Advances in Science, Technology & Innovation

IEREK Interdisciplinary Series for Sustainable Development

Advances in Science, Technology & Innovation (ASTI) is a series of peer-reviewed books based on important emerging research that redefines the current disciplinary boundaries in science, technology and innovation (STI) in order to develop integrated concepts for sustainable development. It not only discusses the progress made towards securing more resources, allocating smarter solutions, and rebalancing the relationship between nature and people, but also provides in-depth insights from comprehensive research that addresses the **17 sustainable development goals (SDGs)** as set out by the UN for 2030.

The series draws on the best research papers from various IEREK and other international conferences to promote the creation and development of viable solutions for a **sustainable future and a positive societal** transformation with the help of integrated and innovative science-based approaches. Including interdisciplinary contributions, it presents innovative approaches and highlights how they can best support both economic and sustainable development, through better use of data, more effective institutions, and global, local and individual action, for the welfare of all societies.

The series particularly features conceptual and empirical contributions from various interrelated fields of science, technology and innovation, with an emphasis on digital transformation, that focus on providing practical solutions to **ensure food, water and energy security to achieve the SDGs.** It also presents new case studies offering concrete examples of how to resolve sustainable urbanization and environmental issues in different regions of the world.

The series is intended for professionals in research and teaching, consultancies and industry, and government and international organizations. Published in collaboration with IEREK, the Springer ASTI series will acquaint readers with essential new studies in STI for sustainable development.

ASTI series has now been accepted for Scopus (September 2020). All content published in this series will start appearing on the Scopus site in early 2021.

Mourad Bezzeghoud • Zeynal Abiddin Ergüler •
Jesús Rodrigo-Comino • Mahesh Kumar Jat •
Roohollah Kalatehjari • Deepak Singh Bisht •
Arkoprovo Biswas • Helder I. Chaminé •
Afroz Ahmad Shah • Ahmed E. Radwan •
Jasper Knight • Dionysia Panagoulia •
Amjad Kallel • Veysel Turan •
Haroun Chenchouni • Attila Ciner •
Matteo Gentilucci
Editors

Recent Research on Geotechnical Engineering, Remote Sensing, Geophysics and Earthquake Seismology

Proceedings of the 2nd MedGU, Marrakesh 2022 (Volume 3)

Springer

Editors
Mourad Bezzeghoud
School of Sciences and Technology
University of Évora
Évora, Portugal

Jesús Rodrigo-Comino
University of Granada
Granada, Spain

Roohollah Kalatehjari
School of Future Environments
Auckland University of Technology
Auckland, New Zealand

Arkoprovo Biswas
Department of Geology, Institute of Science
Banaras Hindu University
Varanasi, Uttar Pradesh, India

Afroz Ahmad Shah
Universiti of Brunei Darussalam
Gadong, Brunei Darussalam

Jasper Knight
University of the Witwatersrand
Johannesburg, South Africa

Amjad Kallel
Sfax National School of Engineering
University of Sfax
Sfax, Tunisia

Haroun Chenchouni
Higher National School of Forests
Khenchela, Algeria

Matteo Gentilucci
School of Science and Technology
University of Camerino
Camerino, Italy

Zeynal Abiddin Ergüler
Kütahya Dumlupınar University
Kütahya, Türkiye

Mahesh Kumar Jat
Malaviya National Institute of Technology
Jaipur, India

Deepak Singh Bisht
Western Himalayan Regional Centre
National Institute of Hydrology
Jammu, India

Helder I. Chaminé
School of Engineering (ISEP)
Polytechnic of Porto
Porto, Portugal

Ahmed E. Radwan
Faculty of Geography and Geology
Institute of Geological Sciences
Jagiellonian University
Kraków, Poland

Dionysia Panagoulia
National Technical University of Athens
Athens, Greece

Veysel Turan
Bingol University
Bingöl, Türkiye

Attila Ciner
Istanbul Technical University
Istanbul, Türkiye

ISSN 2522-8714 ISSN 2522-8722 (electronic)
Advances in Science, Technology & Innovation
IEREK Interdisciplinary Series for Sustainable Development
ISBN 978-3-031-48714-9 ISBN 978-3-031-48715-6 (eBook)
https://doi.org/10.1007/978-3-031-48715-6

This Springer imprint is published by the registered company Springer Nature Switzerland AG
The registered company address is: Gewerbestrasse 11, 6330 Cham, Switzerland

Paper in this product is recyclable.

About the Conference

About MedGU

Steps Toward the Creation of a Mediterranean Geosciences Union (MedGU)

Mediterranean Geosciences Union (MedGU) aims to create a unique federation that brings together and represents the Mediterranean geoscience community specializing in the areas of Earth, planetary, and space sciences.

MedGU will be structured along the lines of American Geophysical Union (AGU) and European Geosciences Union (EGU).

The plan is to establish a large organization for the Mediterranean region that is more influential than any one local geoscience society with the objective of fostering fundamental geoscience research, as well as applied research that addresses key societal and environmental challenges.

MedGU's overarching vision is to contribute to the realization of a sustainable future for humanity and for the planet.

The creation of this union will give the Earth sciences more influence in policy-making and in the implementation of solutions to preserve the natural environment and to create more sustainable societies for the people living in the Mediterranean region. It is hoped that the union will also provide opportunities to Mediterranean geoscientists to undertake interdisciplinary collaborative research. MedGU plans to recognize the work of the most active geoscientists with a number of awards and medals.

Although MedGU has not yet been officially inaugurated, its first annual meeting was organized in November 2021 in Istanbul (MedGU-21). This has provided a forum to achieve a consensus for the formation of this non-profit international union of geoscientists. Membership will be open to individuals who have a professional engagement with the Earth, planetary, and space sciences and related studies, including students and retired seniors.

Nabil Khélifi (MedGU Founder, Germany) and Attila Çiner (MedGU Interim President, Turkey) in collaboration with Abdelaziz Mridekh (MedGU-22 Local Chair, Morocco) have taken an ambitious approach to the launch of the second MedGU Annual Meeting 2022 and hope to develop it in the near future into the largest international geoscience event in the Mediterranean and the broader MENA region. Its mission is to support geoscientists based in this region by establishing a Global Geoscience Congress.

It is expected that hundreds of participants from all over the world will attend this second MedGU Annual Meeting 2022, making it one of the largest and most prominent geosciences events in the region. So far, over 1300 abstracts have been submitted from 95 countries. The meeting's sessions will cover a wide range of topics with more details available on the conference tracks.

This second 2022 Annual Meeting will have a "hybrid" format, with both in-person and virtual participation. Springer, its official partner, will publish the proceedings in a book series (indexed in Scopus) as well as a number of special issues in diverse scientific journals (for more details, see Publications). The official journal of MedGU is Mediterranean Geoscience Reviews (Springer).

Conference Tracks

The scientific committee of the MedGU invites research papers on all cross-cutting themes of Earth sciences, with a main focus on the following 18 conference tracks:

- Track 1. Atmospheric Sciences, Meteorology, Climatology, Oceanography
- Track 2. Biogeochemistry, Geobiology, Geoecology, Geoagronomy
- Track 3. Earthquake Seismology and Geodesy
- Track 4. Environmental Earth Sciences
- Track 5. Applied and Theoretical Geophysics
- Track 6. Geo-Informatics and Remote Sensing
- Track 7. Geochemistry, Mineralogy, Petrology, Volcanology
- Track 8. Geological Engineering, Geotechnical Engineering
- Track 9. Geomorphology, Geography, Soil Science, Glaciology, Geoarchaeology, Geoheritage
- Track 10. Hydrology, Hydrogeology, Hydrochemistry
- Track 11. Marine Geosciences, Historical Geology, Paleoceanography, Paleoclimatology
- Track 12. Numerical and Analytical Methods in Mining Sciences and Geomechanics
- Track 13. Petroleum and Energy Sciences and Engineering
- Track 14. Sedimentology, Stratigraphy, Paleontology, Geochronology
- Track 15. Structural Geology, Tectonics and Geodynamics, Petroleum Geology
- Track 16. Special Session on Astrogeology, Impact Craters and Meteorites
- Track 17. Special Session on climate and sea-level change during the Cenomanian-Turonian Anoxic Event: Synthesis of sedimentological, micropaleontological and geochemical records
- Track 18. Special Session on hydrogeological and climatic risks, their management and the effect of climate change on groundwater quality.

About the Conference Steering Committee

General Committee

Honorary Chair

A. M. Celâl Sengör
Associate Editor, *Mediterranean Geosciences Reviews* (Springer)
Eurasia Institute of Earth Sciences, Istanbul Technical University Istanbul, Turkey

Conference Supervisor

Nabil Khélifi
Senior Publishing Editor
MedGU-21 Supervisor

Program Chair

Mustapha Meghraoui
Editorial Board Member, *Mediterranean Geosciences Reviews* (Springer)
Editor of *Arabian Journal of Geosciences* (Springer)
IPG Strasbourg, France

Publications Chair

Attila Çiner
MedGU (Interim) President
Founding Editor-in-Chief, *Mediterranean Geosciences Reviews* (Springer)
Chief Editor—Tracks 11 and 14, *Arabian Journal of Geosciences* (Springer)
Eurasia Institute of Earth Sciences, Istanbul Technical University, Turkey

Conference Manager

Mohamed Sahbi Moalla
Performer—The Leading Conference Organiser, Tunisia
Journal Coordinator, *Euro-Mediterranean Journal for Environmental Integration* (Springer)
ISET, University of Sfax, Tunisia

Conference Support

Mourad Amer
Founder and CEO of IEREK
Editor of ASTI Series (Springer/IEREK)
IEREK, Alexandria, Egypt

Local Committee

Chair

Abdelaziz Mridekh
Associate Editor, *Arabian Journal of Geosciences* (Springer)
University Ibn Tofail, Kenitra, Morocco

Coordinators

El Hassane Chellai
University Cadi Ayyad, Marrakech, Morocco

Bouabid El Mansouri
University Ibn Tofail, Kenitra, Morocco

Noureddine Laftouhi
University Cadi Ayyad, Marrakech, Morocco

Azzouz Kchikach
University Cadi Ayyad, Marrakech, Morocco

Jean-Louis Bodinier
Program Lead Geology and Sustainable Mining
Mohammed VI Polytechnic University

Members

Abdallah El Hmaidi, Moulay Ismail University, Meknes, Morocco
Abdelhadi El Ouali, Moulay Ismail University, Meknes, Morocco
Aicha Benmohammadi, Ibn Tofail University, Kenitra, Morocco
Ali Essahlaoui, Moulay Ismail University, Meknes, Morocco
Annis Moumen, National Applied School, Kenitra, Morocco
Atika Fahmi, Ibn Tofail University, Kenitra, Morocco
Driss El Azzab, Sidi Mohamed Ben Abdellah University, Fes, Morocco
Hassan Echarfaoui, Ibn Tofail University, Kenitra, Morocco
Jamal Chao, Ibn Tofail University, Kenitra, Morocco
Jaouad Dabounou, Hassan 1 University, Settat, Morocco
Malika Kili, Ibn Tofail University, Kenitra, Morocco
Mohamed Bouhaddioui, National School of Mine, Rabat, Morocco
Mohamed Jaffal, Cadi Ayyad University, Marrakech, Morocco
Mohammed Ouhcine, Ibn Tofail University, Kenitra, Morocco
Moulay Hachem Awragh, Moulay Ismail University, Meknes, Morocco
Mustapha Boualoul, Moulay Ismail University, Meknes, Morocco
Mustapha Boujemaoui, Moulay Ismail University, Meknes, Morocco
Nabil El Mouçaid, Central School of Engineering, Casablanca, Morocco
Souad Haida, Ibn Tofail University, Kenitra, Morocco
Abdullah Sukkar, Department of Geomatics Engineering, Istanbul Technical University, Istanbul, Turkey
Melek Rebai, Performer—The Leading Conference Organiser, Tunisia
M. Bassem Abdelhedi, Performer—The Leading Conference Organiser, Tunisia
Oumayma Abidi, Performer—The Leading Conference Organiser, Tunisia
Toka M. Amer, IEREK—International Experts for Research Enrichment and Knowledge Exchange, Egypt

Advisory Committee

Hans Thybo
President of International Lithosphere Program (ILP)
Editor-in-Chief of *Earth and Planetary Science Letters* (EPSL)
Professor at:

- Eurasia Institute of Earth Sciences, Istanbul Technical University, Turkey
- Center for Earth Evolution and Dynamics, University of Oslo, Norway

A. M. Celâl Sengör
Associate Editor, *Mediterranean Geosciences Reviews* (Springer)
Eurasia Institute of Earth Sciences, Istanbul Technical University Istanbul, Turkey

François Roure
Chief Editor—Track 15
Arabian Journal of Geosciences (Springer)
IFP—Energies Nouvelles, France

Giovanni Bertotti
Associate Editor, *Mediterranean Geosciences Reviews* (Springer)
Geoscience and Engineering, Delft University of Technology, The Netherlands

Abdullah Al-Amri
Founder and Editor-in-Chief
Arabian Journal of Geosciences (Springer)
King Saud University, Saudi Arabia

Akiça Bahri
Director for Africa at the International Water Management Institute (IWMI), Ghana (2005–2010)
Coordinator of the African Water Facility (AWF) at the African Development Bank (2010–2015)
Director of Research at the National Research Institute for Agricultural Engineering, Water, and Forestry (INRGREF), Tunisia (since 2016)
Professor at the National Agricultural Institute of Tunisia (INAT), Tunisia (since 2017)
Awardee of the International Water Association (IWA) Women in Water Prize (2018)
Associate Editor, *Euro-Mediterranean Journal for Environmental Integration* (Springer) (since 2019)
Minister of Agriculture, Water Resources and Fisheries in Tunisia (2019–2020)

Program Committee

Chair

Mustapha Meghraoui
Editorial Board Member, *Mediterranean Geosciences Reviews* (Springer)
Editor of *Arabian Journal of Geosciences* (Springer)
IPG Strasbourg, France

Members

Attila Çiner
MedGU (Interim) President
Founding Editor-in-Chief, *Mediterranean Geosciences Reviews* (Springer)
Chief Editor—Tracks 11 and 14, *Arabian Journal of Geosciences* (Springer)
Eurasia Institute of Earth Sciences, Istanbul Technical University, Turkey

Luc Bulot
North Africa Research Group
The University of Manchester, UK

Sami Khomsi
Georesources Lab, CERTE, University of Carthage, Tunis, Tunisia

Hasnaa Chennaoui Aoudjehane
Meteoritical Society Fellow
Laureate, "Prix Paul Doistau–Émile Blutet" from the French Academy of Sciences
Editor of *Arabian Journal of Geosciences* (Springer)
Professor, Hassan II University of Casablanca, Morocco

Catherine Kuzucuoglu
Associate Editor, *Mediterranean Geosciences Reviews* (Springer)
Research Director Emeritus
CNRS, Laboratoire de Géographie Physique
UMR 8591, Meudon, France

Elena Xoplaki
Chief Editor, *Euro-Mediterranean Journal for Environmental Integration* (Springer)
Justus-Liebig-University Giessen, Germany

Publications Committee

Chair

Attila Çiner
MedGU (Interim) President
Founding Editor-in-Chief, *Mediterranean Geosciences Reviews* (Springer)
Chief Editor—Tracks 11 and 14, *Arabian Journal of Geosciences* (Springer)
Eurasia Institute of Earth Sciences, Istanbul Technical University, Turkey

Members

Zeynal Abiddin Erguler
Chief Editor—Track 8
Arabian Journal of Geosciences (Springer)
Dumlupinar University, Kutahya, Turkey

Amjad Kallel
Chief Editor—Track 4
Arabian Journal of Geosciences (Springer)
Managing and Development Editor, *Euro-Mediterranean Journal for Environmental Integration* (Springer)
ENIS, University of Sfax, Tunisia

Mourad Bezzeghoud
School of Sciences and Technology (ECT)
Insititut of Earth Sciences (IIFA)
University of Évora, Portugal

Hesham El-Askary
Professor of Remote Sensing and Earth Systems Science
Editor of *Arabian Journal of Geosciences* (Springer)
Director Computational and Data Sciences Graduate Programs
Center of Excellence in Earth Systems Modeling and Observations
Schmid College of Science and Technology, Chapman University, USA

Zakaria Hamimi
President of ArabGU
IAGETH VP for Africa and IAGETH National Chapter for Egypt
Editor of *Arabian Journal of Geosciences* (Springer)
Professor, Benha University, Benha, Egypt

Syed E. Hasan, Ph.D.; RG; FGS
Chair, Environmental Characterization and Remediation Technical Working Group, AEG
Department of Earth and Environmental Sciences
University of Missouri-Kansas City, USA

François Roure
Chief Editor—Track 15
Arabian Journal of Geosciences (Springer)
IFP—Energies Nouvelles, France

Anastasia Kiratzi
Professor of Seismology
Faculty of Sciences, Aristotle University of Thessaloniki, Greece

Broder Merkel
Chief Editor—Track 10
Arabian Journal of Geosciences (Springer)
Associate Editor of *Environmental Earth Science* (Springer)
Publisher of Freiberg Online Geoscience (FOG)
Institute of Geology, Technische Universität Bergakademie Freiberg, Germany

Preface

The proceedings of the 2nd Mediterranean Geosciences Union (MedGU-2) held in Marrakesh onsite and online in November 2022 consisted of four volumes titled:

Proceedings Volume 1: Recent Advancements from Aquifers to Skies in Hydrogeology, Geoecology, and Atmospheric Sciences.
Proceedings Volume 2: Recent Research on Sedimentology, Stratigraphy, Paleontology, Geochemistry, Volcanology, Tectonics, and Petroleum Geology.
Proceedings Volume 3: Recent Research on Geotechnical Engineering, Remote Sensing, Geophysics, and Earthquake Seismology.
Proceedings Volume 4: Recent Research on Environmental Earth Sciences, Geomorphology, Soil Science, and Paleoenvironments.

These volumes are based on the accepted conference papers for either oral/poster presentations or selected for online publication during the MedGU-2.

This 3rd volume contains 68 papers that discuss the latest advances in (1) geological and geotechnical engineering, (2) geomechanical studies based on numerical and analytical methods, and (3) geoinformatics and remote sensing. Remote sensing with high-resolution satellite imagery, seismology, geochemistry, theoretical geophysics and related profiles with tomographic images, earthquake geology with times series, radar interferometry and related geodetic-GPS campaigns, and well-logging contribute to scrutinizing the earth's tectonic architecture. Fundamental questions that address the genesis and evolution of our planet are built upon data collection and experimental investigations under physical constitutive laws, which are the conditions for successful scientific research.

The content of these research studies would provide new scientific knowledge to further understand geotechnical engineering-related topics. The proceedings of the MedGU-2 are of interest to all researchers, practitioners, and students in all fields of geosciences.

Évora, Portugal — Mourad Bezzeghoud
Kütahya, Türkiye — Zeynal Abiddin Ergüler
Granada, Spain — Jesús Rodrigo-Comino
Jaipur, India — Mahesh Kumar Jat
Auckland, New Zealand — Roohollah Kalatehjari
Jammu, India — Deepak Singh Bisht
Varanasi, India — Arkoprovo Biswas
Porto, Portugal — Helder I. Chaminé
Gadong, Brunei Darussalam — Afroz Ahmad Shah
Kraków, Poland — Ahmed E. Radwan
Johannesburg, South Africa — Jasper Knight
Athens, Greece — Dionysia Panagoulia
Sfax, Tunisia — Amjad Kallel
Bingöl, Türkiye — Veysel Turan
Khenchela, Algeria — Haroun Chenchouni
Istanbul, Türkiye — Attila Ciner
Camerino, Italy — Matteo Gentilucci
June 2023

Contents

Applied and Theoretical Geophysics

About the Editors

Mourad Bezzeghoud School of Sciences and Technology, University of Évora, Portugal

Mourad Bezzeghoud is Full Professor at the Physics Department of the School of Sciences and Technology and Researcher at the Institute of Space and Earth Sciences, University of Évora. He was in charge of several positions of responsibility in several institutions as, i.e., Founder and Director of the Department for Study and Seismic Monitoring of CRAAG (Algiers, Algeria), Director of the School of Sciences and Technology (University of Évora, Portugal), or Member of the Management Council of the University of Évora. Presently, he is President of the Scientific Council of the School of Sciences and Technology (University of Évora) and Coordinator of the Institute of Space and Earth Sciences (University of Évora). He has published dozens of articles in specialized revues and journals and several books and chapters in book. He works in the area of Earth Sciences (Solid Earth Geophysics) with an emphasis in Seismology. He interacted with more than 200 collaborators in co-authorship of scientific works. In his curriculum Ciência Vitae (FCT, Portugal: https://www.cienciavitae.pt//9715-98D9-D5BB), the most frequent terms in the context of scientific and technological production are: earthquake seismology, seismic source, historical and instrumental seismicity, focal mechanisms, earth's crust structure, seismotectonics, strong ground motion modeling, seismic network, seismic hazard along the Western part of the Eurasia-Nubia plate boundary, and muography as a new tool for geophysics.

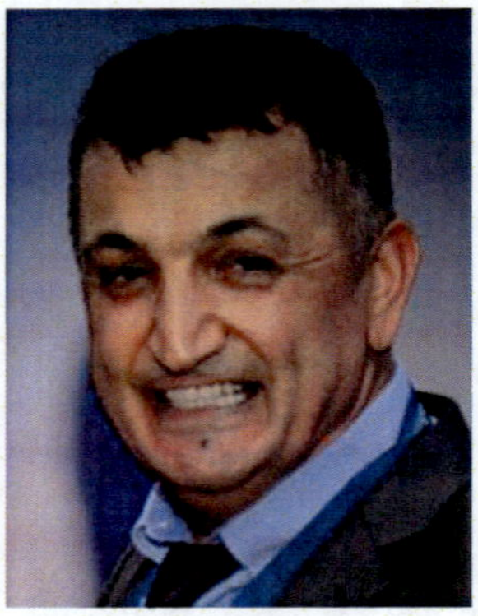

Zeynal Abiddin Ergüler Kutahya Dumlupinar University, Geological Engineering Department, Kutahya, Turkey

Prof. Dr. Zeynal Abiddin Ergüler is Full Professor in the geological engineering department at Kutahya Dumlupinar University (Turkey). Dr. Ergüler holds a B.Sc. (1998), an M.Sc. (2001), and a Ph.D. degree (2007) in Geological Engineering from Hacettepe University (Turkey). His research interests mainly focus on rock mechanics, engineering geology, environmental geology, and soil mechanics. His current investigation is to understand and model the thermo-hydro-mechanical behavior of shale rocks in the area of shale gas production. In addition to performing many types of research and industry-funded projects, he has also taught and supervised undergraduate and graduate students. In 2017, Dr. Ergüler joined the *Arabian Journal of Geosciences* (AJGS) as Editor responsible for evaluating submissions in the fields of rock mechanics, engineering geology, environmental geology, and soil mechanics.

Jesús Rodrigo-Comino University of Granada, Spain

Ph.D. D.Eng. Jesús Rodrigo-Comino is Member of the Department of Regional and Physical Geography at the University of Granada (Spain). He completed his first Ph.D. in Geography at the University of Málaga (Spain) and Trier (Germany) in 2018 and the second one in the engineering of Geomatics and Topography at the Polytechnic University of Valencia (2023). His current research interests include soil geography, regional geography, and land degradation. He coordinates the Terra Lab 2 EGEMAP (Environmental Geography and Mapping) www.egemap.eu.

Mahesh Kumar Jat Malaviya National Institute of Technology Jaipur, India

Dr. Mahesh Kumar Jat is Professor of Water Resources at Malaviya National Institute of Technology (MNIT) Jaipur (India). He is teaching and doing research in the area of Hydrological Modeling, Integrated Water Management, Geospatial Technologies, and Land Use Land Cover Change Modeling. He has published more than 100 research articles in ISI/Scopus journals and conferences (total citations: 2330; h-index—21; i-index 30). He has completed 12 research projects amounting to more than 65 million Indian rupees. He is Member of many national and international professional bodies such as the European Geosciences Union (EGU), ASCE, IWRS, IAH, and IRS. He sits as Member of the board of studies of many institutes in India. He is Associate Editor and Editorial Member in 2 ISI journals. He has delivered 40 keynote and invited talks at international and national conferences. He has edited and reviewed many articles. He has supervised 11 Ph.D. students and 36 M.Tech. students (with a thesis). Professor Jat has widely traveled abroad visiting many countries to present his research findings.

Roohollah Kalatehjari Auckland University of Technology, New Zealand

Dr. Roohollah (Roo) Kalatehjari is Highly Accomplished Academic and Researcher with over 15 years of experience in geotechnical engineering. He is Senior Lecturer in Geotechnical Engineering, Associate Head of School—Learning and Teaching at the School of Future Environments (SoFE) at Auckland University of Technology (AUT), Director of the Advanced Construction Research cluster, and Founder and Director of the Geotechnical Engineering Laboratory. His research focuses on preserving the environment and finding sustainable and environmentally friendly solutions for soil stabilization and slope stability. Roo is also Active Member of several professional engineering and education organizations, including Engineering New Zealand, the New Zealand Geotechnical Society, and the International Society for Soil Mechanics and Geotechnical Engineering. He is Fellow of HEA, UK, and Member of Life Cycle Assessment New Zealand.

Deepak Singh Bisht National Institute of Hydrology, Roorkee, India

Dr. Deepak Singh Bisht is working as Scientist "B" at National Institute of Hydrology, an autonomous society under Ministry of Jal Shakti, River Development and Ganga Rejuvenation, Government of India. He did his Ph.D. and master's from Indian Institute of Technology Kharagpur in the domain of Hydrology and Climate Change. His research interests primarily include, but not limited to, Hydrological Modeling, Urban Drainage, Flood Modeling, Climate Change Impact Analysis, Remote Sensing Applications, and Springshed Management.

Arkoprovo Biswas Banaras Hindu University, India

Dr. Arkoprovo Biswas is Assistant Professor at the Department of Geology, Institute of Science, Banaras Hindu University (BHU), Varanasi. He received his B.Sc. (2002) in Geology from Presidency College, University of Calcutta, M.Sc. (2004) in Geological Science, M.Tech. (2006) in Earth and Environmental Science from IIT Kharagpur, and P.G. Diploma (2009) in Petroleum Exploration from Annamalai University. He joined Geostar Surveys India Pvt. Ltd. as Geophysicist in 2006 and later joined WesternGeco Electromagnetics, Schlumberger, as On-Board Data Processing Field Engineer/Geophysicist in 2007 and served there till 2008. In 2013, he received his Ph.D. in Exploration Geophysics from IIT Kharagpur. Later, he joined the Department of Earth and Environmental Sciences, Indian Institute of Science Education and Research Bhopal, as Visiting Faculty in 2014 and completed his tenure in 2015. He again joined the Wadia Institute of Himalayan Geology (WIHG) Dehradun in 2016 as Research Associate, and later he joined BHU in October 2017. He is Experienced Geophysicist with research interests in Near Surface Geophysics, Integrated Electrical and Electromagnetic Methods, Geophysical Inversion, Mineral, and Groundwater Exploration and Subsurface Contamination. He has published more than 50 papers on theoretical modeling, inversion, and application in practical geoscience problems in peer-reviewed international and national journals and 6 book chapters. He also published two edited books with Springer on *Advances in Modeling and Interpretation in Near Surface Geophysics* and *Self-Potential Method: Theoretical Modeling and Applications in Geosciences*. Dr. Biswas received the Prestigious M. S. Krishnan Medal of the Indian Geophysical Union (IGU), Hyderabad, for the year 2019 and the B. C. Patnaik Memorial Gold Medal from the Society of Geoscientists and Allied Technologists (SGAT), Bhubaneshwar, Odisha, India, in 2019. He is also Life Fellow of the Geological Society of India and International Science Congress Association, Fellow of the Society of Earth Scientists, Life Member of the Indian Geophysical Union, and Active Member of the Society of Exploration Geophysicist, USA. He is Editor of the Journal *Results in Earth Sciences* as well as Associate Editor of the *International Journal of Geophysics*, *Results in Geophysical Sciences*, *Contributions to Geophysics and Geodesy*, *Journal of Earth System Sciences*, *Arabian Journal of Geosciences*, and *Spatial Information Research*. Dr. Biswas is also Member of the International Editorial Advisory Board of *Natural Resources Research* and *Bitlis Eren University Journal of Science and Technology*.

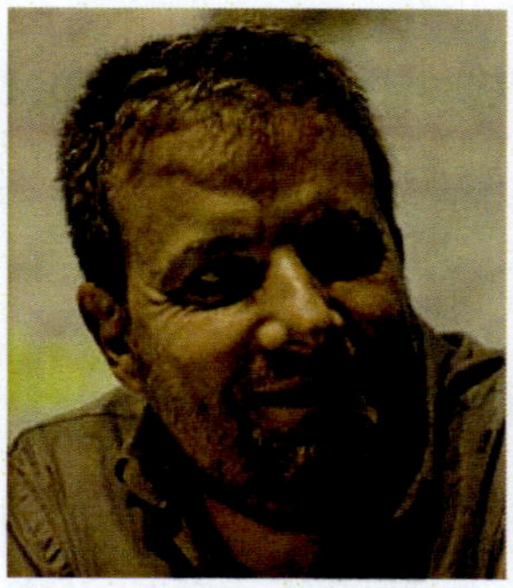

Helder I. Chaminé School of Engineering of Porto (ISEP), Portugal

Helder I. Chaminé is Skilled Geologist (B.Sc., Ph.D., D.Sc.) and Professor of engineering geosciences at the School of Engineering (ISEP) of the Polytechnic of Porto, Portugal, with over 33 years of experience in multidisciplinary geosciences research and practice. Before joining the academy in 2001, he worked for over 13 years on international and national projects for mining hydrogeomechanics and geology, structural geology mapping, applied mineralogy, rock engineering, exploration geology, and groundwater. His major research interests are GIS mapping techniques for applied geology, geotechnics and natural hazards, engineering geosciences and hydrogeomechanics, hard-rock hydrogeology, urban groundwater, water resources, and thermal waters management. He has interests in geomining heritage, geoethics, history of cartography, military geosciences, higher education dissemination, and geoprofessional core values. Presently, he is Head of the Laboratory of Cartography and Applied Geology (LABCARGA|ISEP), Senior Researcher at Centre GeoBioTec|U. Aveiro and Centre IDL|U. Lisbon, as well as belongs to the executive board of the M.Sc. Geotechnical and Geoenvironmental Engineering program (OE+EUR-ACE Label) and the Department of Geotechnical Engineering (ISEP). Furthermore, he was Consultant and or responsible for over 70 projects of applied geology, hydrogeomechanics, slope geotechnics, mining geology, exploration hydrogeology, hard-rock hydrogeology, water resources, urban groundwater, and applied mapping (Mozambique, Portugal, and Spain). He has co-authored over 220 publications in indexed journals, conference proceedings/full papers, book chapters, and technical and professional papers. He co-edited over 16 special volumes (journals and Springer Series Books) and is presently editing topical collections for several international journals. He serves as Associate Editor (*SN Applied Sciences*, *Discover Water*, *Arabian Journal of Geosciences*, *Euro-Mediterranean Journal for Environmental Integration*) and on the editorial or advisory boards (*Mediterranean Geoscience Reviews*, *Geotechnical Research ICE*, *Journal of Geoethics and Social Geosciences*, *Cadernos do Laboratório Xeolóxico de Laxe*, *Revista Geotecnia SPG*, *Geología Aplicada a la Ingeniería y al Ambiente ASAGAI*). He also served as Associate Editor in the *Hydrogeology Journal*. In addition, he has a broad activity as Reviewer for several international journals. In 2021, the journal *SN Applied Sciences* awarded him an outstanding Guest Editor and Editorial Board Member. In addition, he integrates as Moderator or Session Chair in several conferences, workshops, and meetings. He is Vice-President of the Portuguese Chapter of the International Association of Hydrogeologists (IAH) and Deputy Secretary of the Technical Commission of Environmental Geotechnics from the Portuguese Geotechnical Society (SPG). Furthermore, he served as Invited Expert Evaluator of the Bologna Geoscience program for DGES

(Portugal) and Scientific Projects Evaluation for NCST, 2017–2019 (Kazakhstan), and NRF|RISA, 2019 (South Africa), as well as Coordinator of "Geology on Summer/Ciência Viva" program at ISEP (2005–2019) for geoscience dissemination. He has also been active in teaching and supervising many Ph.D., M.Sc., and undergraduate students.

Afroz Ahmad Shah Universiti of Brunei Darussalam, Brunei

Afroz Ahmad Shah is Senior Assistant Professor of Structural Geology at the Department of Geosciences, Universiti of Brunei Darussalam (UBD). He graduated from IIT Kanpur India in 2006 and then completed Ph.D. at James Cook University, Australia, in 2010, a post-doctorate at Earth Observatory of Singapore in 2013, and joined his first academic job as Senior Lecturer of Structural Geology at Curtin Sarawak, Miri, Malaysia, before joining UBD. His research mainly involves the brittle deformation of lithospheric plates, focusing on earthquake-causing faults in South and Southeast Asia. It also includes floods, landslides, and land subsidence hazards. Shah works on earthquake science education and outreach and frequently writes in newspapers, magazines, and blogs. He is National Geographic Explorer, and his work on earthquake hazards in Kashmir has been featured in the National Geographic Magazine. He loves writing and has been involved in writing about geological and climatic hazards.

Ahmed E. Radwan Institute of geological sciences, Jagiellonian University, Krakow, Poland

Dr. Ahmed E. Radwan is Adjunct Professor at the Institute of Geological Sciences of the Jagiellonian University (Poland). Dr. Radwan has academic and industrial experience, since he obtained his Ph.D. in geophysics at Sohag University, Egypt, besides his proficient work in the oil and gas industry as Section Head at the exploration department of the Gulf of Suez petroleum company (Gupco), Egypt. As Post-doctoral Research Scientist, he attended Innsbruck University in Austria in 2019. In 2020, he joined the Jagiellonian University in Poland. Despite his youth, he has received numerous awards from international organizations such as the International Union of Geological Sciences (IUGS), the Geochemical Society (GS), the Clay Minerals Society (CMS), the Austrian Forschungsgemeinschaft (FG), the Narodowa Agencja Wymiany Akademickiej (NAWA), the Austrian Federal Ministry of Education, Science, and Research (BMBWF), and petroleum companies. Dr. Radwan has authored more than 100 papers in highly indexed international peer-reviewed journals, published four book chapters, and presented at numerous international conferences. Dr. Radwan is Associate Editor in *Asian Earth Sciences*, *Marine and Petroleum Geology*, *Geoenergy Sciences and Engineering*, *Petroleum Exploration and Production Technology*, *The Geological Journal*, *Energy Geosciences*, and *Petroleum Research*, in addition to being Editorial Board Member of *Unconventional Resources*. Dr. Radwan is Book Reviewer and Research Article

Reviewer for several publishers and journals, and he organizes many special issues in different journals with the leading publishers. His research interests focused on multidisciplinary research integrating geosciences (geophysics, geochemistry, and geology), petroleum engineering, and reservoir engineering, as follows: (1) geology areas include: petroleum geology, reservoir characterization, sedimentology, facies analysis, depositional environment, diagenesis, paleoenvironment interpretations, subsurface analysis, basin analysis, reservoir quality, fluid flow, fractures, formation evaluation, and unconventional and conventional resources; (2) petroleum engineering (petroleum geomechanics, drilling, fluids, and casing design); (3) reservoir engineering (reservoir geology and geophysics, reservoir damage, production optimization, water flooding, stimulations, fluid flow, and enhanced recovery); (4) the geophysics fields of study (ex. formation evaluation, petrophysics, borehole geophysics, and rock typing); (5) geochemistry fields of study include geochemical characterization, basin modeling, petroleum systems, and isotope analysis; (6) petroleum geomechanics (pore pressure, wellbore stability, in-situ stress orientation, and magnitudes); (7) machine learning applications in the energy industry; and (8) energy storage.

Jasper Knight University of the Witwatersrand, South Africa

I am Geoscientist with research interests in the spatial and temporal variability in morphosedimentary system responses to rapid hemispheric-scale climatic and environmental changes during the late Pleistocene and Holocene. I focus thematically on glaciers, rivers, coasts, and mountains. I focus geographically on Africa, Ireland, northwest USA, Australia, the European Alps, and various places in Asia and South America.

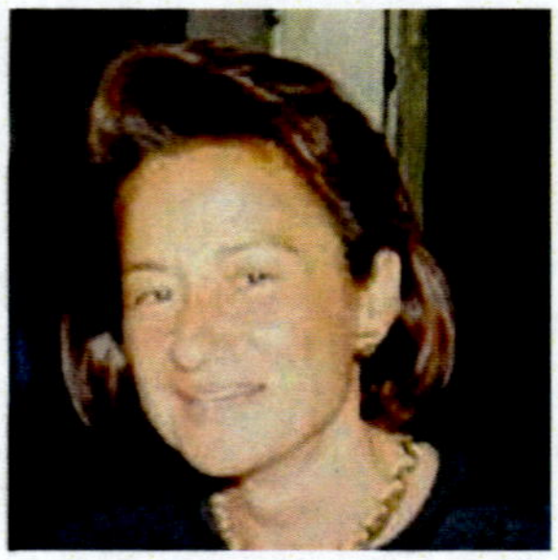

Dionysia Panagoulia National Technical University of Athens, Greece

Dr. Dionysia Panagoulia is Associate Professor at the National Technical University of Athens, Greece, with expertise in hydrology, hydroclimatology, and water systems. She is Author of more than 145 published research works, including being Co-editor of the book *River Basin Management—Under a Changing Climate* and lately has extended her research work to water economics theory and complex time/dark matter approaches. She has over 30 years of research experience in floods and their risk and hazard, extreme events, precipitation, global climate contribution to local climate, climate change, low flows, droughts, maximum/minimum temperatures, sediment transport, groundwater/streamflow interaction, ANN, WEF Nexus, and water economics and management. She has cooperated in joint research with the University of Stuttgart, Germany; CNRS Laboratory of Oceanology and Geosciences Wimereux, France; Wageningen University, Netherlands; and McGill University, Canada. She was/is Member of twelve scientific societies, Reviewer and Guest Editor for thirty-two international journals, and Recipient of the 2018 Outstanding Reviewer Award from Water MDPI.

Amjad Kallel Sfax National School of Engineering, University of Sfax, Sfax, Tunisia

Dr. Amjad Kallel is currently Associate Professor of Environmental Geology at the Sfax National School of Engineers at the University of Sfax, Tunisia. He holds a B.Eng. in Georesources and Environment (1998) from the University of Sfax (Tunisia) and an M.Sc. degree and a Ph.D. degree in Georesources and Environment (2004) from Hokkaido University (Japan). He joined Venture Business Laboratory (VBL) at Akita University, Japan (2005–2006), as Researcher focusing on refining and recycling technologies for the recovery of rare elements from natural and secondary sources. On his return to Tunisia, he worked at the University of Gabes from 2006 to 2011, where he contributed to the elaboration of teaching programs at the Higher Institute of Water Sciences and Technologies of Gabes. Since 2011, he has joined the Sfax National School of Engineers. There, he has also been involved in various research projects related to Environmental Geology and Environmental Geotechnics. Dr. Kallel has co-organized many prestigious workshops, seminars, and international conferences. In 2016, Dr. Kallel joined the *Arabian Journal of Geosciences* (Springer) and the *Euro-Mediterranean Journal for Environmental Integration* (Springer) as Chief Editor and Managing Editor, respectively.

Veysel Turan Bingol University, Turkey

Veysel Turan is currently Associate Professor at Faculty of Agriculture, Bingöl, Türkiye. His research focuses on the interactions among Soil Science, Environmental Remediation, Food Chemistry, and Climate Change. His Google Scholar h-index is 38 and Web of Science ESI top papers is 13 as "highly cited papers" and 4 as "hot paper". He was awarded an Excellence in Reviewing for the *Journal of Environmental Management*, *Resources, Conservation and Recycling* and Best Reviewer Award *Critical Reviews in Environmental Science and Technology* in 2022.

Haroun Chenchouni Higher National School of Forests, Khenchela, Algeria

Dr. Haroun Chenchouni is Associate Professor and Research Scientist (Ecologist) at the Higher National School of Forests (Khenchela, Algeria). He is former Associate Professor at the University of Tebessa (Algeria). He holds a doctorate degree in Ecology and Environment from the University of Batna 2 (Algeria) and an M.Sc. (Magister) in Dryland Ecology from the University of Ouargla (Algeria). He graduated as Engineer in Plant Ecology and Forest Ecosystems from the Department of Biological Sciences (University of Batna, Algeria). His research interests are fairly broad; he uses statistical modeling approaches to understand how natural environments, mainly climatic and edaphic factors, and anthropogenic perturbations influence biological interactions, shape trends in population dynamics, and influence community diversity. He uses various biological models to investigate biological interactions and community ecology of arid and semiarid ecosystems of North Africa. At various universities in Algeria, he teaches forest ecology, biostatistics, and ecological modeling. He has published more than 100 peer-reviewed publications and internationally recognized research papers. He is also involved in national and international research projects. In 2017, he joined the *Arabian Journal of Geosciences* (AJGS) as Associate Editor. Then in 2019, he was assigned as Chief Editor of Topic 2 (biogeochemistry, geobiology, geoecology, geoagronomy) to handle submissions dealing with various fields of biogeosciences, geoecology, climate change, plant and soil science, agricultural and forest environment, and environmental sciences.

Attila Ciner Istanbul Technical University, Turkey

Attila Ciner is Sedimentology and Quaternary Geology Professor at the Eurasia Institute of Earth Sciences at Istanbul Technical University, Turkey. After graduating from the Middle East Technical University in Ankara (1985), he obtained his M.Sc. degree at the University of Toledo, USA (1988) and his Ph.D. at the University of Strasbourg, France (1992). He works on the tectono-sedimentary evolution of basins and Quaternary depositional systems such as moraines, fluvial terraces, alluvial fans, and deltas. He uses cosmogenic nuclides to date these deposits. He primarily focuses on the glacial deposits and landscapes and tries to understand paleoclimatic and paleoenvironmental changes since the Last Glacial Maximum. Lastly, he was part of the Turkish Antarctic Expedition. He spent two months working on the site recognition and decision of the future Turkish scientific research station to be implemented on the continent. He is Founding Editor-in-Chief of *Mediterranean Geoscience Reviews* and Chief Editor of *Arabian Journal of Geosciences*, both published by Springer. He received the Humboldt Foundation Georg Forster lifetime achievement award in 2022. He has published more than 100 peer-reviewed articles and book chapters.

Matteo Gentilucci University of Camerino, School of Science and Technology, Geology division, Camerino, Italy

Matteo Gentilucci had his B.Sc. degree in "Geological Sciences" from University of Camerino, Italy, in 2009 and then earned M.Sc. degree in 2012 from University of Camerino, Italy, in "Geoenvironmental Resources and Risks". He earned a first-level master's degree in GIS for the governance of the territory in 2012 and a Ph.D. in "Sciences and technology: Earth Science" from the University of Camerino in 2017. From February 2017 to October 2018, he is Researcher at the Experimental Geophysical Observatory of Macerata, and in 2018, he became Honorary Fellow in Advanced GIS. In 2018 and 2019, he is Contract Professor of the University of Camerino. Since 2022, he has been Researcher at the University of Camerino and holds courses in meteorology, climatology, and GIS. He is Editorial Board Member of the *Journal of Experimental Sciences*, of the *Journal Sustainability*, of the *Arabian Journal of Geosciences*, and of the International Research Conference Committee. He is Author of numerous articles in international journals. Dr. Matteo Gentilucci is involved in numerous research projects related to climate. The main topics of interest and expertise are: geographic information systems applied to the environment, relationship between climate and phenology, climate change, global warming, and extreme events. https://www.scopus.com/authid/detail.uri?authorId=57201441740.

Geotechnical Engineering

Geo-engineering Education in Civil Engineering Courses: Successful Strategies in ISEP (Porto, Portugal)

José Filinto Trigo, José Cândido Freitas, Agostinho Mendonça, Maria de Fátima Silva, and Tiago Abreu

Abstract

This article intends to present the advantages of including activities of a practical nature, in curricular units of Soil Mechanics and Rock Mechanics, both in undergraduate and master's degrees in Civil Engineering, or in the development of master's theses. The creation of a reduced scale physical model of a reinforced soil structure with tires is presented. The didactic aspects of the development of a work within the service carried by ISEP's Construction Studies Center (NEC) are also described. In addition, a coastal geotechnical study, developed by an undergraduate Civil Engineering student in ERASMUS mobility, is reported. The examples described involved several students and the results achieved, evidenced by the permanent interest they showed throughout the development of their studies and the excellent marks they obtained in their final evaluations, attest to the effectiveness of this teaching strategy with a strong practical component.

Keywords

Small-scale physical model • Finite element numerical modeling • Coastal geotechnics

1 Introduction

The field of theoretical concepts of the fundamental disciplines of geotechnics is of paramount importance for monitoring the subsequent issues of this knowledge area. This expertise, transferred in curricular units of Soil Mechanics and Rock Mechanics, is a task that can be simplified if it is complemented by activities of a practical nature.

On the other hand, conventional university education cannot give the student sufficient experience that has to be gained in practice (Rybak et al., 2017).

This study intends to present the advantages, concerning educational purposes, of including activities of a practical nature, facilitating the transmission of knowledge in this scientific area and, simultaneously, the acquisition of skills that will facilitate the integration and professional development of students. Three initiatives, with distinct characteristics, developed at the 'Instituto Superior de Engenharia do Porto' (ISEP), are presented: (i) creation of a small-scale physical model of a reinforced soil structure with tires, (ii) the development of a work within the service carried by ISEP's Construction Studies Center (NEC), concerning a building that suffered a serious set of damages, associated with instability processes of geotechnical character, (iii) and a coastal geotechnical study, developed by a Civil Engineering student in ERASMUS mobility.

2 Practical Application Cases

2.1 Physical and Numerical Models of a Tire-Reinforced Earth Retaining Structure

End-of-life tires, in some regions of the world, are still a serious problem. Earth retaining structures built with end-of-life tires filled with soil are used in some countries, especially as gravity-type structures. Its application to earth support structures is a known technique, referenced in many practical and academic works (Long, 1993; Medeiros et al., 2000).

J. F. Trigo (✉) · J. C. Freitas · M. de Fátima Silva · T. Abreu
Department of Civil Engineering, School of Engineering (ISEP), Polytechnic of Porto, Porto, Portugal
e-mail: jct@isep.ipp.pt

A. Mendonça
Department of Geotechnical Engineering, School of Engineering (ISEP), Polytechnic of Porto, Porto, Portugal

M. Bezzeghoud et al. (eds.), *Recent Research on Geotechnical Engineering, Remote Sensing, Geophysics and Earthquake Seismology*, Advances in Science, Technology & Innovation,
https://doi.org/10.1007/978-3-031-48715-6_1

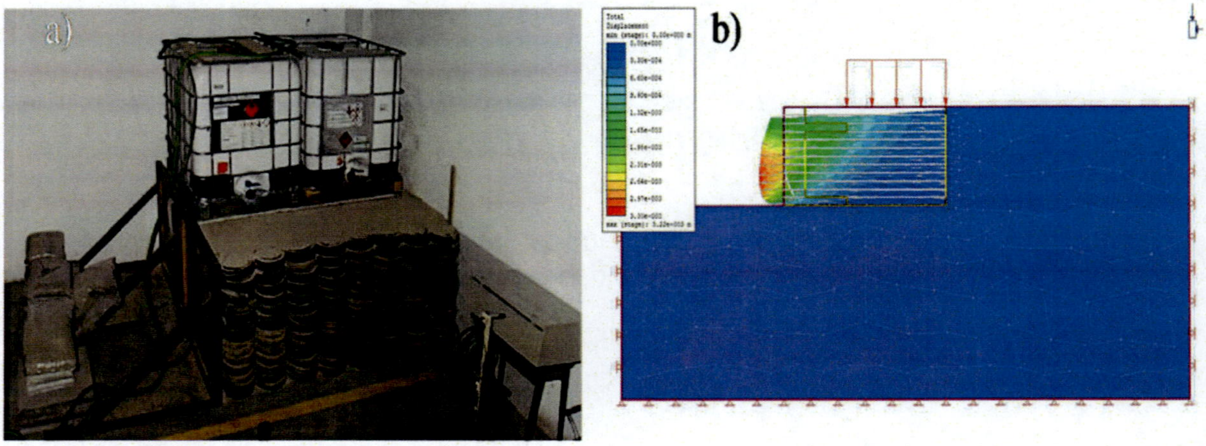

Fig. 1 a Load setup of the physical model. b Numerical modeling

This research aimed to promote student participation in assessing the feasibility of using end-of-life tires in reinforced soil structures. The students started by being involved in the definition of the model geometry and in the laboratory characterization of the landfill soil, which was adopted with granular characteristics and a reduced moisture content to facilitate its analysis in terms of stresses and the monitoring of its behavior in terms of settlements. The laboratory experience included some expedient tests for the characterization of the tire-soil assembly. To predict the behavior of the model, which is more difficult due to its small scale and the need to respect similarity relations, an automatic finite element computer calculation tool was used (Phase2, from Rocscience). The students also had the opportunity to collaborate in the definition of the physical model observation and monitoring plan, which integrated the installation of load cells with a set of Linear Variable Differential Transformer (LVDT) sensors (to control the evolution of the wall displacements). The loading was applied to the landfill surface with tanks that were successively filled with water (see Fig. 1). The students were responsible for controlling this load, collecting and processing a large source of instrumentation data, both concerning ground pressures and displacements at different wall heights.

The entire constructive sequence was numerically modeled by the students (see Fig. 1b) who had the opportunity to compare the two sets of results and to seek numerical model calibration to bring their results closer to the experimental results.

2.2 Structure Inspection and Diagnosis

ISEP's Construction Studies Center conducted a study of a building that experienced high displacements, causing severe structural damages.

A master's student in Civil Engineering followed the entire process of identifying the causes of this accident and the safety level of the building. A geological-geotechnical prospecting campaign of the foundation mass was initially conducted. The student witnessed the accomplishment of this set of tests, of which he only knew the respective theoretical description, collaborating in its interpretation. Inspections were also carried out on the foundations of some pillars, with the manual drilling of wells. The student observed all this inspection work and crossed the information collected with the previously performed in situ tests. The structure of the building was also inspected. Then, the compatibility of the anomaly pattern observed in the structure was verified with the characteristics identified with the geological-geotechnical prospecting carried out and with the displacements experienced by some of the structure foundations.

The student consolidated a set of knowledge transmitted in both Civil Engineering study cycles and acquired new knowledge and skills. This was also the first opportunity to participate in a work of a professional nature, witnessing all the stages involved and contacting the different stakeholders.

2.3 ERASMUS Project

A study was developed for La Garrofera beach for an ERASMUS project. This is a semi-urban sandy beach located in Valencia (Spain) with a vulnerable coastal stretch facing coastal erosion, known by the student who inhabits this region.

This project offered to the student the unique opportunity to deal with a real case, putting the teachings into practice, providing the characterization of the beach as well as the morphology, nature and soil particle sizes, bathymetry,

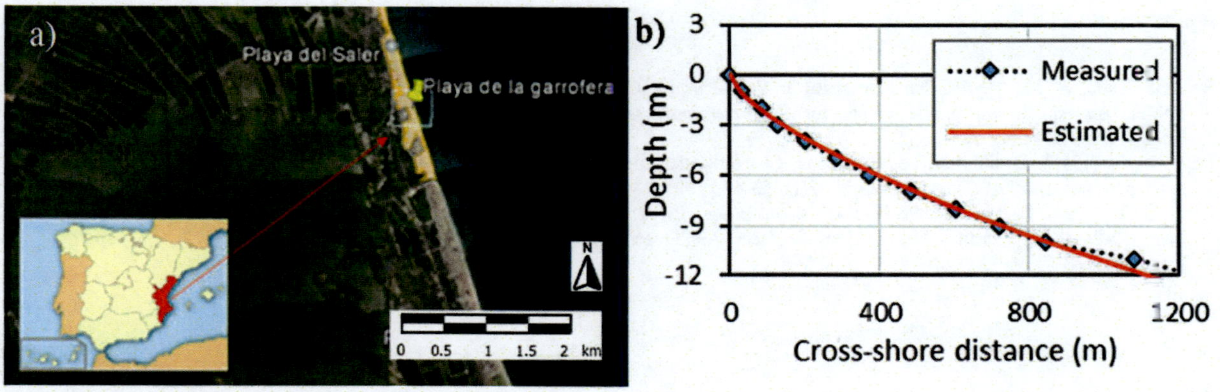

Fig. 2 **a** Localization of La Garrofera beach. **b** Comparison between the measured and estimated values of the cross-shore profile according to Dean model ($d50 = 0.27$ mm)

coastal dynamics and maritime climate. Moreover, this was a chance to learn and to handle the SMC program (Coastal Modeling System), which integrates a series of numerical models and allows to model the study site evaluating the beach evolution.

A study on the existent cross-shore profile was also performed, using three different models to evaluate its equilibrium state (Bernabeu-Tello et al., 2002; Dean, 1991; Yu & Slinn, 2003). The hypothesis behind the equilibrium beach profile is that beaches respond to wave forcing by adjusting their form to an equilibrium or constant shape attributable to a given type of incident wave or sediment characteristic. The importance of this study is directly connected with the management and nourishment of beaches.

The bathymetry of the beach was processed by the student through AutoCAD Software and the convergence to a cross-shore profile was made automatically using Microsoft EXCEL tools. Figure 2 shows an interesting adjustment to the Dean's equilibrium profile (Dean, 1991), considering a median grain size of $d50 = 0.27$ mm. It is noteworthy to mention that the obtained data coming from a cartographic study of the Valencia and Alicante regions, made by the Spanish Environmental Ministry, registers a relevant diameter of the area equal to $d50 = 0.27 \pm 0.4$ mm. Therefore, it can be said that this beach reveals a very good fit to Dean's equilibrium profile. Such adjustment is probably reflected in the beach stability observed in the past decade. This kind of projects provide opportunities for students to work with different tools and to develop their analytical and critical thinking skills that, most likely, will be applied throughout their professional life.

3 Conclusions

Curricular units of Soil Mechanics and Rock Mechanics involve complex knowledge that must be transferred with special care, in a methodical and organized way.

The experience of ISEP demonstrates that this transmission of knowledge is simplified if supported by practical activities, that students find most interesting.

The adoption of physical models with undeniable didactic qualities, associated with the use of software that allows their numerical reproduction, is a teaching strategy with good results proven at ISEP.

The integration of students to collaborate in work requested from ISEP, by public institutions or private companies, is also particularly appreciated by students, as it prepares them for the professional life that lies ahead. These collaborations may also be included in master dissertations or in undergraduate academic papers.

For students in mobility, the development of case studies involving their country and region of origin can also contribute to their increased enthusiasm. The development of these case studies may continue after the return to their school of origin, maintaining contact with ISEP and opening the possibility of fertile research partnerships between that school and ISEP.

The high grades obtained by students in these practical works, as a result of their dedication, are the best parameter to validate the success of this teaching strategy that has been applied at ISEP.

References

Bernabeu-Tello, A. M., Medina-Santamaría, R., & Vidal-Pascual, C. (2002). An equilibrium profile model for tidal environments. *Scientia Marina, 66*(4), 325–335.

Dean, R. G. (1991). Equilibrium beach profiles: Characteristics and applications. *Journal of Coastal Research, 7*(1), 53–84.

Long, N. T. (1993). *Le Pneusol: Recherches—Réalisations—Perpectives* [Ph.D. thesis]. Institut National des Sciences Appliquées, Lyon, France.

Medeiros, L. V., Sayão, A. S., Gerscovich, D. M., & Sieira, A. C. (2000). Reuso de Pneus em Geotecnia. In *Seminário Nacional sobre Reuso/Reciclagem de Resíduos Sólidos Industriais* (pp. 1–19), São Paulo, Brasil.

Rybak, J., Ivannikov, A., Egorova, A., Ohotnikova, K., & Fernandes, I. (2017). Some remarks on experience based geotechnical education. *International Multidisciplinary Scientific GeoConference: SGEM, 17*(1.2), 1003–1011.

Yu, J., & Slinn, D. N. (2003). Effects of wave-current interaction on rip currents. *Journal of Geophysical Research: Oceans, 108*(C3).

Time-Dependent Settlements' Analysis of a Spoil Heap

George A. Papatheodorou, Alexandros I. Theocharis, Nikolaos C. Koukouzas, Christos Roumpos, and Ioannis E. Zevgolis

Abstract

Surface coal mining involves excavating the overburden material, typically dumped in massive spoil heaps; appropriately managing and reclaiming these areas is a global priority. Time-dependent settlements are a crucial challenge to any efficient reclamation, such as renewable energy systems or civil infrastructure. Several studies examined the geotechnical stability of spoil heaps, but very few include characterization of the spoil material, and none a time-dependent settlements' analysis. This work simulated time-dependent settlements as self-load 1D consolidation settlements to model the spoil response and predict their completion time and magnitude. The Finite Element Method was used for the consolidation analysis implementing the Hardening Soil model for the spoil. The calibration of the model parameters was based on a multipoint settlement gauge with two-year measurements from a massive spoil heap of 136 m in Northern Greece. Validation of results has been primarily based on secondary in situ measurements of surface markers spanning a larger period and on in situ and laboratory geotechnical testing. The numerical simulation fitted the multipoint settlement gauge measurements well, while the parameters were within the range of the geotechnical testing results. The simulation results agreed well with the validation measurements, further supporting the model's parameters. The settlement completion time was calculated at 22 years, denoting the heap's massive nature and the material's fine-grained nature. Overall, the consolidation analysis of the spoil heap's time-dependent settlements, serving as a paradigm for similar cases, provides valid lessons and results according to the available measurements and can serve as a design basis, at least on a preliminary evaluation.

Keywords

Waste dumps • Spoil piles • Colliery spoil dumps • Geotechnical engineering • Consolidation

1 Introduction

Extracting coal and lignite (brown coal) near the surface (opencast, open-pit, or other types of surface mining) involves excavating the overburden material. This spoil material is usually transported and dumped in nearby areas forming massive spoil heaps. These vast areas should now be adapted to the post-lignite era within this concept. However, these waste materials pose significant challenges, mainly due to the highly heterogeneous spoil nature. Finding the appropriate ways to manage them efficiently is a global priority challenging engineers, researchers, and communities.

This work investigates the issue of time-dependent settlements, which is a crucial challenge to any efficient reclamation, such as renewable energy systems or civil infrastructure. Several studies examined the geotechnical stability of spoil heaps, but very few include characterization of the spoil material, and none a time-dependent settlements' analysis. This work simulates time-dependent settlements as self-load 1D consolidation settlements to model the spoil response and predict their completion time and magnitude. The Finite Element Method was used for the consolidation analysis implementing the Hardening Soil model for the spoil. The calibration of the model parameters was based on

G. A. Papatheodorou · A. I. Theocharis · N. C. Koukouzas
Centre for Research and Technology Hellas, Chemical Process and Energy Resources Institute, Athens, Greece

C. Roumpos
Mining Engineering and Closure Planning Department, Public Power Corporation, Athens, Greece

I. E. Zevgolis (✉)
School of Mining and Metallurgical Engineering, National Technical University of Athens, Athens, Greece
e-mail: izevgolis@metal.ntua.gr

M. Bezzeghoud et al. (eds.), *Recent Research on Geotechnical Engineering, Remote Sensing, Geophysics and Earthquake Seismology*, Advances in Science, Technology & Innovation,
https://doi.org/10.1007/978-3-031-48715-6_2

a multipoint settlement gauge with two-year measurements from a massive spoil heap of 136 m in Northern Greece (Zevgolis et al., 2021). Validation of results has been primarily based on secondary in situ measurements of surface markers spanning a larger period and on in situ and laboratory geotechnical testing.

2 Materials and Methods

Figure 1a illustrates the heap's construction stages (in red) and the implemented stratigraphy (in black). The soil was divided into four layers according to the settlements' profile. A typical and straightforward way to simulate soil's time-dependent settlements relates to soil consolidation. This process describes the soil volume change during the dissipation of the pore pressures developed during construction, in the present case, during the dumping of the spoils. In that vein, this case's settlements that evolve with time are simulated as self-weight 1D consolidation settlements using the Finite Element Method as implemented in the PLAXIS2D software (version 20.04).

Numerous constitutive models have been developed over the past forty years for modeling the stress–strain behavior of soils. This work employs the Hardening Soil model for all materials, an advanced, commercial standard model. It incorporates advanced elements, including nonlinear response to deviatoric loading, stress-dependent stiffness, independent oedometric response, and different response to primary loading and unloading–reloading conditions. Dry unit weight is equal to 17 kN/m^3 from the experimental results (Zevgolis et al., 2021), and saturated unit weight was considered equal to 20 kN/m^3 for all layers; the friction angle was 25° and the cohesion 18 kPa.

The methodology of calibration and validation of the model and the subsequent analysis was as follows:

Step 1: Estimating the order of magnitude of the significant soil parameters elastic modulus E_{50} and water permeability coefficient. Simplified analysis for the 136 m deep soil formation and calculation of the distribution and temporal evolution of the settlements.
Step 2: Fine-tuning and calibrating the parameters to obtain a settlement profile similar to the measurements from 2015 to 2017.

3 Results

The permeability was calibrated between 3×10^{-6} and 5×10^{-6} cm/s, and the E_{50} was 3.5–6.5 MPa for the upper and the bottom layers and 30 MPa for the third layer. Overall an excellent agreement is declared; however, a discrepancy is observed related to the curvature of the settlements in the upper layer, probably related to the inhomogeneous nature of the spoil not being accurately simulated. Based on the calibrated results, the self-weight consolidation settlements were compared with an independent marker measurement. The marker measures the settlements independently from the multipoint gauge and has been used for further validation; Fig. 2a illustrates this comparison. The difference between the two is mainly attributed to a small spoil heap layer of 4 m not being simulated as the

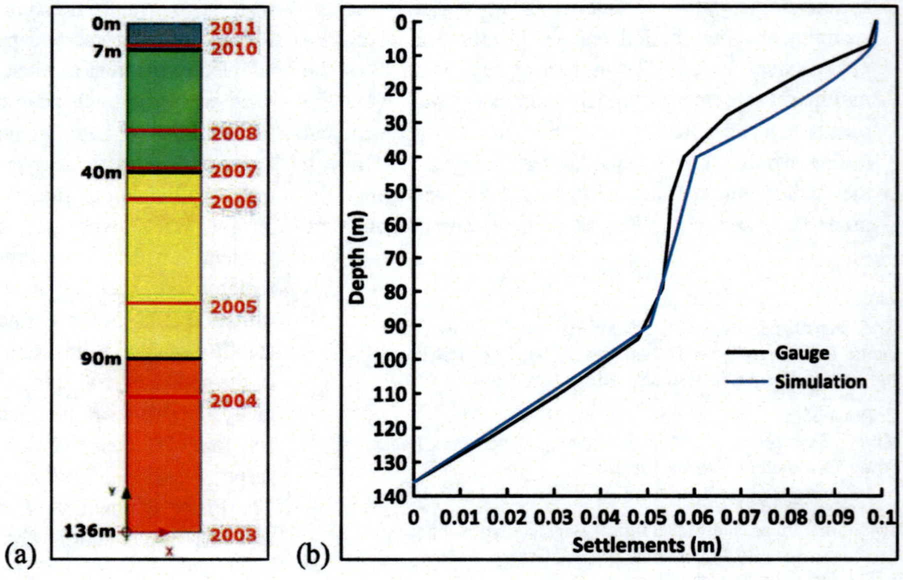

Fig. 1 **a** Soil layers and construction stages; **b** simulation results and settlement profile of multipoint gauge

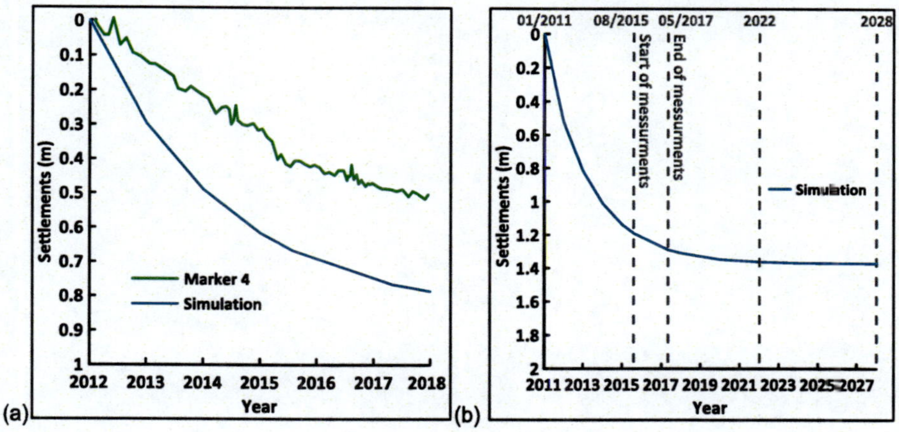

Fig. 2 **a** Simulation settlements versus measured marker settlements; **b** consolidation settlement evolution with time

multipoint gauge did not reach that depth. Nevertheless, the overall trend and the order of settlement it similar for the two cases.

Figure 2b illustrates the general evolution for the whole period from the heap construction until the end of the consolidation. The self-weigh 1D consolidation analysis finished practically in 2022; considering the heap's creation started in 2000, the total settlement procedure needed 22 years.

4 Conclusions

This work simulated time-dependent settlements as self-load 1D consolidation settlements using the Finite Element Method and based on a multipoint settlement gauge from a massive spoil heap in Northern Greece. Validation of results has been based on an independent surface marker's measurements spanning a larger period and on in situ and laboratory geotechnical testing. The numerical simulation fitted the multipoint settlement gauge measurements well, while the parameters were within or at the limits of the geotechnical testing results. Additionally, the simulation results agreed well with the validation measurements, further supporting the model's parameters. The consolidation time of the whole spoil heap was calculated at 22 years, denoting the heap's massive nature and the material's fine-grained nature. Overall, the analysis of the spoil heap's time-dependent settlements, serving as a paradigm for similar cases, provides valid lessons and results according to the available measurements and can serve as a design basis, at least on a preliminary evaluation.

Acknowledgements This work has received funding from the European Union's Research Fund for Coal and Steel under the project SUMAD grant agreement No 847227.

Reference

Zevgolis, I. E., Theocharis, A. I., Deliveris, A. V., Koukouzas, N. C., Roumpos, C., & Marshall, A. M. (2021). Geotechnical characterization of fine-grained spoil material from surface coal mines. *Journal of Geotechnical and Geoenvironmental Engineering*, *147* (7), 04021050.

A Time-Domain Deconvolution Procedure for Soil Deposits with Nonlinear Properties

Luis A. Pinzón, Miguel A. Mánica, Diego A. Hidalgo-Leiva, and Luis G. Pujades

Abstract

In this study, a procedure to perform a time-domain deconvolution in nonlinear elastoplastic materials has been proposed. The aim is to develop accelerograms at the base of a finite element model (FEM) from surface records to perform soil-structure interaction analyses. This procedure is based on finding a relationship between the surface response and the surface target spectrum, modifying the input signal at the base through a spectral matching technique until obtaining a surface response compatible with the target spectrum. To this end, a FEM of a soil deposit, considering its nonlinear properties and ground motion records from the L'Aquila earthquake, was used to validate the methodology; a Eurocode 8 design spectrum was used as the target. It was found that the procedure produces input signals at the base that generate surface spectra similar to the target spectrum, with errors lower than 10% for all considered periods. Results obtained demonstrate the usefulness of the proposed methodology.

L. A. Pinzón (✉)
Scientific and Technological Research Center, Universidad Católica Santa María La Antigua, Panama City, Panama
e-mail: lpinzon@usma.ac.pa

L. A. Pinzón
Sistema Nacional de Investigación, Secretaría Nacional de Ciencia, Tecnología e Innovación, Panama City, Panama

M. A. Mánica
Institute of Engineering, National Autonomous University of Mexico, Mexico City, Mexico

D. A. Hidalgo-Leiva
Earthquake Engineering Laboratory, Universidad de Costa Rica, San Jose, Costa Rica

L. G. Pujades
Department of Civil and Environmental Engineering, Universitat Politècnica de Catalunya, Barcelona, Spain

Keywords

Deconvolution • Dynamic analysis • Finite element • Ground motion • Seismic hazard • Site response • Soil-structure interaction

1 Introduction

A common practice to quantify soil effects on propagated strong motions is to perform one-dimensional site response analyses. This type of analysis can be performed in the frequency-domain and in the time-domain. In both cases, for more realistic results, it is necessary to include the nonlinear properties of materials. In time-domain, the availability of accelerograms fulfilling specific requirements is an issue of broad interest for soil-structure interaction analyses, especially when nonlinear properties are considered (Pinzón et al., 2019, 2020). When numerical dynamic analyses are performed, e.g., using the finite element method, the input signal is required at the base of the model. Nevertheless, acceleration records are usually obtained at the surface and, therefore, it is necessary to perform a deconvolution process to generate the associated motion at the base of the soil deposit. Defining an accelerogram at the base of a model to match a given target motion at the surface is an inverse process in nonlinear analyses that can be solved, approximately, through conventional deconvolution procedures (e.g., equivalent linear method). Nevertheless, results are not always as accurate as expected. In addition, if complex material models are used to characterize the soil in dynamic soil-structure interaction analyses (e.g., using FEM), the input motion derived with an equivalent linear method will not be compatible with the assumed soil behavior. In this study, a simple iterative procedure for time-domain deconvolution in nonlinear elastoplastic materials has been developed. A finite element model in

M. Bezzeghoud et al. (eds.), *Recent Research on Geotechnical Engineering, Remote Sensing, Geophysics and Earthquake Seismology*, Advances in Science, Technology & Innovation,
https://doi.org/10.1007/978-3-031-48715-6_3

PLAXIS 2D (Bentley Systems, 2020) and ground motion records from the 2009 L'Aquila earthquake were used to validate the proposed methodology.

2 Proposed Algorithm

The procedure is based on finding a relationship between the free-field surface response obtained from FEM analyses and the target spectrum at the surface by adjusting iteratively the input signal at the base through a spectral matching technique, until obtaining a surface response compatible with the target motion. The proposed iterative deconvolution procedure is described in Fig. 1.

3 Validation

3.1 Finite Element Model

Since only free-field conditions are considered, the FEM corresponds to a slender column of soil (unit width) with special lateral boundary conditions to account for the semi-infinite lateral extent of the deposit (Zienkiewicz et al., 1988). The soil deposit is characterized as a 30 m deep medium to dense sandy layer, overlying a bedrock. The input motion was applied at the base of the model through a compliant base (Joyner & Chen, 1975) that minimizes reflection of the downward propagating waves. An elasto-plastic constitutive model (Brinkgreve et al., 2007) was

Fig. 1 Algorithm of the proposed approach

1. Set $i = 0$.
2. Set $ACC_i^{\text{base}}(t)$ = initial input motion.
3. Perform FEM site response analysis with $ACC_i^{\text{base}}(t)$ and get $SA_i^{\text{surface}}(T)$.
4. Compute raw correction factor:

   $$F_{i+1}^{\text{raw}}(T) = \frac{SA^{\text{target}}(T)}{SA_i^{\text{surface}}(T)}$$

   IF $F_{i+1}^{\text{raw}}(T) \geq 0.95$ AND $F_{i+1}^{\text{raw}}(T) \leq 1.05$

   $F_{i+1}^{\text{raw}}(T) = 1.0$

   END IF
5. Compute adjusted correction factor:

   $$AF_i(T) = \frac{SA_i^{\text{surface}}(T)}{SA_i^{\text{base}}(T)} \qquad a_{i+1}(T) = 1.025 - 0.4375\, AF_i(T)$$

   IF $a_{i+1}(T) < 0.15$

   $a_{i+1}(T) = 0.15$

   ELSE IF $a_{i+1}(T) > 0.5$

   $a_{i+1}(T) = 0.5$

   END IF

   $$F_{i+1}(T) = 1.0 + \left(F_{i+1}^{\text{raw}}(T) - 1.0\right) a_{i+1}(T)$$
6. Compute new base response spectra as $SA_{i+1}^{\text{base}}(T) = SA_i^{\text{base}}(T) F_{i+1}(T)$.
7. Perform spectral matching to the initial input motion with $SA_{i+1}^{\text{base}}(T)$ as target and get $ACC_{i+1}^{\text{base}}(t)$.
8. Perform FEM site response analysis with $ACC_{i+1}^{\text{base}}(t)$ and get $SA_{i+1}^{\text{surface}}(T)$.
9. Compute error measure E comparing $SA_{i+1}^{\text{surface}}(T)$ and $SA^{\text{target}}(T)$.

   IF $E > TOL$

   Set $i = i + 1$

   GOTO step 4

   END IF

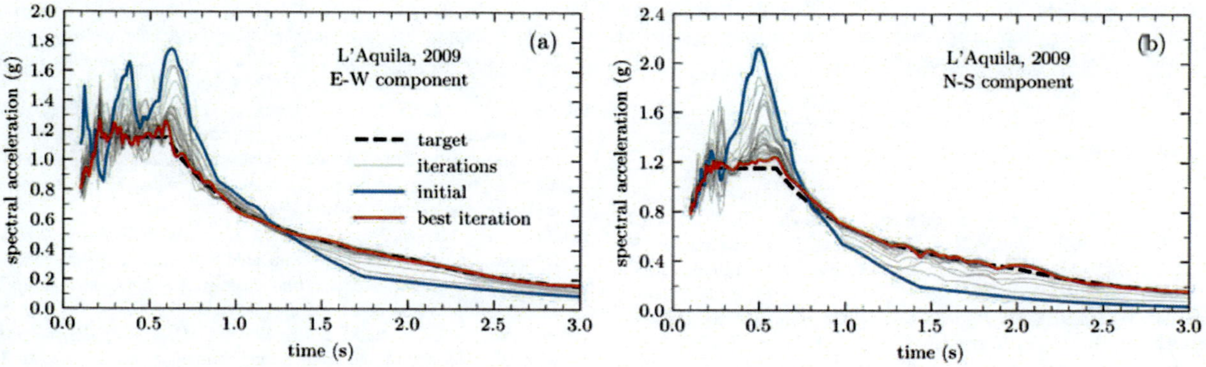

Fig. 2 Surface response spectra resulting from the iterative procedure

adopted to account for the nonlinear response of the soil. The model includes features like small-strain stiffness degradation, a hyperbolic hardening law, and a Mohr–Coulomb type limit surface. This finite element model has been created and analyzed using the PLAXIS 2D software (Bentley Systems, 2020).

3.2 Target Spectrum and Input Motions

To assess the proposed algorithm, an Eurocode 8 (CEN, 2004) design spectrum has been used as a target for the dynamic analyses. The selected design spectrum represents sites with high seismicity ($M_w \geq 5.5$, Type 1) and with dense soil deposits (subsoil class C, $180 < V_{s30} \leq 360$ m/s). A design acceleration of 0.40 g was assumed, which corresponds to a Seismic Zone 1 in Italy (> 0.25 g). A ground motion pair, derived from the horizontal as-recorded components of the L'Aquila (2009) earthquake, was selected as the input motion at the base of the FEM. The records were taken from the PEER Ground Motion Database (Ancheta et al., 2013), considering the site conditions, the target spectrum, and their capacity to significantly excite the soil-structure system. For cases in which the specific site is known, site-specific ground motions should be used according to the seismic hazard of the area.

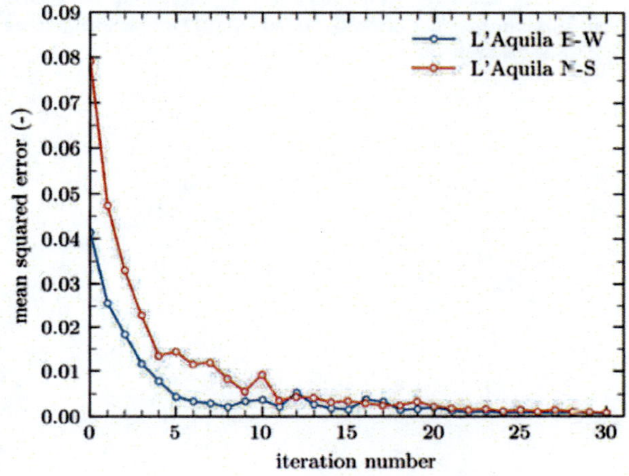

Fig. 3 Evolution of the mean squared error between target spectrum and response spectra resulting from the iterative procedure

4 Results

The procedure described in Sect. 2 was applied to the selected accelerograms. Figure 2 shows the response spectra obtained on the surface of the model after performing 30 iterations. The red line indicates the best iteration, showing the smallest mean square error. In both cases, signals that produce surface spectra similar to the target were obtained, with differences of ± 10% in all considered periods.

In general, the results show progressive improvement and stability with the number of iterations (see Fig. 3). The signals obtained at the iteration with the smallest error preserve a great similarity to the initial records, despite the fact that its response spectrum was modified considerably.

5 Conclusions

Obtained results show the effectiveness of the proposed iterative deconvolution procedure. This methodology can be easily applied to obtain accelerograms on the base of a FEM, from surface records, to match a given surface target spectrum considering the nonlinearity of the ground. This is particularly relevant for application in soil-structure interaction analyses, where complex elastoplastic models are used to characterize the behavior of the soil.

Acknowledgements This research was funded by the National Secretariat of Science, Technology and Innovation of the Republic of Panama (SENACYT) (grant number FID 207-2022).

References

Ancheta, T. D., Darragh, R. B., Stewart, J. P., Seyhan, E., Silva, W. J., Chiou, B. S.-J., Wooddell, K. E., Graves, R. W., Kottke, A. R., Boore, D. M., Kishida, T., & Donahue, J. L. (2013). *PEER NGA-west2 database*. Berkeley.

Bentley Systems. (2020). *PLAXIS CONNECT edition 20.04*. Delft.

Brinkgreve, R. B. J., Kappert, M. H., & Bonnier, P. G. (2007). Hysteretic damping in a small-strain stiffness model. In *10th International Symposium on Numerical Models in Geomechanics* (pp. 737–742). https://doi.org/10.1201/NOE0415440271.ch106

CEN. (2004). *European Standard EN 1998-1:2005 Eurocode 8: Design of structures for earthquake resistance. Part 1: General rules, seismic actions and rules for buildings*. European Committee for Standardization.

Joyner, W. B., & Chen, A. T. F. (1975). Calculation of nonlinear ground response in earthquakes. *Bulletin of the Seismological Society of America, 65*(5), 1315–1336. [Online]. Available: http://www.bssaonline.org/content/65/5/1315.abstract

Pinzón, L. A., Mánica, M. A., Pujades, L. G., & Alva, R. E. (2019). A simplified approach to account for directionality effects on 2D dynamic soil-structure interaction analysis. In *Earthquake geotechnical engineering for protection and development of environment and constructions* (pp. 4490–4497).

Pinzón, L. A., Mánica, M. A., Pujades, L. G., & Alva, R. E. (2020). Dynamic soil-structure interaction analyses considering directionality effects. *Soil Dynamics and Earthquake Engineering, 130*, 106009. https://doi.org/10.1016/j.soildyn.2019.106009

Zienkiewicz, O. C., Bicanic, N., & Shen, F. Q. (1988). Earthquake input definition and the transmitting boundary conditions. In I. S. Doltsinis (Ed.), *Advances in computational nonlinear mechanics. International Centre for Mechanical Sciences (Courses and Lectures)* (pp. 109–138). Springer.

Geological, Physico-Chemical Characterization and Estimation of Clay Reserves Exploited in Pottery in Draa-Tafilalet Region (Morocco)

Toufik Remmal, Halima Jounaid, Fouad Amraoui, Soukaina Bedda, and Assia Laroussi

Abstract

Investigations carried out on pottery in the Draa-Tafilalet (DT) region have shown that this activity is subject to many constraints linked particularly to the depletion of natural resources, lack of workforce, and the quality of the final products. This study contributes to the recognition, characterization, and valuation of clay materials in the DT region in order to control a sustainable supply for local potters. To do this, various analyses are carried out, particularly at the lithological, mineralogical, chemical, and technological characteristics of the clays exploited in pottery, as well as data relating to the estimated reserves in the supply zones. The clays are harvested in different geological contexts from four provinces covering different Saharan structural domains. The analyses generally show a loamy to sandy-loamy composition with a clay fraction not exceeding 14%. Most materials have a plastic character ($15 < PI < 40$). The identified minerals are dominated by quartz, total clay, goethite, calcite and feldspars. The clay fraction is composed of illite, chlorite, kaolinite, smectite, and palygorskite. The collected data are processed in chemical, mineralogical and granulometric diagrams to test the suitability of these materials for use in ceramics. Medium-quality samples were balanced in formulations to improve their pottery making ability. It shows that, despite qualitative fluctuations, DT clays can be used in pottery, subject to adjustments relating to manufacturing processes and firing conditions. The amount of clay consumed annually by potters remains very modest (~ 3000 tons/year) compared to potential clay reserves estimated at 30 million tons.

Keywords

Clay • Ceramics • Valorization • Reserves • Saharan domains • Morocco

1 Introduction

Clays and clay minerals have been widely used for a very long time as main components of raw materials for the manufacturing of various ceramic products. The diversity of their fields of use and properties depends largely on their structure and composition (Reeves et al., 2006). Few studies have been carried out on the quality and potential use of regional clays (Daoudi et al., 2014; El Yakoubi et al., 2006). In DT region, the exploitation of clays in pottery is entirely artisanal, based on empirical knowledge of the local potters. Economically, pottery plays a very important role in employment as it is a significant source of financial gain for the local population. Nevertheless, this sector is not well managed in terms of quantity and quality and is increasingly subject to many constraints related generally to the quality of the produced objects as well as the problems of raw material supply. The purpose of this study is to identify opportunities for valorization and optimization to ensure a quantitative and qualitative availability of these clays which is an essential phase to control a sustainable supply for the potters in the area.

2 Materials and Methods

Current clay mining sites have been listed in different geological contexts spread over four provinces covering different structural domains: Zagora, Errachidia, Tinghir, and Ouarzazate (see Fig. 1). Each site was the subject of a geological reconnaissance supported by lithological and structural sections. 15 clay samples were taken from:

T. Remmal · H. Jounaid (✉) · F. Amraoui · S. Bedda · A. Laroussi
Laboratory of Geosciences Applied to Planning Engineering, Hassan II University of Casablanca, B.P 5366 Maarif, Casablanca, Morocco
e-mail: jounaidhalima@gmail.com

M. Bezzeghoud et al. (eds.), *Recent Research on Geotechnical Engineering, Remote Sensing, Geophysics and Earthquake Seismology*, Advances in Science, Technology & Innovation,
https://doi.org/10.1007/978-3-031-48715-6_4

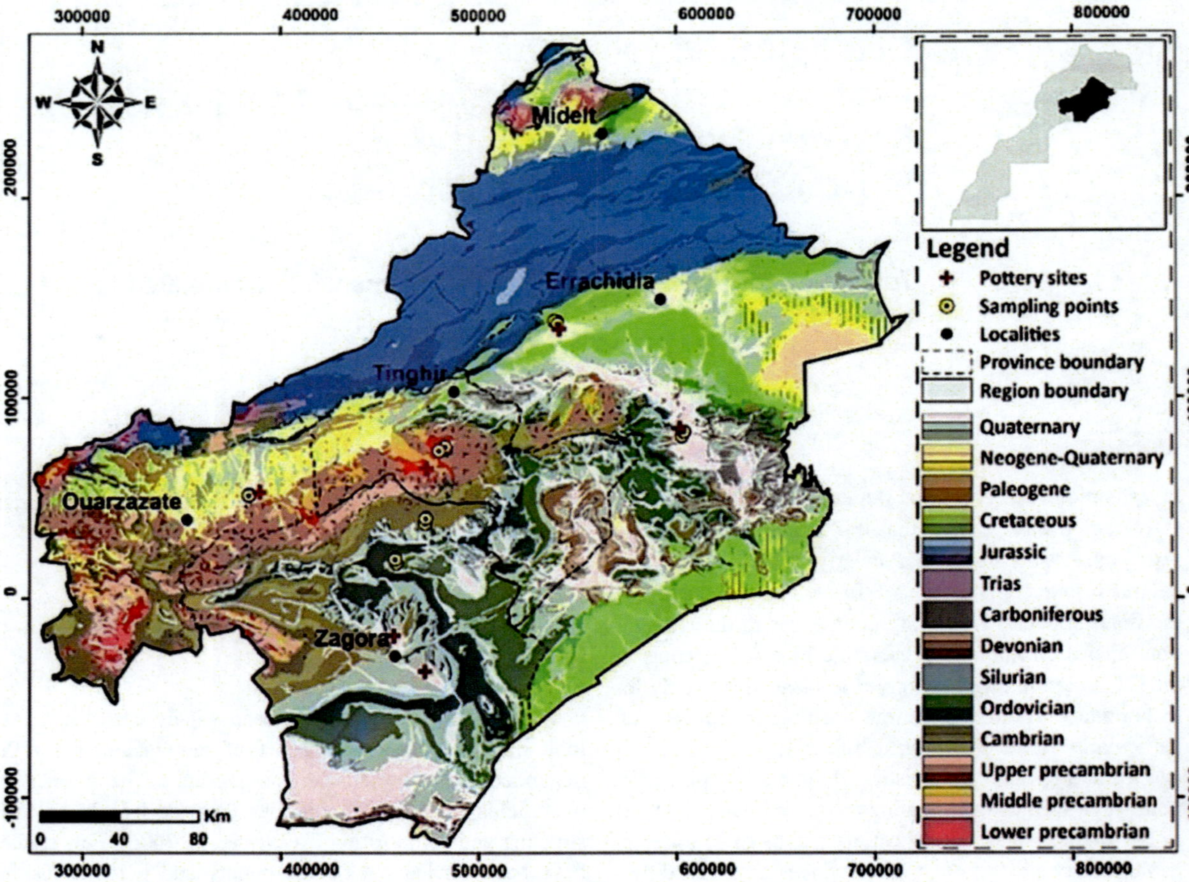

Fig. 1 Geological context of DT region (extracted from the geological map of Morocco at the scale of 1:1,000,000)

Tamegroute (TD1, TD2), Astour (TD3, TD4), Sifa (TD5L, TD5R, TD5S), Tadighoust (TD6J, TD6R, TD6V), Iknioune (TD8B, TD8C, TD9), Skoura (TD10L, TD10TL).

The particle size distribution was carried out using sieving and laser diffraction analysis using a Horiba LA-920 laser analyzer. The Atterberg limits were determined according to the NF P 94-052-1 (1995) and NF P 94-051 (1993) standards. The organic matter content was determined according to XP P 94-047 (1998) standard. The mineralogical analysis was performed by X-ray diffraction using a Bruker D8-Advance diffractometer. The chemical analysis was performed by X-ray fluorescence using a PANalytical Axios spectrometer.

The estimation of the clay resources has been carried out by adopting the sectional estimation method (Stone & Dunn, 1996). The objective is to provide a tonnage estimate for each volume element from a limited sampling. The calculation of the mineral reserves takes into consideration the dimensions of the clay formations. The volume of each section is its area times the average thickness. The obtained volumes are then converted into tons of ore using an estimated density of 1.75.

3 Results

Grain size analyses show a loamy to sandy-loamy composition with a clay fraction not exceeding 14% (Fig. 2a), which gives these materials moderately high porosity and permeability. The plasticity index (PI) deduced from the Atterberg limits shows that most materials have a plastic character ($15 < PI < 40\%$). The organic matter content is relatively low, not exceeding 5%. The minerals are dominated by quartz (10–33%), total clay (8–62%), goethite (8–26%), feldspars (2–15%), and calcite (0–25%). The clay fraction is mainly composed of illite (45–100%) which remains the most dominant mineral in practically all samples and chlorite (0–30%). Other clay minerals are generally

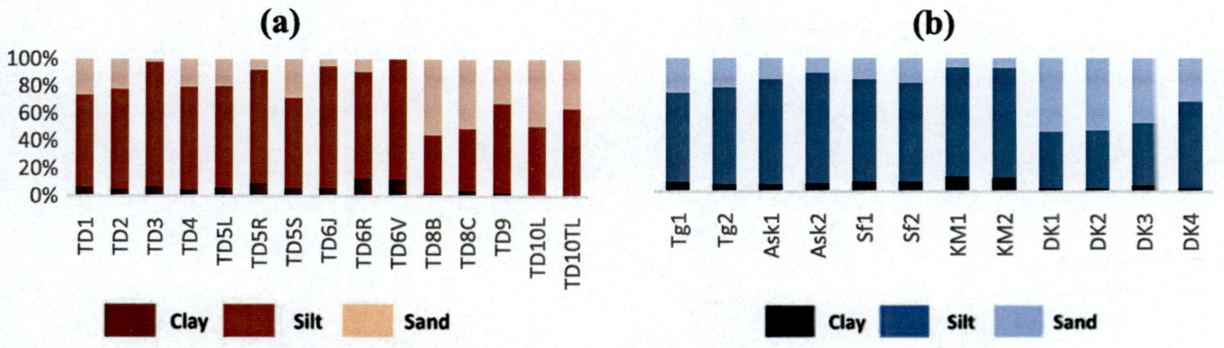

Fig. 2 Grain size distribution of raw clays (**a**) and mixtures (**b**)

poorly represented or even absent, except for some slightly illitic samples where their contents become higher up to 50% for smectite, 40% for kaolinite, 30% for palygorskite, 15% for vermiculite, and 20% for mixed-layer clays. Chemical analyses show that silica remains by far the most represented oxide ($40 < SiO_2 < 67\%$) and it is entirely correlated with the quartz content. The Al_2O_3 content is high in samples rich in clay fraction and/or kaolinite; in both cases, it varies between 15 and 18.5%. The CaO content is disparate (1.9–19%) and varies correlatively with the proportion of calcite in the samples. The contents of alkaline oxides, particularly Na_2O, remain low ($0.17 < Na_2O < 0.35\%$, $1.13 < K_2O < 6.17\%$).

4 Discussion

The ability of clays to be used in the ceramic industry can be tested through multi-criteria diagrams (Fiori et al., 1989; Strazzera et al., 1997), in which the chemical and mineralogical compositions of raw clays and mixtures adopted by the potters are shown according to the reference clays. The details of mixture formulations and compositions are given in Table 1 and Fig. 2b. Two types of reference clays, from the deposits of Fez (FES) and Safi (SAFI), are incorporated into these diagrams.

Overall, the raw clays used by potters are of acceptable quality, due to their mineralogical and chemical composition (see Figs. 3 and 4).

The richness of certain samples in quarto-feldspathic components can lead to a shift outside the scope of the required standard. This gap can, however, be made up for by considering the formulas of mixtures with a new mineralogical distribution which makes it possible to situate the mixtures in the field of adequate or even optimal compositions for the ceramic industry (see Fig. 3b). The mixtures dominated by a single type of clay, particularly those used by the potters of Iknioun (Tinghir), remain of lower quality. In the diagram of Fiori et al. (1989), the raw clays remain globally focused on the domain of the red bodies or deviate

Table 1 Mixture formulations adopted by the artisans of Draa-Tafilalet region

Clay mining site	Mixture formulations	Mixtures references
Tamegroute	100% TD1	Tg1
	100% TD2	Tg2
Astour	75% TD4 + 25% TD3	Ask1
	50% TD4 + 50% TD3	Ask2
Sifa	57% TD5L + 36% TD5R + 7% TD5S	Sf1
	33% TD5L + 33% TD5R + 33% TD5S	Sf2
Tadighoust	57% TD6R + 29% TD6J + 14% TD6V	KM1
	57% TD6R + 43% TD6J	KM2
Iknioun	100% TD8B	DK1
	95% TD8B + 5% TD9	DK2
	90% TD8C + 10% TD9	DK3
	100% TD9	DK4
Skoura	50% TD10L + 50% TD10Tl	Sk

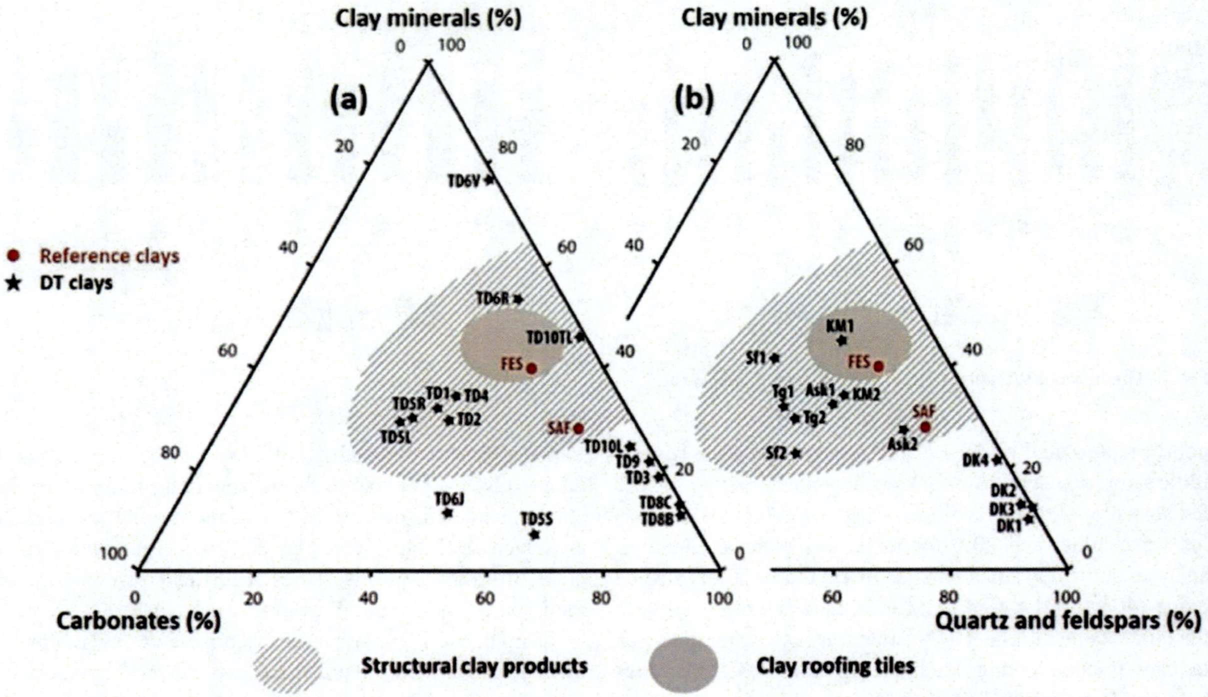

Fig. 3 Ability of raw clays (**a**) and mixtures clays (**b**) to be used in the ceramic industry according to their mineralogical composition

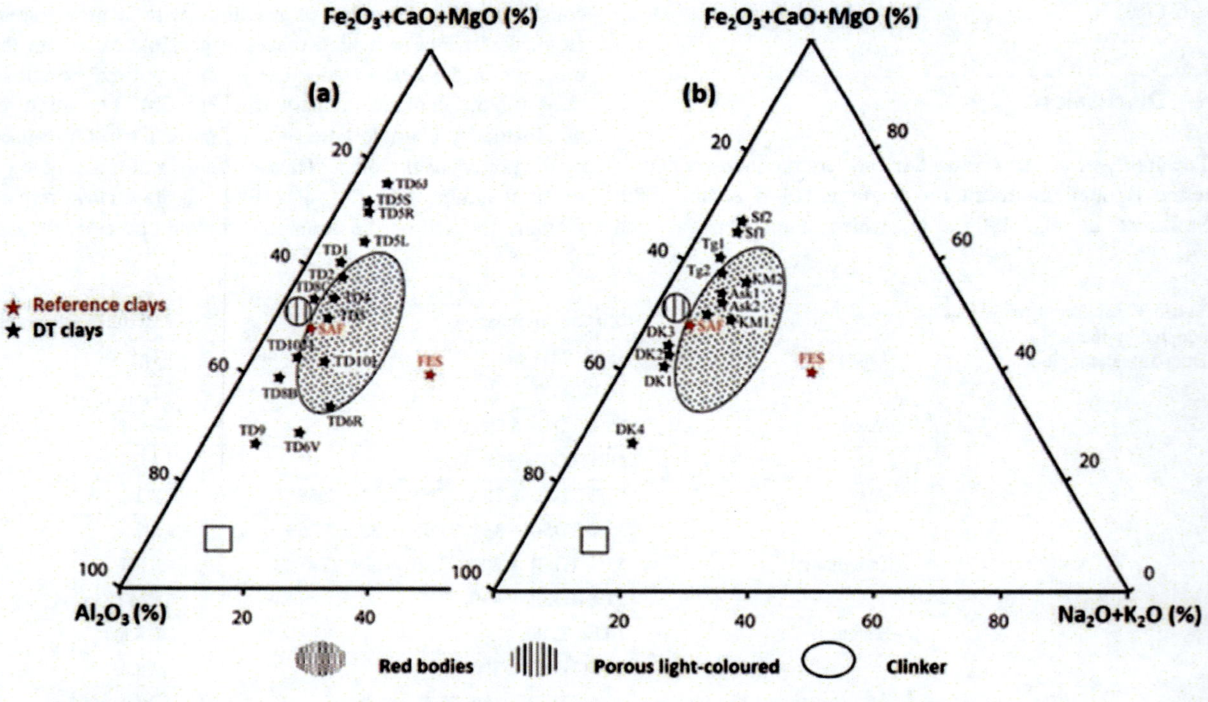

Fig. 4 Ability of raw clays (**a**) and mixtures clays (**b**) to be used in the ceramic industry according to their chemical composition

from it relatively, in the direction of the aluminous and calco-ferro-magnesian poles which reflect the control of the clay phases and the carbonates (see Fig. 4a). On the other hand, and with the exception of rare samples, the pastes made according to the craftsmen's formulation are mostly placed in the area of red bodies in accordance with the quality of the fired pieces produced (see Fig. 4b).

5 Conclusions

The mineralogical and physico-chemical variability of DT clays is linked to the lithological and morpho-structural characteristics of the sampling sites. Overall, the clays studied are suitable for making usual Moroccan pottery (tajines, majmars, kasaâs, jars ...). The formulations adopted by the potters make it possible to optimize the quality of the materials by also acting on the methods of treatment. Finally, the clay reserves are substantial enough (~ 30 million tons) to perpetuate the activity of the sector which remains modest around 3000 tons/year.

References

Daoudi, L., Hicham, E. E., & Latifa, S. (2014). Characteristics and ceramic properties of clayey materials from Amezmiz region. *Applied Clay Science, 102*, 139–147.

El Yakoubi, N., Aberkane, M., & Ouadia, M. (2006). Potentialité d'utilisation d'argiles marocaines de Jbel Kharrou dans l'industrie céramique. *Geoscience, 338*, 693–702.

Fiori, C., Fabbri, B., Donati, G., & Venturi, I. (1989). Mineralogical composition of the clay bodies used in the Italian tile industry. *Applied Clay Science, 4*(5–6), 461–473.

Reeves, G. M., Sims, I., & Cripps, J. C. (2006). *Clay material used in construction*. Geological Society of London.

Stone, J. G., & Dunn, P. G. (1996). *Ore reserve estimates in the real world*. Society of Economic Geologists.

Strazzera, B., Dondi, M., & Marsigli, M. (1997). Composition and ceramic properties of tertiary clays from southern Sardinia (Italy). *Applied Clay Science, 12*(3), 247–266.

Influence of the Amount of Cement on the Drained Shear Strength Parameters of Clay Soil Obtained Through the Direct Shear Test

Sofija Ostojić, Slobodan Živaljević, Zvonko Tomanović, and Radomir Zejak

Abstract

The subject of this paper is the way, in which the amount of cement and the curing time influences the parameters of shearing resistance of clay soil obtained through the direct shear experiment that is routinely conducted in geotechnical laboratories. The direct shear test was performed on silty clay soil samples compacted according to the standard Proctor test, with the compaction degree equal to 95% of the maximum achievable compaction. The shear strength of samples containing 0, 3, 5, 7, and 9% of added cement was tested, and for curing times of 3, 7, and 14 days. The drained shear strength parameters ϕ' and c' were obtained by processing the results in Excel. The drained cohesion increases with increasing the percentage of cement and curing time. The internal friction angle increases with the increase of the percentage of cement up to a certain limit, while a decrease in the internal friction angle value is recorded between 5 and 7% of added cement. In contrast to cohesion, the internal friction angle decreases with the extension of the curing time of the samples.

Keywords

Direct shearing • Clay • Cement • Shear strength

1 Introduction

Many scientific papers have focused on testing of the characteristics of cement-improved soil, with a greater number of authors tested its uniaxial compressive strength (Ho & Chan, 2011; Pakbaz & Alipour, 2012; Tugba, 2015), the shear strength in triaxial compression (Amin, 2015; Asghari et al., 2004; Ho & Chan, 2011), deformation characteristics, as well as parameters important for road construction—the CBR values, while a more limited number of authors tested the shear strength parameters in the direct shear apparatus (Askarani & Pakbaz, 2016; Boroumandzadeh & Pakbaz, 2012). The subject of this research is the influence of the amount of cement on the shear strength parameters of clay soil in a drained direct shear test. Quantification of the impact was carried out through a series of laboratory tests—a drained direct shear test on remolded (pre-compacted) samples of silty clay of low to medium plasticity. A comparative analysis was carried out on samples without added cement, with samples containing 3, 5, 7, and 9% of added cement. All samples were prepared with optimal water content (16%) and compacted by compaction energy according to the standard Proctor test.

2 Materials and Methods

For the purposes of this research, disturbed clay samples were sampled at the location of Đurđevića Tara, municipality of Pljevlja, Montenegro. Based on the conducted identification and classification tests, the soil is classified as silty clay of low to medium plasticity CL-CI according to the British Soil Classification System (BSI, 1981). In the AASHTO classification, the soil is classified as A-6 with a group index of 14. Based on X-ray diffractometry, the minerals present in the untreated sample of the examined soil are quartz (38.2%), muscovite (37.1%), kaolinite (5%), albite (19.6%), and an insignificant percentage of calcite 0.1%.

S. Ostojić
PI Geological Survey of Montenegro, Kruševac bb, 81000 Podgorica, Montenegro

S. Živaljević (✉) · Z. Tomanović · R. Zejak
Faculty of Civil Engineering, University of Montenegro, Džordža Vašingtona bb, 81000 Podgorica, Montenegro
e-mail: slobodanzivaljevic1979@gmail.com

M. Bezzeghoud et al. (eds.), *Recent Research on Geotechnical Engineering, Remote Sensing, Geophysics and Earthquake Seismology*, Advances in Science, Technology & Innovation,
https://doi.org/10.1007/978-3-031-48715-6_5

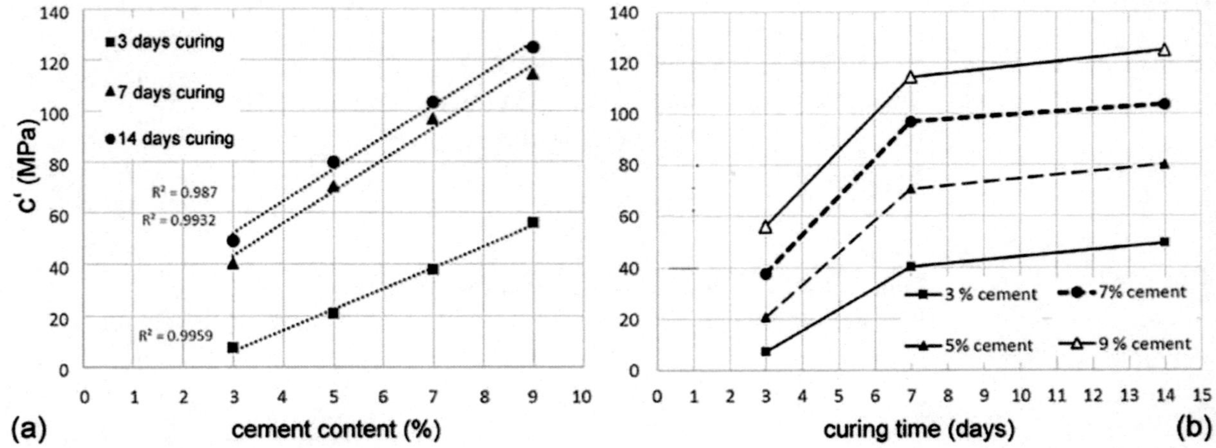

Fig. 1 Effect of cement content and curing time on drained cohesion

Cement is added to pre-dried and mechanically crushed samples in amounts of 3, 5, 7, and 9% by dry weight of the soil. In the process of specimens preparing the cement, CEM II/B-M(V-L) 32.5R was used. After adding cement, the samples were compacted with a compaction energy corresponding to the standard Proctor test. The specimens were 60 mm in diameter and 25 mm height. Given that the dimensions of the samples for the direct shear test are smaller than the samples for the standard Proctor test, a correlation was made that achieved a compaction of approximately 95% of the maximum compaction obtained by the previously performed Proctor test. The average dry unit weight of the samples is $\gamma_d = 17$ kN/m^3. The samples prepared in this way were cured for 3, 7, and 14 days, after which shear strength tests were performed in a drained direct shear test. During the curing period, the samples were protected with PVC foil from the loss of moisture and other influences from the air.

The drained direct shear test was performed in two phases, phase I-consolidation and phase II-shear phase. According to the EN ISO 17892-10, standard horizontal deformation rates were determined based on the consolidation curves obtained in the first phase of the test. Based on the calculation of the required shear rate, it can be concluded that extension of the curing time affects acceleration of consolidation and thus the rate of horizontal deformation. These rates were different for different sample curing times, namely 0.005, 0.015, and 0.05 mm/min for samples cured for 3, 7, and 14 days. During the testing, data were collected which were later processed and analyzed using Microsoft Excel.

3 Results

The results obtained during this research are shown graphically. The diagrams in Figs. 1 and 2 show the obtained values of the drained shear strength parameters, cohesion c' and internal frictional angle ϕ' for the tested samples with different percentages of cement and curing time after adding cement. For untreated specimen (0% cement), values of $c' = 2$ kPa and $\phi' = 15.6°$ are obtained.

4 Discussion

In Fig. 1a, an increase in the value of drained cohesion is clearly observed with an increase in the percentage of cement added to the sample. In addition, an increase in cohesion with an increase in sample curing time is visible (Fig. 1b). In the diagram in Fig. 1a, cohesion values for samples with 3, 5, 7, and 9% cement cured for 3, 7, and 14 days are shown. Increases in the drained cohesion value were noted, so the cohesion value of the uncemented sample is only 2 kN/m^2, while the measured cohesion value for the sample cured for 3 days with 9% of cement was 56 kN/m^2. With each increase in the percentage of added cement, the cohesion value increased, for the case when the samples were cured for 3 days and for increasing the percentage of cement from 3 to 5%, the value of cohesion increased by 180%, while for all subsequent increases, this percentage of increase is smaller, but still significant. Cohesion values increase due to the cementation intergrain bonds and cement hydration. Similar conclusions were reached by Askarani and Pakbaz in their research (Askarani & Pakbaz, 2016).

Contrary to cohesion, the angle of internal friction does not increase constantly with increasing value of added cement. Based on this research and the results presented above, it can be observed that with an increase in the percentage of cement, the angle of internal friction increases to a certain value, after which decreasing values are observed. For samples cured for 3 days, a decrease in the value of the angle of internal friction is observed for the percentage of added cement greater than 7%, while for samples cured for 7 and

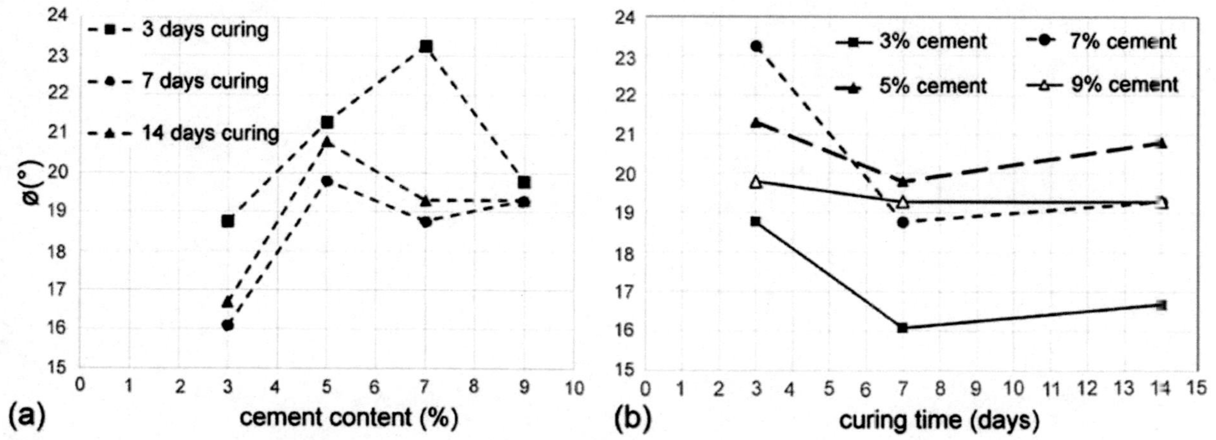

Fig. 2 Effect of cement content and curing time on angle of internal friction

14 days, this decrease can be observed already after a cement content greater than 5% (Fig. 2b). These decreases could be attributed to the linear Mohr–Coulomb approximation of cemented soil nonlinear failure envelope. With an increase in the curing time of the samples, there is a decrease in the value of the internal friction angle of the tested samples. The value of the angle of internal friction shows only a slight increase when the curing time is extended from 7 to 14 days (Fig. 2b).

5 Conclusions

Based on the results of the experimental research conducted within this paper, the following conclusions can be drawn:

- The drained direct shear test was performed on silty clay soil samples compacted according to the standard Proctor test, with the compaction degree equal to 95% of the maximum achievable compaction. The shear strength of samples containing 0, 3, 5, 7, and 9% of added cement was tested for curing times of 3, 7, and 14 days.
- By increasing the percentage of added cement and extending the sample curing time, the value of the cohesion increases.
- The angle of internal friction increases with an increase in the percentage of added cement up to a 5% of added cement for curing times of 7 and 14 days, after which a decrease in value is noted. With an increase in the curing time of the samples, there is a decrease in the value of the internal friction angle of the tested samples. These decreases could be attributed to the linear Mohr–Coulomb approximation of cemented soil nonlinear failure envelope. Future research is planned to include direct shear tests for defining nonlinear failure envelope in a low-normal stress range.

References

Amin, D. (2015). *Triaxial testing of lime/cement stabilized clay: A comparison with unconfined compression tests*. KTH Royal Institute of Technology.

Asghari, E., Toll, D. G., & Haeri, S. M. (2004). Effect of cementation on the shear strength of Teheran gravelly sand using triaxial tests. *Journal of Sciences, Islamic Republic of Iran, 15*(1), 65–71.

Askarani, K. K., & Pakbaz, M. S. (2016). Drained shear strength of over-consolidated compacted soil-cement. *Journal of Materials in Civil Engineering, 28*(05), 04015207.

Boroumandzadeh, B., & Pakbaz, M. S. (2012). Evaluation of effect of cementation on drained shear strength of over consolidated clay soils. *World Applied Sciences Journal, 16*(10), 1375–1379.

BS: 5930. (1981). *British code of practice for site investigations*. BSI.

Ho, M.-H., & Chan, C.-M. (2011). Some mechanical properties of cement stabilized Malaysian soft clay. *World Academy of Science, Engineering and Technology. International Journal of Civil and Environmental Engineering, 5*(2), 76–83.

Pakbaz, M. S., & Alipour, R. (2012). Influence of cement addition on the geotechnical properties of Iranian clay. *Applied Clay Science, 67–68*.

Tugba, E. (2015). Influence of cement treatment on unconfined compressive strength and compressibility of lean clay with medium plasticity. *Arabian Journal for Science and Engineering, 40*(3), 763–772.

The Effect of Microwave Treatment on the Strength of Nigde Marble

Sair Kahraman and Masoud Rostami

Abstract

This study investigates the changes in the strength of Nigde marble irradiated by microwave energy. The uniaxial compressive strength (UCS) and the Brazilian tensile strength (BTS) tests were carried out on the untreated samples and on the samples which were microwave treated at 6 kW for a duration of 5 min. The tests were carried out on both dry, semi-saturated and saturated samples. After the irradiation of the samples by microwave at 6 kW power for 5 min, while the temperature reached about 215 °C for the dry specimens, the temperatures of the semi-saturated and fully saturated specimens raised to about 230 °C and 250 °C, respectively. The strength decrements in the UCS samples are 10.9% and 35.4% for the dry and saturated conditions, respectively. The strength decrements in the BTS samples are 30.4% and 37.7% for the dry and saturated conditions, respectively. The UCS value of the treated dry specimen decreases slightly. On the other hand, the UCS values of the treated samples decline steadily with the increasing saturation degree. The BTS value of the irradiated dry specimen decreases rapidly. The BTS value of the saturated specimens after the treatment declines slowly with the increasing saturation degree that the UCS and the BTS specimens show different behavior under the microwave treatment is probably due to the size effect. Concluding remark is that the microwave treatment has important effect on the strength of Nigde marble.

Keywords

Marble • Microwave treatment • Water saturation • Strength

S. Kahraman (✉) · M. Rostami
Hacettepe University, Ankara, Türkiye
e-mail: sairkahraman@yahoo.com

1 Introduction

The mechanical excavation of rocks has been increasing in rock engineering projects. The low advance rate and the high tool wear are important problems when excavating hard and abrasive rocks. A concrete solution has not been found yet although there are some research studies on this topic. In order to increase cutting rate and decrease wear rate during cutting very hard rocks, recently, the studies on some innovative methods such as microwave-assisted rock breakage have been drawing attention. Several researchers have investigated the effects of microwave treatment on the strength of rocks (Hassani et al., 2016; Kahraman et al., 2020; Motlagh, 2009; Peinsitt et al., 2010; Satish et al., 2006). The aim of these researchers is to provide a base for the possible application of microwave-assisted rock cutting.

Although several researchers have investigated the influence of microwave treatment on the strength of rocks for the possible application in the microwave-assisted rock excavation, there are too limited studies about the effects of microwave treatment on the strength of carbonate rocks. In this study, the effects of microwave treatment were investigated on the mechanical properties of Nigde marble.

2 Experimental Studies

The samples of marble from Nigde area of Turkey were used in the studies. The marble has about 98% $CaCO_3$ content. The NX (54.7 mm) size cylindrical and disc specimens were prepared from the block samples collected from the site. Cylindrical specimens had a height of 136 mm and disc specimens had a thickness of 27 mm.

First the uniaxial compressive strength (UCS) and the Brazilian tensile strength (BTS) values of the untreated specimens were determined. Then, the three UCS specimens were dried with an oven at 105 °C for 24 h. After cooling period, the specimens were treated by microwave at 6 kW

M. Bezzeghoud et al. (eds.), *Recent Research on Geotechnical Engineering, Remote Sensing, Geophysics and Earthquake Seismology*, Advances in Science, Technology & Innovation,
https://doi.org/10.1007/978-3-031-48715-6_6

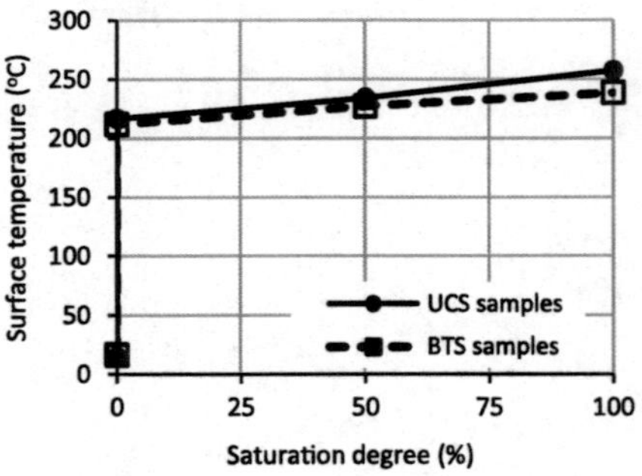

Fig. 1 Correlation between surface temperature and saturation degree

power for 5 min using an industrial microwave oven with a frequency of 2.45 GHz. Other three specimens were saturated with tap water until their weight did not change, then they were treated by microwave at 6 kW power for 5 min. The remaining three specimens were fully saturated and they were left to dry. The semi-saturated conditions (50%) of the specimens were determined by measuring their weights during drying period. After that they were treated by microwave at 6 kW power for 5 min. The surface temperature of each specimen was measured with an infrared gun before the microwave irradiation and immediately after removing the sample from the microwave oven. All surfaces of the samples were scanned by the infrared gun and the average values were used as the final surface temperatures. As soon as the temperatures of the samples were measured, the UCS tests were carried out on the hot specimens. The procedure was repeated for the BTS specimens.

3 Results

After the irradiation of the samples by microwave at 6 kW power for 5 min, while the temperature reached about 215 °C for the dry specimens, the temperatures of the semi-saturated and fully saturated specimens raised to about 230 °C and 250 °C, respectively (Fig. 1). The UCS and the BTS values of the specimens before and after the microwave treatment are given in Table 1 for the different saturation degrees. The strength losses due the microwave treatment were also calculated. The losses in the UCS values are 10.9%, 21.7% and 35.4% for the dry, semi-saturated and fully saturated specimens, respectively. The losses in the BTS values are 30.4%, 34.4% and 37.7% for the dry, semi-saturated and fully saturated specimens, respectively. The BTS loss is 30.4% even for the treatment of the dry specimens. This is due to the fact that small specimens are affected from the irradiation more than that of large specimens.

As indicated in Fig. 2, the UCS value of the treated dry specimen decreases slightly. On the other hand, the UCS

Table 1 Strength losses after the microwave treatment

Saturation degree (%)	UCS (MPa)	UCS loss (%)	BTS (MPa)	BTS loss (%)
	67.3[a]		6.1[a]	
0	59.9	10.9	4.2	30.4
50	52.7	21.7	4.0	34.4
100	43.5	35.4	3.7	37.7

[a] UCS and BTS of the untreated samples

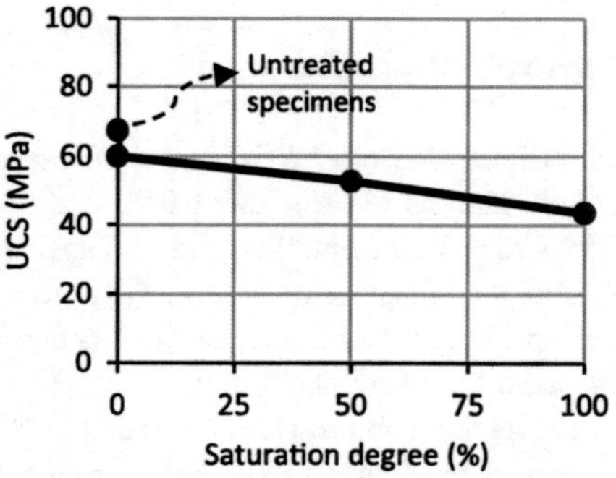

Fig. 2 Correlation between the UCS and saturation degree

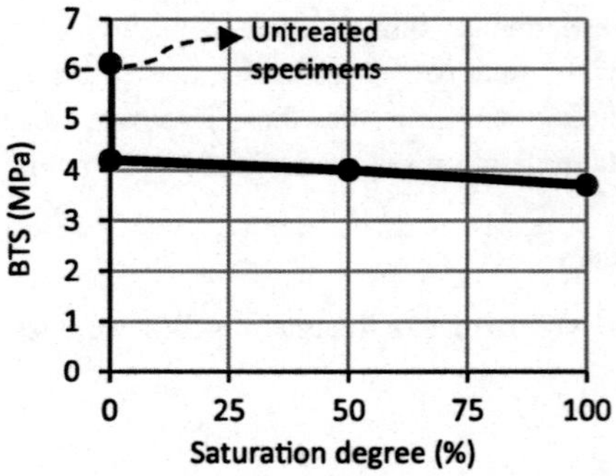

Fig. 3 Correlation between the BTS and saturation degree

values of the treated samples decline steadily with the increasing saturation degree. The BTS value of the irradiated dry specimen decreases rapidly (Fig. 3). The BTS value of the saturated specimens after the treatment declines slowly with the increasing saturation degree that the UCS and the BTS specimens show different behavior under the microwave treatment is probably due to the size effect.

4 Discussion

It was observed that the water saturation did not have much effect on the surface temperature of the UCS and BTS samples, but it was especially effective on the UCS loss. This is because the microwave heats the objects internally and the internal temperature is approximately twice the surface temperature (Hartlieb et al., 2012; Peinsitt et al., 2010).

It was also found that water saturation had less effect on the BTS loss than the UCS loss. This is considered to be because the BTS samples are smaller than the UCS samples, which results in lower interior temperatures.

5 Conclusions

The evaluation of the results shows that microwave treatment significantly decreases the strength of marble. The water saturation increases the degree of influence of microwave treatment. The losses in the UCS values are 10.9%, 21.7% and 35.4% for the dry, semi-saturated and fully saturated specimens, respectively. The losses in the BTS values are 30.4%, 34.4% and 37.7% for the dry, semi-saturated and fully saturated specimens, respectively. It can be concluded that the microwave treatment has important effect on the strength decrement of marble.

Acknowledgements The authors would like to thank the European Union through the ERA-NET Cofund on Raw Materials (ERA-MIN 2) (Reference Number: ERA-MIN-2018_95) and the Scientific and Technological Research Council of Turkey (TUBITAK Grant Number: 119M839) for the financial support of this work.

References

Hartlieb, P., Leindl, M., Kuchar, F., Antretter, T., & Moser, P. (2012). Damage of basalt induced by microwave irradiation. *Minerals Engineering, 31*, 82–89.

Hassani, F., Nekoovaght, P. M., & Gharib, N. (2016). The influence of microwave irradiation on rocks for microwave assisted underground excavation. *Journal of Rock Mechanics and Geotechnical Engineering, 8*, 1–15.

Kahraman, S., Canpolat, A. N., & Fener, M. (2020). The influence of microwave treatment on the compressive and tensile strength of igneous rocks. *International Journal of Rock Mechanics and Mining Sciences, 129*, 104303.

Motlagh, P. N. (2009). *An investigation on the influence of microwave energy on basic mechanical properties of hard rocks* [Dissertation]. Concordia University.

Peinsitt, T., Kuchar, F., & Hartlieb, P. (2010). Microwave heating of dry and water saturated basalt, granite and sandstone. *International Journal of Mining and Mineral Engineering, 2*, 18–29.

Satish, H., Ouellet, J., & Raghavan, V. (2006). Investigating microwave assisted rock breakage for possible space mining applications. *Mining Technology, 115*(1), 34–40.

Effectiveness of Vetiver Grass in *Char* Land Protection

Mohammad Shariful Islam and Dipa Sarker

Abstract

Numerous land protection techniques are available but all of them are not always suitable due to various factors such as resource limitations, cost-effectiveness and environmental impacts. This research work aims to focus on the protection of *char* land (small river islands) which are highly erodible and unstable due to high tidal flow and flood actions. Bioengineering technique is effective in increasing bed roughness. In this context, vetiver grass (*Chrysopogon zizanioides*) has been selected for its special characteristics such as fast growth, submergence tolerance and long root system. Even though there has been studies on the roughness coefficient (*n*) of vegetated soil, there is still gap in study on *char* land. For this purpose, a small-scale physical model-based experiment and theoretical studies were conducted. A rectangular wooden model with a dimension of 3.0 m × 0.7 m 0.6 m filled with *char* soil and a mild bed slope of 1:60 was constructed. Vetiver tillers were planted with a spacing of 20 cm c/c in both directions. Model tests were conducted after 140 days of vetiver plantation varying the discharge in the range of 0.000111–0.000138 m^3/s to determine *n*. The vetiver root system grew to a depth of 50 cm and anchored the bed soil. Vetiver shoots grew up to 127 cm and retarded the flow velocity and also increased the surface roughness and flow depth. Theoretical relationships have been established using Manning's roughness equation, and Strickler's formula for determining the Manning's Roughness Coefficient, *n* of the *char* land for a discharge range of 0.002–0.004 m^3/s. Manning's roughness coefficient was found to be in the range of 0.03–0.07 for different roughness heights. It is concluded that vetiver grass plantation is effective in *char* land protection and reclamation by increasing roughness, retaining soil and reducing erosion.

Keywords

Char land • Land reclamation • Roughness coefficient • Vetiver grass

1 Introduction

The term *Char* means riverine island or mid-channel island, which emerges as a result of continuous accretion in the river bed commonly surrounded by surface water sources (Islam et al., 2014). Accretion of land is subject to continual change, erosion and reformation due to natural hazards like flood, river flow, wave action which wash out the land surface. In most of the cases, *chars* are developed with sandy soil, which is unstable and easily erodible. The *char* land is unique because of its formation, accretion process, soil type and stability. These sandy *chars* have adverse effects on the livelihood of local people. So, as a cost-effective solution, the local people practice bioengineering using local materials like *dholkolmi*, *chaliya*, *binna* or vetiver, *murta*, *aizol/ koros*, etc., with bamboo along with other conventional solutions like boulders, sand-cement mixture, cement concrete blocks, geo-bag, retaining wall, etc. Past research showed vetiver grass (*Chrysopogon zizanioides*) is effective in erosion control, slope stabilization, against wave action and land development due to its unique characteristics of root (Islam & Islam, 2022; Truong, 2000). However, vetiver is intolerant to shading, requires regular watering, trimming and weeding and may also need fertilizer depending on the soil nutrient for proper growth (Truong & Hengchaovanich, 1997). So, with proper maintenance, application of vegetation can have a good result and be cost-effective, flexible,

M. S. Islam (✉) · D. Sarker
Bangladesh University of Engineering and Technology, Dhaka, 1000, Bangladesh
e-mail: msharifulislam@ce.buet.ac.bd

M. Bezzeghoud et al. (eds.), *Recent Research on Geotechnical Engineering, Remote Sensing, Geophysics and Earthquake Seismology*, Advances in Science, Technology & Innovation,
https://doi.org/10.1007/978-3-031-48715-6_7

applicable in remote areas and adaptable with local geo-environmental condition. Studies also showed that, vetiver hedge can trap fine to coarse sediments from surface runoff (Truong & Loch, 2004). In Bangladesh, vetiver is being used on road slope, coastal and flood embankment, village mound protection and landslide mitigation (Islam et al., 2021). However, the application of vetiver grass in *char* land is yet to be studied. The objective of this study is to determine the roughness coefficient of vetiver grass embedded soil from *char* land for its protection against soil erosion and land degradation as well as its reclamation.

2 Materials and Methods

Soil: Soil sample was collected from Kodaillar *Char*, Narayanganj (23.685° N, 90.685° E). Silt and clay content of the soil is 72.8% and 23.2%, respectively, and FM is 0.60. Coefficient of permeability and friction angle varies between 1.03×10^{-2} and 9.29×10^{-5} cm/s and 28–36°, respectively. The pH is 6.7 and nutrient contents such as organic matter, nitrogen, potassium, phosphorus, sulfur, boron and zinc are 0.57%, 0.03%, 0.79 meq/100 g, 29.12 ppm, 12.70 ppm, 0.14 ppm and 0.20 ppm, respectively, which is not adequate for proper growth of vegetation.

Vetiver Grass: Naturally grown, good quality vetiver grass with stiff and erected stem and finely structured root system was collected from Pubail, Gazipur (23.93° N, 90.47° E) and transported to the model site with appropriate arrangement.

Theoretical Approach for Roughness Coefficient, *n*: A number of empirical methods are used for estimating *n*. Although much research has been done on determining Manning's roughness coefficient, *n*, for different vegetated soil, very little has been conducted concerning that for *char* land. A theoretical relationship has been established using Manning's and other's roughness coefficient equation (Table 1).

Physical Model Approach for Roughness Coefficient, *n*: Two rectangular wooden models (3 m × 0.7 m × 0.6 m) with a bed slope of 1:60, namely Bare *Char* Model (BCM) and Vetiver *Char* Model (VCM) were constructed in BUET premises. The temperature, humidity and sunlight of this area ranged between 18–32 °C, 62–81% and 7.5–9.4 h/day, respectively, during the study period. The average annual rainfall is 1875 mm. The models were filled up with *char* soil (moisture content: 5–7%), compacted in three layers (compaction energy: 40 kN m/m^3, density: 10.74 kN/m^3). Among the two models, one was covered with triple tiller vetiver grass at a spacing of 20 cm c/c on both direction, whereas the other one was in bare condition. At the time of plantation, root length was 7 cm and shoot length was 23 cm. After 140 days, shoot and root length grew up to 127 cm and 50 cm, respectively. Number of tillers increased up to 5–6/point.

A container was placed at upstream of the model which acted as an inlet tank. A pipe was connected to the tank to supply flow into the model. The water in the tank was supplied through a pipe from the overhead tank and a specific head (0.45 m) was maintained by continuous supply. The bottom of the water tank was placed near the surface of the model and a metal sheet was used to ensure uniform flow. A valve was set at the upstream end to adjust and a flow meter was connected to the pipe to measure the flow discharge. Three flow discharges were considered and time laps for each of the flow were up to 30 min. Flow discharge was adjusted by the flow meter. At that time, flow depth was measured at downstream. Average velocity was determined manually for each of the cases. At the end, soil erosion for each condition was measured. Figure 1 shows the schematic diagram of experimental set up for VCM.

3 Results

Under theoretical consideration for flow discharges 0.002, 0.003 and 0.004 m^3/s; Manning's roughness coefficient for *char* soil has been found to be 0.011. For vegetation with different effective roughness heights (k_s), *n* increases significantly ranging from 0.03 to 0.05. From Fig. 2a, flow depth increases with flow discharge and it is higher in vegetated condition compared to bare condition. With vegetation, flow depth increases with increasing effective

Table 1 List of equations used for the theoretical approach (Sarker, 2022)

Author	Equation	Description
Strickler (1923)	$n = 0.047\, d^{1/6}$	d = grain diameter in mm
Subramanya (1982)	$n = 0.047\, d_{50}^{1/6}$	d_{50} = diameter of the bed material corresponding to 50% finer in mm
Manning's (1889)	$\bar{u} = \frac{1}{n} R^{2/3} S^{1/2}$	$\bar{u}$ = mean velocity of flow (m/s), R = hydraulic radius (m), S = channel slope, n = Manning's roughness coefficient

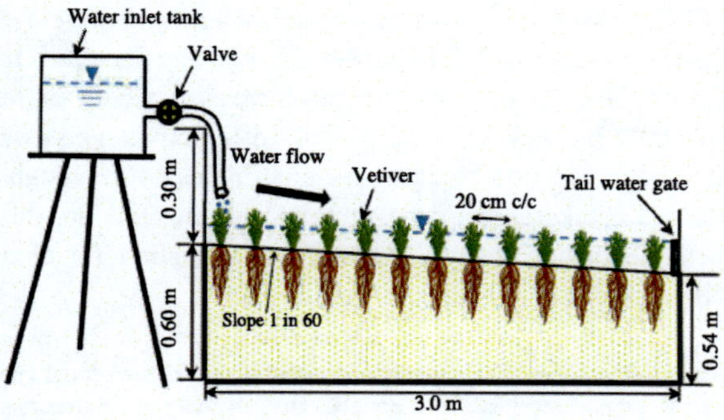

Fig. 1 Schematic diagram of experimental setup for Vetiver *Char* Model (VCM)

roughness height. From Fig. 2b, $\overline{u}$ increases with flow discharge and it is higher in bare condition compared to vegetated condition. With vegetation, mean velocity, $\overline{u}$ decreases with increasing effective roughness height.

In physical experiment (Fig. 2c), flow discharges at a range of 11.0×10^{-5}–14.0×10^{-5} m^3/s and k_s 1.27 m have been used. At these rates, the average flow velocity is 0.18–0.27 m/s for BCM and 0.11–0.16 m/s for VCM. Flow depth is 0.01–0.02 m for BCM and 0.06–0.07 m for VCM. For BCM, flow discharge increases the depth of flow and corresponding average velocity also increases simultaneously. When vetiver is added as a roughness element, flow

Fig. 2 **a** Flow depth-discharge relationship and **b** mean velocity-discharge relationship from theoretical estimation, **c** flow depth-discharge and average velocity-discharge relationship from model test for flow through bare condition and with vetiver

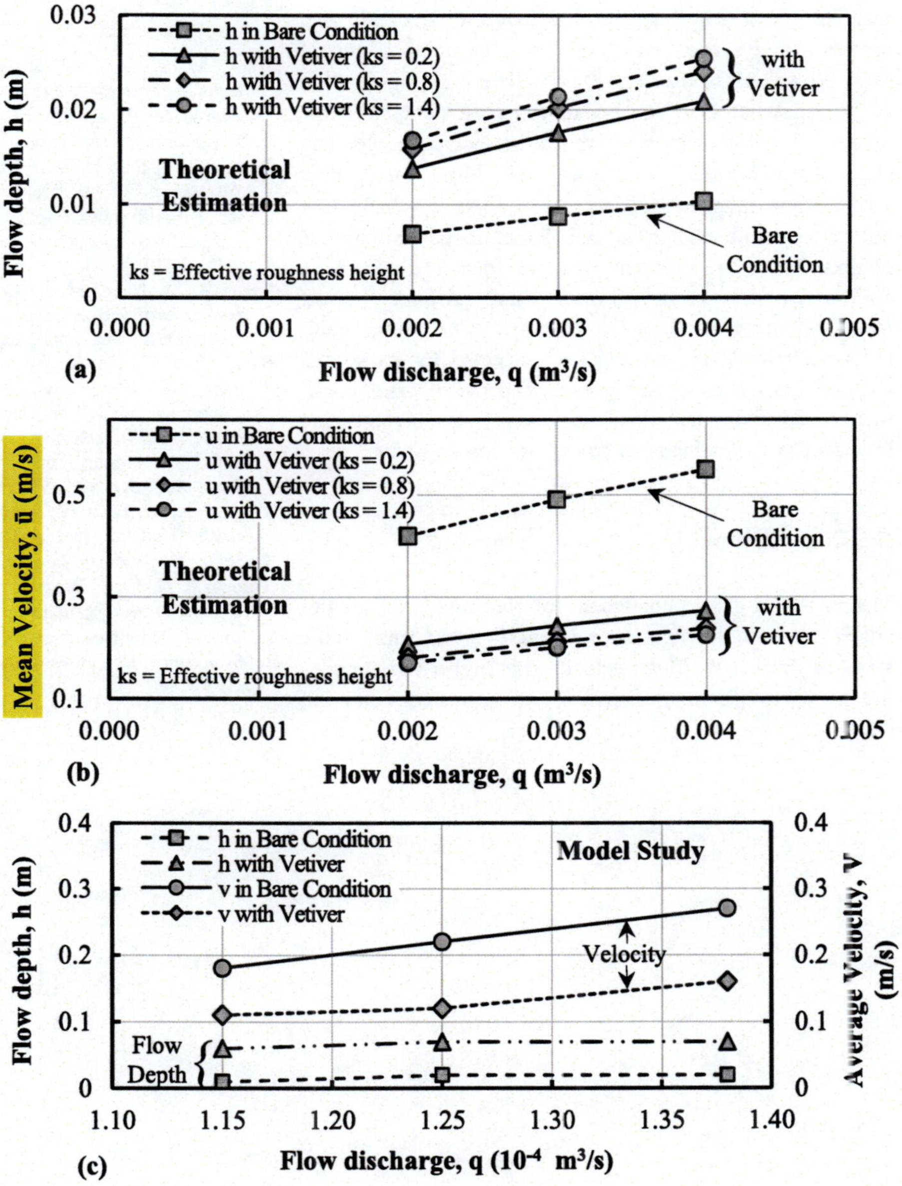

depth increases slightly but average velocity decreases compared to bare condition.

4 Discussion

It is observed that vetiver growth is well satisfactory in nutrient deficient *char* soil. It means that the application of vetiver grass as a bioengineering tool can be effective and sustainable in *char* land protection and reclamation.

From the theoretical calculation for specific flow discharges and increasing effective roughness heights, the $\overline{u}$ of flow decreases and thus Manning's roughness coefficient increases in case of vegetated conditions. During experimental conditions when vetiver is added as a roughness material for a specific flow, discharge flow depth also increases and average velocity decreases up to a certain limit. However, for bare conditions, there is no significance of bed material alone on roughness coefficient and flow velocity. So, for both cases, vegetation acts as a blocking material which minimizes surface erosion by reducing flow velocity and initiates land accretion. Also, it satisfies the criteria for sedimentation under different effective roughness heights. However, different types of local reeds such as *Saccharum pontaneun* (*Kansh*) and *Arundo donux* (*Nolkhagra*) are practiced but vetiver is effective for stabilizing char land. Vetiver has been selected for its special characteristics such as fast growth, capable of extreme climatic condition, submergence tolerance, strong stems and long root system which has strong binding capacity.

5 Conclusions

Manning's roughness coefficient for *char* soil is 0.011. For vegetation, with different (k_s), n increases significantly in the range of 0.03–0.05. Mean velocity $\overline{u}$ is higher in bare condition (0.41–0.55 m/s) than that of vegetated condition (0.17–0.27 m/s). In the physical experiment, average velocity increases in BCM (0.18–0.27 m/s) and decreases in VCM (0.11–0.16 m/s). Flow depth increases in both BCM (0.01–0.02 m) and VCM (0.06–0.07 m). Vetiver grass on *char* bed acting as a blocking material increases the roughness and decreases the flow velocity and erosion. So, this bioengineering technique is an effective solution for *char* land protection.

Acknowledgements The authors acknowledge the infrastructural and financial support received from Bangladesh University of Engineering and Technology (BUET), Dhaka-1000, Bangladesh, for carrying out the research work.

References

Islam, M. S., Latif, M. B., & Islam, T. (2021). Bioengineering technique for protecting submersible roads in Haor districts of Bangladesh (Chap. 21). In *Research development in geotechnics, geo-informatics and remote sensing*. Springer Nature.

Islam, M. S., Sultana, S., Saifunnahar, M., & Miah, M. A. (2014). Adaptation of char livelihood in flood and river erosion areas through indigenous practice: A study on Bhuapur riverine area in Tangail. *Journal of Environmental Science and Natural Resources, 7*(1), 13–19.

Islam, T., & Islam, M. S. (2022). Submergence and wave action resilience of vetiver grass protected fly ash amended soil slopes. *Geotechnical and Geological Engineering, 40*, 3643–3668.

Sarker, D. (2022). *Analyzing roughness co-efficient of vetiver planted char land using physical model* [M.Sc. engineering thesis]. Department of Civil Engineering, BUET, Dhaka-1000, Bangladesh.

Truong, P. (2000). Vetiver grass system: Potential applications for soil and water conservation in Northern California. In *Stiff Grass Technology Seminar*, Woodland (pp. 562–571).

Truong, P., & Loch, R. (2004). Vetiver system for erosion and sediment control. In *13th International Soil Conservation Organization Conference*, Paper No. 247, Brisbane, Australia (pp. 1–6).

Truong, P. N. V., & Hengchaovanich, D. (1997). Application of the Vetiver Grass system in land stabilization, erosion and sediment control in civil construction. In *Proceedings of Southern Region Symposium*, Department of Main Roads, Queensland.

Check for updates

Predictive Algorithm for Landslide Using MATLAB Simulink

Diego Villalobos, Gustavo Gatica, Pierre-Yves Descote, Luis F. Robledo, Ivo Fustos, Vicente Tapia, Wenbin Jian, and Wen Nie

Abstract

The study of earth movements and their occurrence factors contributes positively to the reduction of damages resulting from disasters caused by landslides, falls or alluvium. The investigation of early warning systems for landslide events that is developed for this case uses main variables for the generation of movement. In this case, the main conditioning factor is the slope, and the triggering factor is the precipitation, measured through the daily precipitation history of the mountain town of San Jose de Maipo, Chile. Using fuzzy logic rules, an algorithm capable of establishing logical relationships between variables with a prediction accuracy ratio of over 88% is developed, which implies that the applied model is efficient at the time of detecting a real warning signal.

Keywords

Landslide • Early warning systems • Fuzzy logic • Predictive algorithm • Simulink • Geological modeling • Geological hazard

D. Villalobos · G. Gatica · P.-Y. Descote (✉) · L. F. Robledo · V. Tapia
Facultad de Ingeniería, Universidad Andrés Bello, Santiago, Chile
e-mail: pierre.descote@unab.cl

I. Fustos
Department of Civil Engineering, University of La Frontera, Temuco, Chile

W. Jian · W. Nie
Department of Geotechnical and Geological Engineering, Fuzhou University, Fuzhou, PR China

School of Resources and Environmental Engineering, Jiangxi University, Ganzhou, PR China

Spatial Information Technology and Big Data Mining Research Center, Chengdu, PR China

1 Introduction

Massive landslides, and particularly landslides, are a geological hazard that has meant potential damage throughout history in certain regions of the planet (Cruden et al., 1996). This studying case located in the mountainous province of the Metropolitan Region, Chile. An area that is highly exposed to the occurrence of these natural phenomena, due to the high precipitation and a steep and mountainous geomorphology. The GIS analysis tools mainly provide us with the resulting susceptibility for the studied areas (Achour et al., 2017 Manchar et al., 2018), but in simulation environments using the slope as the main conditioning factor and meteorological factors as the triggers of the phenomenon, this study proposes to develop and validate a prediction algorithm for the occurrence of landslides, to foresee future movements and avoid human losses and damages in the communities (UNISDR, 2012).

2 Methodology

Daily rainfall, El Niño, OAA and terrain slope data are used and entered into a predictive algorithm based on fuzzy logic (Ferson & Hajagos, 2004), in a Mamdani type system composed of a fuzzifier, in which the fuzzy sets and their membership of the data universe are processed through control rules, developed in MATLAB Simulink (Wang, 2015). This system replicates the behavior of the dynamic system that conditions and originates slips based on the main variables.

3 Results

3.1 Membership Functions

To use the fuzzy logic tool (Fustos et al., 2020; Wang, 2015), intervals are created by Goodman's algorithm (Goodman, 1965) using multinomial probability models to

M. Bezzeghoud et al. (eds.), *Recent Research on Geotechnical Engineering, Remote Sensing, Geophysics and Earthquake Seismology*, Advances in Science, Technology & Innovation,
https://doi.org/10.1007/978-3-031-48715-6_8

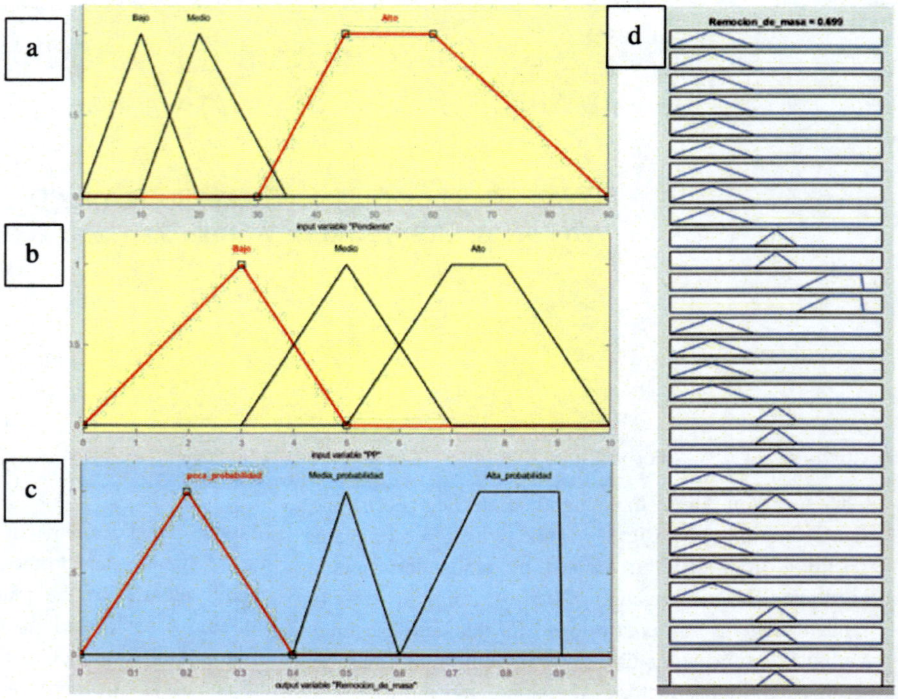

Fig. 1 Membership functions of: **a** slope, **b** precipitation, **c** landslide and **d** fuzzy logic rules for landslide

analyze the membership functions. In this work, the confidence intervals allow defining the uncertainty in the estimation of the relative frequencies or proportions of the multinomial distributions. Probability boxes are also used to represent the uncertainty due to the precision of the multinomial probability models (Ferson & Hajagos, 2004; Ferson et al., 2003; Hall & Lawry, 2004). After defining the intervals for each parameter, rules are created for each membership function given by trapezoidal type functions under the Gaussian bell, applied to the parameters obtained by Goodman's algorithm (Goodman, 1965). Figure 1 illustrates the membership functions for the main variables and for the sliding phenomenon.

3.2 Fuzzy Logic Rules and Graphics

The rules established for the model correspond to an IF AND THEN system, in which all combinations between variables are made with the additional commands NOT, AND, and OR (Fig. 1, right). As a result, the variables have 3 states within the model, (low, intermediate, and high probability state) for the occurrence of a removal. The graphs (Fig. 2) are the result of the behavior of each variable according to the rules established for the model, simulating the slope (*x* axis) versus daily scale precipitation (Fig. 2a), slope versus Antarctic Oscillation (AAO) and slope versus El Niño phenomenon (Fig. 2b, c). The *z* axis in the graphs illustrates the probability of landslide occurrence in the study area, where the cold blue tones represent a low probability of occurrence (slope angles close to zero and low precipitation/influence of AAO and El Niño phenomenon), while the warm yellowish tones show a higher probability of landslide occurrence associated mainly to the increase of the slope and the amount of fallen water associated to the precipitation and meteorological phenomena of greater intensity, showing the relationship that these variables have according to the rules in which the model was configured.

3.3 ROC Curve

The Receiver Operating Characteristic curve (ROC curve) (Cruden et al., 1996) is known as the sensitivity rate, which is expressed through a graph between false positives (*x* axis) and false negatives (*y* axis). It is used to compare models and identify which one gives better performance as a classifier and in turn, the area under the curve (AUC) can be used as a summary of the quality of the model, measuring the accuracy in the prediction of binary events, that occur or not. The choice of an indicator, in this case an algorithm, implies a compromise in relation to the rate of correct predictions and false predictions, but because the signals come from continuous variables, they are not binary for this case, so the signals must be calibrated (Fig. 3), so that the operator defines a threshold above which any indicator is considered as valid.

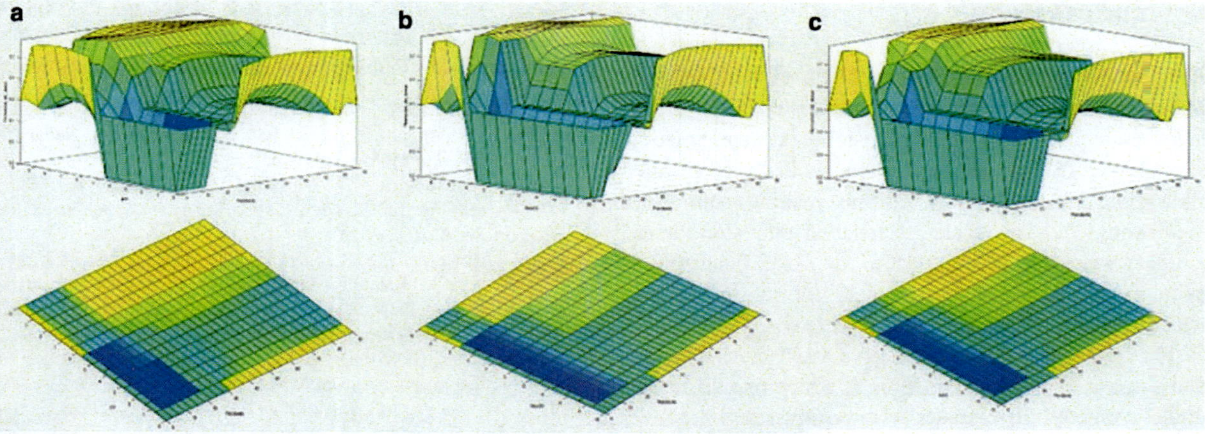

Fig. 2 **a** Slope versus daily precipitation; **b** slope versus Antarctic Oscillation (AAO) and **c** slope versus El Niño phenomenon

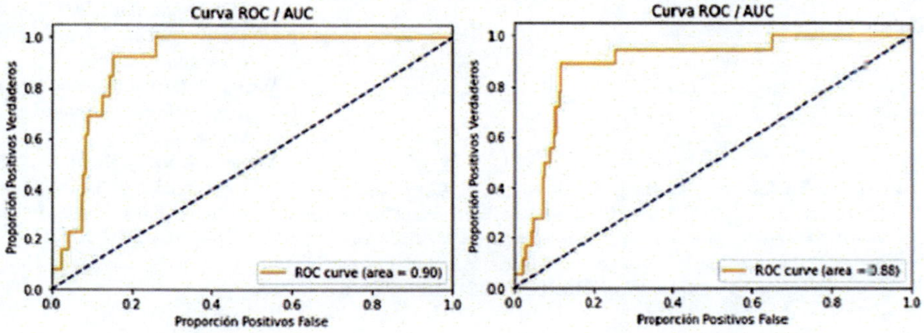

Fig. 3 Calibration ROC curve and validation ROC curve

4 Discussion

There is a wide range of factor that condition the physical stability of a terrain in general (Yuvaraj & Dolui, 2021), but the study of landslides in this area, the slope is the most important variable in high mountain conditions as well as simplifying the modeling of the problem. The use of climatic factors and daily precipitation as triggers of events is due to the historical relationship that exists between climatic cycles and heavy rainfalls with the generation of alluvium and landslides. Using the analytical and variable membership graphs, it can be inferred that intervals *A* and *D* (Table 1) are the scarcest data within the records, which makes it probabilistically ambiguous, while intervals *B* and *C* present greater certainty when simulating a scenario in which the risk increases proportionally to the precipitation on a daily scale. The slope has an impact on the model when exceeding 50°, where, added to the precipitation, it shows an increase in the probability of occurrence of movement (Fig. 2a).

In the ROC curve analysis, for the calibration phase of the algorithm, 30% of the research data and 70% of simulated data were applied in which the failures were analyzed to modify certain parts of the code, where there is more ambiguity for the model. The results of the ROC curve show results within the range 0.85–0.90, meaning that the model can differentiate well between negative and positive signals, choosing the correct classification.

Table 1 Risk intervals

	A	*B*	*C*	*D*
Very low	0.0	0.0	0.05	0.2
Low	0.0	0.2	0.2	0.4
Middle	0.2	0.4	0.4	0.6
High	0.4	0.6	0.75	0.9
Very high	0.75	0.9	> 0.9	> 0.9

In the validation phase, 100% of the data are applied to calculate the ROC curve and create a confusion matrix showing the true positives and true negatives of the dataset, as well as the false negatives and positives to calculate the logistic regression already implemented together with the AUC value. The ROC curve analyzed for the research case is within the range 0.85–0.95, making the correct classification choice for the algorithm very likely.

5 Conclusions

The precipitation and slope conditions of the Cajon del Maipo territory are modeled and evaluated, forcing their logistic coupling through a probability system based on fuzzy logic. The logistic regressions applied present a feasible yield (AUC) over 80%, thus making non-random predictions. The daily scale precipitation estimator obtained a higher weighting with respect to the slope, indicating the existence of a hydrometeorological control in the area, which is above the geomorphological factor.

The ROC curve model used gave satisfactory results by delivering a ROC curve above 0.88, which implies that the model correctly differentiates false positives and negatives, generating an adequate logical response for the early warning analysis that may be originated. Thus, the use of logistic models and precipitation simulations for the study and control of landslides in the study area is considered applicable.

References

Achour, Y., Boumezbeur, A., Hadji, R., Chouabbi, A., Cavaleiro, V., & Bendaoud, E. A. (2017). Landslide susceptibility mapping using analytic hierarchy process and information value methods along a highway road section in Constantine, Algeria. *Arabian Journal of Geosciences, 10*(8). https://doi.org/10.1007/s12517-017-2980-6

Cruden, D. M., Varnes, D. J., Turner, A. K., & Schuster, R. L. (1996). *Landslides: Investigation and mitigation* (Special Report 247) (675 pp.). Transportation Research Board, National Research. National Academy Press.

Ferson, S., & Hajagos, J. G. (2004). Arithmetic with uncertain numbers: Rigorous and (often) best possible answers. *Reliability Engineering and System Safety, 85*.

Ferson, S., Kreinovick, V., Ginzburg, L., & Sentz, F. (2003). *Constructing probability boxes and Dempster–Shafer structures* (Report SAND2002-4015).

Fustos, I., Abarca-del-Río, R., Mardones, M., González, L., & Araya, L. R. (2020). Rainfall-induced landslide identification using numerical modelling: A southern Chile case. *Journal of South American Earth Sciences, 101*.

Goodman, L. A. (1965). On simultaneous confidence intervals for multinomial proportions. *Technometrics, 7*(2).

Hall, J. W., & Lawry, J. (2004). Generation, combination, and extension of random set approximations to coherent lower and upper probabilities. *Reliability Engineering and System Safety, 85*.

Manchar, N., Benabbas, C., Hadji, R., Bouaicha, F., & Grecu, F. (2018). Landslide susceptibility assessment in Constantine region (NE Algeria) by means of statistical models. *Studia Geotechnica et Mechanica, 40*(3), 208–219. https://doi.org/10.2478/sgem-2018-0024

UNISDR. (2012). *Towards a post-2015 framework for disaster risk reduction*. Geneva.

Wang, C. (2015). *A study of membership functions on Mamdani-type fuzzy inference system for industrial decision making*. Lehigh University.

Yuvaraj, R. M., & Dolui, B. (2021). Statistical and machine intelligence based model for landslide susceptibility mapping of Nilgiri district in India. *Environmental Challenges, 5*, 100211. https://doi.org/10.1016/j.envc.2021.100211

Innovative Solution for Retaining Walls Anchorage

Michael Doubrovsky and Virzhiniia Oganesian

Abstract

High soil retaining walls as well as deep water berths of solid construction need to be anchored in order to provide structures' stability and required bearing capacity. Traditional anchor systems for such facilities in many cases are resource demanding and costly. An innovative (patented) design of the anchor system is worked out. Anchor bearers are made of "comb" type as several small anchor plates fixed along the rigid core. Anchor force is taken by all plates of the comb simultaneously. Distance between anchor plates on the rigid core may be regulated to achieve the maximal positive effect from the point of view of anchor system bearing resistance. Other adjustable parameters are angle between the plate and the core axis and plate sequence (if plates are of different height). Study of the innovative anchor system included both physical (in laboratory conditions by use of soil box in scale of 1:20) and numerical modelling. Based on the executed study, the basic factors that determine the bearing capacity of both the anchor device and the whole structure (the depth of the anchor bearers, the angle of anchor plates' inclination, the number of plates, the height of the various plates in the "comb" and the order of their location in the device) are analysed. As it was demonstrated by our research, proposed design approach provides increasing bearing capacity both of the anchorage and of the soil retaining structure in whole (keeping the same material consumption for the anchorage as at the traditional solutions with one large anchor plate or with anchor sheet pile wall). On the other hand, proposed structure may provide decreasing of the material consumption (keeping the same structure's bearing capacity as at the known facilities).

Keywords

Anchor system • Retaining wall • Sheet piling • Physical modelling • Numerical modelling

1 Introduction

In civil engineering, broad application found rigid retaining walls of sheet pile type with anchored front elements. Anchor devices and, in particular, anchor plates are very important structural elements of such quay walls. Today, a relevant issue of design and construction of sheet pile walls is the development of new anchor systems of a high bearing capacity with minimal costs of materials and resources.

A retaining wall with an improved anchoring system in the form of a "comb", developed by the authors (Doubrovsky & Oganesian, 2012) (see Fig. 1), contains a front wall 1, backfill 2 behind the wall and anchor rods 3 with anchor bearers located in the backfill behind the soil sliding wedge. Anchor bearers are made of a "comb" type with anchor plates 5 fixed along a rigid core 4. The distance between adjacent anchor plates is determined to provide a condition: the plate closest to the front wall does not cross the boundaries of the soil extrusion wedge in front of the plate more distant from the front wall (soil extrusion wedges are built at an angle of 45° + $\varphi/2$ to the vertical, where φ is the angle of internal friction of the soil).

The advantages and design features of innovative anchor system, as well as the principle of its operation, are described in our articles (Doubrovsky & Oganesian, 2013; Doubrovsky & Oganesyan, 2013). The sheet pile wall 1 with the cap 6 works under the pressure of the backfill soil 2, like a beam plate, the upper end of which is pivotally attached to

M. Doubrovsky (✉)
Coastal and Marine Engineering Research Institute,
32004 Technion City, Haifa, Israel
e-mail: m.doubrovsky@gmail.com

V. Oganesian
Danube Commission, Budapest, 1068, Hungary

M. Bezzeghoud et al. (eds.), *Recent Research on Geotechnical Engineering, Remote Sensing, Geophysics and Earthquake Seismology*, Advances in Science, Technology & Innovation,
https://doi.org/10.1007/978-3-031-48715-6_9

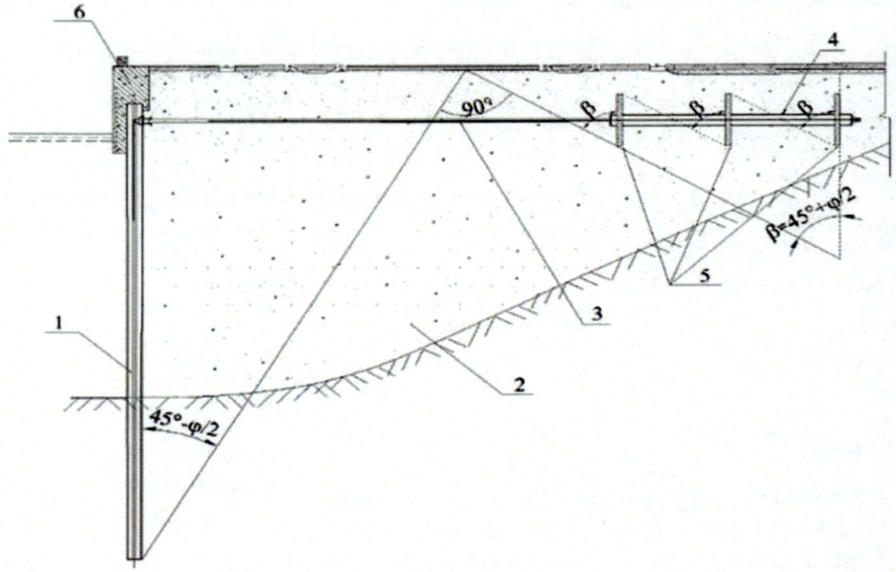

Fig. 1 Retaining wall with innovative anchor system

the anchor device, and the lower end is clamped in the ground. Wall 1, the movement of the upper end of which is limited by the anchor device, under the pressure of the backfill soil 2, bends and tends to turn around the point of attachment of the anchor rod 3. This is prevented by the bottom soil resistance located in front of the wall. Anchor rod 3 transfers the force from the ground pressure to the anchor system of a "comb" type. All the forces are taken simultaneously by all plates of the "comb" 5.

The location of anchor plates on a rigid core allows to fix the most favourable distances between the plates. In particular, this makes it possible to maximize the use of the reactive pressure of the soil in front of the plates. In traditional high-pitched anchor slabs placed in trenches, the lateral surface of the upper part of the anchor plate gives a smaller contribution to the bearing anchor capacity compared to its lower part.

2 Experimental Studies and Discussion

2.1 Numerical Analysis of Anchor System

In the numerical analysis of the work of the "comb", three design considerations of its positions in depth were evaluated: at the level of the upper, middle and lower thirds of the height of the anchor plate of the traditional design. The results are given in Tables 1 and 2.

Thus, with the same material consumption, the bearing capacity of the proposed solution (n = 2, 3, 6) is greater and grows with the depth of the anchor device (Table 1). On the other hand, with the same bearing capacity of the traditional structure and the new technical solution (E_1 = 630 kN, Table 2), the material consumption for the anchor device of a "comb" type decreases with an increase in the number of plates. Consequently, when reducing material consumption 1.5 times, the required value of the bearing capacity is provided, or there is the possibility of significant increase of bearing capacity for high retaining walls (e.g. in case of deep-sea berths).

2.2 Experimental Research and Analysis

At Odessa National Maritime University, authors have carried out experimental model research of the proposed anchor device with both vertical (perpendicular to the anchor rod axis) and inclined anchor plates.

Earlier in the works of Yakovlev et al. (2004), Budin (1974), Tsagarelli (1963), experimental and numerical studies of anchor plates were considered, but no exhaustive answer was given to the question of how the angle of inclination of the anchor plate affects its bearing capacity. At the same time, in most cases, the results of calculations using the Coulomb methods and the Theory of the Limit Stress State of Soil give significantly different results not only from each other, but also in comparison with experimental data.

In our research, experiments with an anchor device of a "comb" type with vertical and inclined plates were carried out for the first time. For comparison, under similar conditions, the work of an anchor device of a traditional design was also considered.

Table 1 Comparison of bearing capacity of retaining wall anchor devices

Number of plates in anchor device, *n*	Height of anchor plates, *m*	Depth of the top of the plate, *T* (m)	Bearing capacity of anchor devices with the same material consumption (κN)	Increment of the bearing capacity of anchor device E_i/E_1
1	3.0	2.0	E_1 = 630	1.0
2	1.5	3.5	E_2 = 765	1.21
3	1.0	4.0	E_3 = 810	1.29
6	0.5	4.5	E_4 = 855	1.36

Table 2 Comparison of retaining wall anchor devices material consumption

Number of plates in anchor device, *n*	Depth of the top of the plate, *T* (m)	Cumulative area of plates ($м^2$) with the same bearing capacity of anchor device	Decline of the area of anchor device E_i/E_1
1	2.0	S_1 = 3.0	1.0
2	3.5	S_2 = 2.26	1.33
3	4.0	S_3 = 2.07	1.45
6	4.5	S_4 = 1.92	1.56

The following initial data was chosen for the experiment:

- height of the single anchor plate (in traditional solutions) $H = 0.3$ m;
- height of the plates of the "comb"-type anchor device $h = 0.1$ m;
- number of plates in the "comb"-type anchor device-3;
- length of plates in layout $L = 0.5$ m;
- distance between plates of the anchor device $l = 0.2$ m;
- depth of the single anchor plate $T = 0.15$ m from the top of the experimental box.

In order to determine the optimal position of the "comb", three positions in terms of depth of the proposed anchor device were considered:

- the "upper" position—$T = 0.15$ m from the top of soil surface in box;
- the "middle" position—$T = 0.25$ m;
- the "lower" position—$T = 0.35$ m.

The model of a "comb" was made of two types, for each of which a series of experiments were carried out in three positions at different depths:

- vertical plates are positioned under the angle of 90° to anchor rod;
- inclined plates are positioned under the angle of 10, 20, 30° to vertical axis (Table 3).

As a result of the research of the influence of the inclination angle of anchor device plates on its performance, a family of curves for each type of anchor device was obtained, described by a two-parameter function:

$$f(T,E) = f_1(T) * ef_2(T), \quad (1)$$

where

$$f_1(T) = a_1 + b_1 * T + c_1 * T^2 \quad (2)$$

$$f_2(T) = a_2 + b_2 * T + c_2 * T^2 \quad (3)$$

Table 3 Results of experimental research of the bearing capacity of anchor systems

No	Anchor device type	Bearing capacity (κN)	Bearing capacity increment (%)
1	Traditional anchor device with the single anchor plate	3.0	–
2	Innovative anchor device of a "comb" type (vertical plates)	3.9	30
3	Anchor device with inclined anchor plates (10°)	4.15	35
4	Anchor device with inclined anchor plates (20°)	4.25	42
5	Anchor device with inclined anchor plates (30°)	4.7	57

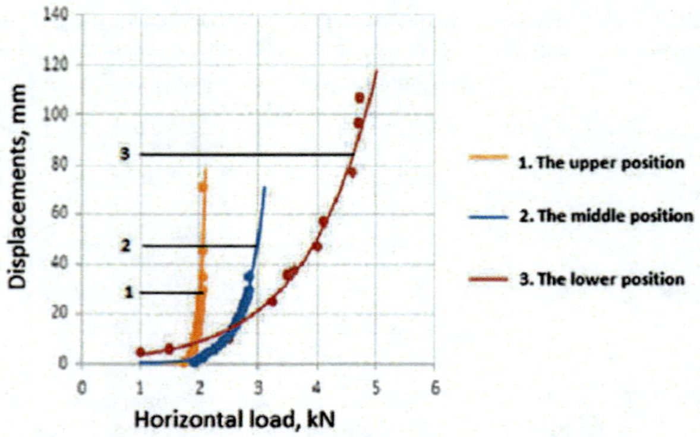

Fig. 2 Dependence of the displacements of a "comb" with the plates inclination angle 30° towards the backfill on the depth of the anchor device and the applied horizontal load

$f_1(T)$, $f_2(T)$	parameters of the function $f(T, E)$;
T	independent variable, set by the experimenter, describing the depth of the anchor device;
E	independent variable, set by the experimenter, describing horizontal load applied to anchor device;
a_1, b_1, c_1, a_2, b_2, c_2	two-parameter equation ratios, calculated by the Least Squares Method.

Curves describing the most advantageous constructive type of "comb" are shown in Fig. 2. For plates with inclination angle of 30° towards the backfill, when putting the ratios values calculated by the Least Squares Method into formula (1) obtaining:

$$f(T,E) = \left(3.2 - 34.14 * T + 85.54 * T^2\right) * e^{42.5 - 293.9 * T + 385.5 * T^2} \quad (4)$$

Fixing the plates on a rigid core facilitates the installation of the anchor device, since there is no need to use an auxiliary rigid mounting traverse, which is typical for anchor devices with flexible anchor rods. The entire "comb" (anchor plates fixed to the core) can be placed in the design position by a crane during one operation.

Besides, depending on the design solution and ground conditions on the construction site, the "comb" can be placed horizontally or obliquely, and the plates themselves in the "comb" can be fixed to the core perpendicular or at an angle to ensure the necessary reaction of the backfill soil.

Comb rigid elements with short anchor plates are more reliable and service-strong in case of dynamic loads than traditional flexible tie rods with high plates.

As considered anchor system of quay walls or high retaining walls in most practical cases are located in sandy soils (without rheological processes), long-term performance was not considered at this research stage.

3 Conclusions

As shown by the research performed (physical and numerical modelling), the considered anchor device of a "comb" type provides a lower material consumption (with the same bearing capacity as in traditional solutions). On the other hand, the proposed design can provide a higher bearing capacity of the structure (with the same material consumption as in the known solutions), thereby improving the performance of the quay wall. From a technological point of view, there are some advantageous in comparison with traditional approach. In particular, there is no need to use temporarily auxiliary rigid mounting traverse and to provide a trench for anchor plates stability. So, studied solution makes it possible to reduce cost of installation work and construction materials of the structure as a whole.

References

Budin, A. (1974). *Sheet pile walls*. Stroyizdat.

Doubrovsky, M., & Oganesian, V. (2012). Quay wall. Patent 105067, E02B 3/06 (2006.01) E02D 29/02 (2006.01), Ukraine.

Doubrovsky, M., & Oganesian, V. (2013). The improvement of operational characteristics of anchorage bearing of sheet pile wall berths. *Collection of Scientific Works, Issue 3*(38), Poltava.

Doubrovsky, M., & Oganesyan, V. (2013). Innovative solutions in the field of geotechnical construction and coastal geotechnical engineering under difficult engineering-geological conditions of Ukraine. In *Proceedings of the 18th International Conference on Soil Mechanics and Geotechnical Engineering*, Paris.

Tsagarelli, Z. (1963). Experimental research of the pressure of a granular medium on retaining walls inclined towards the backfill. In *Soil mechanics, foundations and footings*, Moscow.

Yakovlev, P., Gotman, A., & Kurmaev, R. (2004). *Interaction of structures with soil and pile foundations*. Astroprint.

Some Tasks of Increasing and Identifying the Reserves of the Bearing Capacity of Anchor Fastenings of Offshore Fixed Platforms

Latif F. Aslanov and Firidun L. Aslanov

Abstract

With the development of great depths of the sea and the need to drill deep wells from stationary platforms, the issue of using an anchor fastening design with a large bearing capacity became acute, and for this purpose, a combined type design was proposed. The design feature is that a metal pipe driven into the upper layer of weak soil of the sea bottom, through which a well is drilled to the required depth, forms an ordinary bored pile in the lower part, followed by filling the entire internal cavity with cement mortar to the level of the platform, inside which the platform is installed tubular metal anchor. The purpose of this design was to reduce the length of the anchor fastening in the ground with the same bearing capacity due to additional pouring with cement mortar. At the same time, to assess the bearing capacity of such a structure, a calculation model is used, which consists of the fact that the cement slurry, before hardening, produces lateral pressure on the walls of the wells, similar to hydraulic pressure, and the reaction from the soil is taken equal to this pressure, and it remains unchanged in magnitude even when the cement slurry hardens. As for the bearing capacity of the butt, the solution of the problem obtained for a round stamp pressed into an elastic–plastic body is used for it. Taking into account also the bearing capacity of the clogged part of the structure, calculation formulas were obtained to determine the bearing capacity of the combined anchor fastening. According to the regulations on environmental protection, drilling works are carried out with seawater and a solution made from natural clay, the resulting slurry does not contain harmful toxic substances and it is allowed to be thrown to the bottom of the sea as it does not pollute the environment. Pouring this solution to the bottom of the sea does not violate the requirements of the legislation on protecting the economic importance of water bodies and the natural local conditions inhabited by living organisms from an ecological point of view.

Keywords

Stationary platforms • Offshore • Pile • Bearing capacity

1 Introduction

To assess the bearing capacity of the structure, a calculation model is used, which consists in the fact that the cement slurry, before hardening, produces lateral pressure on the walls of the well, similar to hydrostatic pressure, and the reaction from the soil is taken equal to this pressure, and it remains unchanged in magnitude even when the cement slurry hardens. As for the bearing capacity of the butt, the solution of the problem obtained for a round stamp pressed into an elastic–plastic body is used for it. Taking into account also the bearing capacity of the clogged part of the structure, calculation formulas were obtained for determining the bearing capacity of the combined anchor fastening in the form:

at $\varphi_{\text{soil}} \leq 20$

$$N = \frac{\gamma_{\text{p}} utg\varphi_{\text{soil}}}{2}(l^2 + 2Ll) + A(\gamma_{\text{soil}} l_0 \eta_1 + c_{\text{soil}} \eta_2) + N_{\text{c}}. \quad (1)$$

at $\varphi_{\text{soil}} \geq 20$

$$N = \frac{\gamma_{\text{p}} utg\varphi_{\text{soil}}}{2}(l^2 + 2Ll) + AB_{\text{q}} \gamma_{\text{soil}} l_0 (1 + \sin\varphi_{\text{soil}}) + N_{\text{c}}. \quad (2)$$

In both cases, the condition $\gamma_{\text{p}}(L + l) \leq R_{\text{soil}}$.

The required length of the internal anchor is determined from the condition of ensuring the strength of the cement stone of the unreinforced section from the equation:

L. F. Aslanov (✉) · F. L. Aslanov
Azerbaijan University of Architecture and Construction, Ayna Sultanova St. 5, AZ1073 Baku, Azerbaijan
e-mail: latif.aslanov@bk.ru

M. Bezzeghoud et al. (eds.), *Recent Research on Geotechnical Engineering, Remote Sensing, Geophysics and Earthquake Seismology*, Advances in Science, Technology & Innovation,
https://doi.org/10.1007/978-3-031-48715-6_10

$$N = \frac{\gamma_p utg\varphi_{soil}}{2}\left(l^2 + 2Ll\right) + AR_{(c.s)u}; \quad (3)$$

where L—is the height of the pouring of cement mortar, counted from the end of the clogged pipe to the level of pouring; l—is the length of the bored pile, counted from its end to the end of the clogged pipe; l_0—length of anchoring in the ground; l_1—the length of the inner anchor, counted from the end of the clogged pipe; N_c—bearing capacity of the clogged part of the structure; φ_{soil}, c_{soil}—respectively, the angle of internal friction and soil cohesion; η_1, η_2, B_q—coefficients depending on the angle of internal friction of the soil.

To evaluate the effectiveness of the proposed combined design of anchor fastening and the calculation model, tests were carried out on models close to full-scale ones. Sharafi et al. (2016) analyzed a vertical monopile under static lateral load. Subanantharaj Palammal and Senthilkumar (2018) studied laboratory model tests of vertical and groups of piles under vertical and transverse. Wang et al. (2018) developed an approach to modeling, in the study Yuan et al. (2017) a special model installation was manufactured, in Yuan et al. (2019) a model test was carried out. The tabular values of the bearing capacity coefficients are borrowed from Berezantsev (1960). Under the dynamic impact of the load on the anchor fastenings, not only the structure, but also the surrounding soil is subjected to oscillatory motion. Due to the inertial property of the soil, the nature of the work of the anchor fastening changes, and in some cases, it can significantly affect the bearing capacity. The problem in this formulation was solved by the author. In Aslanov (2015a, 2015b, 2022), we have considered the application of large section bored piles with a "hard core" to offshore structures. To determine the reaction of the soil due to the oscillation of the structure, a plane problem is considered, due to which it is reduced to a hydrodynamic problem. At the same time, the obtained differential equations for the cases of vertical and horizontal impact of dynamic load on the anchor fastening structures are:

$$\mathrm{EA}\frac{\partial^2 z}{\partial x^2} + (m + \alpha_0 uR\rho_{soil})\frac{\partial^2 z}{\partial t^2} + \beta uR\rho_{soil}\omega\frac{\partial z}{\partial t} = p_b \sin \omega t; \quad (4)$$

$$\mathrm{EJ}\frac{\partial^2 z}{\partial x^4} + (m + \alpha_0 uR\rho_{soil})\frac{\partial^2 z}{\partial t^2} + \beta uR\rho_{soil}\omega\frac{\partial z}{\partial t} + k_0 Dz = p_b \sin \omega t; \quad (5)$$

$$\begin{cases} \alpha = \frac{1}{\nu R}\frac{I_1(I_1 - xI_0) + N_1(N_1 - xN_0)}{(I_1 - xI_0)^2 + (N_1 - xN_0)^2} \\ \beta = \frac{1}{\nu R}\frac{N_1(I_1 - xI_0) + I_1(N_1 - xN_0)}{(I_1 - xI_0)^2 + (N_1 - xN_0)^2} \\ x = \nu R; \nu = \frac{\omega^2}{c^2} \end{cases} \quad (6)$$

where $J(x)$, $N(x)$—are Bessel functions; m, EI, EA—mass and rigidity characteristics, respectively, in the vertical and horizontal directions; k_0, ρ, c—coefficient of the bed, density and speed of sound propagation in the ground; P_b, P_q—disturbing external vertical and horizontal loads; u, R—perimeter and radius of the cross section of the anchor fastening; ω—oscillation frequency. Taking into account that the stiffness of the bond is much less than the rigidity of the anchor, the condition of displacement or failure will be fulfilled at the moment when the elastic forces of the bond reach the value of the soil friction force. Such a calculation model made it possible to obtain the ultimate load perceived by the anchor fastening:

$$p = \tau_{soil} ul\xi; \quad (7)$$

$$\begin{cases} \xi = \frac{\sqrt{\left(1 + \frac{\omega^2}{c^2k^2}\right)^2 + (F\omega)^2}}{1 + \alpha_m\frac{\omega^2}{c^2k^2}} \\ \frac{M_0}{\overline{M} + \overline{m}}kl = -tgkl_i \\ \alpha_m = \frac{\overline{m}l}{M + \overline{m}l}; \\ \overline{m} = \alpha uR\rho_{qr}; \\ \overline{M} = \frac{M}{0.243l} \end{cases} \quad (8)$$

where F is the damping modulus of ground vibrations; α—coefficient determined by formula (6), and for most equipment, $\alpha = 1$ can be taken; k is the coefficient determined by formula (8); M—is the mass of the structure and equipment per anchor; τ—soil friction force; M_0—is the mass of the equipment; ω—frequency of vibrations of equipment; l—is the length of the anchor fastening in the ground. Formulas (7) and (8) were obtained in the case of a single-layer soil, and in the case of a multi-layer soil, it is necessary to use the formula

$$p_0 = u\sum \tau_{(soil)i} l_i \xi_i \quad (9)$$

Here, index i shows the soil layer ($i = 1, 2, 3 \ldots$). The practical use of these calculation formulas has shown that the bearing capacity of anchor fastening under the conditions of the dynamic impact of drilling equipment is significantly reduced and this reduction in some cases reaches 50%. Criteria for the limit state of anchor fastening are obtained, taking into account the dynamic parameters of drilling equipment. To communicate the anchorage with length l to move at a speed v in the soil, it is necessary to expend power W. In this case, the equilibrium condition can be written in the following form:

$$(N - p)v = W \quad (10)$$

where N is the load taken by the anchorage (bearing capacity);

P is the total weight of anchoring and equipment.

2 Settings or Methods or Materials and Methods or ..., Etc.

The calculation model is used to solve the problem. In this case, the end effect of anchor fastening is additionally taken into account; a method for assessing strength and stability, taking into account the stress–strain characteristics of the structures of offshore fixed platforms; a method for assessing the bearing capacity of anchor fastening of offshore fixed platforms, taking into account the design features under conditions of static and dynamic impact; a method for predicting the settlement of supporting blocks of offshore fixed platforms during construction in silty soils; constructive solutions and anchor fastening devices that allow the design of a stationary platform to be used repeatedly; the purpose of this design was to reduce the length of the anchorage in the ground with the same bearing capacity due to additional pouring with cement mortar.

3 Results

The load-bearing capacity of anchoring on the ground for static and dynamic loads has been studied, and calculation formulas have been developed that characterize the conditions for the construction of these structures and the operation of drilling equipment. The strength of the core of cement stone in a cage for bending and compression was studied using these results when calculating the anchoring of stationary offshore platforms.

4 Discussion

In practice, the oscillation frequencies are much less than the speed of sound propagation in the ground, then the values are less than the speed of sound propagation in the ground, then the values determined by formula (6) at the value $vR \rightarrow 0$, $\alpha_0 = 1$ and $\beta = 0$. In this case, Eqs. (4) and (5) become much simpler. As the obtained equations show, to take into account the inertial properties of the soil, it is enough to sum the inertial mass of the anchor fastening with the added mass of the soil. Sample testing showed that the discrepancy between the results of calculation formulas and experiments is no more than 8%. Thus, it has been established that by taking the appropriate characteristics of the pipe and cement stone, it is possible to reduce the consumption of metal in the design of anchor fastening. In this case, not only savings in the use of scarce metal are achieved, but also the production of work on the anchoring device is greatly facilitated. The question of the nature of the change in the strength of the cement stone core in the cage along the height of the anchoring is also considered.

5 Conclusions

A method has been developed for predicting the settlement of the support blocks of stationary platforms during installation on soft and silty soils, and the influence of support shoes in assessing the installation stability of a stationary platform has been studied. It has been established that the bearing capacity of the anchorage, the formation of which proceeded with additional pouring of cement mortar (loading) is 1.5 ÷ 1.7 times higher than the same anchorage, but made without loading. The analysis of the research results showed that taking into account the shear forces of the soil helps to identify the margin of safety, and taking into account the influence of the longitudinal force when calculating horizontal loads increases the values of the bending moment and the depth of anchoring.

References

Aslanov, L. F. (2015a). Wave interaction of offshore structure and shelf soil through large section piles with a 'hard core' on the half-space model. *Neftyanoe Khozyaystvo, 2*, 78–81.

Aslanov, L. F. (2015b). Interaction between large cross-sections bored piles with 'hard core' under dynamic loads and shelf soils. *Scientific Bulletin of National Mining University, 5*, 21–25.

Aslanov, L. F. (2016). Reflected wafes from bored or CFA piles of large section in the offshore soils. *Neft. khozyaystvo, 7*, 112–116.

Aslanov, L. (2022). Optimization of the calculation of the piles of fixed offshore platforms. In H. El-Askary, Z. A. Erguler, M. Karakus, & H. I. Chaminé (Eds.), *Research developments in geotechnics, geo-informatics and remote sensing. CAJG 2019. Advances in Science, Technology & Innovation*. Springer. https://doi.org/10.1007/978-3-030-72896-0_32

Berezantsev, V. G. (1960). *Strength analysis of bases of structures*. Gosstroiizdat (in Russian).

Sharafi, H., Maleki, Y. S., & Fard, M. K. (2016). Three-dimensional finite difference modeling of static soil-pile interactions to calculate p-y curves in pile-supported slopes. *Arabian Journal of Geosciences, 9*, 5. https://doi.org/10.1007/s12517-015-2051-9

Subanantharaj Palammal, J., & Senthilkumar, P. K. (2018). Behavioural analysis of vertical and batter pile groups under vertical and lateral loading in sand. *Arabian Journal of Geosciences, 11*, 706. https://doi.org/10.1007/s12517-018-4032-2

Wang, Y. X., Guo, P. P., Dai, F., Li, X., Zhao, Y. L., & Liu, Y. (2018). Behaviour and modelling of fiber reinforced clay under triaxial compression by using the combining superposition method with the energy based homogenization technique. *International Journal of Geomechanics*. https://doi.org/10.1061/(ASCE)GM.1943-5622.0001313

Yuan, B., Sun, M., Wang, Y., et al. (2019). Full 3D displacement measuring system for 3D displacement field of soil around a laterally loaded pile in transparent soil. *ASCE International Journal of Geomechanics, 19*(5), 04019028.

Yuan, B., Xu, K., Wang, Y., Chen, R., & Luo, Q. (2017). Investigation of deflection of a laterally loaded pile and soil deformation using the PIV technique. *ASCE International Journal of Geomechanics*. https://doi.org/10.1061/(ASCE)GM.1943-5622.0000842

Check for updates

Displaying Water Table Levels, Flow Direction for Predicting Construction Techniques Using Geographic Information as a Tool: Case Study of Kumba (South West, Cameroon)

Bahel Benjamin, Sigue Cyrille, Ngwem Bayiha Blaise, Bepuaka Ekuka, Ndive Molua Martin, Yamb Emmanuel, Owona Sébastien, and Bock Hyeng Alain

Abstract

The rapid economic growth of the town present the matter of water issue as very important problem in terms of human life, construction life, agriculture, etc. … This study is to predict techniques of foundation construction through the displaying of the water table at the flow direction in the town of Kumba and GIS. It is characterized by a significant research question which is the level of fall and rise in groundwater levels within the town of Kumba and this influence on choice of types of foundation in construction. This study is directed to decision-makers and technicians of the construction field to develop policies facilitating the supervision when building construction foundation by informing about water level depth and its flowing direction in the town. To achieve this, depths of static water levels were measured in over 200 randomly selected hand-dug wells in Kumba, after their geolocation and data were collected during the dry season (November and March 2017) and during the rainy season (between April and October 2017). Data were analyzed and treated using Microsoft Excel and GIS software us as Golden Surfer, Global Mapper, and ArcGIS. The results show us the water table contours of the city in dry and rainy seasons which vary between 26.1 and 2.0 m. Quarters such as Kumba-Mbeng and Meboka Streets have water levels ranging between 0 and 4 m in the rainy season and deeper above 5 m in the dry season. This variation of water level in the areas of Kumba may threaten foundation construction that why in those areas shallow foundation are not advice if the constructed structures is important only deep foundation is suitable. In Kumba Station, Mile 1, Bulletin Street (Fongong Quarter), and parts of Fiango show that water table is to deep water and proper for the shallow foundation but very hard for water supply through borehole. Groundwater flow direction was revealed to be toward the south and southeastern parts of Kumba. The significance of the study is to propose to the technician the direct application on the field of chosen types of foundations according to the quarter and proposed groundwater supply possibilities.

Keywords

GIS • Prediction • Water table level • Kumba • Foundation

B. Benjamin (✉) · S. Cyrille · B. Ekuka · N. M. Martin
Department of Topography and Real Estate Management, HTTTC, University of Buea, Kumba, Cameroun
e-mail: bahelbenjamin@yahoo.fr

N. B. Blaise · Y. Emmanuel
Department of Civil Engineering, ENSET, University of Douala, Douala, Cameroon

B. Benjamin
Department of Civil Engineering, ENSPD, University of Douala, Douala, Cameroon

O. Sébastien
Department of Earth Sciences, Faculty of Sciences, University of Douala, Douala, Cameroon

B. H. Alain
School of Technology, North Carolina A&T State University, 1601 East Market Street, Greensboro, NC 27411, USA

1 Introduction

The spatial prediction of the water table can be used for many applications related to civil works (foundations, excavations, irrigation in agriculture, etc.) also in urban and environmental management activities. The depth of the water table can be measured in existing wells to determine the effects of season, climate, or human impact on groundwater. A number of studies have been conducted to determine the depth of groundwater level in other parts of the world and their link with the climate, environmental impact. Water table depth has significance for agricultural,

M. Bezzeghoud et al. (eds.), *Recent Research on Geotechnical Engineering, Remote Sensing, Geophysics and Earthquake Seismology*, Advances in Science, Technology & Innovation,
https://doi.org/10.1007/978-3-031-48715-6_11

construction, and domestic potentials, and different flowing methods have been adopted, to avoid accidents difficulties. Ground flow water models have used (Akoachere et al., 2018, 2019) to monitor seasonal geochemical variations, the water quality, and groundwater aquiferous formations in Kumba. The saturated hydraulic conductivities of aquifers in Kumba were studied in 2016 and Darcy apparent (Akoachere & Ngwesse, 2016, 2017). They are many flows when rains in Kumba, but those flows are not only the facts of water rains also the influence of the water table level of the areas. In other words, some quarters of Kumba are under wetting areas each time it rains.

The objective of our work is to determine the displaying water table and the flow direction for the prediction of construction foundation techniques in Kumba. To obtain this work objective some specifics objectives must be attend as: establis the map location of the 200 boreholes used for the experimentation, evaluate the water evolution on each boreholes to obtain water table and map then mapping of the water direction of the underground water flowing and the proposed types of foundations appropriate for the water level areas. The significance of the study is to help the lawmakers and technicians to suggest the appropriate construction in each areas of the town and the types of foundations to be realize.

2 Materials and Methods

2.1 Materials

The choice of the materials was directly related to answering research. A map of Kumba was used during the reconnaissance stage of this work. The boreholes and well dug were located with a GPS remote (global positioning system) Garmin GPSMAP 64 to collect heights of points and geographical coordinates in the study area. A 50 m ribbon tape with a plumb bob attached to it was used to measure depth of well water and recorded in a filed notebook. A computer, Microsoft Excel, and some GIS tools such as ArcGIS, Global Mapper, and Golden Surfer 12 were equally used to analyze and display spatial information in the form of maps (Fig. 1).

3 Result

3.1 Elevation Map

Figure 2 presents the elevation of well points. Areas with high elevations were found to have deeper wells and higher depth-to-static water levels in both seasons.

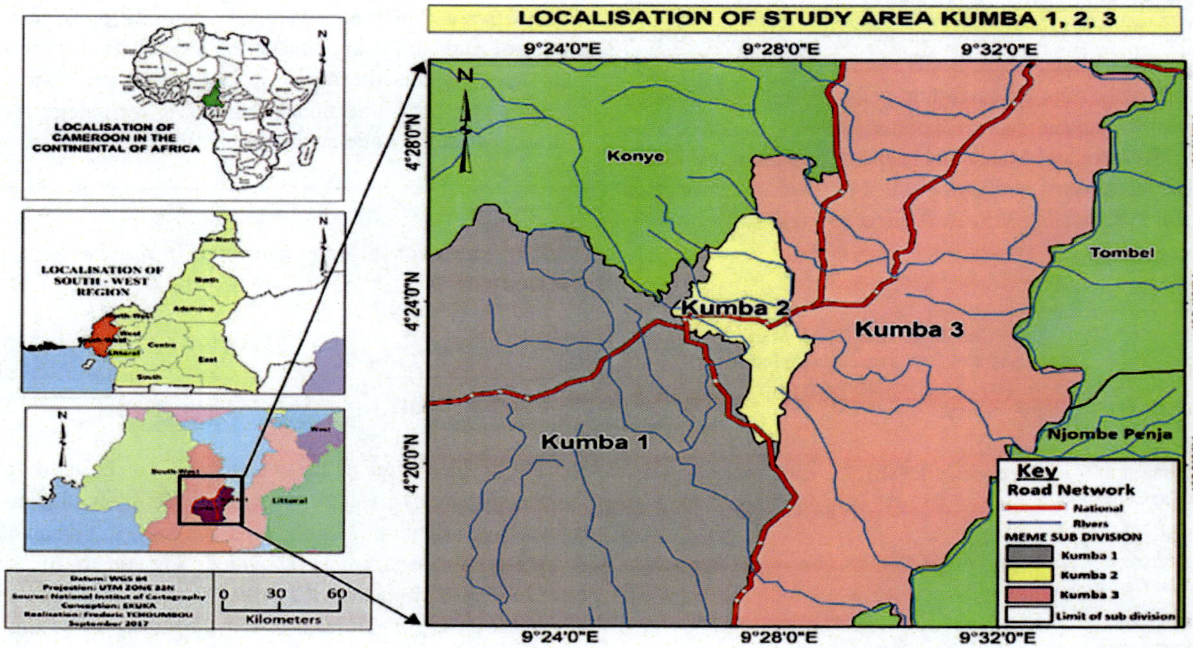

Fig. 1 Location map of Kumba in Cameroon

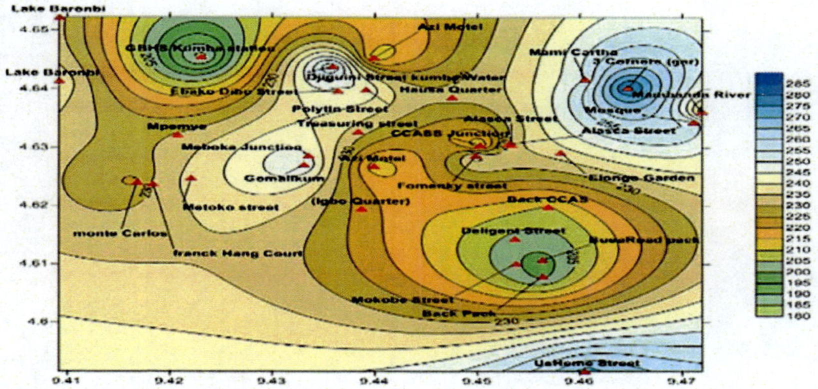

Fig. 2 Contour map showing the elevation of well points (above sea level) in the study area

3.2 Depth-to-Static Water Level

See Fig. 3.

3.3 Depth of Water Table in Relation to Building Construction

See Fig. 4.

3.4 Different Types of Foundations Used on the Field

See Figs. 5 and 6.

4 Discussion

The choice of shallower than depth foundation is very complex and more costly when it is wrongly design. These show the construct of a foundation is not only due to the low bearing capacity of the soil of an area but the level of water table. Figure 4a illustrates the spatial variation of water table level in Kumba, particularly between April and August. Areas such as Buea road have very shallow (0–4 m). As such, foundations in these areas should be well constructed, following the rigorous design methods and appropriates types of foundation and raised to withstand and control humidity when buildings in this areas are on swampy one.

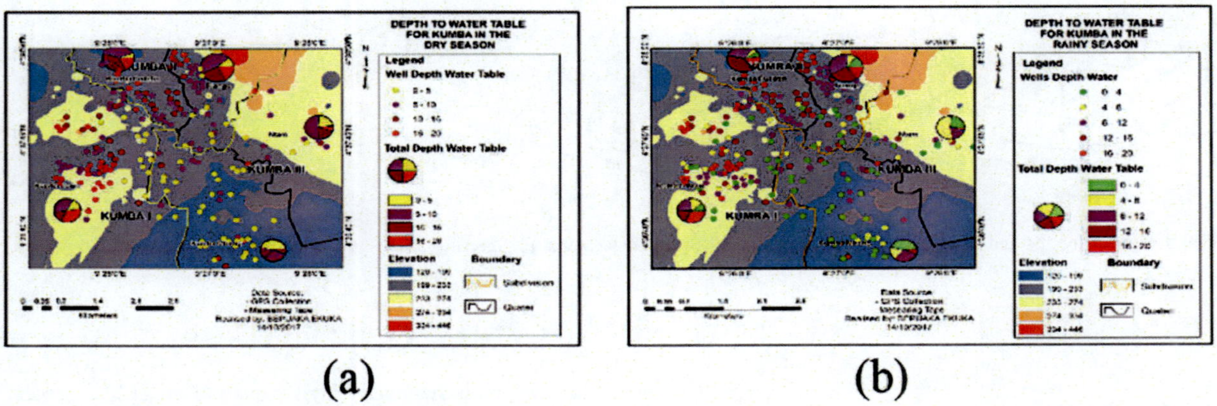

Fig. 3 Depth-to-static water table level map: **a** for dry season, **b** rainy season

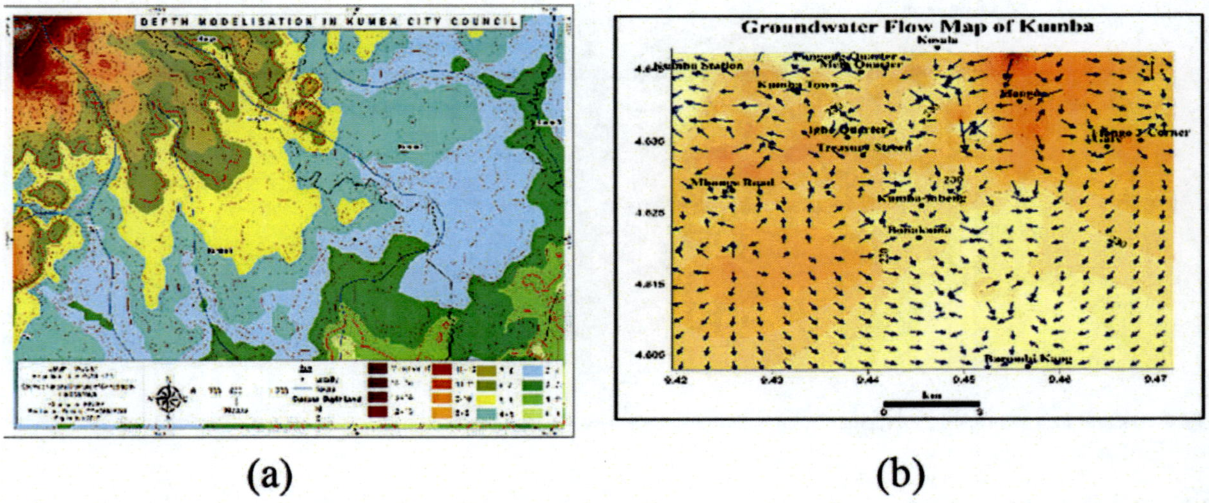

Fig. 4 a Depth to water table map of Kumba, b groundwater flow direction

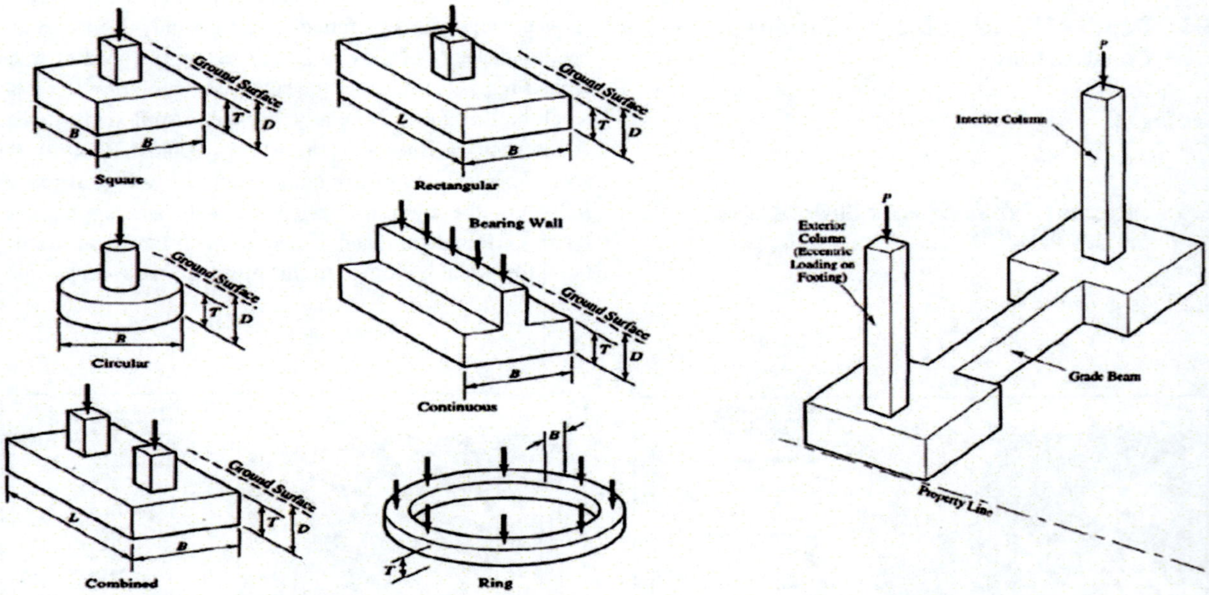

Fig. 5 Spread footing and strap footing foundation. *Source* SNU Geotechnical Engineering 2019

5 Conclusions

The result shows us a radial flow of water from the Northern and Northwestern parts of the study area. This is in line with results gotten in 2018 by Akoachere et al. It is also realise that water discharges in high elevation areas like Station, Mbonge Road Fongong quarter on with shallow foundation is proper and in low elevated areas like Buea Road proper for deep foundation (Fig. 4a, b). Topography plays a significant role in the flow of groundwater and the variation of water table in Kumba.

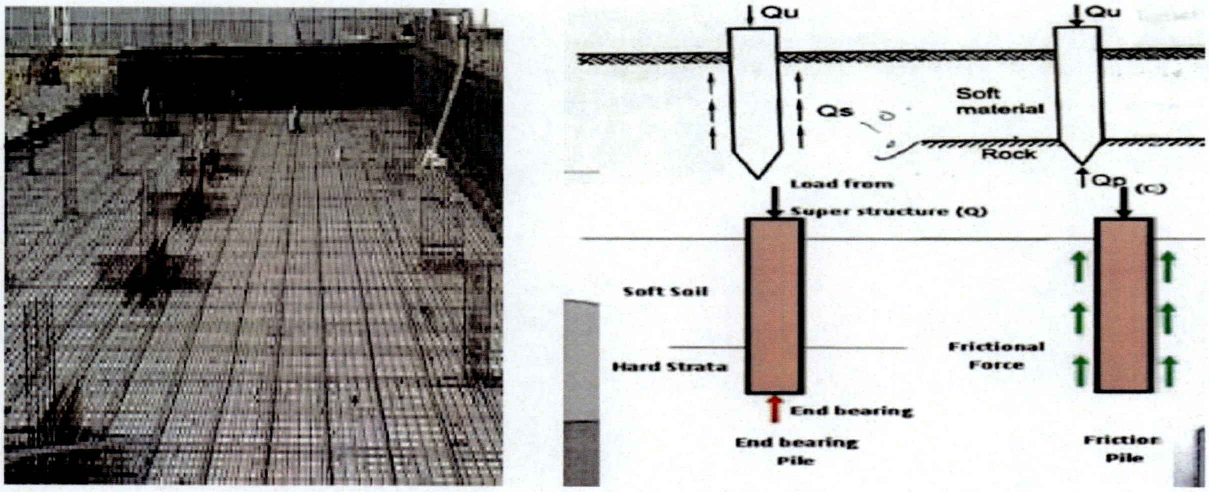

Fig. 6 Raft and deep foundation for heavy load swamping soil. *Source* SNU Geotechnical Engineering 2019

References

Akoachere, R. A., Hosono, T., Eyong, T. A., Ngassam, M.-C. P., Nkongho, R. N., & Okpara, S. O. (2019). Trace metals in groundwater of Kumba and environs in Cameroon. *Open Access Library Journal, 6*, e5824. https://doi.org/10.4236/oalib.1105824

Akoachere, R. A., & Ngwese, Y. M. (2016). Saturated hydraulic conductivities and high yield zones in the phreatic aquiferous formations in Kumba-Cameroon: Determined from slug-in tests in dug wells. *Journal of Hydrogeology & Hydrologic Engineering, 5*, 4.

Akoachere, R. A., & Ngwesse, Y. M. (2017). Darcy and apparent velocities of groundwater in phreatic aquiferous formations in Kumba-Cameroon: Determined by use of trigger-tube tracer test method in dug wells. *Journal of Hydrogeology and Hydrologic Engineering, 6*, 1.

Akoachere, R. A., Ngwesse, Y. M., Egbe, S. E., Eyong, T. A., Edimo, S. N., & Tambe, D. B. (2018). Groundwater monitoring in gneisso-basaltic fractured rock aquiferous formations of Kumba, South West Region Cameroon: Seasonal variations in the aqueous geochemistry and water quality. *Journal of Geoscience and Environment Protection, 6*(11). https://doi.org/10.4236/gep.2018.611003

The Optimal Approach to Valuing Mud Stored in Dam Reservoirs—Case Study of Morocco

Said Mohafid, Laila Stour, and Ali Agoumi

Abstract

Since dams are built on rivers generally to store water, they also store solid materials from floods. These materials settle and consolidate to create enormous quantities of mud, hampering the normal functioning of the structures and threatening the safety of the dam and surrounding populations. In addition to preventive solutions reducing this siltation phenomenon, dam managers often opt for desilting reservoirs as a curative solution. This solution produces enormous quantities of mud to be stored in landfills so far from the watercourse and dam reservoir, causing enormous environmental damage and considerable loss of land. Given the increasing scarcity and rising costs of various raw materials, the utilization of this mud as a valuable resource in diverse sectors including civil engineering, agriculture, crafts, and environmental applications has become imperative. An optimal approach to recover the mud extracted from dam reservoirs is designed and presented in the present article. This approach, which is the result of our research on mud valorization experiences around the world, is based on the definition of the characteristics to be studied, the tests to be carried out on the mud and the limit values to be observed, and on a diagram of the procedures to be followed in the valuation process. To show the canvas for using this approach, the present research work summarizes a case study carried out in the Moroccan context. This case study is based on the inventory of the experiences of some Mediterranean basin countries to define the possible uses of mud and on the laws and regulations of Morocco to choose the laboratory tests to accomplish and the limit values not to be exceeded. As a result, a guide for valuing mud extracted from dam reservoirs has been developed. It is composed of three elements: A grid of tests to accomplish on mud and limit values to observe, a matrix of characteristics to study for each intended use, and a diagram of the optimal approach to follow. It is a framework that Moroccan dam managers can follow to gain part of the cost of desilting operations. The guide can be adapted to other countries by replacing the limit values of the standards used in Morocco with those of the country in question.

Keywords

Dam • Mud-silting • De-siltation • Valorization • Environment

1 Introduction

Once deposited in dam reservoirs, the sediments consolidate in the form of juxtaposed layers, trapping more or less coarse formations called mud. The mud deposited in dam reservoirs remains harmful because they degrade the quality of the stored water (pollution, eutrophication, and turbidity) and presents a threat to the safety of the dam and the neighboring populations because the mud deposited increases the load applied to the dam, shut the outlets bottom designed to ensure to emptying reservoir in the event of failure of the structure, decrease the useful capacity of the reservoir, and get lost a large part of the budget invested in the construction of the dam. As long as it is deposited, the mud remains harmful, and dam managers always seek to get rid of it, whether by flushing laden water during floods or by desilting operations.

S. Mohafid (✉) · L. Stour
Process and Environment Engineering Laboratory, Faculty of Sciences and Techniques of Mohammedia, Hassan II University of Casablanca, B.P. 146, Mohammedia, Morocco
e-mail: said.mohafid@hotmail.fr

A. Agoumi
Civil Engineering, Hydraulics, Environment, and Climate Change Laboratory, Hassania School of Public Works, Casablanca, Morocco

M. Bezzeghoud et al. (eds.), *Recent Research on Geotechnical Engineering, Remote Sensing, Geophysics and Earthquake Seismology*, Advances in Science, Technology & Innovation,
https://doi.org/10.1007/978-3-031-48715-6_12

Despite the advantages of this operation, it also has the following disadvantages:

- It is a very expensive operation,
- It requires large land to be occupied as landfills for the extracted vases,
- The millions of cubic meters of the evacuated mud pollute the environment.

The possible use of this mud in the industry will surely compensate for some of the disadvantages of desilting operations.

2 Potential Sectors of the Possible Use of Mud

The appearance of mud stored in dam reservoirs as well as their enormous quantities (of the order of tens of millions of cubic meters per dam) naturally lead us to think of using them in industries and sectors consuming large quantities of raw materials. It is for these reasons that we have pushed our research into the following sectors:

- Civil engineering sector: Backfilling roads and the development of platforms (Seklaoui, 2016), backfilling trenches, manufacturing concrete (as fillers or as expanded aggregates), producing clinker for cement (Prax, 1979), and manufacturing terracotta bricks (Bonnet & Gaillard, 2001);
- Agriculture: Fertilization (or fertilizer effect) by providing nutrients for cultivated plants, amendment by improving soils or crop supports through their physical (cohesion, porosity, etc.) and/or chemical (pH, cation exchange capacity, etc.) and/or biological (microbial biomass, mineralizable nitrogen properties …), ecological rehabilitation of disturbed areas (old industrial wastelands and recent deconstruction, slag heaps, or newly developed areas) which requires reconstituting a substrate favorable to revegetation (Bataillard et al., 2017);
- Environmental protection: Water tightness layer for a lagoon or landfill (Benasla, 2015), soil amendment (Levacher et al., 2006);
- Craftworks: Ceramic products and tiles (Guerraoui et al., 2008) and glass manufacturing (Olivier, 1978).

3 Characteristics to Consider

The usability of a vase by the sectors and activities mentioned above is determined by several decisive characteristics, including clayey or clay contents, granularity or particle size analysis, mineralogical characteristics, organic matter content, water content, mechanical characteristics, chemical characteristics, as well as the presence of traces of metals and organic compounds.

4 Laboratory Tests to Be Performed and Limit Values

The potential sectors for the use of mud mentioned above, as well as the characteristics to consider, have been listed in the experiences of some Mediterranean countries concerning the valorization of mud retained by dams. The laws and regulations of each country make it possible to define the tests to execute to determine the characteristics of the materials and the limit values not to be exceeded to be usable in each sector. As an example, the characteristics to observe to using mud in road works are clayey, granularity, organic matter content, and water content.

A variety of laboratory tests are to be conducted, including the dry sieving test after washing, the methylene blue value test, the Atterberg limits tests, the Casagrande cup test, the steaming method, the chemical method, and the calcination method. These tests encompass a comprehensive analysis of the mud's characteristics, ranging from particle size distribution assessment to plasticity determination, as well as chemical composition. By employing this diverse range of laboratory tests, a thorough understanding of the mud's properties can be obtained, providing valuable insights for various research purposes and engineering applications.

Laboratory tests to accomplish for identifying the aforementioned characteristics can be classified into three categories, namely:

- Physical tests that include among others: Limits of Atterberg, Casagrande cup, roller method, dry sieving method after washing, sedimentometry, and calcination method.
- Chemical tests that include methylene blue test, chemical method of determining the organic matter content …
- Radiation and spectroscopy tests include X-ray diffraction, fluorescence X, inductively coupled plasma-optical emission spectrometry (ICP-OES), and inductively coupled plasma-mass spectrometry.

Limit values to be observed for each laboratory test are determined by the laws and regulations used in the country concerned. These limit values depend on the quality of the materials and also on the requirements of each sector of activity. An example of the limit values to be observed is given in the next paragraph relating to the case study.

5 The Optimal Approach to Valuing Mud Stored in Dam Reservoirs (Case Study of Morocco)

Through our extensive research, we have developed an optimal and pragmatic approach for the valorization of mud retained in dam reservoirs. This approach centers around the creation of a tailored guide specific to the relevant country. Serving as a comprehensive resource, this guide offers valuable insights into the assessment of valuing mud. By utilizing this guide, decision-makers will be better regarding the use of mud. This guide is composed of the following three components:

1. The test grid that outlines the required tests and limit values to be observed. This grid is presented as a table, with each row representing different activities and the columns providing information on the sector, activity, mud use, characteristics, desired value, and tests to be conducted.
   As an example of a few lines of this grid, for the use of mud as a form layer in road works relating to Civil Engineering, we define the tests to be carried out and the limit values as shown in Table 1.
2. The matrix of the main characteristics to be studied in order to optimize the choice of tests according to the aforementioned grid. It presents the activities related to the different sectors where the use of mud is possible on the one hand, as well as the specificities to be considered on the other hand. This matrix is a useful tool to guide the selection and prioritization of appropriate laboratory tests according to the targeted activity.
3. The Procedure Flowchart: In order to provide a clear and structured approach for utilizing this guide, we outline the sequential procedure as follows:

(1) Choose a dam based on factors such as the availability of extractable mud in the reservoir mud deposition and storage areas, and proximity to the desired reuse area.
(2) Select the sectors of activity through market research, supply and demand analysis, technical and economic studies, and feasibility and profitability assessments.
(3) Refer to both local databases (in paper or digital format) on the chosen dam's sediments and the geographic information system (GIS) database. These databases mainly concern quantitative data and the results of laboratory tests previously carried out on the mud extracted from the dams.
(4) Conduct laboratory tests in a phased manner, depending on the targeted activity:
  a. Define the specific tests for first and second phases.
  b. Collect relevant samples for testing.
  c. Perform trials for the first phase.
  d. Analyze the results obtained from the first phase trials.
  e. Proceed with the second phase of trials if the results from the first phase are conclusive.
  f. Analyze the results obtained from the second phase trials.
(5) Decision-Making: Based on the test results, make one of the following decisions:
  a. Retain the selected dam for the valorization of its mud.
  b. Reject the idea of utilizing the stored mud from the selected dam.
  c. Update the databases and the GIS database by entering the obtained results.

We summarize the process diagram in the provided flowchart (Fig. 1).

Table 1 Required tests and threshold values for the utilization of mud as a subsurface layer in road construction

Characteristics	Test to be carried out	Limit values
Granularity	• Dry sieving after washing • Sedimentometry	Maximum grain size D_{max} < 50 mm
		Passing 80 μm sieve of 35% minimum
Clayey	The methylene blue value test (MBV)	2.5 < MBV < 8 (g/100 g)
	• The limits of Atterberg • The Casagrande cup • The roller method	12% < I_p < 40%
Water content	Steaming method	*W* (%) Fine soil conditions *A*1, *A*2, *A*3
Organic matter content	Chemical method or calcination method	3% < OM < 10%

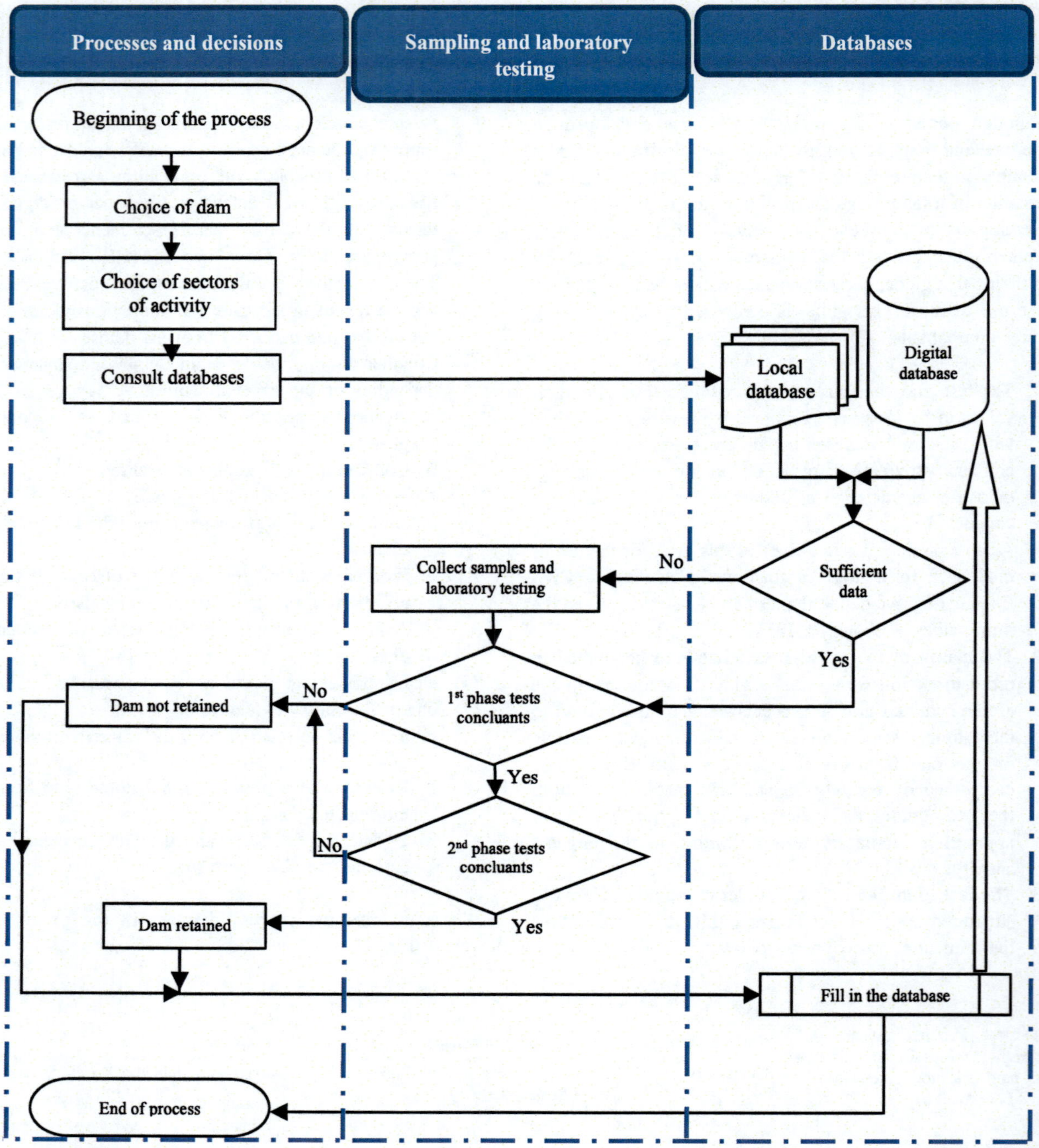

Fig. 1 Flowchart of valuing mud process

6 Conclusion

The framework for valuing mud retained in dam reservoirs, which has been presented in this article, offers dam managers an optimal approach to recover a portion of the desilting operation costs. This approach enables more sustainable and economically viable management of dam reservoirs, as it promotes the utilization of mud as a valuable resource rather than considering it as a waste material. Furthermore, by integrating this framework into their decision-making processes, dam managers can contribute to the overall efficiency and profitability of their operations, while also considering the environmental and economic benefits of mud valorization.

In this approach, we do not perceive mud as harmful material to the environment and dangerous for populations, but rather as a raw material for different industries. The application of this approach to the case of Morocco by producing a detailed guide for valuing mud will be enriched by a concrete and in-depth study of the guidelines of this guide on one or more dams in Morocco. Samples of the mud will be taken from dam reservoirs for laboratory testing, the results of which will be the subject of our future publications.

References

Bataillard, P., Chevrier, B., & Hoang, V. (2017). Valorisation à terre des sédiments de dragage: Retour d'expérience en France et à l'international, BRGM/RP-67329-FR (p. 45).

Benasla, M. (2015). *Caractérisation de la vase de dragage du barrage de l'Oued Fodda et valorisation en tant que matériau de construction* [Thèse doctorale]. Université des Sciences et de la Technologie d'Oran Mohamed.

Bonnet, J. P., & Gaillard, J. M. (2001). *Céramique silicatées, Matériaux et processus céramiques* (pp. 113–140). Hermes Science.

Guerraoui, F., Zamama, M., & Ibnoussina, M. (2008). Caractérisations minéralogique et géotechnique des argiles utilisées dans la céramique de SAFI (Maroc). *African Journal of Science and Technology (AJST), Science and Engineering Series, 9*(1), 1–11.

Levacher, D., Colin, D., Carina Perroni, A., Duan, Z., & Sun, L. (2006). Recyclage et valorisation de sédiments fins de dragage à usage de matériaux routiers. In *IXème Journées Nationales Génie Civil—Génie Côtier*, Brest (pp. 603–612).

Olivier, E. (1978). *Technologie des matériaux de construction T2*. Entreprise Moderne Edition.

Prax, A. (1979). *Mémento substances utiles (Matériaux de carrière) clinker portland, matière premières utilisés pour la fabrication du ...; 79 SGN 148 MTX* (pp. 27–28).

Seklaoui, O. (2016). *Valorisation des sédiments du barrage d'El Mardja Sidi Abed: Etude technico-économique* [Thèse de doctorat] (119 pp.). Université Mouloud Mammeri de Tizi-Ouzou.

Characterization of Moroccan Phosphogypsum Toward a Road Engineering Application

Sarra Meskini, Toufik Remmal, Moncef Benmansour, and Azzeddine Samdi

Abstract

Around 30 Mt of phosphogypsum are generated annually in Morocco. It is a by-product from the phosphoric acid industry where the majority is stockpiled by Jorf Lasfar and Safi plants. The purpose of this study is to characterize Moroccan phosphogypsum on the physico-chemical, mineralogical, morphological, geotechnical, and environmental level in order to assess its suitability as a road material. To this end, several techniques have been implemented, namely: X-ray fluorescence and X-ray diffraction (XRF/XRD), laser granulometry, inductively coupled plasma mass spectrometry (ICP-MS), scanning electron microscopy (SEM), proctor and oedometer tests as well as radioactivity and leaching tests. The results showed that Moroccan phosphogypsum has a pH of about 3.45, a fine grain size (< 200 μm) marked by the presence of parallelepipedic crystals. Chemical and mineralogical analyzes showed that phosphogypsum is mainly composed of dihydrate gypsum (86%), but contains impurities such as quartz, heavy metals, and radioelements from phosphate rock, in addition to some acid residues due to the transformation process. These impurities constitute the main source of pollution linked to the storage of phosphogypsum insofar as heavy metals can migrate from the heap to the groundwater and the radionuclides can decompose and pose risks to human health. From a geotechnical point of view, phosphogypsum is a low-density material with compressive behavior and low resistance to immersion due to the solubility of gypsum which further increases owing to its acidic nature. Although the radiological indices relating to phosphogypsum allow its use in road engineering, this is not possible given its sensitivity to water variations which could impact its bearing capacity as well as the risk of leaching of contaminating elements. Thus, for phosphogypsum to be valorized in road engineering, it requires a treatment in order to stabilize its behavior, in particular the improvement of its resistance to compression and immersion. Lime–fly ash treatment is a promising solution for a use as a road base material.

Keywords

Phosphogypsum • Pavement • Radioactivity • Impurities • Chemical composition

1 Introduction

Phosphogypsum (PG) is the by-product from the transformation of phosphate rock into phosphoric acid following the dihydrate process. This transformation entails the passage of impurities such as heavy metals and radionuclides, initially present in the ore, to phosphogypsum which hinders its large-scale valorization, in particular for the production of construction materials and the amendment of agricultural soils. Road engineering is considered as a viable and promising route for the effective elimination of phosphogypsum as it consumes large volumes of construction materials. In this context, several investigations have been conducted, the results of which have demonstrated the weakness of phosphogypsum toward compression and immersion. This study is a contribution to the characterization of phosphogypsum on the physico-chemical,

S. Meskini (✉) · T. Remmal
Laboratory of Geosciences Applied to Urban Development Engineering, Faculty of Sciences Ain Chock, University Hassan II of Casablanca, Casablanca, Morocco
e-mail: Sarra.meskini-etu@etu.univh2c.ma

M. Benmansour
National Center for Energy Sciences and Nuclear Techniques, Rabat, Morocco

A. Samdi
Laboratory of Physico-Chemistry of Inorganic Materials, Faculty of Sciences Ain Chock, University Hassan II of Casablanca, Casablanca, Morocco

M. Bezzeghoud et al. (eds.), *Recent Research on Geotechnical Engineering, Remote Sensing, Geophysics and Earthquake Seismology*, Advances in Science, Technology & Innovation,
https://doi.org/10.1007/978-3-031-48715-6_13

morphological, mineralogical, geotechnical, and environmental level in order to assess its suitability for use in road engineering in the Moroccan context.

2 Materials and Methods

The phosphogypsum sample, under characterization, was freshly collected from the Jorf Lasfar phosphoric acid plant, El Jadida, Morocco. In order to determine its chemical composition, in major and trace elements, X-ray fluorescence spectroscopy and inductively coupled plasma mass spectrometry analyzes were performed, respectively. The mineralogical composition of PG was determined through X-ray diffraction analysis. The physico-chemical characterization concerned the determination of its pH, water content, real and apparent densities, and grain size distribution. The morphology of phosphogypsum was determined by mean of scanning electronic microscope observations, performed in secondary electrons mode, while the geotechnical characterization was carried out via a compaction modified proctor test and an oedometer test. The environmental impact of phosphogypsum was assessed through a radiochemical test (gamma spectrometry) and leaching test (TCLP procedure), followed by a metallic trace elements dosage by atomic absorption spectrometry.

3 Results and Discussion

3.1 Chemical and Mineralogical Compositions

The chemical composition of the phosphogypsum sample is given in Table 1, revealing a major composition of gypsum (71% $CaSO_4$) with the presence of low amounts of silica (9.65% SiO_2). Mineralogical analysis showed that it is a dihydrate gypsum ($CaSO_4$, $2H_2O$), accompanied by quartz. Table 1 highlights also the presence of phosphoric oxide (1.62% P_2O_5), which reflects the existence of residual phosphoric acid H_3PO_4 and a considerable amount of impurities due to the phosphate rock nature and/or the transformation process type. In fact, the introduction of these impurities, during the chemical attack, can be attributed to the substitution of calcium or sulfate ions with other elements having the same size and charge (Ennaciri et al., 2020). These impurities, as determined by ICP-MS trace analysis, represent 0.2% of the elements not identified by XRF. The detected elements are alkaline earth metals (Ba, Be, and Sr), toxic metalloids (As and B), metals (Co, Cr, Cu, Mo, Ni, Zn, and Nb), heavy metals (Pb, Cd), and rare earth elements (Y) with concentrations varying between 0.2 and 243 ppm.

3.2 Physico-chemical and Morphological Properties

The phosphogypsum sample is a yellowish moist powder with a water content of 27% and a pH of 3.45. Its reel and apparent densities are in the order of 2410 kg/m^3 and 740 kg/m^3, respectively. The grain size analysis confirmed its fine nature insofar as the size of its particles varies between 2 and 315 μm, with a median diameter d_{50} of 58 μm, and a majority (69%) having a diameter less than 80 μm. The SEM micrographs highlighted a tabular form of the phosphogypsum crystals with a heterogeneous size (20–200 μm length and 5–30 μm width).

3.3 Geotechnical Properties

3.3.1 Compaction Characteristics

Results from the compaction proctor test on the phosphogypsum sample are plotted in Fig. 1. The optimal water content is 15.8% corresponding to a maximal dry density of 14.82 kN/m^3. The flattened aspect of the proctor curve means that the dry density is little influenced by water variation, insofar as a variation of 9% only varies the dry density by 0.2 and maintains it in the range 14.6–14.8 kN/m^3.

3.3.2 Consolidation Test Results

The compressibility curve of PG (Fig. 2) highlights a strong compressible behavior, marked by significant settlements for low loading stresses. This resulted in a compression index (C_c) of 0.295 and a pre-consolidation stress of (σ'_c) 400 kPa which classify the phosphogypsum sample as a fairly highly compressible material.

3.4 Environmental Behavior

3.4.1 Radiochemical Properties

Results from gamma spectrometry highlighted a high activity of the radionuclides from the uranium decay chain

Table 1 Chemical composition of PG (major elements)—XRF

Oxides	Al_2O_3	Fe_2O_3	CaO	K_2O	MgO	Na_2O	P_2O_5	TiO_2	SiO_2	SO_3	L.O.I	Others
%	0.77	0.15	38.53	0.05	0.21	0.55	1.62	0.03	9.65	32.76	14.36	1.32

L.O.I loss on ignition

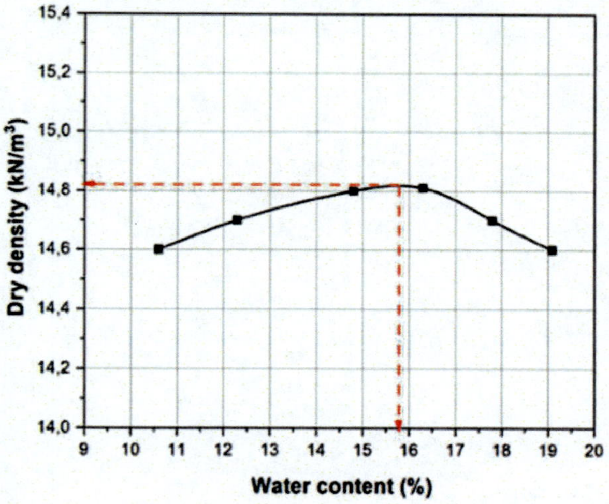

Fig. 1 Proctor curve of PG

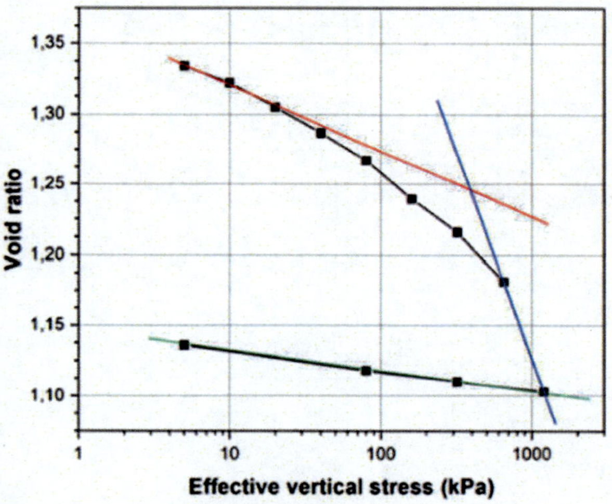

Fig. 2 Odometer curve of PG

U-238, reflecting their predominance in the phosphogypsum sample, in particular Ra-226 (Meskini, 2022). Compared to the phosphogypsum activities reported in the literature (Kuzmanović et al., 2020; Zrelli et al., 2018), Moroccan material shows rather high values for Ra-226 and of the same order for Th-232 and K-40. These values exceed the threshold limits allowing phosphogypsum to be used indoor, yet are acceptable for a use in road engineering.

3.4.2 Leaching Characteristics

The chemical analysis of the recovered solution at the end of the leaching test revealed the passage of some trace elements from phosphogypsum to water at concentrations varying between 0.002 and 0.290 mg/l. The most leached metal is zinc, followed by lead, nickel, then copper, and chrome in lesser amounts. According to the obtained concentrations, phosphogypsum does not represent a significant risk of contamination of water following the leaching of metallic trace elements it contains since these levels remain well below the limit thresholds set by the US-EPA (1992).

4 Conclusion

The aim of this study was to characterize Moroccan phosphogypsum on the physico-chemical, mineralogical, geotechnical, and environmental level in order to assess its suitability for use in road engineering. Results showed that phosphogypsum does not present a risk of contamination of groundwater since the concentrations of heavy metals are below the limit thresholds and its radioactivity is acceptable for use in road sub-layers. However, from a geotechnical point of view, phosphogypsum has a low density and a strong compressive behavior that could impact its bearing capacity. Its use in the raw state is therefore not possible, and hence, a stabilizing treatment is required.

References

El Zrelli, R., Rabaoui, L., Daghbouj, N., Abda, H., Castet, S., Josse, C., van Beek, P., Souhaut, M., Michel, S., Bejaoui, N. & Courjault-Radé, P. (2018). Characterization of phosphate rock and phosphogypsum from Gabes phosphate fertilizer factories (SE Tunisia): High mining potential and implications for environmental protection. *Environmental Science and Pollution Research, 25*, 14690–14702. https://doi.org/10.1007/s11356-018-1648-4

Ennaciri, Y., Zdah, I., El Alaoui-Belghiti, H., & Bettach, M. (2020). Characterization and purification of waste phosphogypsum to make it suitable for use in the plaster and the cement industry. *Chemical Engineering Communications, 207*, 382–392. https://doi.org/10.1080/00986445.2019.1599865

Kuzmanović, P., Todorović, N., Forkapić, S., Petrović, L. F., Knežević, J., Nikolov, J., & Miljević, B. (2020). Radiological characterization of phosphogypsum produced in Serbia. *Radiation Physics and Chemistry, 166*. https://doi.org/10.1016/j.radphyschem.2019.108463

Meskini, S. (2022). *Le phosphogypse dans le domaine routier: Valorisation et potentiel d'utilisation*. University Hassan II of Casablanca.

US-EPA. (1992). *Methode 1311: Toxicity characteristic leaching procedure* (pp. 1–35).

New Protection Technique for Flexible Pavement Structure

Tahar Ayadat

Abstract

This paper concerns the development of a new protection technique to improve simultaneously the waterproof or the impermeability, the thermal insulation, and the bearing capacity of flexible pavement structures. The technique consists on the introduction of a layered system composed of three different materials into the pavement structure, in this case: Foamglas material, XPE foam laminated Alufoil, and geotextile Terram 2000. The new protection technique is believed to limit frost penetration in the frost susceptible subgrade soil, reduce significantly road network damages associated to differential heaving and bearing capacity loss during seasonal temperature variations, restrict water infiltration, and thus reduce considerably the associated damage and the rehabilitation costs. Furthermore, a procedure was developed for the determination of the thicknesses different layers of the improved flexible pavement structure.

Keywords

Flexible • Pavement • Protection • Technique • Foamglas • Terram • Alufoil

1 Introduction

In many regions around the world, seasonal temperature variations cause road network damages associated to differential heaving and bearing capacity loss during spring (Vaitkus et al., 2019). As a result, deterioration of the ride quality, cracking of the asphalt concrete layer and finally, increased rutting and pothole formation are likely to occur (among others, Ragnoli et al., 2018; Tamrakar, 2019). This leads to increased maintenance and higher rehabilitation costs. This problem is compounded by the infiltration of water in the pavement structure (Adlinge & Gupta, 2005). The primary source of water in pavements is precipitation (rain). This water can enter the pavement structure through several ways, including infiltration from shoulders and ditches. In order to improve the durability of the pavement, thermal insulation is commonly used as a preventive measure to limit the frost penetration in frost sensitive subgrade soil. However, the effectiveness of this type of pavement improvement is still inconclusive. Thus, new alternative methods should be employed to solve the problem of pavement deterioration effectively. The methods employed should combine simultaneously water proof, thermal insulation, ground reinforcement, and erosion control. The objective of this paper is the development of a new technique to improve simultaneously the impermeability, the thermal insulation, and the bearing capacity of flexible pavement structures. The technique consists of the introduction of a layered system, composed of three different materials, into the pavement structure, in this case: Foamglas material, XPE foam laminated Alufoil, and geotextile Terram 2000 (i.e., Thermal insulation material, waterproof material, and ground stabilization or reinforcement material, respectively). These three thin layered materials are placed or installed in the interface between the base and sub-base layers of the flexible pavement.

2 Description of the New Pavement Structure

The flexible pavement structure comprises a surface layer, which comprises asphaltic concrete; a base layer, wherein the surface layer is adjacent and above the base layer; a sub-base layer, wherein the base layer is adjacent and above the sub-base layer; and a subgrade layer, wherein the

T. Ayadat (✉)
Prince Mohammad Bin Fahd University, P.O. Box 1664
Al-Khobar, 31952, Saudi Arabia
e-mail: tayadat@pmu.edu.sa

M. Bezzeghoud et al. (eds.), *Recent Research on Geotechnical Engineering, Remote Sensing, Geophysics and Earthquake Seismology*, Advances in Science, Technology & Innovation,
https://doi.org/10.1007/978-3-031-48715-6_14

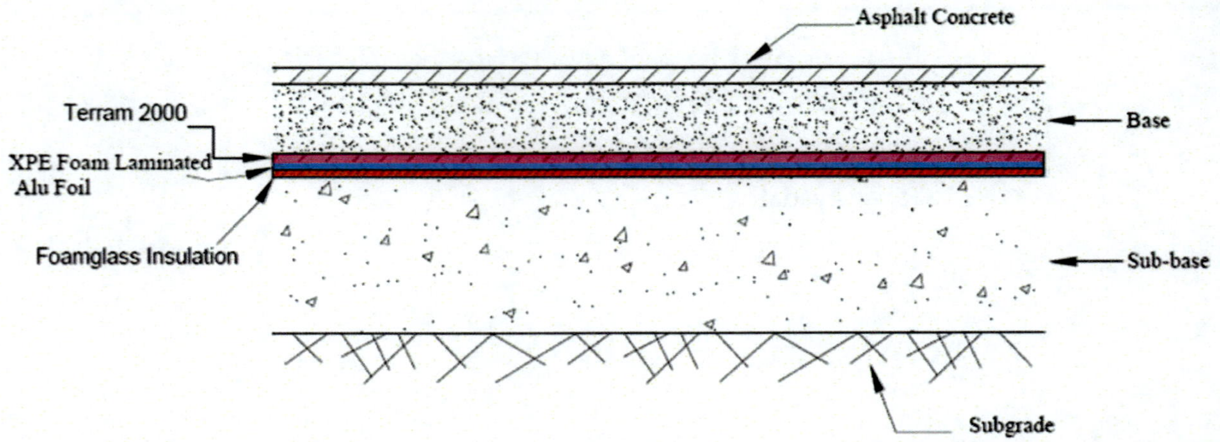

Fig. 1 Cross-section of the improved pavement structure

sub-base layer is adjacent and above the subgrade layer; wherein: a layered system composed of first, second, and third materials different from each other as an interface layer between the base layer and the sub-base layer; the first material of the system is configured to provide the structure with ground stabilization or reinforcement properties; the second material of the system is configured to provide the structure with waterproofing or impermeability properties; the third material of the system is configured to provide the structure with thermal insulation properties (Fig. 1).

The first material is a geotextile fabric selected from the group consisting of polypropylene and polyethylene. It has the following properties:

- a tensile strength of approximately 14.5 kN/m;
- a tensile elongation of approximately 60%;
- a CBR puncture resistance of approximately 2750 N;
- a cone drop of approximately 26 mm;
- a pore size of approximately 65 μm; and
- a permeability of approximately 55 l/m^2.

The second material is a waterproof heat insulation material selected from the group consisting of XPE foam and laminated Alufoil; the first material is adjacent and above the second material in the structure. It has the following properties:

- a thermal conductivity of approximately 0.03 W/mK;
- a reflectivity of 95–97%;
- a temperature resistance of approximately 40–100 °C;
- a heat insulating ability of approximately 34.5–56.6 °C;
- a water vapor transmission rate of less than 1 g/m^2/day;
- a tensile load of at most 44 N in a machine direction and at most 99 N in a transverse direction;
- a grams-per-square meter value of approximately 180–410 g/m^2; and
- a density of approximately 431 g/m^2.

The third material is a Foamglas insulation material, such as Foamglas T3 + Slab XL, and the second material is adjacent and above the third material in the structure. It has the following properties:

- an operating temperature range of approximately − 200 to 430 °C;
- a thermal conductivity of 0.036 W/mK; and
- a compressive strength of at least 500 kPa.

It is worthy to note that the first material further provides waterproof properties; the second material further provides thermal insulation properties; the third material further provides both waterproof and reinforcement properties.

The unique features of the new protection technique when compared to the existing solutions are that it provides a superior protection technique. The combination of the multi-functions of the different materials in the layered system limits frost penetration in the frost susceptible subgrade soil, reduces significantly road network damages associated to differential heaving and bearing capacity loss during seasonal temperature variations, restricts water infiltration, and thus reduces considerably the associated damage and the rehabilitation costs. The new technique preserves the balances between technological, environmental, and social costs.

3 Theoretical Development

The method proposed in this paper for the design of the new pavement structure provides a procedure and recommendations regarding the determination of total thickness of the

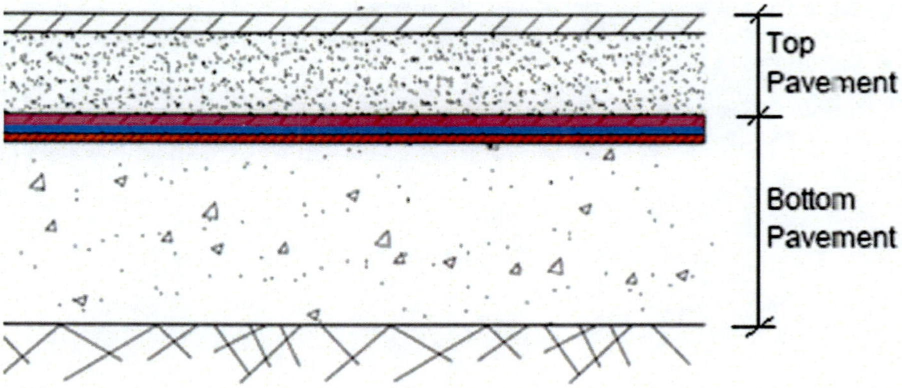

Fig. 2 Components of the improved flexible pavement structure

pavement structure as well as the thickness of the individual structural components. Based on the AASHTO guide for the design of pavements structures, the new pavement structure was considered as two superposed pavement components (Fig. 2), in this case: top pavement (asphalt concrete + base) and bottom pavement (protection + sub-base). Accordingly, the structural number SN (indicative of the of the total pavement thickness required) is determined by the following equation.

$$\mathrm{SN} = (a_1 \times D_{\mathrm{hma}} + a_2 \times m_{\mathrm{base}} \times D_{\mathrm{base}})_{\mathrm{Top\,Pavement}} + (a_3 \times t_{\mathrm{protection}} + a_4 \times m_{\mathrm{subbase}} \times D_{\mathrm{subbase}})_{\mathrm{Bottom\,Pavement}}$$

where

- a_1 Coefficient of the asphalt concrete layer relative strength
- a_2 Coefficient of the base layer relative strength
- a_3 Coefficient of the protection layer relative strength
- a_4 Coefficient of the sub-base layer relative strength
- m Modifier for more than normal moisture (drainage coefficient)
- D Layer thickness (in.)
- t Thickness of protection layer.

The design monographs (AASHTO, 1993) are used to estimate the different SN values. Then, the required SN are converted to actual thickness of surfacing, base, and sub-base by means of appropriate layer coefficients representing the relative strength of the construction materials. Furthermore, the coefficients of the different layer strengths and the drainage coefficients are deduced form the design guide (AASHTO, 1993). However, the coefficient of the protection layer (a_3) is estimated to be 0.50 by considering a resilient elastic modulus (M_R) of the protection materials around 9.5×10^5 psi. The value of M_R was determined using the following equation (Heukelom & Klomp, 1962):

$$M_R = 1500 \times \mathrm{CBR}$$

where

- CBR California bearing ratio (deduced from the properties of Foamglas material and the geotextile Terram 2000).

4 Conclusion

A new protection technique was proposed in the present work to improve simultaneously the waterproof or the impermeability, the thermal insulation, and the bearing capacity of flexible pavement structures. Conversely to the prior art references, the technique consists on the introduction of a reinforcing layered system composed of three different materials into the pavement structure, in this case: Foamglas material, XPE foam laminated Alufoil, and geotextile Terram 2000. The combination of the multi-function of the layered system enhances the performance of the flexible pavement in respect to waterproofing, thermal insulation, and bearing capacity. Separately, the reinforcing materials of the layered system provide different protections, simultaneously. Furthermore, a procedure was developed for the determination of the thicknesses different layers of the improved flexible pavement structure.

References

AASHTO. (1993). *Guide for design of pavement structures*. American Association of State Highway and Transportation Officials.

Adlinge, S. S., & Gupta, A. K. (2005). Pavement deterioration and its causes. *IOSR Journal of Mechanical and Civil Engineering*, 9–15.

Heukelom, W., & Klomp, A. J. G. (1962). Dynamic testing as a means of controlling pavements during and after construction. In *Proceedings of the International Conference on the Structural Design of Asphalt Pavements* (pp. 667–679), University of Michigan Ann Arbor.

Ragnoli, A., De Blasiis, M. R., & Di Benedetto, A. (2018). Pavement distress detection methods: A review. *Infrastructures, 3*, 58, pp. 1–19. https://doi.org/10.3390/infrastructures3040058

Tamrakar, N. K. (2019). Overview on causes of flexible pavement distresses. *Bulletin of Nepal Geological Society, 36*, 245. ISSN 2676-1394.

Vaitkus, A., Žalimienė, L., Židanavičiūtė, J., & Žilionienė, D. (2019). Influence of temperature and moisture content on pavement bearing capacity with improved subgrade. *Materials (Basel), 12*(23), 382. https://doi.org/10.3390/ma12233826

Slake Durability of Granitic Rocks in Wet and Dry Conditions

Lekan Olatayo Afolagboye, Abel Ojo Talabi, and Olubunmi Oluwadare Owoyemi

Abstract

Weathering processes due to heating and cooling, wetting and drying cycles, salt decay, and acidic rains (atmospheric pollution) can affect the durability and long-term behavior of rock masses and aggregates. The resistance of rocks to weathering can be described using a durability parameter referred to as slake durability index. The main purpose of this study is to evaluate the influence of weathering processes simulated by heating and cooling and wetting and drying on the durability of fine-grained and porphyritic granites rocks from Ado-Ekiti, southwestern Nigeria. Petrographic and physical tests were carried out on the rocks using appropriate international standards. The slake durability test followed the procedure stipulated by ASTM D4644 standard was performed. The test cycles were conducted in wet (with distilled water) and dry (without water) conditions. The granites contain minerals such as quartz, microcline, hornblende, biotite, and muscovite. The fine-grained granite has lower water content, water absorption capacity, and porosity compared to porphyritic granite. The rate of mass loss decreases as the number of slaking cycles increased. This rate of mass loss is found to be higher at the early cycles than the final cycles. The maximum mass loss was observed in samples used for wet slake durability test. Slake durability indices of the granitic rocks decreased as number of slaking cycles increases and is higher in dry test than in wet test. Comparison between the results of dry and wet slake durability tests show that the effect of water on the rate of rock degradation differs in both rocks. The porphyritic granite is sensitive to water in terms of their slake durability compared to fine-grained granite. This may be attributed to the variation in physical properties of the granitic rocks.

Keywords

Slake durability • Weathering • Granite • Petrographic characteristics • Slaking cycle

1 Introduction

The properties of rock masses or aggregates are greatly affected by the hot, humid, and rainy climate of the tropical regions. Rocks in these regions are influenced by weathering processes such as heating and cooling and wetting and drying. As opined by Momeni et al. (2015), the most significant gradual natural destructive process is weathering due to heating and cooling, wetting and drying, and freezing and thawing, catalyzed by the presence of water and salt as a principals decay agents, as well atmospheric pollution (acidic rain). Slake durability represents an important engineering parameter (Gökceoglu et al., 2000) and denotes "the degradability of rocks due to the process of physical and chemical decay as in exfoliation, disintegration, hydration, oxidation, and abrasion". Over the past years, there have been several studies on the slake durability and slaking properties of different rocks (Arman et al., 2021; Ghobadi & Momeni, 2011). These rocks were subjected to different wetting and drying cycles to assess the combined influence of pH of slaking fluid, the number of slaking cycles, and mineralogical composition on the disintegration behavior and durability of rocks (Gupta & Ahmed, 2007; Momeni et al., 2017; Yagiz, 2018).

Understanding the deterioration of rocks after repeated heating–cooling cycles is also important in places experiencing large daily temperature variations (thermal shock weathering) due to alternate heating and cooling of the rock

L. O. Afolagboye (✉) · A. O. Talabi
Department of Geology, Ekiti State University, Ado-Ekiti, Nigeria
e-mail: lekan.afolagboye@eksu.edu.ng

O. O. Owoyemi
Department of Geology, Kwara State University, Malete, Nigeria

M. Bezzeghoud et al. (eds.), *Recent Research on Geotechnical Engineering, Remote Sensing, Geophysics and Earthquake Seismology*, Advances in Science, Technology & Innovation,
https://doi.org/10.1007/978-3-031-48715-6_15

surface because of solar radiation. This article aims to predict the effect of the weathering process on the durability of granitic rocks from Ado-Ekiti, SW Nigeria. This aim was achieved by carrying out Slake Durability Tests (SDT) in wet and dry conditions on samples of Fine-Grained Granite (FGG) and Porphyritic Granite (PG) from Ado-Ekiti, SW, Nigeria. A comparison between the dry and wet SDT results is used to elucidate the possible impact of water on the alteration of the granitic rocks.

2 Materials and Methods

Rock samples used in this study were collected from granitic batholiths in Ado-Ekiti, Nigeria. These granitic batholiths are part of the rocks of the Precambrian basement complex (PBC) of southwestern Nigeria, which underlies most parts of Ado-Ekiti. The granitic batholith comprises porphyritic granite, medium-grained granite, and fine-grained granite (Afolagboye et al., 2016). Rock samples were obtained from two varieties of granite found in Ado-Ekiti: fine-grained granite (FGG) and porphyritic granite (PG). The FGG and PG samples were taken, respectively, from their respective outcrops and a quarry site. About 400 * 400 * 600 mm block of each granitic rock samples was collected in the field. The selection of rock samples was based on freshness, weathering state, and homogeneity in macroscopic attributes such as fractures. Before performing slake durability tests (SDT), physical, textural, and mineralogical properties of the FGG and PG rock samples were determined using ISRM Standard (2007).

The mineralogical properties of the granitic rocks were observed by preparing thin sections and studying under a polarizing microscope and the images analyzed. The modal composition was obtained by point counting method. The physical and index tests including dry density, water absorption capacity, water content, porosity, point load test (PLT), as well Schmidt Rebound Hammer (SRH) test. The SHR test was held perpendicular to the surface of the block samples, and the readings were taken by a single operator. PLT was carried out at least three times, and the average values recorded.

On the granitic rock samples, two series of SDT were performed. The first series were undertaken under wet conditions ($T°$ of water = 20 °C) following the procedure stipulated by ASTM D4644 (2008). In order to evaluate the long-term durability of granitoids by monitoring mass loss and basing on the work of Momeni et al. (2017), eight slaking cycles were applied instead of two stipulated in this standard. The second series of SDT was like the first series in procedure except that the slaking was conducted under dry conditions, i.e., the test was conducted without slaking fluid in the trough. The rock fragments retained in the drum (after approximately 10 min of rotation) in both dry and wet SDT tests were heated at 110 °C for twenty-four hours for each slaking cycle, cooled, and their mass determined. In both dry and wet SDT, the mass loss was calculated following the formulate indicated by ASTM D4644 (2008).

3 Results and Discussion

3.1 Petrography

Two PG and two FGG samples from granitic batholiths were investigated petrographically for their mineralogical composition. The petrographic characteristics of the granitic rocks reflect the mineralogical characteristics and textural variations observed in hand specimens and on the outcrops. The PG is comprised of quartz (12–18%), feldspars (microcline, 16–23%; orthoclase, 18–21%; plagioclase, 20–26%), altogether making up about 54–63% of the total composition. Hornblende (6–9%), biotite (8–12%), and muscovite (1–2%) constitute the minor mineral components. The fine-grained granite comprises quartz (40–46%), microcline (6–7%), hornblende (32–35%), and biotite (5–10%), altogether making up about 89–93%.

3.2 Physical and Index Properties

The physical and index properties of the PG and FGG are presented in Table 1. The average water content, water absorption capacity, and porosity of the FGG are lower compared to those of the PG. On the other hand, the point load strength (PLS) and SRH values of FGG are higher compared to those of PG. The variation in properties of the two granitic rocks may be due to the variation in texture and degree of compactness of the rocks.

3.3 Slake Durability Test

Figure 1a shows the effects of the drying–cooling and wetting–drying cycles on the variation of mass loss of the granitic samples during the SDT. In both sets of tests, the rate of mass-loss decreased as the number of slaking cycles increased. The rate of mass loss is higher in the initial cycles than in the final cycles. In addition, the rate of mass loss is higher in wet SDT than in dry SDT. The greatest mass loss was observed in PG samples (wet SDT), while the least was observed in FGG samples (dry SDT).

Figure 1b shows the effect of the number of slaking cycles on the slake durability index in the two series of SDT. In the two-test series, slake durability indices of the granitic rocks decreased as the number of slaking cycles increased.

Table 1 Physical and index properties of the granitic rocks

Rock type		W (%)	WAC (%)	η (%)	ρ (g/cm^3)	PLS (MPa)	SRH
PG	Range	0.27–0.5	0.51–0.98	0.45–0.56	2.6–2.64	7.84–9.81	50–56
	Mean	0.37	0.71	0.52	2.61	9.02	52.80
FGG	Range	0.06–0.1	0.39–0.53	0.089–0.67	2.53–2.65	14.71–16.67	50–52
	Mean	0.078	0.49	0.34	2.57	14.76	50.4

w water content; *WAC* water absorption capacity; η porosity; ρ density; *PLS* point load strength; *SRH* Schmidt rebound hammer

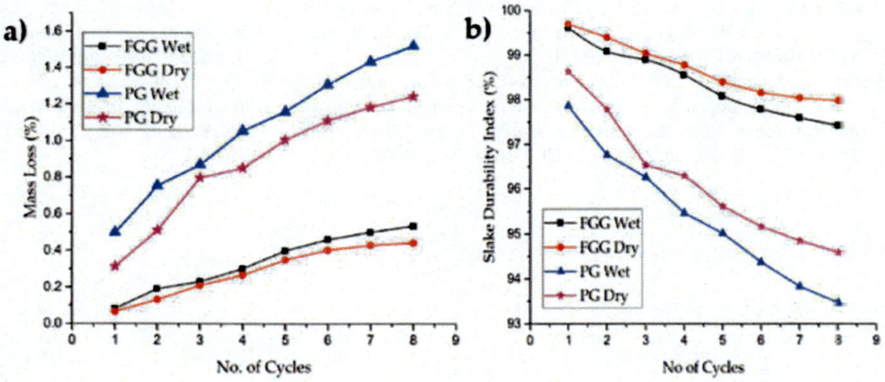

Fig. 1 Effect of slaking cycles on **a** mass loss, **b** slake durability index of the FGG

The slake durability index is higher in the dry test than in the wet test. Like mass loss, the rate at which the durability index decreased during the initial cycles was faster than during the end cycles. The FGG samples possessed a higher slake durability index in all the cycles and test series.

Comparison between the dry and wet SDT results suggests that the effect of water on the rate of rock degradation differs for the different granitic rock types. The PG is sensitive to water in terms of slake durability. For FGG, the obtained SDI values from the dry and wet testing were relatively similar, indicating that water has little influence on their rate of degradation. When compared with the durability classification provided by Broch and Franklin (1972), the most durable appeared to be the FGG, which has extremely high durability according to the classification. The PG, on the other hand, has very high to extremely high durability. The difference in the slake durability (in wet and dry tests) may be attributed to the slight difference in texture, degree of compactness, strength, and porosity of the rocks.

4 Conclusions

This study deals with the effect of the weathering process on the durability of on samples of fine-grained and porphyritic granite from Ado-Ekiti, SW, Nigeria, by using the SDT in wet and dry conditions. Petrologically, the granitic rocks are not significantly different in terms of observed mineral constituents but differ in terms of texture. The average water content, porosity, and water absorption of the FGG are lower compared to those of the PG. The mass loss against the number of slaking cycles curve reveals that the rate of mass loss decreases as the number of slaking cycles increases. Maximum mass loss was observed in samples used for wet SDT. Slake durability indices of the granitic rocks decreased as the number of slaking cycles increased. The slake durability index is higher in the dry test than in the wet test. The FGG samples have a higher slake durability index in all the cycles and test series. A comparison between the results of dry and wet SDT shows that the effect of water on the rate of rock degradation differs in both rocks. The PG is sensitive to water in terms of its slake durability compared to FGG. The PG with coarse texture, high porosity, and low strength is relatively more susceptible to degradation when in contact with water. The variations in material properties of the granitic rocks may be responsible for the difference in sensitivity to water.

References

Afolagboye, L. O., Talabi, A. O., & Akinola, O. O. (2016). Evaluation of selected basement complex rocks from Ado-Ekiti, SW Nigeria, as source of road construction aggregates. *Bulletin of Engineering Geology and the Environment, 75*, 853–865. https://doi.org/10.1007/s10064-015-0766-1

Arman, H., Abu Saima, M., Abdelghany, O., & Paramban, S. (2021). Comparative study on degradability characteristics of evaporitic and carbonate rocks from Al Ain, United Arab Emirates. *IOP Conference Series: Earth and Environmental Science, 906*, 012130. https://doi.org/10.1088/1755-1315/906/1/012130

ASTM D4644_08. (2008). *Standard method for slake durability of shales and similar weak rocks*. ASTM International.

Broch, E., & Franklin, J. A. (1972). The point-load strength test. *International Journal of Rock Mechanics and Mining Sciences, 9*, 669–676. https://doi.org/10.1016/0148-9062(72)90030-7

Ghobadi, M. H., & Momeni, A. A. (2011). Assessment of granitic rocks degradability susceptive to acid solutions in urban area. *Environment and Earth Science, 64*, 753–760. https://doi.org/10.1007/s12665-010-0895-6

Gökceoglu, C., Ulusay, R., & Sönmez, H. (2000). Factors affecting the durability of selected weak and clay-bearing rocks from Turkey, with particular emphasis on the influence of the number of drying and wetting cycles. *Engineering Geology, 57*, 215–237. https://doi.org/10.1016/S0013-7952(00)00031-4

Gupta, V., & Ahmed, I. (2007). The effect of pH of water and mineralogical properties on the slake durability (degradability) of different rocks from the Lesser Himalaya, India. *Engineering Geology, 95*, 79–87. https://doi.org/10.1016/j.enggeo.2007.09.004

ISRM (International Society for Rock Mechanics). (2007). The complete ISRM suggested methods for rock characterization, testing and monitoring: 1974–2006. In R. Ulusay & J. A. Hudson (Eds.), *Suggested methods prepared by the commission on testing methods* (p. 628). Kozan Ofset.

Momeni, A., Hashemi, S. S., Khanlari, G. R., & Heidari, M. (2017). The effect of weathering on durability and deformability properties of granitoid rocks. *Bulletin of Engineering Geology and the Environment, 76*, 1037–1049. https://doi.org/10.1007/s10064-016-0999-7

Momeni, A., Khanlari, G. R., Heidari, M., Bagheri, R., & Bazvand, E. (2015). Assessment of physical weathering effects on granitic ancient monuments, Hamedan, Iran. *Environmental Earth Sciences, 74*, 5181–5190. https://doi.org/10.1007/s12665-015-4536-y

Yagiz, S. (2018). The effect of pH of the testing liquid on the degradability of carbonate rocks. *Geotechnical and Geological Engineering, 36*, 2351–2363. https://doi.org/10.1007/s10706-018-0467-1

Comprehensive Investigation of the Role of the Geometrical, Topological, and Mechanical Properties of Discrete Fracture Network, Rock Type, and In-Situ Stress on Over-Excavation Around a Tunnel of El Teniente Mine

Amin Hekmatnejad, Francisco Muñoz, Jorge Prado, Carolina Saavedra, Eduardo Roja, Benoit Crespin, and Bastián Aguilar

Abstract

Over-excavation in underground opening is one of the most significant geo-mechanical hazards in mining and civil projects with potential consequences including cost overruns, time delays, increased rehabilitation, grade dilution, equipment damages, injuries, loss of human lives, and social or environmental consequences. The main objective of this study is to detail study the role of non-controllable parameters, namely rock type, geometrical, topological, kinematical and geo-mechanical properties of the discontinuity network, and in-situ stress on over-excavation around a tunnel of El Teniente Mine. To this end, we investigate the role of each of these parameters on the over-excavation around tunnel using cause–effect approach. We performed extensive sensitivity analysis using Monte Carlo simulation techniques of discrete fracture network and carry out the rigid block failure analysis to capture contribution of the non-controllable parameters on the kinematic state of the rock blocks. It is worth to mention that we assume that the constant value of in-situ stress around tunnel due to low coefficient of variation of the in-situ stress around tunnel (less than 5%). The rock type does not show a significant impact on the over-excavation. The mutual interaction between the induced stress around tunnel and discontinuities shows a significant correlation with observed over-excavation around tunnel. We used the microseismic data to study the potential correlation between over-excavation, fracturing process, and the non-controllable factors. According to the results of this study, the pre-existing discontinuities and their mutual interaction with induced stress around tunnel are the key factors on variational trend of the over-excavation along tunnel at El Teniente mine.

Keywords

Over-excavation • Discrete fracture network • Initial degree of active damage • Analysis of rock block instability • Induced stress field • Sensitivity analysis

A. Hekmatnejad (✉)
Escuela de Ingeniería Química, Pontificia Universidad Católica de Valparaíso, Valparaíso, Chile
e-mail: amin.hekmatnejad@pucv.cl

F. Muñoz
Department of Mathematical Engineering, Faculty of Physical and Mathematical Sciences, University of Chile, Santiago, Chile

J. Prado
Center for Mathematical Modeling, University of Chile, Santiago, Chile

C. Saavedra · B. Aguilar
School of Mining Engineering, University of Talca, Camino Los Niches Km. 1, Curicó, Chile

E. Roja
Gerente Corporativo de Geociencias, CODELCO, Santiago, Chile

B. Crespin
UMR CNRS 7252, University of Limoges, XLIM/ASALI, 87000 Limoges, France

1 Introduction

The essential part of underground mining is the development of a huge amount of underground opening spaces (e.g. 3000 km tunnels at El Teniente mine, Chile—source: Codelco). Studies indicate that underground excavation and tunnel failure costs in major civil and mining engineering projects represent 10–30% of the total production costs (Brzovic & Villaescusa, 2007).

The over-excavation defines as unplanned excavated parts of a tunnel (outside of the theoretical tunnel design) due to redistribution of the local stress field and induced damage by blasting. The factors influencing the

M. Bezzeghoud et al. (eds.), *Recent Research on Geotechnical Engineering, Remote Sensing, Geophysics and Earthquake Seismology*, Advances in Science, Technology & Innovation,
https://doi.org/10.1007/978-3-031-48715-6_16

over-excavation can be divided into two main groups: controllables including excavation process and tunnel geometry and design and non-controllables including rock mass properties and in-situ stress. Hudson (2013) reported that the over-excavation is influenced mainly by the rock mass properties and the tunneling process is less significance.

The main objective of this study is to investigate the potential role of the discontinuity system, rock type, stress field, and tunnel shape on the likelihood of the occurrence of over-excavation and its severity.

2 Methodology

It is considered to investigate the potential cause–effect relationship of P_{32}, $C.V._{\text{ver-Total}}$, active ratio of discontinuities, stress field, and rock type on the severity of the over-excavation. It is used the universal discontinuity index (UDi) (Hudson, 2013) to consider the mutual role of discontinuities and stress field as indication of the initial degree of active damage as Eq. (1)

$$\text{UDi} = P_{32} \times [1 + w_1 \text{C.V.}_{\cdot\text{ver-tot}} + w_2(N_{\text{AF}}/N_{\text{TF}}) + w_3(\text{C.V.}_{\cdot\text{tot-}\sigma 1})] \quad (1)$$

where UDi is the universal discontinuity index. P_{32} is the volumetric fracture intensity. $\text{C.V.}_{\cdot\text{ver-tot}}$ is the circular variance with respect to the vertical direction. $\text{C.V.}_{\cdot\text{tot-}\sigma 1}$ is the total circular variance with respect to the direction of a vector which forms an angle of $45° + \varphi_w/2$ with maximum principal stress. N_{AF} is the number of active fractures. N_{TF} is the total number of fractures within rock mass. w_1, w_2, and w_3 are the weighting factors depending on rock engineering design.

3 Results

3.1 Analysis the Role of Discontinuity Parameters, Stress Filed, Rock Type, and UDi Along Tunnel

The objective of this section is to investigate the potential influence of the discontinuity system parameters including the number of fracture sets, volumetric fracture intensity (P_{32}), circular variance, and UDi on the observed over-excavation along the tunnel. Figure 1 shows the variation of each factor against over-excavation along the tunnel. To include the role of the stress field and its interaction with discontinuity system, causing the rock strength anisotropy

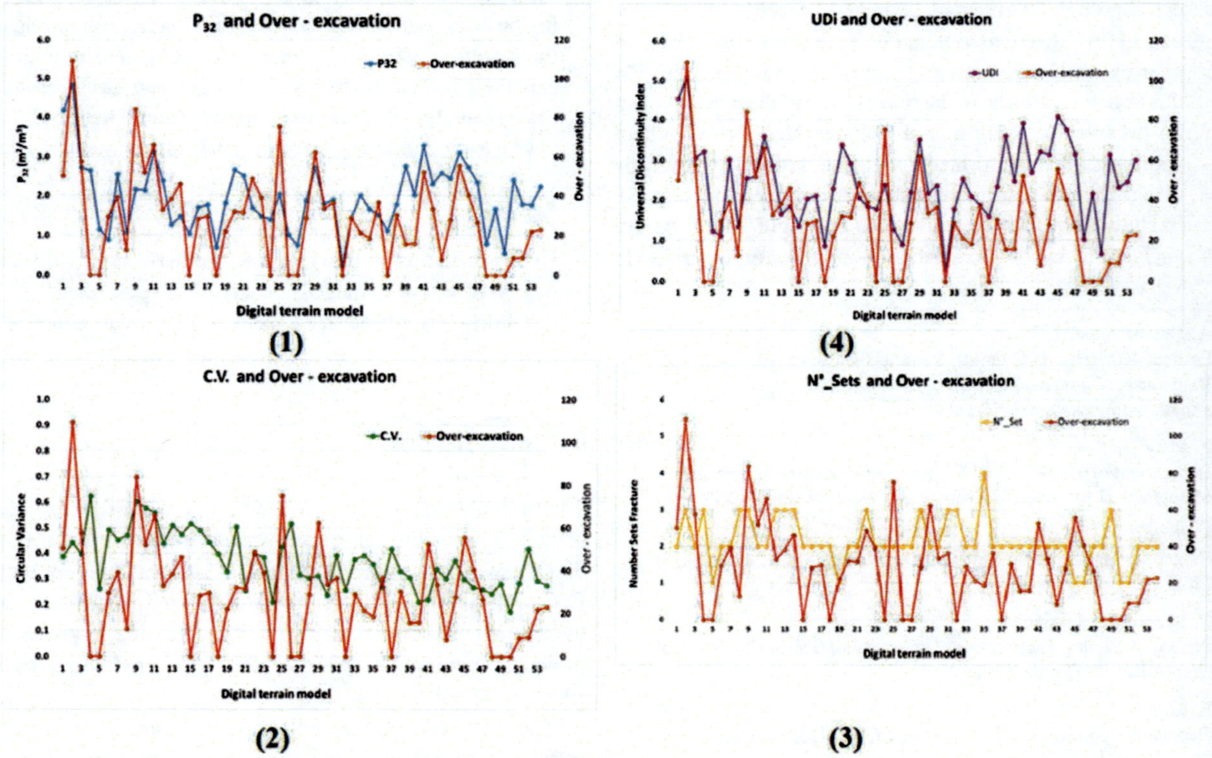

Fig. 1 1 Combo plot of P_{32} versus over-excavation, 2 combo plot of circular versus over-excavation, 3 combo plot of number of fracture sets versus over-excavation, 4 combo plot of UDi versus over-excavation

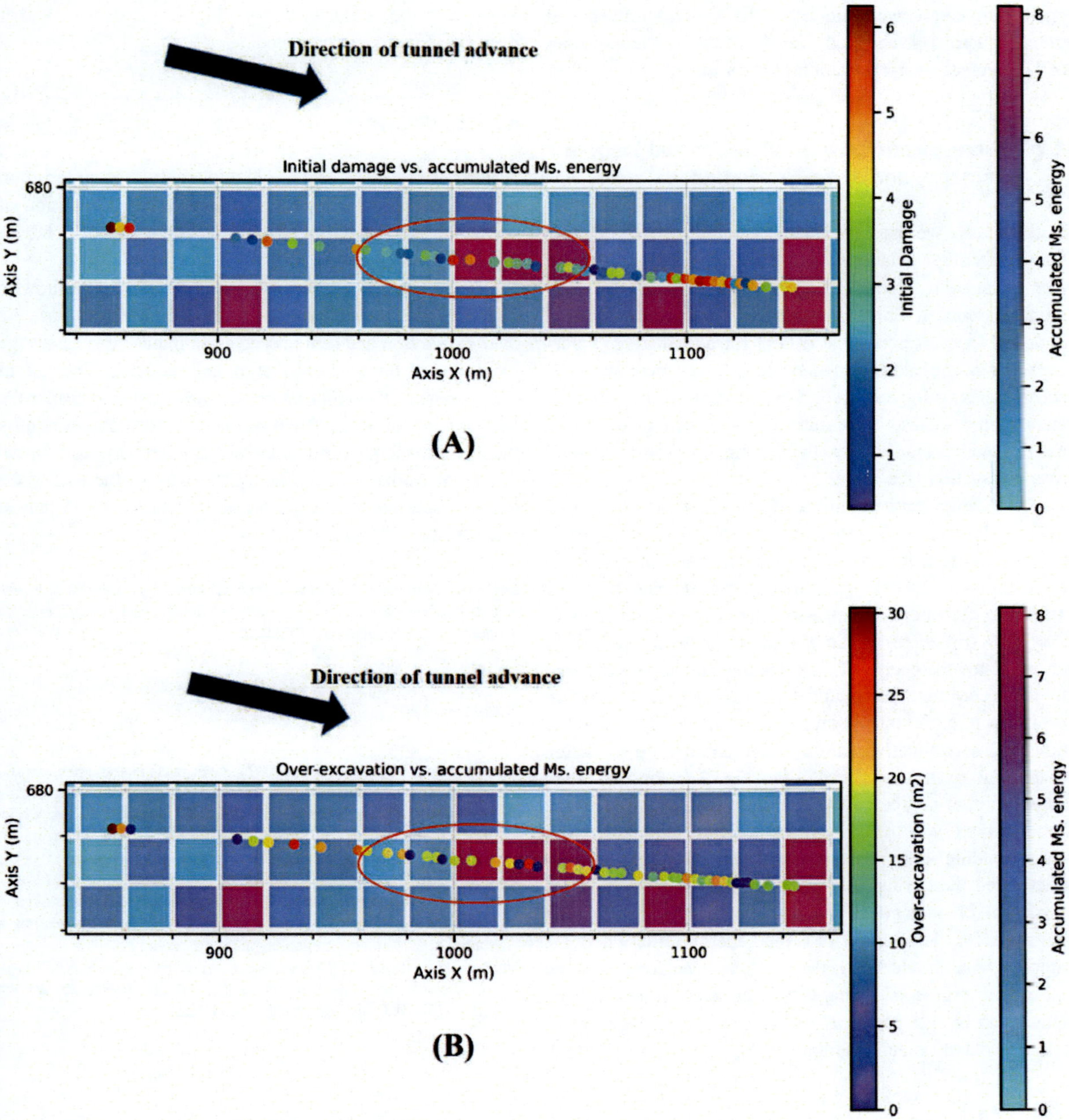

Fig. 2 Map of the initial damage versus metrics of microseismic events. **A** Initial damage versus accumulated energy, **B** over-excavation versus accumulated energy

and heterogeneity, we estimate the UDi for each sequence of tunnel. Figure 1(4) shows the results of the estimated values of UDi versus over-excavation along a tunnel.

3.2 Microseismic Data Analysis, Initial Degree Damage, and Over-Excavation

In this study, we used the microseismic dataset to analysis their potential relationship with the degree of the over-excavation along a tunnel. We limited our study domain around a tunnel under study. The metric which we obtained from microseismic dataset for quantitative characterization of the microseismic events is the accumulated total seismic energy for each cell. We plot each of these parameters against UDi as a measure of initial damage degree of each tunnel advance and their corresponding values of over-excavation (see Fig. 2).

The relation between this metric and over-excavation is not trivial but its explainable. For example, in the area within the red ellipsoid in Fig. 2A on the first parts of tunnel (in direction of tunnel advance), the accumulated energy is low as the initial damage degree is low; however, with passing of $x = 1000$ (m) there are an abrupt jump on the values of accumulated the energy. This phenomenon can be explained by the according to the multistage model of failure, in which every fracture of J rank is prepared by smaller defects $J - 1$ rank and accumulative nature of the damage up to failure point. For example, according to Fig. 2A, for the lower value's initial degree of damage (less than 3.5, within the zone marked with red ellipse) the accumulated energy with corresponding cells is low (less than 4) which can be interpreted the during hole period of tunnel excavation the evolution of damage is low and the rock mass has high potential for the storage of strain energy. But when the tunnel advances into $x = 1000$ (m) there is an abrupt change the initial degree of damage and the accumulated energy which can be interpreted as an evolution of damage due to fracturing and potential failure of the rock mass at tunnel. This can be confirmed by Fig. 2B as the higher over-excavations are observed.

4 Conclusion

According to the results of this study, the interaction between the initial degree of damage and concentration of stress are key parameters on the degree of the failure (over-excavation).

The initial degree of damage based on UDi rating shows a high degree of correlation with the intensity of the over-excavation and can be used as an indicator to assess the likelihood of the over-excavation and its depth. The role of the geometry of vicinity tunnels is important to consider the local mining stiffness. The result of this study shows that by the combination of the microseismic activity and initial degree of damage the remaining capacity of the rock mass can be assessed which is critical to classify the J ranked microseismic for warning.

Acknowledgements The first author acknowledges the funding project of Fondecyt Iniciacion No 11221093 and Basal Grants Center for Modeling ACE210010 and FB210005.

References

Brzovic, A., & Villaescusa, E. (2007). Rock mass characterization and assessment of block-forming geological discontinuities during caving of primary copper ore at the El Teniente mine, Chile. *International Journal of Rock Mechanics and Mining Sciences, 44*, 565–583.

Hekmatnejad, A., Rojas, E., Saavedra, C., & Crespin, B. (2022). Presentation of the Universal Discontinuity index (UDi) system and its application to predict the geometry of over-excavation along a tunnel at El Teniente mine, Engineering geology.

Hudson, J. A. (2013). *A review of rock engineering systems (RES) applications over the last 20 years*. Paper presented at the ISRM SINOROCK 2013, Shanghai, China.

Rock Mass Joint Sets Identification Through Stereographic Projection and Unsupervised Learning: A Comparative Study

Kaoutar Clero, Said Ed-Diny, Mohammed Achalhi, Mouhamed Cherkaoui, Tarik Soror, Said Rziki, Sanaa El Fkihi, Andronic Boanarijesy, and Mohamed Nadour

Abstract

Geological and geotechnical studies have been usually based on experimentation, analytical analysis, graphical representations, and numerical modeling. However, with the development of the artificial intelligence field and the sharp increase of its applications, using machine learning and deep learning methods has gained a significant importance in these studies. Within this context, this paper presents, first, an overview about recent research articles that studied the application of machine learning approaches to rock mass studies. Second, a case study of rock mass joint sets identification at Draa Sfar mine in Morocco is presented. Based on a database of 400 measurements of dips and dips directions for different rock mass discontinuities such diaclases and schistosity, a comparative study between stereographic projection as a classical method and K-means clustering as an unsupervised machine learning method is established, to identify the mine's main joint sets. The results show that the clustering of the joints sets through both two methods is concordant. However, the use of stereographic projection for joints sets identification does not need prerequisites while K-means clustering presents the limitation of requiring the initial number of clusters as an input. In this case, the use of pre-analysis methods, such as the elbow method or the silhouette score, is mandatory. In addition, both of the methods need sufficient number of inputs in order to give accurate values of joint sets' spatial orientation characteristics. To sum-up, our comparative study shows the advantages of classical modeling for rock mass joint sets identification, but also introduces the potential and the limitations that have one of the clustering algorithms within this topic. The challenge now is to train and test more sophisticated artificial intelligence methods that will allow advanced analysis of geotechnical and geological data, to confirm if these tools can partially or entirely substitute classical modeling.

Keywords

Rock mass • Joints sets • Stereographic projection • Artificial intelligence • Unsupervised learning

1 Introduction

During the last years, the use of artificial intelligence methods has known an important increase in many fields, including earth sciences. In fact, many research articles and reviews on geological and geotechnical topics that were published in the last years discuss the potential of using machine learning or deep learning tools to complete or substitute the classical methods usually used in these disciplines.

As examples of machine learning application within these topics, different regression models were used in Wang et al. (2021), Dehghan et al., (2010), Matin et al., (2018), El-Sebakhy et al., (2012) to estimate some geomechanical parameters that are usually defined through experimentation, such as the uniaxial compressive strength, the elasticity modulus, the permeability… The obtained results have shown high prediction accuracies. Besides, classification

K. Clero (✉) · A. Boanarijesy · M. Nadour
Moroccan Foundation for Advanced Science, Innovation and Research (MASCIR), Mohamed Al Jazouli Street, Madinat Al Irfane, Rabat, Morocco
e-mail: k.clero@mascir.ma

K. Clero · S. Ed-Diny · M. Achalhi · M. Cherkaoui
National School of Mines of Rabat (ENSMR), Hadj Ahmed Cherkaoui Avenue, Agdal, Rabat, Morocco

T. Soror · S. Rziki
R&D and Project Management, REMINEX Engineering, BP 469, Marrakech, Morocco

S. E. Fkihi
Rabat IT Center, ADMIR Lab, National School of Computer Science and Systems Analysis (ENSIAS), IRDA Team, Université Mohammed V, Rabat, Morocco

M. Bezzeghoud et al. (eds.), *Recent Research on Geotechnical Engineering, Remote Sensing, Geophysics and Earthquake Seismology*, Advances in Science, Technology & Innovation,
https://doi.org/10.1007/978-3-031-48715-6_17

models were also applied to rock mass studies, such as in Klyuchnikov et al. (2019), where some machine learning algorithms were used for rock type identification through binary classification and showed high performance.

As a special focus, several research articles present the use of machine learning methods for rock mass discontinuity analysis. In Chen et al., (2022), sixteen machine learning classifiers were applied to digital images captured on a rock tunnel face to classify areas into discontinuities (trace) and intact rock areas (non-trace). In addition, new methods combining several machine learning algorithms for discontinuity identification through 3D point cloud data were established in Kong et al. (2020), Chen et al., (2016). The methods allowed therefore identifying the rock mass discontinuities, clustering the joint sets, characterizing the orientation of the discontinuity, and estimating the trace length and the joint spacing. However, the limitation of the methods used in these contributions is the need for scanning data (e.g., 3D point clouds) to train the models. Access to these scanning or imaging inputs is not always easy and affordable, while discontinuity orientation measurements are often accessible through manual surveys. Thus, in this paper, we present another approach based on numerical data obtained from structural measurements that were collected on the site of the Draa Sfar underground mine in Morocco to identify rock mass joint sets using a comparative study between stereographic projection and K-means clustering algorithm.

2 Dataset and Processing Method

Draa Sfar is a deep underground mine located in Morocco, at 16 km on the Northwest of Marrakech city. The mine produces primary metals (Cu, Pb, and Zn). Figure 1 presents the mine's geographical location (Ait Lemkademe et al., 2022).

The main lithostratigraphic units in the mine are rhyodacite igneous rock, sandstone pelites and shales, pyroclastic meta-tuffs with rhyodacite elements, polymetallic massive sulfide ore body, and black and calcareous pelites and shales (Salama et al., 2018), as presented in Fig. 2. Besides, the main joint sets in the mine are the schistosity set S1, sub-vertical diaclases J1, and sub-horizontal diaclases J2 (Clero et al., 2022).

In this context, this study aims to compare the potential of stereographic projection and K-means clustering algorithm in identifying the joint sets presented above through a comparative analysis.

The input dataset that was collected on the site of Draa Sfar underground mine contains 400 measurements of dips

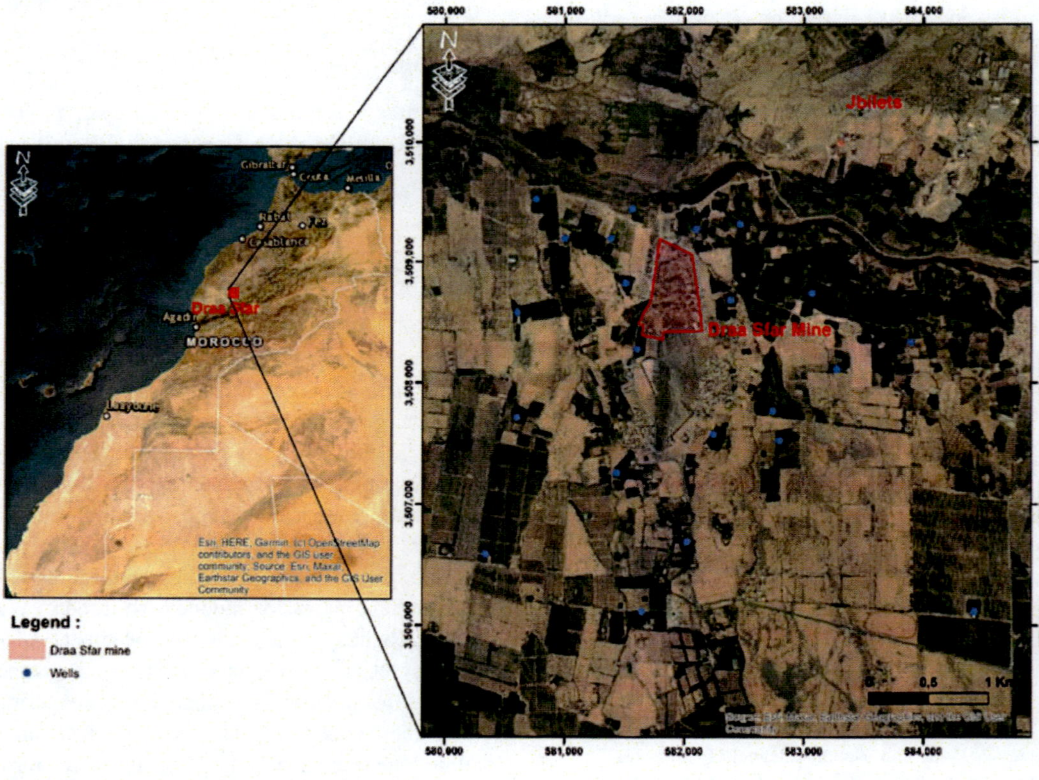

Fig. 1 Geographic situation of Draa Sfar mine (Ait Lemkademe et al., 2022)

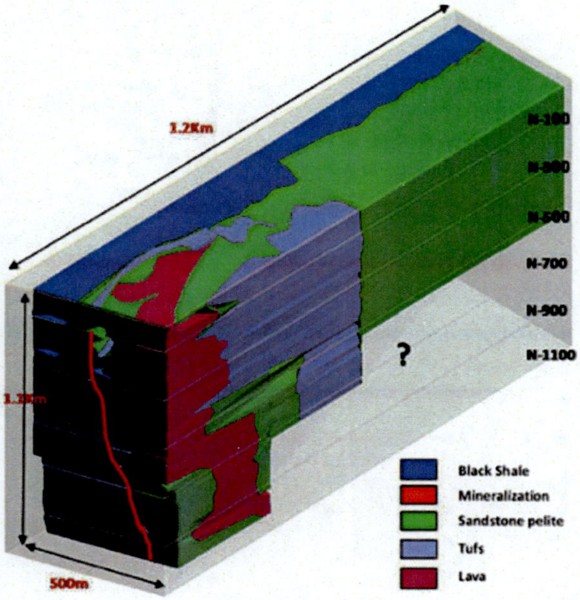

Fig. 2 Lithostratigraphy of Draa Sfar mine (Ait Lemkademe et al., 2022)

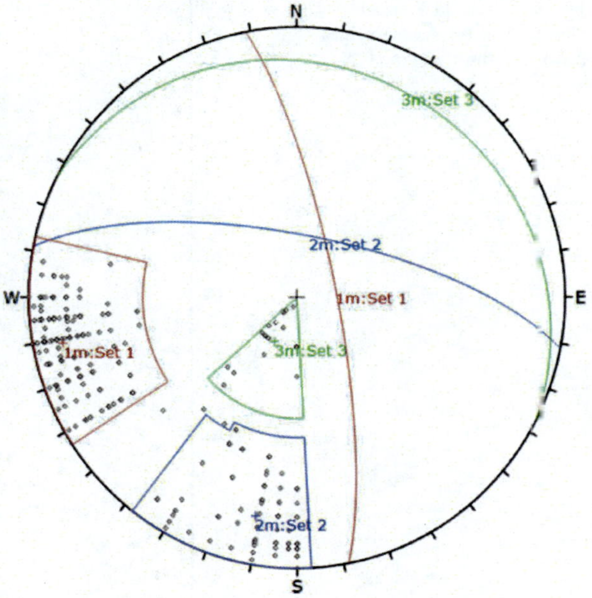

Fig. 3 Scatter plot of the discontinuities spatial data and their clustering into three groups on the stereographic projection

and plane directions for different rock mass discontinuities that are present in the mine. To compare the clustering of the joint sets through both stereographic projection and K-means clustering, first, a pre-processing step is applied to the dataset, to convert the plane direction to the dip-direction values, using Eq. (1).

$$\text{Dip Direction} = |90 - \text{Plane Direction}| \quad (1)$$

Second, we established, on one hand, the stereographic projection of the obtained results on DIPS software to visualize the joint sets clustering. As a reminder, stereographic projection is a classical method for analyzing orientation-based geological data, which allows the projection of three-dimensional components of a sphere on a two-dimensional plane and identifying the main characteristics of the joint sets.

On the other hand, we developed an algorithm based on K-means clustering method using Python, and we applied it to the same data. This unsupervised machine learning model is a centroid-based and distance-based algorithm. Thus, distance is calculated to assign points, representing data, to each cluster, defined by its centroid.

3 Results

The orientation-based analysis through stereographic projection presents three clusters, as shown in Fig. 3. Thus, we distinguish three joint sets that have different spatial orientations and different densities. However, K-means clustering method requires the initial number of clusters as an input. To predict this number, there are some methods that can be used, such as the elbow method or the silhouette score.

Applying these two pre-analysis methods to our dataset gives concordant results presenting three clusters as the optimal number of joint sets. Figure 4 shows the clusters representation generated by applying a 3-means clustering algorithm to our dataset. The number of points in each cloud is different. Thus, we recognize here also a difference of each joint set's dominance.

4 Discussion

Through both methods above, the rock mass discontinuities are clustered into three groups. The main characteristics of each set are presented in Table 1.

Stereographic projection and K-means clustering show a concordant clustering of the discontinuities into three joint sets. This is relatively adequate with on-site observations, where three main joint sets are reported. In fact, in this mine, there are three joint sets related to the schistosity set (presenting a dip of 80 and a dip-direction of 100), sub-vertical diaclases (presenting a dip of 75 and a dip direction of 000), and sub-horizontal diaclases (presenting a dip of 00 and a dip direction of 000). These joint sets can be respectively approximated to set 1, 2, and 3 identified through our analysis. However, we notice that there is a slight difference between the clustering results and the real values reported,

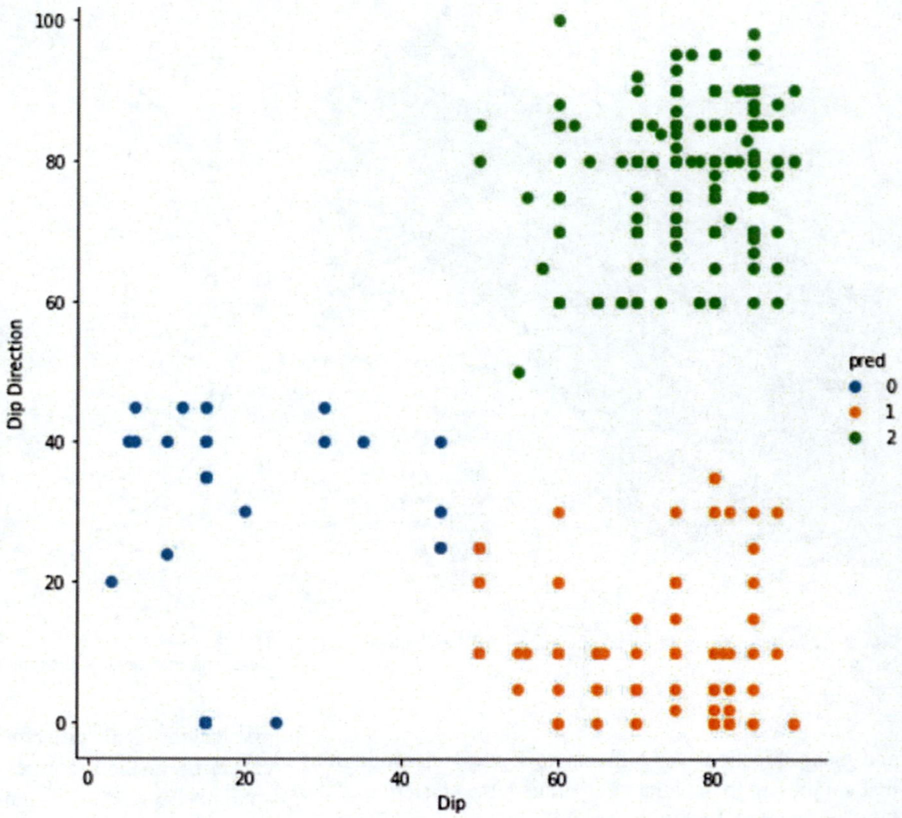

Fig. 4 Results of K-means clustering of the discontinuities presenting three joint sets

Table 1 Comparison of stereographic projection and K-means clustering results

	Mean dip values			Mean dip-direction values		
	Set 1	Set 2	Set 3	Set 1	Set 2	Set 3
Stereographic projection	78	71	15	79	11	27
K-means clustering	77	71	19	78	10	27

especially for sub-horizontal diaclases. This can be explained by the lack of samples related to this set in the database. In fact, the retro-analysis of the clustering results shows that only 28 samples belong to set 3. Therefore, more measurements are required as inputs to obtain more accurate results.

Besides, although the results of the stereographic projections and K-means clustering are concordant, one of the main limitations of the K-means clustering method is that it requires the clusters' number as an input, while the stereographic projection directly gives the information about the main joint sets.

5 Conclusions

As a conclusion, this study shows that the use of stereographic projection and K-means clustering for rock mass joint sets identification present many similarities but not the same prerequisites. Therefore, the use of K-means unsupervised learning can be considered as a complement to the stereographic projection but cannot substitute it for joint sets identification unless the number of sets is initially known. Thus, there are two options:

- If the number of joint sets is already known, K-means clustering model can be used to cluster the measurements into the separate sets and define their main characteristics (e.g., dip, dip direction). In this case, this model can be very useful especially when not having access to stereographic projection tools, especially non-open-access ones.
- If only raw measurements are accessible without a previous idea about the number of sets, stereographic projection is more adequate to be used for an inclusive analysis.

In both of the cases, the more measurements are accessible the more the results are accurate and representative of the mine's real structural characteristics.

Acknowledgements We would like to thank the MANAGEM Group and its subsidiary CMG for allowing conducting research on the Draa Sfar site as an industrial partner in this project. As a reminder, this publication is part of the work undertaken by different partners composed of Moroccan Foundation for Advanced Science, Innovation and Research (MASCIR), REMINEX Engineering, R&D and Project Management, MANAGEM group, National School of Mines of Rabat (ENSMR), University Cadi Ayyad (UCA), and National School of Computer Science and Systems Analysis (ENSIAS). This research is conducted within the framework of the "Intelligent Connected Mine" project, which has been supported by the Moroccan Ministry of Higher Education, Scientific Research and Innovation, the Digital Development Agency (DDA), the National Center for Scientific and Technical Research of Morocco (CNRST) through the Al-Khawarizmi project in addition to MANAGEM group and MASCIR supports for this project.

References

Ait Lemkademe, A., Michelot, J. L., Benkaddour, A., Hanich, L., & Heddoun, O. (2022). Origin of groundwater salinity in the Draa Sfar polymetallic mine area using conservative elements (Morocco). *Water, 15*(1), 82.

Chen, J., Huang, H., Cohn, A. G., Zhang, D., & Zhou, M. (2022). Machine learning-based classification of rock discontinuity trace: SMOTE oversampling integrated with GBT ensemble learning. *International Journal of Mining Science and Technology, 32*(2), 309–322.

Chen, J., Zhu, H., & Li, X. (2016). Automatic extraction of discontinuity orientation from rock mass surface 3D point cloud. *Computers & Geosciences, 95*, 18–31.

Clero, K., Ed-Diny, S., Cherkaoui, M., Soror, T., Rziki, S., Achalhi, M., El Fkihi, S., & Boanarijesy, A. (2022). A review of geotechnical instabilities identification and monitoring at deep underground mines: Case of Draa Sfar mine in Morocco. *International Journal of Civil Infrastructure, 5*, 51–59.

Dehghan, S., et al. (2010). Prediction of uniaxial compressive strength and modulus of elasticity for Travertine samples using regression and artificial neural networks. *Mining Science and Technology (china), 20*(1), 41–46.

El-Sebakhy, E. A., et al. (2012). Functional networks as a new data mining predictive paradigm to predict permeability in a carbonate reservoir. *Expert Systems with Applications, 39*(12), 10359–10375.

Klyuchnikov, N., et al. (2019). Data-driven model for the identification of the rock type at a drilling bit. *Journal of Petroleum Science and Engineering, 178*, 506–516.

Kong, D., Wu, F., & Saroglou, C. (2020). Automatic identification and characterization of discontinuities in rock masses from 3D point clouds. *Engineering Geology, 265*, 105442.

Matin, S. S., et al. (2018). Variable selection and prediction of uniaxial compressive strength and modulus of elasticity by random forest. *Applied Soft Computing, 70*, 980–987.

Salama, L., Mouguina, E. M., El Bachari, E., Rddad, L., Outhounjite, M., Essaoudi, M., Maacha, L., & Zouhair, M. (2018). Numerical heat and fluid flow modeling of the Hercynian Draa Sfar polymetallic (Zn–Pb–Cu) massive sulfide deposit, Central Jbilets, Morocco. *Arabian Journal of Geosciences, 11*(24), 1–19.

Wang, Y. T., et al. (2021). Machine learning approaches to rock fracture mechanics problems: Mode-I fracture toughness determination. *Engineering Fracture Mechanics, 253*, 107890.

Valorization Potential of a Greek Coal Spoil Heap by Civil Linear Infrastructure

Alexandros I. Theocharis, Ioannis E. Zevgolis, Nikolaos C. Koukouzas, and Christos Roumpos

Abstract

Coal surface excavations create wastes, typically dumped in nearby areas forming massive spoil heaps. Several reclamation options have been proposed, including constructing linear civil infrastructure. Evaluations on this reclamation type have not been presented in the literature as the necessary geotechnical characterization of the spoil material is generally lacking, and various literature, design standards, and regulations are included. In this work, the potential of constructing a railroad or a highway on a massive spoil heap in Greece is assessed based on the geotechnical characterization of the spoil material. The soil body is highly heterogeneous due to the extraction and deposition processes related to the heap's construction. Thus, the soil mass was considered one unified spoil material with significant uncertainty for a preliminary assessment. The representative material of the spoil heap is high plasticity silt (MH for USCS). The mean values of the constraint modulus (E_s) for 100–200 kPa, the compression (c_c), and recompression (c_r) indices were determined through 1D consolidation tests as 2.7, 0.216, and 0.37 MPa, respectively. The most significant property of the subgrade's suitability for road or railroad construction is stiffness, quantified through an elastic modulus (e.g., the resilient modulus). Each country's regulations specify the minimum subgrade stiffness requirements for railroads; spoil stiffness is significantly smaller than the minimum requirements. Moreover, based on soil classification, worldwide standards qualitatively characterize the highway subgrade and its potential; high plasticity silt—the representative spoil material—is generally unwanted as a subgrade material for highways. Additionally, considering the resilient modulus and the CBR, the spoil results are far below the minimum requirements. Overall, the spoil material is too soft to withstand linear infrastructure without ground improvement.

Keywords

Waste dumps • Spoil piles • Colliery spoil dumps • Land reclamation • Sustainability • Circular economy • Geotechnical engineering

1 Introduction

Extracting coal or lignite near the surface (opencast, open-pit, or surface mining) involves excavating the overburden material covering the deposits. This material, called spoil waste, is typically transported and dumped in nearby areas forming massive spoil heaps, also called waste dumps, spoil piles, or colliery spoil dumps. Zevgolis et al. (2021) systematically quantified the uncertainty of a massive spoil heap's geotechnical properties in Greece. Laboratory results based on an extended investigation of a spoil material originating from lignite mines were compiled and thoroughly analyzed. The landowner is considering alternatives for the future use of the spoil heap. This work examines two reclamation scenarios of this massive heap; it aims to answer whether heaps of lignite mines can be used for civil linear infrastructure projects. Specifically, two different construction scenarios were evaluated a highway and a railroad. Criteria based on literature, design standards, and regulations were employed to determine the spoil's suitability for each scenario.

A. I. Theocharis · N. C. Koukouzas
Chemical Process and Energy Resources Institute, Centre for Research and Technology Hellas, Athens, Greece

I. E. Zevgolis (✉)
School of Mining and Metallurgical Engineering, National Technical University of Athens, Athens, Greece
e-mail: izevgolis@metal.ntua.gr

C. Roumpos
Mining Engineering and Closure Planning Department, Public Power Corporation, Athens, Greece

M. Bezzeghoud et al. (eds.), *Recent Research on Geotechnical Engineering, Remote Sensing, Geophysics and Earthquake Seismology*, Advances in Science, Technology & Innovation,
https://doi.org/10.1007/978-3-031-48715-6_18

2 Railroad

The components of a traditionally ballasted railroad can be divided into those of the superstructure and the substructure (Fig. 1). The substructure consists of the trackbed and the subgrade and influences the superstructure's resiliency, contributing to the rail's elastic deflection under wheel loading. The subgrade is critical for the railroad embankment, being the foundation of the entire railway.

Regulations and standardized procedures govern railroad design. Most of them aim to determine the thickness and material properties of the trackbed layers. They all account for the subgrade and its suitability, mainly through the soil's classification. High plasticity soils with more than 15% of fines—the spoil's characteristic material—are placed in the worst class and cannot serve as subgrade without any reinforcement. Moreover, the most significant property of the subgrade's suitability is stiffness quantified through an elastic modulus. Alamaa (2016) summarized several standards regarding railroad construction in a comparative study; typical minimum Young's modulus values are from 30 to 60 MPa for a soil formation to be acceptable.

Based on Zevgolis et al. (2021), the spoil material exhibits a mean Young's modulus of 1.7 MPa. Moreover, $E = 10$ MPa based on the Mohr–Coulomb model and $E_{50} = 5$ MPa for the hardening soil model calibration (Theocharis et al., 2020). All these values are significantly smaller than the minimum requirements.

3 Highway

The design and construction of a road is a multi-parameter project. The essential elements of such a design are the topography, the pavement's construction materials, and the soils, called subgrade, below the pavement. The pavement is founded on the subgrade and consists of several layers that ensure the road's serviceability through the appropriate transfer of the stresses. Stresses that develop in a well-designed pavement are well below the failure strength of the pavement materials. The fundamental engineering property included in road design is stiffness. The stiffness of the pavement layers and subgrade defines the stress and strain distribution. Several measures can be used for characterizing this stiffness; soil classification is primarily used along with the California bearing ratio (CBR) and the resilient modulus (M_R, Fig. 2).

Based on soil classification, worldwide standards characterize the subgrade and its potential. The spoil material does not fall in a specific category since it is a mixture of soils. However, this spoil's characteristic material has been identified mostly as high plasticity silt (MH) according to the USCS (Howard, 1986; Zevgolis et al., 2021) or as A7-5 on AASHTO (AASHTO, 2008). Soils belonging to these classes are ranked as poor, and they are undesirable in road construction. Another important parameter used to measure the performance of the subgrade is the CBR value. It is a comparative measure of the soil's shear resistance and has been widely correlated with stiffness and used extensively to assess a subgrade's suitability. Based on design standards, CBR values are commonly used for highways, airports, parking lots, and other pavement constructions. The minimum requirements in CBR vary slightly in different regulations and manuals. In Greek regulations, a minimum value of CBR = 3 is required. The spoil material has not been tested to determine its CBR; however, based on empirical relations and the mean values of Zevgolis et al. (2021), CBR ≈ 0.1–0.50. This value is significantly lower than the minimum requirements of any standard. Finally, the resilient modulus (M_R) is an elasticity modulus of the same order of magnitude as Young's modulus and a critical property in pavement design (Fig. 2). The M_R has also been widely correlated with several geotechnical properties and particularly the CBR, leading in the present case to a resilient modulus of $M_R \approx 4$–7 MPa. However, for $M_R < 31$ MPa, the subgrade generally requires improvement. The expected value for the spoil is far below the typical values or the minimum requirement.

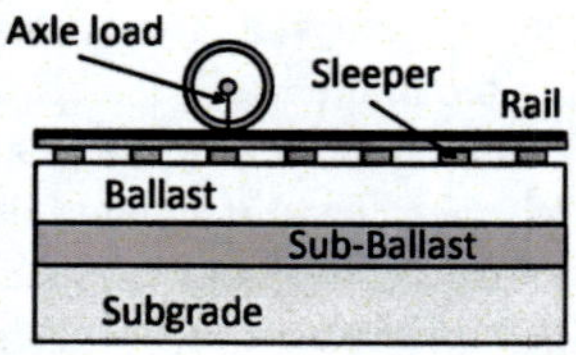

Fig. 1 Simplified components of a typical ballasted track (Burrow et al., 2007)

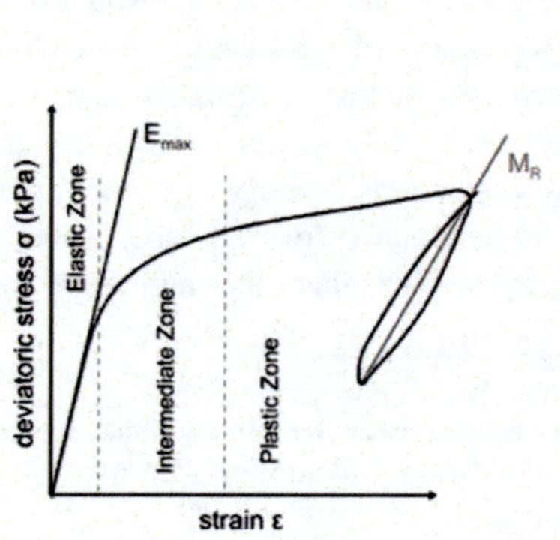

Fig. 2 M_R definition and relationship with Young's modulus during cyclic loading (Sas et al., 2017)

4 Conclusions

In this work, two alternatives were examined regarding the reclamation of a significant Greek spoil heap, a highway and a railroad. The spoil material under consideration is a mixture of primarily silts and clays, exhibiting high plasticity and low stiffness. The mean Young's modulus is around 1.7 MPa ranging from 5 to 10 MPa, according to calibrations on monotonic triaxial tests. These values are very low and indicate a soft soil. Furthermore, the spoil material is highly heterogeneous, proposing uncertainties larger than most natural soils. Highways and railroads generally require stiff soils to ensure that large and differential settlements do not jeopardize serviceability. The stiffness is quantified through the CBR value (indirectly), Young's modulus, and the resilient modulus M_R. Since no experimental results exist regarding the CBR or the M_R, they were calculated for this work using well-established empirical relationships. Based on the results, the spoil material does not meet the minimum requirements of the typical standards, and thus, it cannot serve as a subgrade without proper treatment. Overall, the spoil material cannot be considered adequate, and ground improvement techniques must be implemented.

Acknowledgements This work has received funding from the European Union's Research Fund for Coal and Steel under the project SUMAD grant agreement No. 847227.

References

AASHTO. (2008). *LRFD bridge design specifications*. American Association of State Highway and Transportation Officials.

Alamaa, A. (2016). *High-speed railway embankments—A comparison of different regulations* (Master of Science, KTH Royal Institute of Technology, Stockholm).

Burrow, M., Bowness, D., & Ghataora, G. (2007). A comparison of railway track foundation design methods. *Proceedings of the Institution of Mechanical Engineers Part F-Journal of Rail and Rapid Transit—PROC INST MECH ENG F-J RAIL R, 221*, 1–12.

Howard, A. K. (1986). Soil classification handbook: unified soil classification system. Denver, Colorado: Geotechnical Branch, Division of Research and Laboratory Services, Engineering and Research Center, Bureau of Reclamation.

Sas, W., Gluchowski, A., Gabryś, K., Soból, E., & Szymański, A. (2017). Resilient modulus characterization of compacted cohesive subgrade soil. *Applied Sciences (Switzerland), 7*.

Theocharis, A. I., Zevgolis, I. E., & Koukouzas, N. C. (2020). Validation of constitutive models for spoil material of brown coal mines. In *Proceedings of 3rd Conference of the Arabian Journal of Geosciences (CAJG)*.

Zevgolis, I. E., Theocharis, A. I., Deliveris, A., V., Koukouzas, N. C., Roumpos, C., & Marshall, A. M. (2021). Geotechnical characterization of fine-grained spoil material from surface coal mines. *Journal of Geotechnical and Geoenvironmental Engineering 147* (7), 04021050.

Numerical Studies on Underground Thermal Energy Storages

Julian Schleicher, Islam Marzouk, Matthias Rebhan, and Franz Tschuchnigg

Abstract

The ever-increasing global energy consumption leads to a rise in demand for renewable energy sources and storage solutions. Nevertheless, in several cases energy is not sufficiently stored and dissipates unused (e.g., waste heat from air conditioning systems). This results in increasing the air temperature, especially in urban areas. Underground thermal energy storage (UTES) systems can be used to utilize underground soil to store unused energy for use when needed (e.g. district heating). The objective of this paper is to investigate the implementation of a UTES system in the 2D finite element software PLAXIS. Furthermore, the modelling approach used for thermally insulating the storage tank by using improved diaphragm walls was evaluated. In case of a UTES system, it is necessary to model a heating phase in which the storage stores energy for later use. For this purpose, different modelling approaches were investigated. Each analysis aimed to heat the storage reservoir to 90 °C. Afterwards, the difference in temperature distribution in the soil was investigated. Moreover, the influence of the thermal properties of the diaphragm walls on the temperature distribution was reviewed. Concrete with a density of 2400 kg/m^3 was compared to a lightweight concrete developed at the Institute of Structural Concrete at Graz University of Technology. The thermal properties of the lightweight concrete were determined based on laboratory tests. The lightweight concrete had a thermal conductivity of 1.026 W/m K, which led to a significant reduction in heat losses from the storage.

Keywords

Numerical modelling • Underground thermal energy storage system (UTES) • Concrete thermal properties • Temperature effects

1 Introduction

The aim of the presented research was to evaluate the possibilities of modelling an underground thermal energy storage system in PLAXIS 2D (Bentley Systems, 2020). As a result, different modelling approaches for the heating phase were compared to assess the differences and to identify the most suitable approach. Subsequently, the impact of the thermal properties of the diaphragm walls—used to enclose the storage see Fig. 1—on the heat losses and the influence on the adjacent soil and groundwater was investigated.

2 Methods

2.1 Model in PLAXIS 2D

The model used for calculations is shown in Fig. 1. The model has a width of 300 m and a depth of 70 m. The storage is 150 m wide and 25 m deep. The diaphragm walls (red) have a width of 1 m and a depth of 30 m.

The walls are 5 m embedded into the silty clay (brown area in Fig. 1). The blue area in Fig. 1 represents a sandy gravel layer. The thermal properties of the soil were assumed based on a literature review (Falkner, 2014; He et al., 2020; Rees et al., 2000; Thomas & Rees, 2009) and are shown in Table 1. The groundwater table is located at 2 m below the surface.

J. Schleicher · I. Marzouk (✉) · M. Rebhan · F. Tschuchnigg
Institute of Soil Mechanics, Graz University of Technology, Foundation Engineering and Computational Geotechnics, 8010 Graz, Austria
e-mail: islam.marzouk@tugraz.at

M. Bezzeghoud et al. (eds.), *Recent Research on Geotechnical Engineering, Remote Sensing, Geophysics and Earthquake Seismology*, Advances in Science, Technology & Innovation,
https://doi.org/10.1007/978-3-031-48715-6_19

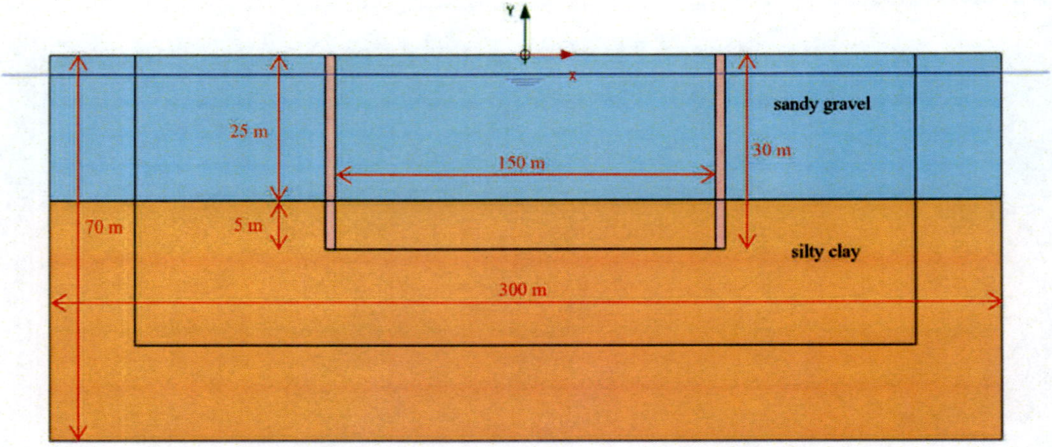

Fig. 1 UTES model in PLAXIS 2D

Table 1 Soil properties used for calculations

Soil	Thermal conductivity λ_s (W/m K)	Specific heat capacity c_s (J/kg K)	Initial void ratio e_{init} (–)
Sandy gravel	4.00	800	0.50
Silty clay	3.00	1400	0.50

2.2 Cross-Sections for Evaluation

Since the temperature field in the subsurface is a complex problem that cannot be captured with a single cross-section, Fig. 2 shows some of the cross-sections relevant for the evaluation. For the results presented below, the temperature distribution next to the diaphragm wall outside the storage was of interest. A distinction is made between vertical (A–E) and horizontal (1–5) sections. Cross-sections A–E are located at 0.5, 2, 5, 10, and 20 m distance from the diaphragm wall, respectively, while cross-sections 1–5 are located at 2, 5, 10, 20, and 30 m below the ground surface, respectively.

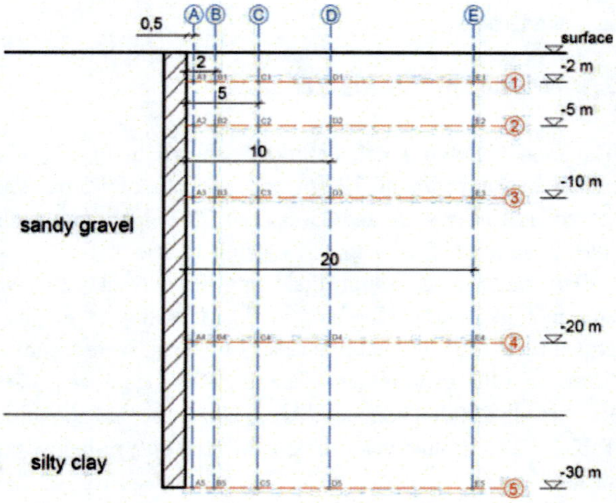

Fig. 2 Cross-sections for evaluation

2.3 Calculation Details

To simulate an underground thermal energy storage, thermal boundary conditions are defined. PLAXIS 2D (Bentley Systems, 2020) offers two possibilities either line-based thermal flow boundary conditions or cluster-related thermal conditions. As the main aim was to simulate a fully heated storage over a calculation time of 150 days, mainly line-based thermal conditions were applied inside of the storage. These were set to 90 °C to simulate a constant storage temperature.

3 Results

3.1 Comparison of Modelling Approaches

To investigate the modelling of the heating phase of the UTES system, three different approaches were considered as follows:

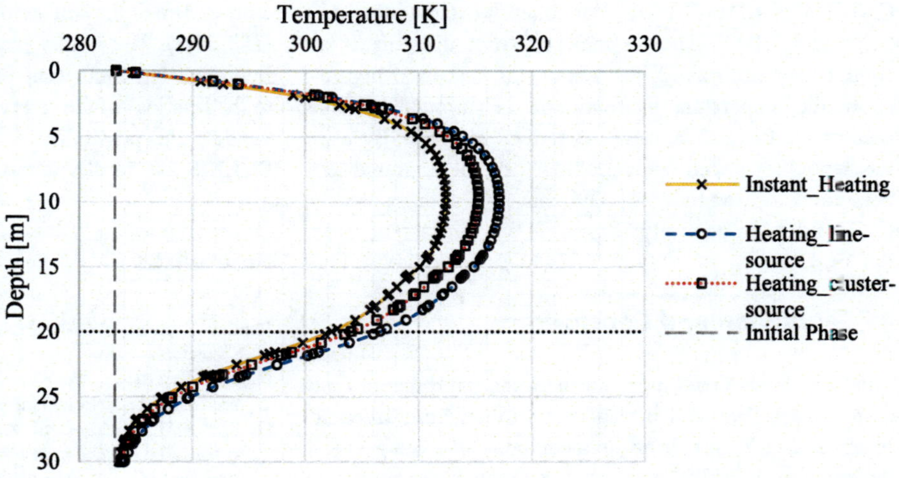

Fig. 3 Heating phase studies—temperature distribution after 150 days

- Instant heating: 90 °C cluster condition, instant heating
- Heating line source: 90 °C line-based condition, linear time-dependent heating
- Heating cluster source: 90 °C cluster condition, linear time-dependent heating.

For linear time-dependent heating, a heating period of 93 days (~ 3 months) was assumed in which the temperature of the storage increased from 10 °C (initial soil temperature) to 90 °C. In this case the storage loses heat during the heating phase which leads to an increase in the temperature of the surrounding soil on the outside of the UTES. As a result, the temperature of the surrounding soil is expected to be higher compared to the approach of instant heating. After the heating phase, the storage temperature remains at 90 °C for 150 days. With this modelling procedure, it was possible to determine the effects of the different modelling approaches, e.g. the influence of the heat losses during the heating phase. The temperature distributions after 150 days at cross-section B are exemplary shown in Fig. 3.

3.2 Impact of Thermal Properties of the Diaphragm Wall

An essential part of the investigation was to evaluate the influence of the thermal properties of the diaphragm wall on the heat loss of the storage, i.e. the temperature distribution in the subsoil. Consequently, a newly developed concrete with a thermal conductivity of 1.026 W/m K (named LC) was compared to a classic concrete with a thermal conductivity of 2.0 W/m K (named NC). Both types of concrete have a heat capacity of 1000 J/kg K. These parameters are based on a literature review (ÖNORM EN 12524, 2000;

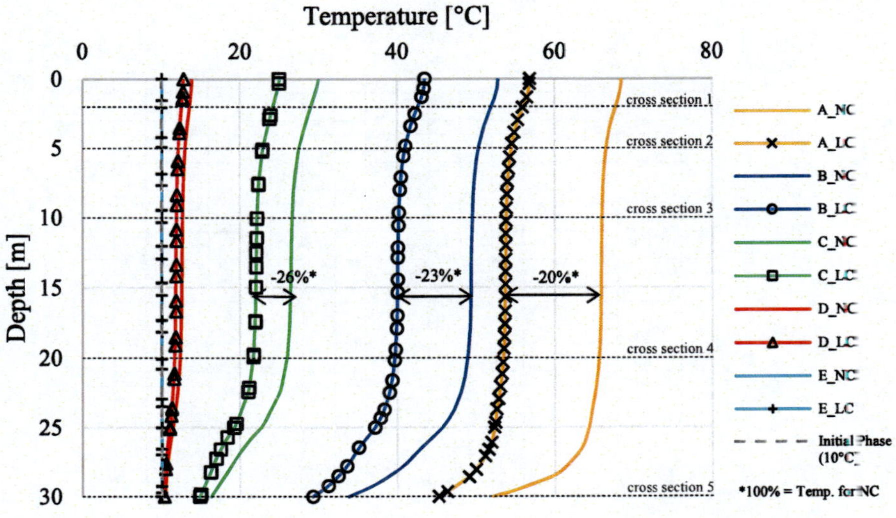

Fig. 4 Vertical temperature distribution in the surrounding subsoil

ÖNORM B 8110-7, 2013) and intensive lab testing (see Schleicher, 2022). The thermal properties of concrete are significantly influencing the amount of heat loss. Figure 4 shows the temperature distribution in the surrounding soil at sections A, B, C, D, and E as described in Fig. 2. By using the newly developed concrete "LC", the underground temperature decreased by 26%, 23%, and 20% at cross-sections C, B, and A, respectively.

4 Discussion and Conclusion

Figure 3 shows the comparison between the different modelling approaches, which resulted in a maximum difference of about 2–3 K in subsoil temperature. Concerning Fig. 4, the cross-sections further away from the diaphragm wall (D, E) are hardly influenced by the heat loss of the reservoir. In contrast, cross-sections A–C show significant differences between the two concrete types. The use of the newly developed concrete "LC" leads to a reduction of the underground temperature by up to 26% (cross-section C). Cross-sections B and A show a reduction of 23% and 20%, respectively.

In summary, the investigations concerning the modelling of the heating phase showed minor influences on the temperature distribution in the subsurface. Although a time-dependent heating phase with a cluster source is the more realistic approach, no significant differences in the temperature of the soil are observed after the calculation time (150 days). The results presented in Fig. 4 demonstrate the importance of the thermal properties of concrete with respect to the heat loss of the reservoir. To confirm the results and to validate the possibility of modelling UTES systems with PLAXIS 2D, further investigations are necessary.

References

Bentley Systems. (2020). *PLAXIS 2D: Reference manual*. Bentley Systems.

Falkner, S. (2014). *Modellierung und Simulation von thermischen Speichern*. Technische Universität Wien.

He, H., Li, M., Dyck, M., Si, B., Wang, J., & Lv, J. (2020). Modelling of soil solid thermal conductivity. *International Communications in Heat and Mass Transfer, 116*, 104602.

ÖNORM B 8110-7. (2013). *Wärmeschutz im Hochbau - Teil 7: Tabellierte wärmeschutztechnische Bemessungswerte*. Austrian Standards Institute, Wien.

ÖNORM EN 12524. (2000). *Baustoffe und -produkte Wärme- und feuchteschutztechnische Eigenschaften: Tabellierte Bemessungswerte*. Austrian Standards Institute, Wien.

Rees, S., Adjali, M., Zhou, Z., Davies, M., & Thomas, H. (2000). Ground heat transfer effects on the thermal performance of earth-contact structures. *Renewable and Sustainable Energy Reviews, 4*, 213–265.

Schleicher, J. (2022). Auswirkungen eines thermisch verbesserten Konstruktionsbetons auf Underground Thermal Energy Storage (UTES) Systeme (Master's thesis, Graz University of Technology).

Thomas, H. R., & Rees, S. W. (2009). Measured and simulated heat transfer to foundation soils. *Geotechniques, 4*, 365–375.

Geoinformatics

Geospatial Predictive Analytics Model for Urban Impervious Surface Detection: A Study on North Central Province, Sri Lanka

Indra Mahakalanda, Sandun Dassanayake, Dineth Chandrasiri, Shalitha Abeysingha, Ruvishan Vithanachchi, and Nipun Tharuka

Abstract

Impervious surfaces change the natural hydrology due to lower levels of water infiltration, increasing stream peak flows and flood risks. Concentrated storm water runoff over the landscape can contribute to pollutants and contamination of drinking water, streams and aquifers. In the past decade (2010–2020), the North Central Province of Sri Lanka has experienced a series of anomalously severe flash flood events during annual monsoon rain from December to January. While regional paddy production has experienced successes and failures, the failures have dominated due to adverse climate conditions. This study aims to develop supervised machine learning-based geospatial analytics models to classify spatial and temporal impervious surface cover changes. Following the literature on remote sensing in conjunction with machine learning, we deploy Google Earth Engine-based machine learning algorithms under the localised climate zone (LCZ) classification workflow approach to predict the imperviousness of surfaces in the northern part of Sri Lanka during 2013–2020. The ground truth for the training data set is established via Google earth images and field survey data extracted from urban areas such as the Anuradhapura and Polonnaruwa districts of the North Central Province. Random forest (RF) and classification and regression tree (CART) classifications were used to train and test the data extracted from the Landsat imageries. CART classification gives promising results. Performance measures (F1 scores) for impervious, vegetation, water, agriculture and bare lands are 0.71, 0.96, 0.96, 0.91 and 0.91, respectively. The predictive model with a pixel density analysis conducted at the lowest level of local administrative divisions appears practically and conceptually appealing to aggregated and disaggregated urban systems.

Keywords

Impervious surface • Google Earth Engine • CART • Random forest

1 Introduction

Rapid urbanisation increases impervious surface cover, preventing soil water infiltration. This causes urban heat island effect (Ranagalage et al., 2020; Senevirathne et al., 2021), flash floods, groundwater depletion and agricultural land depletion in tropical countries with agricultural economies. Urban planning and resource allocation could benefit from a holistic assessment of impervious surface cover spatiotemporal variations. This study addresses the lack of mechanisms for periodic estimation of impervious cover land use in Sri Lanka. Remotely sensed data helps sustainably manage impervious surfaces.

Land cover classification (LULC) images show impervious surface cover boundaries. However, long-term LULC classifications using remote sensing data require processing and managing massive amounts of unstructured satellite image data. Google Earth Engine (GEE) and other cloud computing resources help us overcome computational challenges. Kannangara et al. (2022) downloaded and classified surface waterbody vegetation cover over six years (2016–2021) for four seasons using GEE. Classifying two vegetation cover frequencies. Water, vegetation, and non-vegetation (other) classes cycle annually. Seasonal vegetation dynamics. The study helps policymakers optimise urban surface water body environmental resource management. The GEE can process remotely sensed data globally. This requires current Landsat data, classification tools and lots of

I. Mahakalanda (✉) · S. Dassanayake · D. Chandrasiri · S. Abeysingha · R. Vithanachchi · N. Tharuka
University of Moratuwa, Katubedda, Sri Lanka
e-mail: indram@uom.lk

M. Bezzeghoud et al. (eds.), *Recent Research on Geotechnical Engineering, Remote Sensing, Geophysics and Earthquake Seismology*, Advances in Science, Technology & Innovation,
https://doi.org/10.1007/978-3-031-48715-6_20

computation. Using multitemporal Landsat data in GEE, the study (Gordon et al., 1984) maps settlements. Global local climate zone (LCZ) mapping is in demand (Patel et al., 2015). The authors in Patel et al. (2015); Onojeghuo et al., (2018) show that the GEE platform can be employed to analyse LCZ maps using supervised classification methods such as classification and regression tree (CART), random forest (RF) and support vector machine (SVM).

This study employs GEE to classify changes in impervious surface cover in Sri Lanka's North Central Province using two machine learning-based classification algorithms: RF and CART. The proposed framework comprehensively explains urban impervious surface area classification and quantification. In land-use and land-cover (LULC) classification, the SVM produces less variability with different training sets. The GEE-classified high-resolution spatial and temporal datasets of impervious-related land use can be used to determine the optimal land use for urban growth and resource balance.

2 Methods

2.1 Study Area and Data Set

The North Central Province of Sri Lanka was selected as the study area for this investigation, the rectangle encompassing coordinates 7° 39′ 1″ N, 79° 57′ 6″ E to 8° 55′ 6″ N, 81° 20′ 25″ E. The North Central Province is composed of two main districts, Polonnaruwa and Anuradhapura, and is notable for its rich cultural heritage, including numerous archaeological sites featuring brick works and monolithic structures. The area is characterised by fertile lands in the dry zone, featuring a substantial vegetation cover and numerous water bodies, including man-made and natural ponds, tanks, and reservoirs. To generate the Normalized Difference Built-up Index (NDBI) of the study area, the study utilised the Landsat Operational Land Imager (OLI) spectral bands 2, 3, 4, 5 and 6, covering the period from 2013 to 2020.

2.2 Supervised Learning Classification Workflow

Earth observation data processing, digitisation, preprocessing sample data for model training and validation and classification algorithm application comprise the study. An open-source GIS software (QGIS) was used to digitise sample data from the Sri Lanka Survey Department Website for Anuradhapura and Polonnaruwa districts. The GEE removed cloud masking, performed mathematical operations and cropped the earth observation data to the Region of Interest (ROI). The classification algorithm was trained and validated using sample datasets with a 7:3 ratio of training to validation polygon sets, using 1400 and 600 pixels for each LCZ extracted from the stratified random sample data set [following guidelines in Mann et al. (2009)]. This study used Landsat 8 data from the World Urban Database and Access Portal Tools (WUDAPT) database and RF and CART models for supervised machine learning.

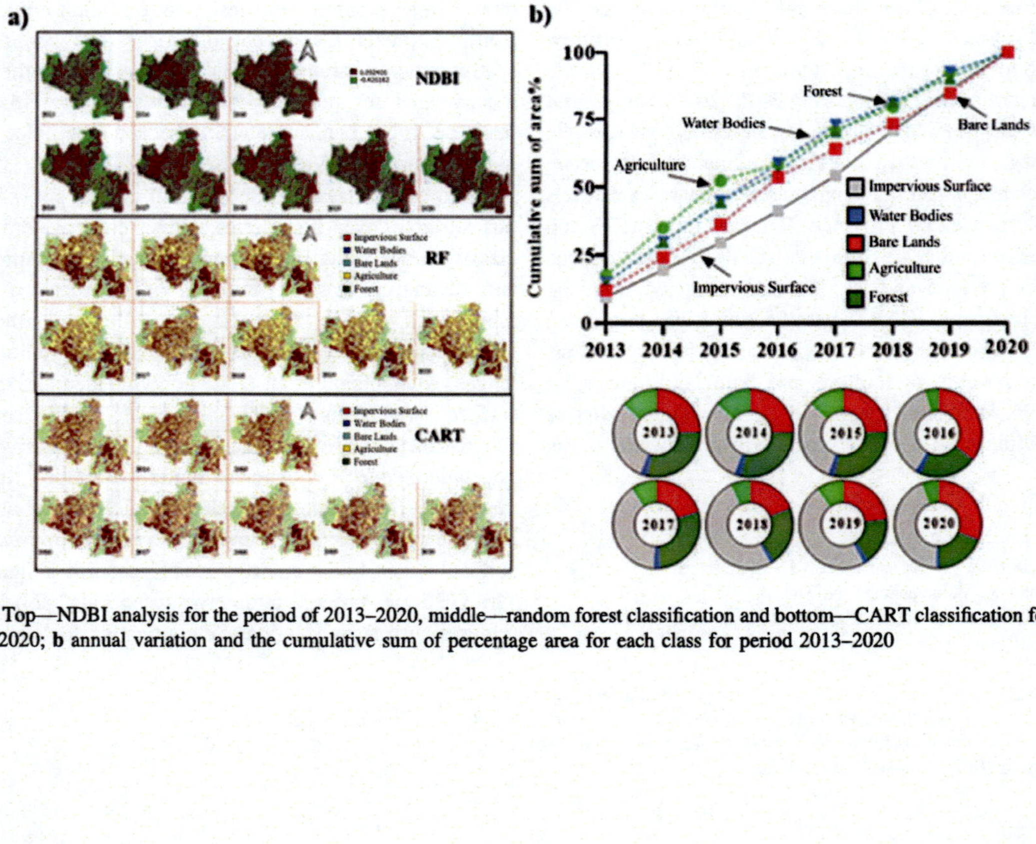

Fig. 1 **a** Top—NDBI analysis for the period of 2013–2020, middle—random forest classification and bottom—CART classification for the period of 2013–2020; **b** annual variation and the cumulative sum of percentage area for each class for period 2013–2020

3 Analysis and Discussion

Earth observation data are categorised by resolution, electromagnetic spectrum, energy source, imaging medium and bands. Satellite resolution affects classification precision. NDBI uses the near-infrared (NIR) and shortwave infrared (SWIR) band identifiers to focus on impervious surfaces.

Figure 1 (top panel) shows NDBI values from 2013 to 2020. This ratio minimises terrain illumination and atmospheric effects. Positive NDBI values indicate built-up features and negative values impervious surfaces. Positive NDBI values indicate impervious surface growth. The middle and bottom panels of Fig. 1 show the RF and CART classifications. The results show that low-density impervious surfaces are overestimated, and high-density surfaces are underestimated. In the bottom panel of Fig. 1, CART classification accuracy is above 80%. CART gives precision and recall measures (*F*1 scores) for impervious, vegetation, water, agriculture, and bare land from 0.71 to 0.91. The CART classification correctly identified vegetation and water surfaces, but impervious surfaces need more work. However, it is worth noting that the North Central Province is a significant paddy cultivating region, and the harvesting period can last for three to four months. The aggregation of satellite data over this time frame can lead to under-representation (or over-representation) of vegetation in different time periods and harvesting seasons, affecting the accuracy of the classification results.

Figure 1 demonstrates the changes in impervious surface cover from 2013 to 2020. The results show that the impervious surface cover increased from 2015 to 2018 and then decreased after 2018. Additionally, the period from 2015 to 2018 recorded a decrease in agricultural land cover.

4 Concluding Remarks

The NDBI indicates a spatial variation of the impervious surface cover area in the North Central Province from 2013 to 2020. The CART shows a more significant promise for the ML-based remote sensing image classification to detect the temporal variation of the impervious surface changes. GEE-classified high-resolution spatiotemporal datasets can determine urban growth and resource balance land use. The results can be applied in environmental protection and sustainable future city development strategies.

References

Gordon, A. D., Breiman, L., Friedman, J. H., Olshen, R. A., & Stone, C. J. (1984). Classification and regression trees. *Biometrics*.

Kannangara, K. A. T. T., Shoukie, M. B., Nayomi, M. F. A., Dassanayake, S. M., Jayawardane, C., & Anjula, A. B. N.(2022). Spatiotemporal growth dynamics of invasive plant distribution in Bolgoda Lake, Sri Lanka: a GIS based approach. In *Proceedings of the 2022 Moratuwa Engineering Research Conference (MERCon)* (pp. 1–6). IEEE..

Mann, G., McDonald, R., Mohri, M., Silberman, N., & Walker, D. D. (2009). Efficient large-scale distributed training of conditional maximum entropy models. In *Advances in Neural Information Processing Systems 22—Proceedings of the 2009 Conference*

Onojeghuo, A. O., Blackburn, G. A., Wang, Q., Atkinson, P. M., Kindred, D., & Miao, Y. (2018). Mapping paddy rice fields by applying machine learning algorithms to multitemporal Sentinel-1A and Landsat data. *International Journal of Remote Sensing, 39*(4), 1042–1067.

Patel, N., Angiuli, E., Gamba, P., Gaughan, A., Lisini, G., Stevens, F., et al. (2015). Multitemporal settlement and population mapping from Landsat using Google Earth Engine. *International Journal of Applied Earth Observation and Geoinformation, 35*, 199–208.

Ranagalage, M., Ratnayake, S., Dissanayake, D., Kumar, L., Wickremasinghe, H., Vidanagama, J., et al. (2020). Spatiotemporal variation of urban heat islands for implementing nature-based solutions: A case study of Kurunegala, Sri Lanka. *ISPRS International Journal of Geo-Information, 9*(7), 461.

Senevirathne, D., Jayasooriya, V., Dassanayake, S., & Muthukumaran, S. (2021). Effects of pavement texture and color on Urban Heat Islands: An experimental study in tropical climate. *Urban Climate, 40*, 101024.

Analysis of the Structural Health Response of Beirut Port Silos Under Seismic Loading

Sahar Ismail, Wassim Raphael, Emmanuel Durand, and Fouad Kaddah

Abstract

The analysis of the structural health response of the Beirut port silos, two years after August 4, 2020 explosion, is presented in this chapter. This work evaluates the structural response of the remaining standing silos caused by the blast loading using 3D laser scan measurements, performed every couple of months, and live monitoring of the silos' tilt changes carried out using four ultraprecise triaxial inclinometers installed in strategic locations on the North and South silos' blocks. The tilted silos are then modelled using Abaqus to test the status of the silos under different earthquake loadings. Therefore, the objective of the performed study is not only to assess and monitor the real condition of the remaining standing silos, but also to check their status and prevent their catastrophic failure under seismic conditions. The results indicate that till March 2022; the North block silos have been rotating vertically by a rate of 0.85 mm/day since the day of the explosion while the South block got stabilized. This rate varies with bad weather and sharp changes in temperature. Moreover, the triaxial inclinometers' results indicate that the displacements of the monitored silos have increased in the last couple of months to reach 45 cm on silo#84 in July 2022. The obtained results show that the remaining standing silos will exhibit serious damage and possible failure in case of moderate intensity earthquakes.

Keywords

Blast loading • Seismic loading • Numerical analyses • Abaqus explicit • Concrete damage plasticity

S. Ismail (✉) · W. Raphael · F. Kaddah
Saint Joseph University of Beirut, Beirut, 17-5208, Lebanon
e-mail: sahar.ismail@net.usj.edu.lb

E. Durand
Amann Engineering, 1212 Geneva, Switzerland

1 Introduction

In view of the Beirut port silos' damage caused by August 4, 2020 explosion, the structural status health of these silos was monitored and assessed. The explosion occurred at warehouse number 12 at Beirut port (Fig. 1). It caused more than 200 deaths and 7400 injuries, as well as more than $15 billion in damage. The explosion destroyed two rows of Beirut port silos, that were built in 1968 and reinforced in 2000, as well as two silos of the remaining standing third row. Since the day of the blast, the silos have been tilting. Therefore, in the work presented in this study, the structural health/displacement of the standing silos since the day of the blast is presented and discussed. This study is the first to monitor and publish the displacements of the remaining standing silos two years after the explosion whereas other studies performed on the subject are limited to estimating the blast magnitude (Ismail et al., 2021; Rigby et al., 2020; Valsamos et al. 2020). Moreover, using Abaqus, the behavior of the silos in case of occurrence of moderately intense earthquakes is simulated.

2 Structural Health Monitoring of Beirut Port Silos

The structural health of Beirut port silos was conducted via 3D laser scan measurements and installed triaxial inclinometers. The 3D laser scan measurements were conducted in five phases: September 2020, November 2020, March–April 2021, July 2021, and March 2022 using two equipment: the LEICA BLK 360 Imaging scanner and the Z + F 5010X Imager scanner (Leica BLK360 Imaging Laser Scanner, 2021; Z+F IMAGER 5010X, 2021). Every scan measurement acquired one billion 3D points to cover every centimeter square of the silos. Note that 3D scan measurement technology, used since the 1980s, is a nondestructive and noncontact technology that allows the generation of

M. Bezzeghoud et al. (eds.), *Recent Research on Geotechnical Engineering, Remote Sensing, Geophysics and Earthquake Seismology*, Advances in Science, Technology & Innovation,
https://doi.org/10.1007/978-3-031-48715-6_21

Fig. 1 The Beirut port silos, **a** before and **b**, **c** after August 4, 2020 blast (The Washington Post, 2020; TBS, 2020)

high-resolution and 3D digitized images of an object. It has been used in several domains such as in the monitoring soil erosion in Turkey (Yakar & Yilmaz, 2011; Yilmaz et al., 2012). In July 2021, four ultraprecise autonomous Long Range Low Power ("LoRa") triaxial inclinometers and their 4G LoRa router were installed on strategic location on the North and South block silos to monitor the live displacement of the tilted silos. Two inclinometers "2DDC8" and "2DFFA" were installed on silos #63 and 84/West side, one inclinometer "2DFAC" on silo #75–82/East-explosion side on the North block and one inclinometer "2E032" on silo #89/West side on the South block. The inclinometers send information every two minutes: "roll", "tilt" and temperature of sensor "contact temperature" with an ultra-precise measure 1/100,000 of angular degree.

3 The Numerical Model

Using Abaqus explicit, three-dimensional model consisting of the remaining standing Beirut port silos was simulated to analyze the behavior of the structure following moderately intensity seismic events. The simulated earthquakes: El-Centro (1940) $M_w = 6.9$, Loma Prieta $M_w = 6.9$, and Lebanon (2008) $M_w = 5.1$ were used since they possess PGA values around 0.3 g close to the adopted PGA value in Lebanon; Lebanon is located in zone 3 having a PGA value of approximately 0.25 g for a 475-year return period and 0.35 g for a 950-year return period (Huijer et al., 2011). Hence, first the silos were displaced based on the 3D scan and inclinometers' data to simualte their tilting till March 2022, then they were hit at the bottom by the different seismic events. The geometric model was based on the real plans and projects' data: 14, 8.5 m diameters, 48 m height silos having a 29 cm wall thickness. 139,880 4-doubly curved shell, reduced integration, hourglass control, finite-membrane strain elements were used to model the reinforced concrete silos.

The concrete silos, constructed with a special water-resistant prescribed type of cement (concrete exposed to chloride) based on the 1960s plans and design, having a mean compressive strength of 30 MPa, were heavily reinforced. On average, the horizontal and vertical reinforcement are formed of 14-mm reinforced steel diameter every 180 cm and 12-mm reinforced steel diameter every 300 cm, respectively. Note that the rebars' layer option was used to define the layers of steel reinforcements. Moreover, the concrete and steel were tied using the tie command and the base of the structure was assumed fixed. As for material, the concrete damage plasticity model while considering strain rate effects was adopted to model the nonlinear behavior of the reinforced concrete and the elastic-perfectly plastic material was adopted to model the behavior of the steel reinforcements. Table 1 presents the used material and Fig. 2 shows the stress–strain relation of the reinforced concrete in compressive and tensile region-concrete damage plasticity model with high strain rate effect.

Table 1 Material properties

Steel properties			
ρ (kg/m^3)	E (GPa)	ν	σ_y (MPa)
7850	206	0.3	448

Concrete properties							
ρ (kg/m^3)	ν	Ψ (°)	Eccentricity	$fb0/fc0$	K	Viscosity parameter	E (GPa)
2400	0.2	30	0.1	1.16	0.67	0	43

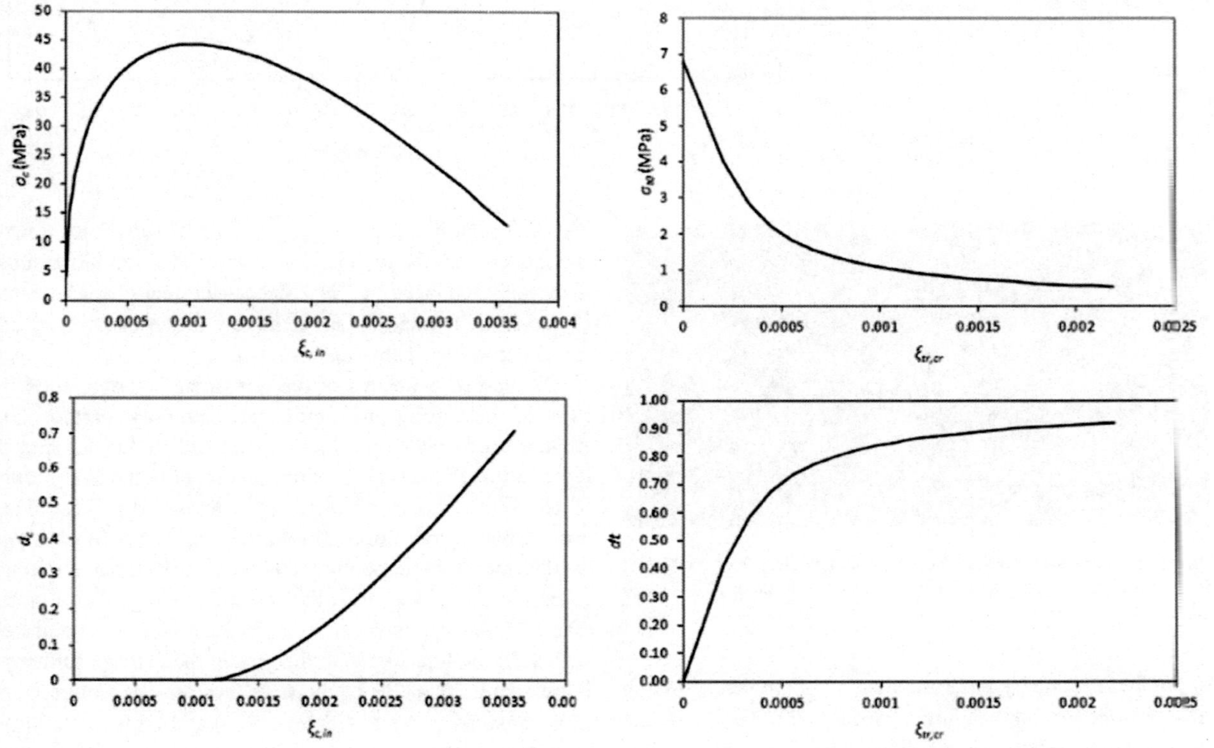

Fig. 2 Stress–strain relation of the reinforced concrete in compressive and tensile region-concrete damage plasticity model with high strain rate effect

4 Results and Discussion

As detailed in Fig. 3 that presents the horizontal displacements of the remaining standing silos, the 3D scan results indicate that following the blast, the South block silos tilted 3–20 cm to the West side and then they got stabilized. However, the North block silos tilted 4–26 cm towards the West side, and then, they shifted their displacement towards the East/crater side to rotate on average by 0.85 mm/day with a maximum and minimum rate of 2 and 0.2 mm/day. In March 2022, the North block silos tilted 8–43 cm towards East/crater side. Note that the maximum displacement is noted on silo#84, located next to the South block silos. The reason behind the stabilization of the South block silos could be caused by the destruction of its first two silos and the office building located next to it during the explosion, absorbing an important part of the blast energy and shifting it away from the foundation. Whereas the blast energy attacked the pile foundation of the North block silos, damaged it, and led to the overall structural instability manifested by the silos' displacements. Note that the rate of silos' inclination is affected as well by the temperature, weather, sun orientation, etc. For example, an acceleration in the displacements of the North block silos was noted during January and February 2022; during periods of heavy rain Lebanon experienced. The displacement at top of silo#84 was equal to 22 cm first of November 2021, it increased to

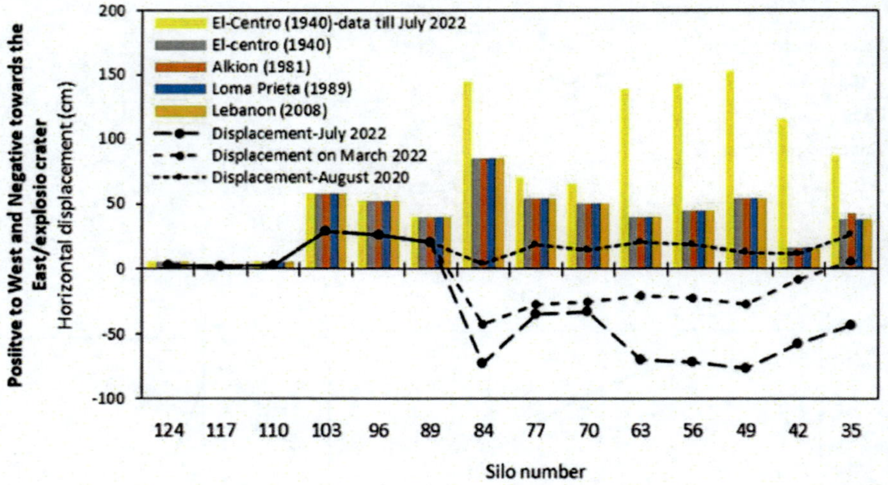

Fig. 3 Maximum horizontal displacements at top of the silos

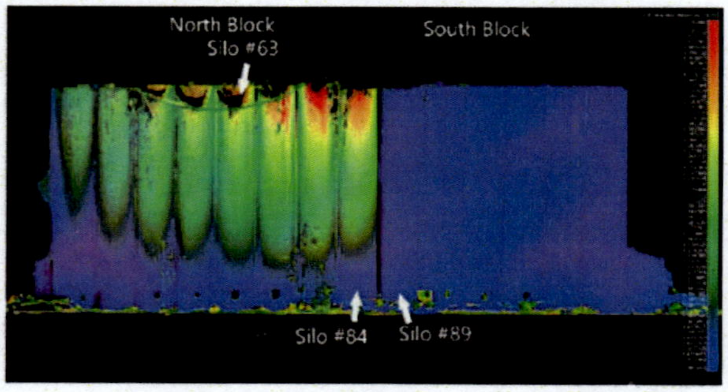

Fig. 4 3D laser scan measurement comparison (in m) of North and South block Beirut port silos: September 2020 and March 2022-Hamra side

30 cm first of January 2022 and then to 43 cm first of March 2022 East/crater side. Figure 4 presents the 3D laser scan measurement comparison of North and South block silos between September 2020 and March 2022.

The inclinometers installed on silos #82-75, 63, 84, and 89 (Fig. 4), validate the 3D scan results and note the changes in silos' displacements followed by the bad weather and sharp changes in temperature. The inclinometers' results indicate that in July 2022, silos #63, 84, and 89 were tilted 23.7, 45.2, and 19.5 cm toward the East/crater side. Thus, from March (last 3D scan measurement) to July 2022, the rate of North block silos' movement varied between 0.14 and 0.57 mm/day, proving that the rate of silos' movement is not linear and the foundation beneath the North block silos is severely damaged. Table 2 details the real displacements of silos #63, 84, and 89 since the day of the blast based on the 3D scan and inclinometers' data.

To analyze the status of the remaining standing silos in case of occurrence of moderately intensity earthquake, Abaqus was used. The results, presented in Figs. 3 and 5 indicate that the seismic loadings cause an immediate tilting of the South and North block silos toward the West side, toward the seismic load. The South and North block silos displacements increase by 100–600% under the different seismic events. Moreover, the seismic loadings cause more than 40% damage in 60% of the silos' surface. Note that the higher the earthquake PGA, the greater the damage surfaces. For example, 65–82% of the silos' surface were more than 20% damaged under El-Centro (1940) and Lebanon (2008), respectively. As a result, any moderetely intensity earthquake will cause severe damage in the remaining standing silos.

5 Conclusions

In the work presented in this study, the structural response of Beirut port silos since August 4, 2020 explosion was presented based on 3D scan and inclinometers' data. Moreover,

Table 2 Real horizontal displacements at top of silos #63, 84, and 89

Silo #	September, 2020 (Phase 1 and 2)	April, 2021 (Phase 3)	July 2021 (Phase 4)	March 2022 (Phase 5)	April 2022	May 2022	June 2022	July 2022
63	20	− 5	− 110	− 200	− 217.1	− 227.1	− 232.0	− 236.6
84	4	− 26	− 320	− 426	− 441.5	− 449.4	− 450.7	− 451.2
89	20	20	200	200	199.47	195.6	194.0	194.7

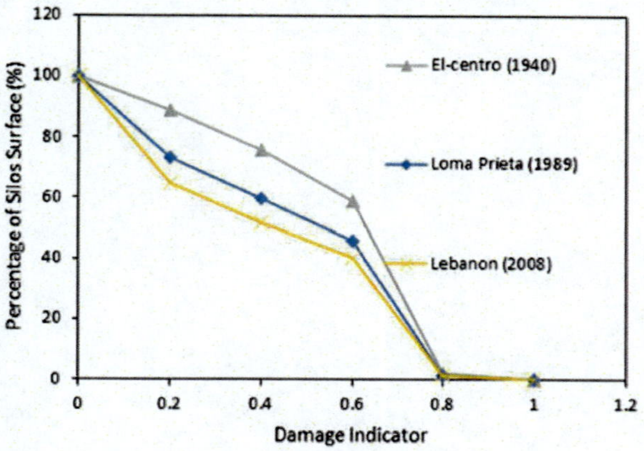

Fig. 5 The cumulative surface damage rate curves of the different simulated cases [damage variables range from 0 (no damage) to 1 (destruction)]

the status of the remaining standing silos in case of occurrence of moderately intensity earthquake was simulated via Abaqus and discussed. The results indicated that the South block has been stable since the day of the blast while the North block has been tilting and its movement is linked to the deterioration of the pile foundation. Moreover, the rate of silos' inclination is affected by the temperature, bad weather, sun orientation, etc. In March 2022, the North block silos have been rotation on average by 0.85 mm/day and in July 2022, silos #63, 84, and 89 were tilted 23.7, 45.2, and 19.5 cm toward the East/crater side. Finally, the numerical results proved that severe damage and possible silos' collapse will occur in case of any moderately intensity seismic event.

References

Huijer, C., Harajli, M., & Sadek, S. (2011). Upgrading the seismic hazard of Lebanon in light of the recent discovery of the offshore thrust fault system. Lebanese *Science Journal, 1*(2).

Ismail, S., Raphael, W., Durand, E., Kaddah, F., & Geara, F. (2021). Analysis of the structural response of Beirut port concrete silos under blast loading. *Archives of Civil Engineering, 67*(3), 619–638.

Leica BLK360 Imaging Laser Scanner. (2021). https://leica-geosystems.com/products/laserscanners/scanners/blk360.

Rigby, S., Lodge, T., Alotaibi, S., Barr, A., Clarke, S., Langdon, G., & Tyas, A. (2020). Preliminary yield estimation of the 2020 Beirut Explosion using video footage from social media. *Shock Waves*.

The Washington Post. (2020). After the blast: Satellite images show destruction in Beirut. *The Washington Post*. https://www.washingtonpost.com/world/2020/08/05/beirut-explosions-satellite-images-destruction/?arc404=true. Accessed August 5, 2020.

Valsamos, G., Larcher, M., & Casadei, F. (2020). Beirut Explosion 2020: A case study for a large-scale urban blast simulation. *Safety Science, 137*, 105190.

Yakar, M., & Yilmaz, H. M. (2011). Determination of erosion on a small fairy chimney. *Experimental Techniques, 35*(5), 76–81.

Yilmaz, H. M., Mutluoglu, O., Kavurmaci, M. M., & Yurt, K. (2012). Monitoring of soil erosion in Cappadocia region (Selime-Aksaray-Turkey). *Environmental Earth Sciences, 66*(1), 75–81.

Z+F IMAGER 5010X. (2021). https://www.zf-laser.com/Z-F-T-Cam.150.0.html?&L=1

Susceptibility Mapping of Wildfires Using XGBoost, Random Forest and AdaBoost: A Case Study of Mediterranean Ecosystem

Furkan Bilucan, Alihan Teke, and Taskin Kavzoglu

Abstract

The Mediterranean Region of Turkey has simultaneously witnessed intense wildfires throughout the summer seasons, resulting in many fatalities and injuries, linked to global warming and climate change. In the region, Antalya is one of the most vulnerable provinces to forest fire activities due to its climatic and anthropogenic conditions. In this chapter, three ensemble-based machine learning algorithms, including Extreme Gradient Boosting (XGBoost), Random Forest (RF), and AdaBoost (AB) were applied to model wildfire susceptibility of the Antalya province situated in the southern coastal zones of Turkey. For this aim, an inventory map was initially constructed by using the publicly available MODIS data retrieved from NASA's Fire Information for Resource Management System between 2001 and 2020. Also, a total of 13 causative factors (temperature, precipitation, wind speed, elevation, slope, aspect, plan curvature, profile curvature, curvature, TRI, NDVI, LULC, and TWI) were selected by considering the main characteristics of the study area and previous studies conducted in this region. According to the results, the XGBoost algorithm produced the highest accuracy of 85.4%, followed by the RF (84.6%) and AB (78.9%). Considering the estimated AUC values, XGBoost and RF outperformed the AB algorithm by about 6%. McNemar's statistical significance test results showed that RF and XGBoost algorithms produced similar performances, but their results significantly differ from that of the AB algorithm. According to the Shapley additive explanations (SHAP) strategy, elevation was found the most effective parameter for the occurrence of wildfire events, while curvature was the least effective one. SHAP analysis also showed that topographic (e.g., elevation and aspect) and climatic (e.g., temperature, precipitation, and wind speed) parameters had a larger impact on wildfire susceptibility. When all susceptibility maps were thematically interpreted, approximately 20% of the study area was covered by a very high fire susceptibility category. To be exact, the highest potential forest fire risk was mainly situated around the center and shoreline zones of the Antalya province. Taking all the results into consideration, producing a reliable wildfire susceptibility map could be used as an early warning system by decision-makers to manage, prevent, and mitigate potential fire risks and protect wildlife.

Keywords

Wildfire susceptibility mapping • Shapley additive explanations • Ensemble machine learning • XGBoost • Random forest

1 Introduction

Forests are regarded as vital natural resources that play a key role in protecting the ecological balance, wildlife habitat, and regulating the carbon cycle of the environment. Wildfires, considered as one of the most hazardous and harmful disasters, are unforeseen, unwanted, and frequently uncontrolled phenomena that damage forests, rural vegetated, and agricultural lands. In recent years, wildfires have considerably increased in Europe and the Mediterranean, although they were not prevalent in the past. According to the 2021 technical report of the European Commission's Joint Research Centre (JRC), the occurrence of wildfires in 43 countries caused to burn of 1,113,464 ha area. When it comes to Turkey, the total burnt area from 612 fires was 206,013 ha, which is the highest amount of recorded wildfire across Europe, the Middle East, and North Africa in 2021 (San-Miguel-Ayanz et al., 2021). The occurrence of wildfires can be triggered by many varied reasons including lightning, sparks from rock falls, volcanic eruptions, climate,

F. Bilucan (✉) · A. Teke · T. Kavzoglu
Department of Geomatics Engineering, Gebze Technical University, Gebze, 41400, Turkey
e-mail: f.bilucan2020@gtu.edu.tr

M. Bezzeghoud et al. (eds.), *Recent Research on Geotechnical Engineering, Remote Sensing, Geophysics and Earthquake Seismology*, Advances in Science, Technology & Innovation,
https://doi.org/10.1007/978-3-031-48715-6_22

topography, and human activities. Human activities that are intentional (i.e., arsonist) or unintentional (i.e., accidental or negligent reasons) have an essential role, especially in populated areas (Leuenberger et al., 2021). Therefore, producing an accurate wildfire prediction model and assessing the causative factors of wildfires have an essential role in efficiently managing wildfire activities (Zhang et al., 2021). In this scope, wildfire susceptibility maps, which forecast the spatial probability of wildfire zones, are of great importance for the reduction and prevention of wildfire risk. The main purpose of this study is to produce accurate and reliable wildfire susceptibility maps for Antalya province by using three ensemble machine learning algorithms, namely Extreme Gradient Boosting (XGBoost), Random Forest (RF), and AdaBoost (AB). In accordance with the purpose of the study, 13 wildfire causative parameters were selected based on the main characteristics of the study area. The predictive performance of the three ensemble machine learning algorithms was analyzed by considering overall accuracy (OA) and area under curve (AUC) score. In addition, McNemar's statistical test was employed to confirm whether the differences between the performances of algorithms were significant or not. Shapley Additive exPlanations (SHAP) strategy was also performed to locally designate the contribution of the conditioning factors.

2 Study Area and Dataset

Antalya province, extending in the east–west direction, is situated geographically between 29° 9′ and 32° 37′ E longitude, 37° 20′ and 36° 5′ N latitude, and encompasses a land of approximately 20.600 km^2 (Fig. 1). The city, situated on Anatolia's southwest coast and bordered by the Taurus Mountains, is the fifth-largest populous city with a population of 2,619,832 in 2022. In recent years, the occurrence of wildfires has become increasingly destructive in Turkey due to meteorological, topographical, and anthropogenic conditions. In addition, climate change, warmer and drier conditions have a direct effect on materials that increase fire ignition and flame speed. In this study, MODIS data retrieved from NASA's Fire Information for Resource Management System between 2001 and 2020 data were employed to model a susceptibility map for the city using three ensemble robust machine learning algorithms. Given that the objective is to produce the wildfire susceptibility maps, 13 causative factors which allow the model to specify areas with risk of fire, were analyzed. The factors were grouped into three categories including topographic variables (e.g., elevation, aspect, slope, profile curvature, plan curvature, and curvature, TRI, and TWI), environmental variables (e.g., LULC and NDVI), and climatic conditions (e.g., temperature, precipitation, and wind speed).

3 Results and Discussion

In this chapter, three ensemble machine learning algorithms were applied and 13 conditioning factors were inputted into the predictive models. Ensemble machine learning algorithms, which are one the most effective tools for generating wildfire susceptibility map, have become very popular (He et al., 2021; Iban & Sekertekin, 2022). According to the results, the XGBoost algorithm produced the highest OA of 85.4%, followed by the RF (84.6%) and AB (78.9%). Similarly, AUC analysis revealed that the XGBoost (91.7%) and RF (91.4%) methods outperformed the AB (88.4%). Furthermore, McNemar's test was conducted to specify the significance of

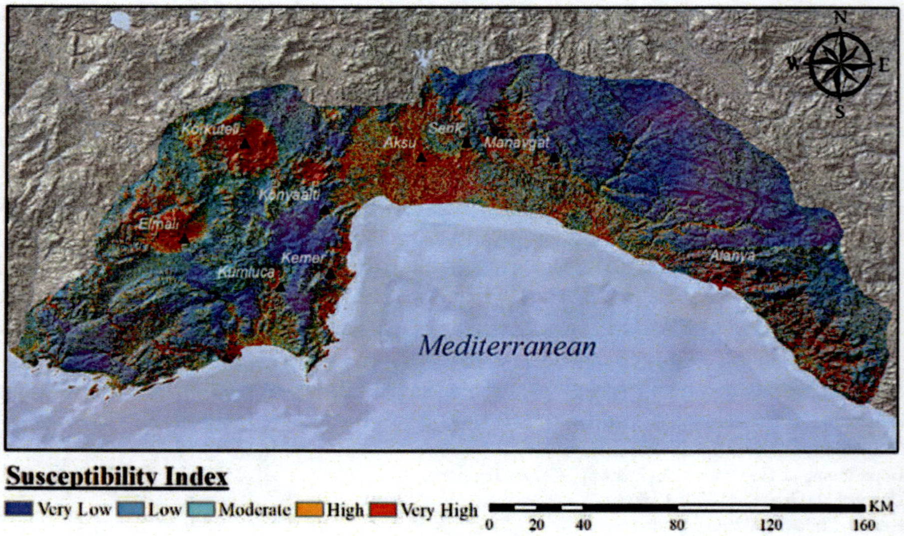

Fig. 1 Landslide susceptibility map of Antalya province produced by XGBoost algorithm

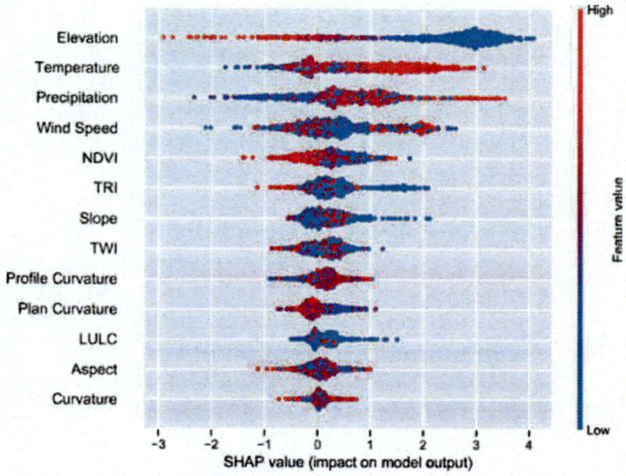

Fig. 2 Summary plot with beeswarm representation for each factor using SHAP values

differences in predictive performances. The test results indicated that the differences between the performances of all algorithms were statistically significant at a 95% confidence interval, except for XGB and RF algorithms.

Given the produced wildfire susceptibility maps with the highest OA, it is observed that one-fifth of the working area was categorized under the very high-risk class. More specifically, the interior and shoreline zones of Antalya province have the highest potential wildfire risk (Fig. 1). Detecting the importance of wildfire causative factors is one of the most critical point through process of producing a wildfire susceptibility map.

Therefore, the SHAP method was applied to investigate the importance of the causative parameters in constructed XGB model, locally. In summary plot (Fig. 2) each point represents the Shapley value for each sample and causative factor and red points correspond to high values and high Shapley values. It was observed that elevation had the highest impact on the model, followed by temperature and precipitation (Fig. 2). Similarly, previous studies proved that the elevation had important influence in modeling the wildfire susceptibility map (Arca et al., 2020; Bem et al., 2018). Also, it was found that temperature is the second most effective causative factors for occurrence of wildfire and it can be clearly said that higher values of temperature have a high positive contribution on the occurrence probability of wildfire. This finding supports the fact that higher temperature values increase the risk of wildfire. Conversely, curvature derivatives were found to be the least crucial wildfire conditioning factors. This fact also has been supported by research which were employed ensemble machine learning algorithm (Yue et al., 2023).

4 Conclusions

In this work, three ensemble machine learning approaches, namely XGBoost, RF, and AB, were applied to create wildfire susceptibility maps of the Antalya province which is one of the most threatened and wildfire-prone cities in the Mediterranean Region of Turkey. Some important conclusions can be drawn from the results of the current study. First, the results noticeably indicated that the XGBoost and RF were superior to the AB method by about 6% in terms of OA, which is a statistically significant difference. Second, topographical and meteorological variables had a higher influence on the produced predictive models. Finally, visual interpretation of the susceptibility map pointed out that the very low susceptibility classes were mainly concentrated in the northeast sections of the province while the coastal areas and middle of the study area were generally categorized as very high susceptible intervals.

References

Arca, D., Hacısalihoğlu, M., & Kutoğlu, ŞH. (2020). Producing forest fire susceptibility map via multi-criteria decision analysis and frequency ratio methods. *Natural Hazards, 104*, 73–89.

de Bem, P. P., de Carvalho Júnior, O. A., Matricardi, E. A. T., Guimarães, R. F., & Gomes, R. A. T. (2018). Predicting wildfire vulnerability using logistic regression and artificial neural networks: A case study in Brazil's Federal District. *International Journal of Wildland Fire, 28*(1), 35–45.

He, Q., Jiang, Z., Wang, M., & Liu, K. (2021). Landslide and wildfire susceptibility assessment in southeast Asia using ensemble machine learning methods. *Remote Sensing, 13*(8), 1572.

Iban, M. C., & Sekertekin, A. (2022). Machine learning based wildfire susceptibility mapping using remotely sensed fire data and GIS: A case study of Adana and Mersin provinces, Turkey. *Ecological Informatics, 69*, 101647.

Leuenberger, M., Parente, J., Tonini, M., Pereira, M. G., & Kanevski, M.: Wildfire susceptibility mapping: Deterministic vs. stochastic approaches. *Environmental Modelling & Software, 101*, 194–203.

San-Miguel-Ayanz, J., Durrant, T., Boca, R., Maianti, P., Libertà G., Artes Vivancos, T., Jacome Felix Oom, D., Branco, A., De Rigo, D., Ferrari, D., Pfeiffer, H., Grecchi, R., & Nuijten, D. (2022). *Advance report on wildfires in Europe, Middle East and North Africa 2021*. Publications Office of the European Union. ISBN 978-92-76-49633-5

Yue, W., Ren, C., Liang, Y., Liang, J., Lin, X., Yin, A., & Wei, Z. (2023). Assessment of wildfire susceptibility and wildfire threats to ecological environment and urban development based on GIS and multi-source data: A case study of Guilin, China. *Remote Sensing, 15*(10), 2659.

Zhang, G., Wang, M., & Liu, K. (2021). Deep neural networks for global wildfire susceptibility modelling. *Ecological Indicators, 127*, 107735.

Sentinel-1 DInSAR for Monitoring Active Landslide "Fish-Fish" Northeast Bulgaria

Mila Atanasova and Hristo Nikolov

Abstract

At the Northern Black Sea coast of Bulgaria, large numbers of active landslides are located. The situation has changed in the last 30 years since large technogenic load was put on it caused by construction activities in that area. For this reason, tracking activation of the landslides and surface movements in them is a critical task that must be carried out. This is the reason to perform a pilot investigation for a single landslide using synthetic aperture radar (SAR) from ESAs' Sentinel-1 mission. As main advantages of said data pointed out should be the uninterrupted time series, the unrestricted access, and the compatibility with other data operating in the radar C-band.

In the recent years, the landslide found in the residential area "Fish-fish" has been the subject of our research by making terrain observations, geodetic and geologic investigations. In the last years, two surveys by unmanned aerial system (UAS) were done spanned in one and half year period focused at studying the deformations that took place in its area. The first one has the task to create precise digital elevation model of the zone, while with the second the surface motions were registered. For this research, SAR data were used to establish the magnitude of the geodynamic processes in the area of the landslide for the period 2015–2022. To complete this task, seven uninterrupted time series covering the time span from late autumn to early spring were created using the small baseline approach (SBAS). The period was limited to the said one, because in the researched area large parts are vegetated, which causes spatial decorrelation in the interferograms. As a result of our research produced were 42 interferometric images using SAR data in SLC format allowing creation of deformation maps for the zone of "Fish-fish" landslide. The analysis of the obtained results confirmed that the applied technique provides sound information and is solid basis for systematic monitoring of the area.

Keywords

Landslides processes • DInSAR • Time series SAR data in SLC format • Small baseline approach (SBAS)

1 Introduction

In the Northeast Black Sea coastal area of Bulgaria, one of the landslides that have been active in the last years is found in residential zone Fish-Fish close to the town of Balchik. Its modern development and recent activations at large extent are due to the anthropogenic activities (mainly construction of homes) that have been carried out during the last three decades in this area.

The surface deformations that took place in the zone of the landslide located on the steep slope slopes of the plateau that face the sea are part from an ancient larger and relatively stabilized landslide The active landslide processes at this site were first registered in year 2000 (according to local inhabitants) and as probable cause for their activation a water mains accident was pointed out. In year 2010, it developed further increasing the affected area up to 5600 m^2. After that it is difficult to provide the exact size of the landslide area due to presence of high and dense vegetation, and gradual earth masses collapses along the steep coast. According to the competent local authorities (Geozastita Varna Ltd., 2018), this landslide activated twice in the last 6 years destroying houses and other infrastructural objects (see Fig. 1). For this site, again the main reason for landslide

M. Atanasova (✉)
National Institute of Geophysics, Geodesy and Geography (NIGGG)—Bulgarian Academy of Sciences, Sofia, 1113, Bulgaria
e-mail: mila_at_zl@abv.bg

H. Nikolov
Space Research and Technology Institute (SRTI)—Bulgarian Academy of Sciences, Sofia, 1113, Bulgaria

M. Bezzeghoud et al. (eds.), *Recent Research on Geotechnical Engineering, Remote Sensing, Geophysics and Earthquake Seismology*, Advances in Science, Technology & Innovation,
https://doi.org/10.1007/978-3-031-48715-6_23

Fig. 1 Recent developments of Fish-Fish landslide (photograph taken by authors in April 2019)

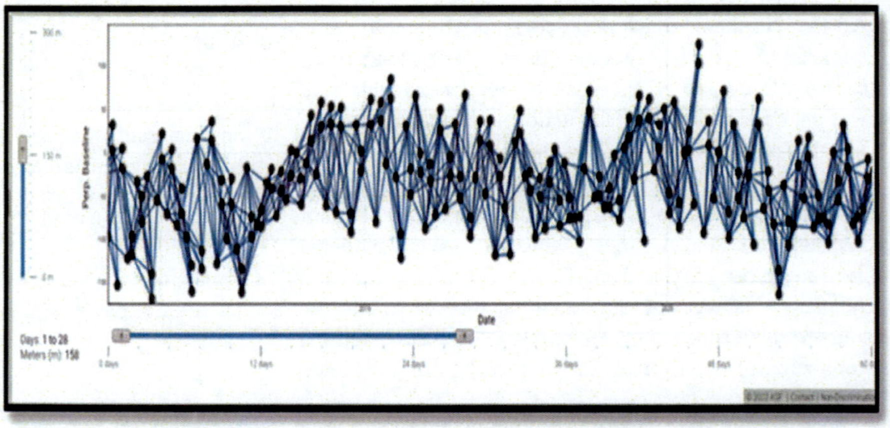

Fig. 2 The interferometric pairs that formed the time series for the period 2019–2022

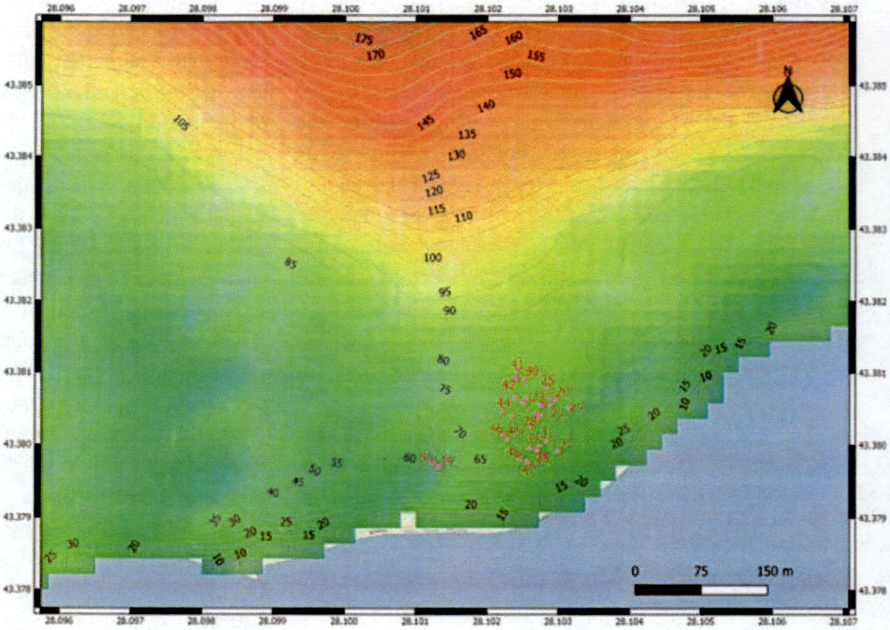

Fig. 3 The digital elevation model produced by the first UAS

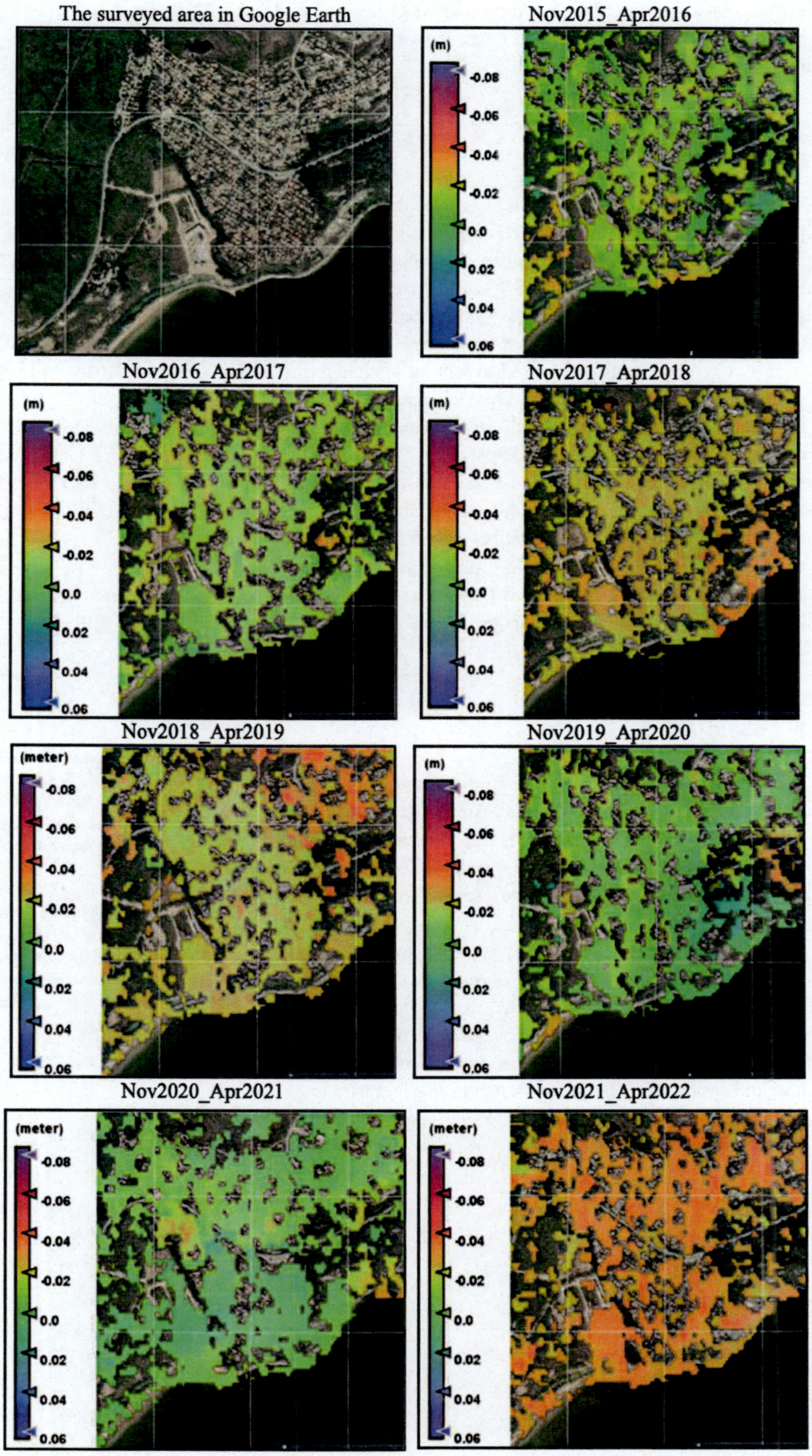

Fig. 4 Maps of SAR-derived displacements for the investigated periods

activations is the rise of the underground water level caused by different sources—torrential rains, fast snow melting, human activities.

2 Method

To assess the surface motions, it is suggested the consecutive DInSAR method to be used, which combines short baselines and short time period, not more than one month. It needs to be mentioned that the SBAS (short baseline subsets) approach is widely used by the geosciences community (Mora et al., 2002). It works best for natural environments over a large scale and can be used to look at gradual terrain changes over time. SBAS requires as input a series of interferograms and as final result creation of single image that shows the surface displacements (ASF SAR Data, https://search.asf.alaska.edu/). To detect the landslides activations in the researched area, SAR images that form a time series from orbit 36 of the Sentinel-1 mission for winter periods for the time span 2015–2021 together with the shortest perpendicular baseline B⊥ were used and downloaded from Copernicus Open Access Hub as in Duysak and Yiğit (2022) (Fig. 2).

3 Results

Two surveys by UAS focused at studying the deformations that took place in the area of Fish-fish landslide area were made in one and half year period. The first one has the task to create precise digital elevation model (DEM) (see Fig. 3) of the zone while with the second one surface motion were registered.

For this research, SAR data were used to establish the magnitude of the geodynamic processes in the area of the landslide for the period 2015–2022. To complete this task, seven uninterrupted time series covering the time span from late autumn to early spring were created using the small baseline approach (SBAS). As shown in Fig. 4, raster image of "Fish-fish" (a) and maps of SAR-derived displacements for the periods indicated.

4 Discussion

The produced displacements from time series from SAR data for the period 2015–2022 reveal that only small amount of surface points can be considered as persistent scatterers. This based on the requirement that the coherence for each interferogram have to be greater than 0.3 for the all scenes. As shown in the Fig. 3, the displacements vary between the studied intervals, which allow conclusion to be drawn that the behavior of the investigated ground motions is not linear in time.

5 Conclusions

Despite the short period that was studied the results obtained from this research are consistent, show same movement rate in the area of the collapse, have high coherence values and thus the produced information is considered to be reliable and comparable.

Acknowledgements This study was supported by Bulgarian National Science Fund under contract KP-06-OPR 06/1.14.12.2018.

References

ASF SAR Data. https://search.asf.alaska.edu/

Duysak, H., & Yiğit, E. (2022). Investigation of the performance of different wavelet-based fusions of SAR and optical images using Sentinel-1 and Sentinel-2 datasets. *International Journal of Engineering and Geosciences, 7*(1), 81–90. https://doi.org/10.26833/ijeg.882589

Geozastita Varna Ltd. (2018). *Annual report*. Available at: https://www.mrrb.bg/bg/godishen-doklad-za-2017-g-na-durjavnoto-drujestvo-za-geozastita-varna/. Retrieved February 3, 2022

Mora, O., et al. (2002). A new algorithm for monitoring localized deformation phenomena based on small baseline differential SAR interferograms. *International Geoscience and Remote Sensing Symposium (IGARSS), 2*, 1237–1239.

Nearshore Bathymetry of Figueira da Foz, Portugal, Derived from Sentinel-1 SAR Satellites

Tiago Abreu, Diogo Santos, Paulo A. Silva, and Paulo Baptista

Abstract

Remote sensing imagery enables large coastal areas to be mapped effectively. In particular, in low-lying sandy coasts exposed to high energetic wave conditions, where rapid and significant morphological changes can occur, the development of satellite-derived bathymetry methodologies represent a cost-efficient technique to estimate water depths. This work presents a satellite-derived bathymetry map of Figueira da Foz coastal stretch in Portugal, using Sentinel-1 synthetic aperture radar (SAR) data. The results are obtained with the application of a wavelet methodology where the nearshore bathymetry is inferred from the observed swell wave patterns captured by the satellite imagery. The results are compared with data extracted from Coastal Nautical Charts and from the Portuguese COaStal Monitoring Program (COSMO). The results indicate a good agreement to map water depths up to 30 m. However, errors increase close to the shore and are practically of the same order of the morphologic feature elevations such as longshore bars or rip channels. Improved results require higher imagery resolution to provide more accurate estimates.

Keywords

Satellite imagery • Bathymetry • Wave-based model • Wavelet transform • Dispersion relation

T. Abreu (✉)
School of Engineering—Polytechnic of Porto (ISEP-IPP), 4249-015 Porto, Portugal
e-mail: taa@isep.ipp.pt

D. Santos · P. A. Silva · P. Baptista
University of Aveiro, Campus de Santiago, 3810-193 Aveiro, Portugal

T. Abreu · P. A. Silva · P. Baptista
CESAM, University of Aveiro, Campus de Santiago, 3810-193 Aveiro, Portugal

1 Introduction

Low-lying sandy coasts exposed to the high-energetic wave conditions of the North Atlantic Ocean, as the western coast of Portugal, present significant morphological changes in time. The nearshore bathymetry needs to be regularly updated, but conventional bathymetric campaigns, as performed by onboard single or multibeam sonars, although highly accurate, cannot be performed, due to safety reasons during wave energetic events (Jawak et al., 2015).

A good complement to obtain bathymetric information in wave-dominated areas can be through remote sensing imagery. The results derived are rather estimated than directly measured and present a lower accuracy when compared to reference single or multibeam hydrographic surveys. Nonetheless, the International Hydrographic Organization (IHO) recognizes the development of Satellite Derived Bathymetry (SDB) as one of the most cost-effective technologies allowing to map large and remote areas (Pe'eri et al., 2013).

This work presents a satellite-derived bathymetry map of Figueira da Foz coastal stretch, located in the Atlantic West coast of Portugal, using Sentinel-1 synthetic aperture radar (SAR) data. The results are obtained with a wavelet methodology proposed by Santos et al., (2020, 2022a) where the nearshore bathymetry is inferred from the surface wave field captured by the satellite imagery. The results are compared with data extracted from Coastal Nautical Charts and from the Portuguese COaStal Monitoring Program (COSMO—https://cosmo.apambiente.pt/).

2 Methods and Data

2.1 Study Area

Figueira da Foz coastal region is located on the west coast of Portugal, being exposed to the wave climate of the North Atlantic Ocean (Fig. 1a). This region comprises sandy

M. Bezzeghoud et al. (eds.), *Recent Research on Geotechnical Engineering, Remote Sensing, Geophysics and Earthquake Seismology*, Advances in Science, Technology & Innovation,
https://doi.org/10.1007/978-3-031-48715-6_24

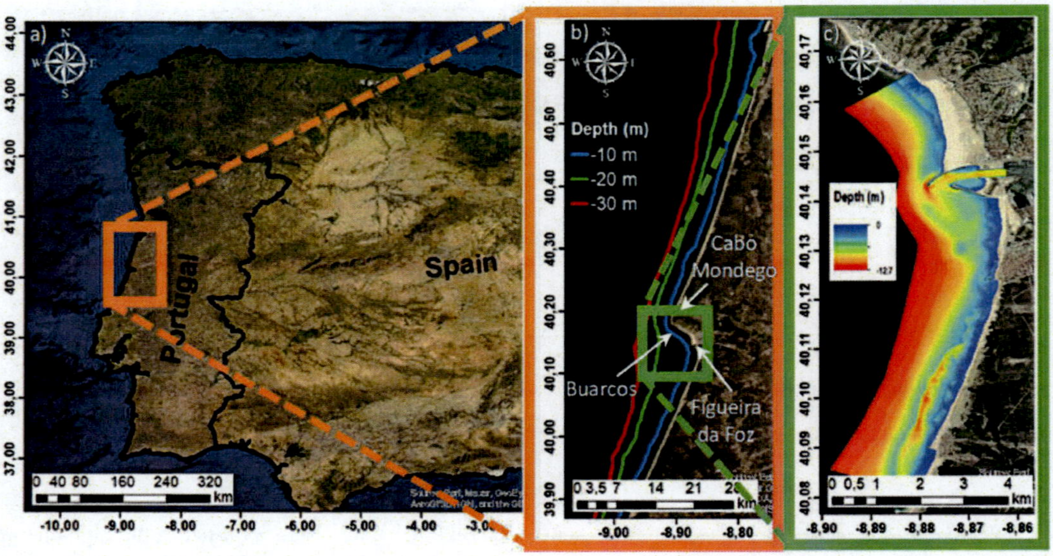

Fig. 1 a Location of Figueira da Foz and the nearshore bathymetry extracted: b from the Coastal Nautical Charts; c and from the COSMO Program

beaches and the Mondego estuary-inlet. To the north, between Cabo Mondego and Buarcos, the coastal stretch reflects rocky features, which transition southward, between Buarcos and Figueira da Foz, to a semi-enclosed sandy coast. South of the jetties, which protect the entrance to the estuary, the coast returns to sandy features.

The bathymetric data provided by the Portuguese Hydrographic Institute (IH) and by the Portuguese Environment Agency (APA) are shown in Fig. 1b, c, respectively. The IH isobaths of 10, 20 and 30 m report to 2013 and were extracted from the Coastal Nautical Charts (www.hidrografico.pt). For shallower depths, bathymetric data were measured at Figueira da Foz harbor neighboring area through the COSMO Program conceived and developed by APA. These measurements were made in August 2020.

2.2 Bathymetry Retrieval

This work uses three Sentinel-1 synthetic aperture radar (SAR) images acquired in 2018, 2019 and 2020 to infer the bathymetry at Figueira da Foz. The images have a spatial resolution of 10 m. The wavelet spectral analysis (WT) proposed by Santos et al. (2020) allows estimation of wavelength changes of the sea surface in the satellite image and to map the shallow marine environment. The estimated water depths rely on the inversion of the wave number dispersion relationship, depending on the local (L) and deep-water wavelengths (L_0). L_0 was obtained from the peak wave period, T_p, measured in Alfredo Ramalho buoy—Leixões. At the time of satellite image acquisition in 2018, 2019 and 2020 the values were T_p = 14.3, 15.6 and 13.9 s, respectively.

3 Results

The estimated bathymetry of Figueira da Foz coastal region using the image processing methodology proposed by Santos et al.'s (2020) is shown in Fig. 2a. The comparison with the data extracted from the Coastal Nautical Charts was assessed by Santos et al. (2022a) for the isobaths of 10, 20 and 30 m, respectively presenting mean absolute errors of 1.7, 2.0 and 3.2 m.

Figure 2b extends the comparison of the depth estimates for shallower depths (< 12.5 m), considering the multi-beam measurements of the COSMO Program (Fig. 1c). The map differences in the Figueira da Foz harbor neighboring area show that the major differences occur when shallow morphologic features are present (e.g., sandbars or rip channels).

4 Discussion

Considering a depth of closure for this study site smaller than 20 m, it is expected that, for deeper Coastal Nautical Chart isobaths, bathymetric changes are minimal. Small differences are confirmed with estimates for the 10 m isobath, despite errors that might occur by adopting an overall constant offshore wavelength in the dispersion relation. Still, errors increase close to the shore and are practically of the same order of the morphologic feature elevations. Since the depth

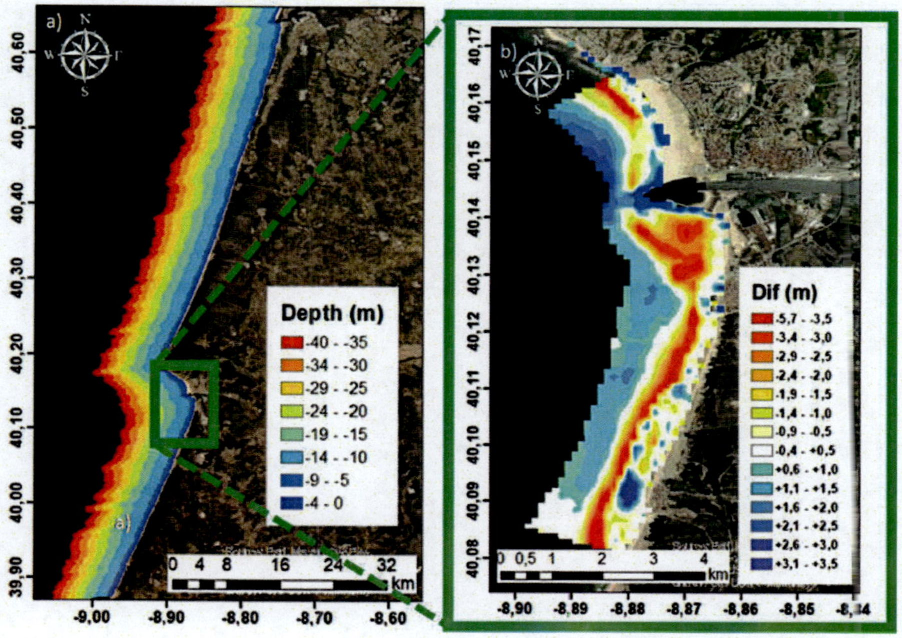

Fig. 2 **a** Estimated bathymetry of Figueira da Foz using WT and **b** bathymetric differences between the depth estimates and COSMO's measured data

estimates consider a satellite spatial resolution of 10 m and the wavelengths in the image are not well detectable close to the shoreline, the bathymetric results are interpolated between the coastline and the first 350 m, where wave breaking occurs. Thus, single images are unable to properly characterize small shallow water features. Improvements could be obtained using successive images that allow the consideration of different wave periods (Santos et al., 2022b).

5 Conclusions

This work explores an image processing methodology that makes use of the wavelet spectral analysis to retrieve the seabed morphology of Figueira da Foz. Despite the good agreement to map greater water depths, SDB are not suitable to extract smaller bathymetric features in shallower water as the evolution of longshore bars or of rip channels require higher imagery resolution or a sequence of static images of the water surface.

Acknowledgements This research was funded by Direção Geral da Política do Mar, through project NAVSAFETY of the Fundo Azul program. Thanks are due to FCT/MCTES for the financial support to CESAM (UIDP/50017/2020+UIDB/50017/2020+LA/P/0094/2020), through national funds.

References

Jawak, S. D., Vadlamani, S. S., & Luis, A. J. (2015). A synoptic review on deriving bathymetry information using remote sensing technologies: Models, methods and comparisons. *Advances in Remote Sensing, 4*, 147–162.

Pe'eri, S., Azuike, C., & Parrish, C. (2013). Satellite-derived bathymetry a reconnaissance tool for hydrography. *Hydro International, 17*, 16–19.

Santos, D., Abreu, T., Silva, P. A., & Baptista, P. (2020). Estimation of coastal bathymetry using wavelets. *Journal of Marine Science and Engineering, 8*, 772.

Santos, D., Abreu, T., Silva, P. A., Santos, F., & Baptista, P. (2022b). Nearshore bathymetry retrieval from wave-based inversion for video imagery. *Remote Sensing, 14*, 2155. https://doi.org/10.3390/rs14092155

Santos, D., Fernandez-Fernandez, S., Abreu, T., et al. (2022a). Retrieval of nearshore bathymetry from Sentinel-1 SAR data in high energetic wave coasts: The Portuguese case study. *Remote Sensing Applications: Society and Environment, 25*, 100674. https://doi.org/10.1016/j.rsase.2021.100674

Monitoring and Assessment Drought Dynamics in High and Middle Moulouya, Morroco Using Remote Sensing and Meteorological-Based Indices

Ali Salem, Mustapha Amiri, Yasir Abduljaleel, Ehab Mohammad Amen, Zana Fattah Ali, Ahmed Awad, and Mohamed Ghzal

Abstract

Agricultural drought has become a major global problem due to its severe impact on rain-fed crop yield and indirect impact on employment and per capita income. Evaluating agricultural drought risk effectively is critical to properly manage current water resources and ensure food security. The current study aims to evaluate the effectiveness of GIS and remote sensing tools for monitoring the spatial and temporal extent of agricultural drought by applying different drought indices for the lands at the high and middle Moulouya basin. The annual weather data from 1978 to 2017 was used to analyze the extent of meteorological drought using a standardized precipitation index (SPI) across the basin. Remote sensing images in 1984, 1995, and 2022 from Landsat 4–5 Thema Tique Mapper (TM) and Landsat 9 operational land image (OLI) were utilized to monitor agricultural drought using the NDVI-based vegetation condition index (VCI). The extreme drought in upper Moulouya occurred during 1983/1984 with an average SPI value of − 1.97. The largest areal extent of agricultural drought conditions was recorded in 2022 compared to 1984 and 1995. Results show that VCI tracks agricultural droughts more effectively than the SPI indices employed in this study. This identification will help to build an effective planning and drought management tool for the agriculture sector over time.

Keywords

Agricultural drought • Drought indices • Remote sensing • GIS • Moulouya basin

A. Salem (✉)
Structural Diagnostics and Analysis Research Group, Faculty of Engineering and Information Technology, University of Pécs, Boszorkány ut 2, 7624 Pecs, Hungary
e-mail: alisalem@gamma.ttk.pte.hu

Civil Engineering Department, Faculty of Engineering, Minia University, Minia, 61111, Egypt

M. Amiri · M. Ghzal
Geomatics and Soil Management Laboratory, Faculty of Arts and Humanities, Université Mohammed Premier Oujda, 60000 Oujda, Morocco

Y. Abduljaleel
Department of Civil and Environmental Engineering, Washington State University, Richland, WA 99354, USA

E. M. Amen
Natural Resources Research Center (NRRC), Tikrit University, Tikrit, Iraq

Z. F. Ali
Doctoral School of Earth Sciences, University of Pécs, Ifjúság útja 6, 7624 Pécs, Hungary

Z. F. Ali
Faculty of Education, Department of Geography, Koya University, Koysinjaq, 46011, Iraq

A. Awad
Ministry of Water Resources and Irrigation (MWRI), Giza, 11925, Egypt

1 Introduction

In arid and semi-arid areas, droughts not only adversely impact crops' growth and food production but also the available water for people's activities. Drought is described as a natural phenomenon that has a negative impact on land and water resources and disrupts hydrological balance owing to precipitation levels dropping below average (Benson et al., 1997; Salem et al., 2023). Climate change has had a substantial influence on the spatiotemporal status of drought. The lands at the middle and upper of Moulouya basins experienced significant levels of drought, resulting in substantial changes in the surrounding environmental, economic, and social fields (Amiri et al., 2021, 2022; Salem et al., 2021). Drought also impacts the steppe formation, especially through the arid spaces of the south and center of the region where Alfa and Sagebrush dominate (Iahs et al., 2021).

M. Bezzeghoud et al. (eds.), *Recent Research on Geotechnical Engineering, Remote Sensing, Geophysics and Earthquake Seismology*, Advances in Science, Technology & Innovation,
https://doi.org/10.1007/978-3-031-48715-6_25

The vegetation condition index (VCI), which is calculated using remote sensing data, is a common tool for drought assessment. Researchers have also employed the satellite-based vegetation indices, such as the normalized difference vegetation index (NDVI)-based VCI to detect and monitor the agricultural drought (Domenikiotis et al., 2004; Kuri et al., 2014; Zambrano et al., 2016). The Standardized Precipitation Index (SPI), too, is a highly valuable tool to assess the probability distribution of meteorological dryness based only on available precipitation data (preferably more than 30 years) (Morsy & Hadı, 2022). The present study aims to assess agricultural drought over the high and middle Moulouya basin using VCI and SPI over the period (1978–2017).

2 Materials and Methods

2.1 Study Area

Among the lands that locate at the middle and high of Moulouya basin, Morocco, the study area was chosen with a surface area of about 54,000 km^2 (Fig. 1).

2.2 Procedure for Calculating (VCI)

As scaling vegetative health based on long-term NDVI variability throughout the research period, the VCI is a more direct indicator of agricultural drought (Domenikiotis et al., 2004). NDVI reflects the vegetation condition through the ratio of response in visible (BAND1-BAND2-BAND3) band and near-infrared (BAND3-BAND4) of advanced very high-resolution Radiometers (AVHRR) of NOAA. Weekly NDVI data of May for the period 1985–2022 were used to produce VCI images for the study area. The VCI is obtained using the following equation:

$$VCI = 100 * \frac{NDVI_j - NDVI_{min}}{NDVI_{max} - NDVI_{min}} \quad (1)$$

$NDVI_{max}$ and $NDVI_{min}$ reflect the maximum and minimum NDVI of each pixel determined for each month, respectively, while j is the current month's index. The VCI value is expressed as a percentage and ranges from 1 to 100. Values of (50–100%), (35–50%), and ($\leq$ 35%) represents above-normal vegetation conditions, drought situations, and severe drought conditions, respectively. The remote sensing images in three different years (1985, 1995, and 2022) from Landsat 4–5 Thematique Mapper (TM) and Landsat 9 operational land image (OLI) were downloaded freely online from the U.S. Geological Survey Platform (https://earthexplorer.usgs.gov).

2.3 Procedure for Calculating SPI

Mckee et al. (1993) propose a standardized precipitation index to assess extreme and anomalous precipitation. Long-term rainfall data (1978–2016) in seven stations were obtained from the Moulouya Hydraulic Basin Agency, and SPI was calculated to determine the spatial and temporal extent and intensity of a meteorological drought event (McKee et al., 1993).

3 Results and Discussions

3.1 Drought Monitoring Through SPI

The SPI is a very popular meteorological drought index which has been frequently used by decision makers for measuring

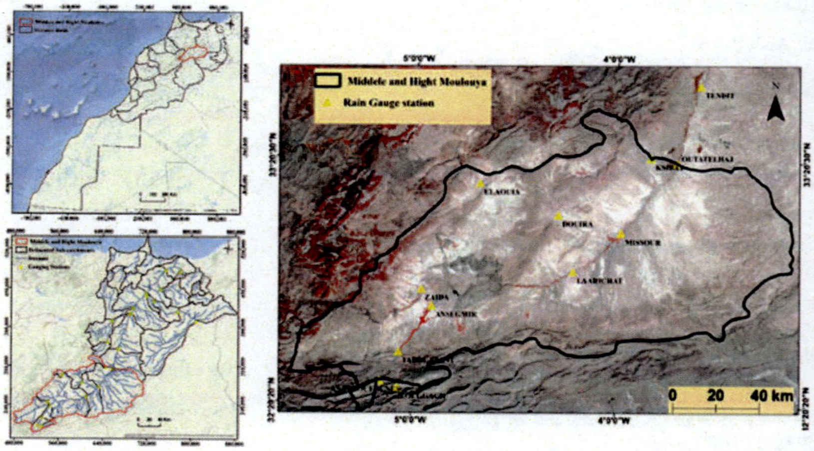

Fig. 1 Location of the study area

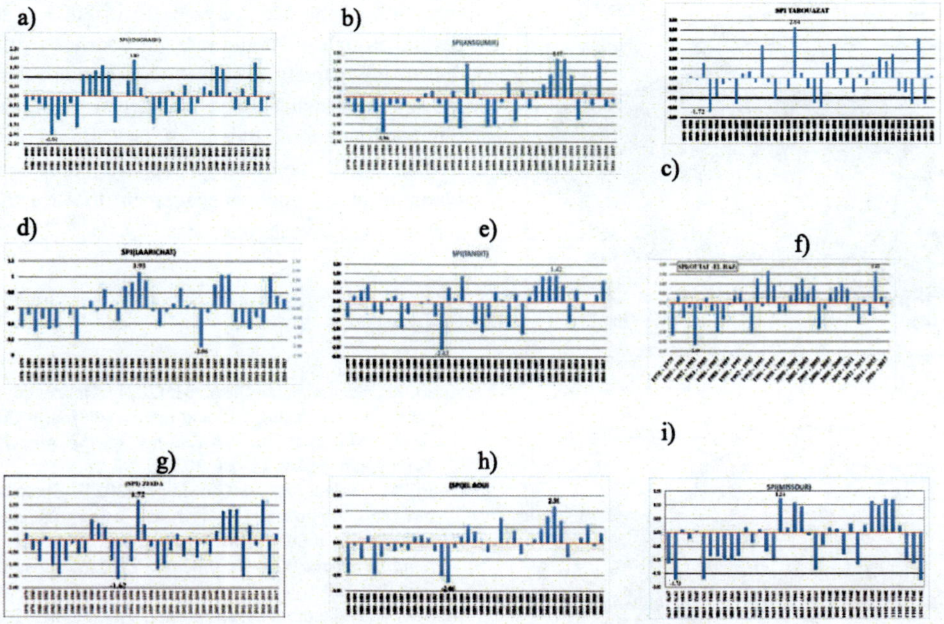

Fig. 2 SPI at annual timescale for 9 stations in Moulouya basin

and monitoring the intensity of meteorological drought events.SPI is used to determine the spatio-temporal extent of long term historical droughts (Dutta et al., 2015). Figure 2 shows the temporal distribution of the average annual precipitation for 9 stations over the study period. For all stations, the temporal analysis indicate drought incidences with a slight temporal variation. The extreme drought happened during 1983/1984 in the stations "Loughagh (SPI = 1.97) and Ansgmir (SPI = 1.96)" (upper Moulouya). For the stations "Tabouaazat and Outat el Hadj", extreme drought was observed once with SPI values of − 1.72 and − 2.19 in the year 1982–1983, respectively. Similar findings of the most recent notable droughts of 1980–1981 to 1985–1986; 1991–1992 to 1994–1995; and 2000–2001 to 2002–2003 have been reported by Ouassou et al. (2007). These dry years exacerbated the chronic deficit of water flow. The average surface water inflow, estimated to be around 19 billion m^3, was decreased to 10 billion m^3 in 1980–1985, 4.9 billion in 1992–1993, and 5.3 billion in 1994–1995 (Dutta et al., 2015).

3.2 Spatio-temporal Variation of VCI

Figure 3 depicts the spatial distribution of the annual indices VCI for the Middle and upper Moulouya basin in 1985, 1995, and 2022. Results show that moderate drought conditions occured across a broad region of Moulouya basin during 2022 with a surface area of 14,194.33 km^2.

The drought condition increased significantly between 1985 and 1995, increasing from 973.72 to 4982.10 km^2 correspondingly. Similar to a study of Ouassou et al. (2007), the agricultural seasons of 1944–1945, 1982–1983, 1994–1995, and 1999–2000 were among the driest in the recent climatological series due to widespread and significant rainfall deficits. The severe drought condition class is detected in positive annual temporal changes between 1985 and 1995 (460.3–1005.95 km^2), as a result of increasing the surface irrigation area of the tree apple in upper Moulouya. However from 1995 to 2022 the drought condition decreased from 4982.1 to 14,194.33 km^2 because of successive years of drought, which has a significant impact in agriculture activities and the water ressources, particularly in the plain area. The most recent drought, which lasted from November 2015 to spring 2016, resulted in a 3% drop in economic growth. This was due to decreased agricultural output, primarily in cereal production, and a lack of mitigation measures, such as crop insurance. The citrus and olive value chains experienced output declines and increased sensitivity to water shortages as a result of the 2015–2016 drought (Verner et al., 2018). It also demonstrates and justifies the utility of remote sensing and GIS techniques in detecting drought-related stress in rain-fed crops. Unlike meteorological data from sparsely distributed meteorological stations, the remote sensing-based index VCI can successfully delineate the spatiotemporal extent of agricultural drought (Dutta et al., 2015).

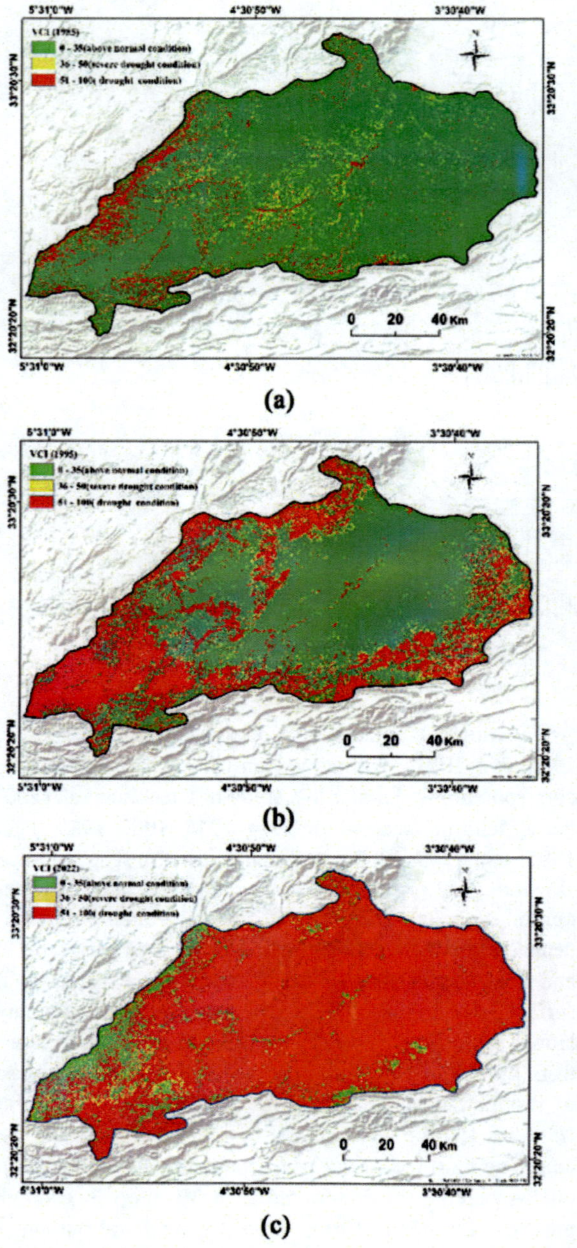

Fig. 3 Inter-annual temporal variation of VCI across the middle and upper Moulouya basin: **a** 1985; **b**1995 and **c** 2022

4 Conclusion

The present study aims to employ the Standard Precipitation Index (SPI) and the vegetation condition index (VCI) to determine the spatiotemporal extent of agricultural drought in the Moulouya basin, Morroco. The VCI results showed that 89% of the Moulouya basin in 2022 is suffering from severe drought-related stress. In contrast to meteorological data from sparsely dispersed meteorological stations; VCI, the remote sensing-based index, can successfully delineate the spatiotemporal extent of agricultural drought. These findings will potentially support constructing effective planning and drought management tool for the agricultural sector in the investigated area.

References

Amiri, M., Salem, A., & Ghzal, M. (2021). Delineation of groundwater potential zones using GIS and remote sensing in Central Moulouya Basin, Morocco. In *Proceedings of the MedGU-21, Istanbul, Turkey*, November 25–28, 2021.

Amiri, M., Salem, A., & Ghzal, M. (2022). Spatial-temporal water balance components estimation using integrated GIS-based Wetspass-M model in Moulouya Basin, Morocco. *ISPRS International Journal of Geo-Information, 11*, 139. https://doi.org/10.3390/ijgi11020139

Benson, G. J., Dambe, D., Darnhofer, T., Gommes, R., Mwongela, G. N., & Pedglev, D. E., Perarnaud, V. (1997). Extreme agrometeorological events

Domenikiotis, C., Spiliotopoulos, M., Tsiros, E., & Dalezios, N. (2004). Early cotton yield assessment by the use of the NOAA/AVHRR derived vegetation condition index (VCI) in Greece. *International Journal Remote Sensing, 25*, 2807–2819.

Dutta, D., Kundu, A., Patel, N. R., Saha, S. K., & Siddiqui, A. R. (2015). Assessment of agricultural drought in Rajasthan (India) using remote sensing derived vegetation condition index (VCI) and standardized precipitation index (SPI). *The Egyptian Journal of Remote Sensing and Space Sciences, 18*, 53–63. https://doi.org/10.1016/j.ejrs.2015.03.006

Iahs, P., Sbai, A., Mouadili, O., Hlal, M., Benrbia, K., & Mazari, F. Z. (2021). Water erosion in the Moulouya watershed and its impact on Dams' Siltation (Eastern Morocco), pp. 127–131 (2021).

Kuri, F., Murwira, A., Murwira, K. S., & Masocha, M. (2014). Predicting maize yield in Zimbabwe using dry dekads derived from remotely sensed vegetation condition index. *International Journal of Applied Earth Observation and Geoinformation, 33*, 39–46.

McKee, T. B., Doesken, N. J., & Kliest, J. (1993). The relationship of drought frequency and duration to time scales. In *Proceedings of the 8th Conference on Applied Climatology, Anaheim, CA, USA*. Meteorological Society, Boston, pp 179–184 (1993)

Morsy, S., & Hadı, M. (2022). Impact of land use/land cover on land surface temperature and its relationship with spectral indices in Dakahlia Governorate, Eqypt. *International Journal of Engineering and Geosciences, 7*(3), 272–282.

Ouassou, A., Ameziane, T., Ziyad, A., & Belghiti, M.(2007). Application of the drought management guidelines in Morocco. *Options Méditerraneenes Series B, 58*, 343–372 (2007)

Salem, A., Abduljaleel, Y., Dezső, J., et al. (2023). Integrated assessment of the impact of land use changes on groundwater recharge and groundwater level in the Drava floodplain, Hungary. *Science Report, 13*, 5061.

Salem, A., Amiri, M., Ghzali, M., & Dezső, J. (2021). Influence of climate and land use changes on the water resource of Missour Oasis, Morocco. *International Multidisciplinary Scientific GeoConference: SGEM, 21*(3.2), 115–123 (2021)

Verner, D., Treguer, D., Redwood, J., Christensen, J., McDonnell, R., Elbert, C., Konishi, Y., & Belghazi, S. (2018). *Climate variability, drought, and drought management in Morocco's Agricultural Sector*. World Bank (2018)

Zambrano, F., Lillo-Saavedra, M., Verbist, K., & Lagos, O. (2016). Sixteen years of agricultural drought assessment of the BioBío Region in Chile using a 250 m resolution vegetation condition index (VCI). *Remote Sensing, 8*, 530. https://doi.org/10.3390/rs8060530

Contribution of Automatic Processing of LiDAR Data to Accelerate Topographic Map Updating

Abdenbi Zagaoui, Mohammed Ettarid, and Abdelilah Tahayt

Abstract

The process of topographic maps updating in the Moroccan Agency for Land Registration, Cadaster and Cartography (ANCFCC) is essentially performed by manual 2D or 3D plotting based on aerial or satellite images. In this chapter, we use data from airborne LiDAR to automatically extract some map features for the purpose of speeding up the map updating process. We processed LiDAR data with both ArcGIS and QGIS Grass software to automatically generate some map data layers like contour lines and hydrography streams. The obtained features were compared by overlapping other data such as orthoimages, manually stereoplot map and contour lines generated after automatic photogrammetric correlation in the same area. The comparison of all datasets showed that LiDAR can bring a significant advantage in some areas like shaded areas of aerial images or areas of complex topographic structures like cliffs as well as in dense forest zones or even relatively urban areas where classic stereoplotting often presents difficulties. However, automatic processing of LiDAR data requires adapted parameters to avoid poorly or over detailed maps, particularly with hydrographic features even if the automatically obtained hydrographic streams present the advantage of being ordered following the stream size. We conclude that automated LiDAR data processing is of high value in terms of accelerating map updating and the extracted features like contours and hydrographic streams are of better quality than traditional methods.

A. Zagaoui (✉)
ANCFCC, El Jadida, Morocco
e-mail: zaga.abdou@gmail.com

A. Zagaoui · A. Tahayt
GEORISK Team, FSTT, Abdelmalek Essaady University, Tetouan, Morocco

M. Ettarid
IAV Hassan II, Rabat, Morocco

Keywords

LIDAR • Map updating • Topography • Hydrography • Contour lines • GIS • ANCFCC

1 Introduction

The process for topographic map updating in the Moroccan Agency for Land Registration, Cadaster and Cartography (ANCFCC) is mainly based on aerial and satellite images used in conventional photogrammetric processes, including stereo preparation, aerial triangulation and interactive 3D stereo plotting in most cases. This process is tedious for the operators and also time-consuming (https://www.ancfcc.gov.ma/nos-métiers/cartographie/processus-production/). The idea behind the present work is to assess the possibility of using ANCFCC LiDAR data together with automated tools for topographic feature extraction (Awrangjeb et al., 2015) such as hydrographic streams (Cantemir et al., 2016) and contour lines (Petrovič & Podobnikar, 2011) to reduce map updating time and effort.

Tools from QGIS free software and commercial ARCGIS software (https://desktop.arcgis.com/fr/arcmap/10.3/tools/spatial-analyst-toolbox/how-flow-accumulation-works.htm) were used for hydrographic stream and contour line generation. The quality of the obtained features was assessed based on comparison with the same features obtained from the classical photogrammetric process and using digital orthoimages as background.

2 Materials and Methods

To automatically generate hydrographic streams and contour lines map layers, a LiDAR based DTM with 5 m resolution in tiff format was used. The DTM was interpolated from a 0.3 pts/m^2 LiDAR point cloud after calibration

M. Bezzeghoud et al. (eds.), *Recent Research on Geotechnical Engineering, Remote Sensing, Geophysics and Earthquake Seismology*, Advances in Science, Technology & Innovation,
https://doi.org/10.1007/978-3-031-48715-6_26

and classification to remove non-ground points using Terrascan software. The point cloud was acquired in 2014 over the area of Al Hoceima on the Mediterranean coast of Morocco using an Airborne Leica ALS70-HA LiDAR sensor in four East-West flight lines with 20% minimum overlap.

The generated hydrographic streams and contour lines were compared with the other sources of data including:

- Topographic map of Al Hoceima at 1:25 000 scale edited in 2015.
- DTM with 5 m resolution generated with automatic correlation followed by manual edition using 2016 aerial images.
- 20 cm resolution orthoimages generated from 2016 aerial images.

Contour lines obtained automatically from the LiDAR DTM using ARCGIS and QGIS were visually compared with each other and then compared with hydrographic streams from 1:25 000 topographic maps to check for consistency, then with manually stereoplotted contour lines from the same map, as well as with interpolated contours from the DTM that was generated with automatic correlation.

Hydrographic streams were extracted several times from the LiDAR DTM with a different value for watershed minimum size. The values 100, 500, 750 and 1000 pixels for the watershed parameter were tested. Hydrographic streams obtained using LiDAR DTM from both ARCGIS Spatial analyst toolbox and QGIS GRASS were compared with each other, and then compared to the hydrography from the 1:25 000 topographic map.

3 Results

3.1 Results Regarding Contour Lines

Contour lines generated from LiDAR data using QGIS and ArcGIS have a near-perfect overlap. When compared with hydrographic streams from the 1:25 000 map, the consistency was very good especially in the contours with V-shaped summits. Comparing LiDAR contours with automatic correlation contours and manually plotted contours (Fig. 1) reveals that the two first contours are generally matching quite well. Manually plotted contours however show some important and nonsystematic differences from the two other contours.

In shaded areas as well as on cliffs, LiDAR contours were by far better than contours from manual plotting or from automatic correlation DTM. In forested areas, LiDAR contours fit better with the ground than contours from manual plotting. In urban areas, even if contours from the LiDAR DTM were less smooth they fitted well with those from the correlation DTM.

3.2 Results Regarding Hydrographic Streams

Hydrographic streams generated from the LiDAR DTM were very similar using either ARCGIS or QGIS GRASS. They were generated in.shp format with streams structured depending on their size (Fig. 2).

Even though the network of hydrographic streams generated from the LiDAR DTM coincides geometrically with streams from the 1:25 000 map, LiDAR generated streams are denser when small values of the parameter "minimum

Fig. 1 Overlap of contour lines from LiDAR DTM (in cyan) with contour lines from the correlation DTM (in yellow), and contours manually stereoplotted (in brown). Red arrows show some areas of big differences

Fig. 2 Network of hydrographic streams automatically generated with QGIS using 100 pixels as minimum watershed size, overlapped the background orthoimage. Stream thickness and color vary depending on stream size

watershed size" (e.g. 100 pixels) are chosen. LiDAR streams become less dense when larger values (like 1000 pixels) are selected. Stream density in the generated layer decreases as values for minimum watershed size increases from 500 to 750 pixels.

4 Discussion

The difference between contour lines from LiDAR and manual plotting can be explained by the lower accuracy of manual 3D plotting of contour lines. This is confirmed by the good matching of contour lines from the LiDAR with those obtained from the correlation DTM which was manually edited based on stereo images that were precisely oriented.

The quality of the interpolated contours depends on the DTM used as input. When the LiDAR point cloud classification is of poor quality, this reduces the quality of the DTM and therefore the accuracy of derived contour lines, especially in forests and urban areas, resulting in less smooth contour lines. The advantage of LiDAR contours compared to manual plotting or contours interpolated from automatic correlation to the DTM appears in areas like cliffs or shaded areas where the ground is invisible on images but well covered with LiDAR data, as well as in dense forest zones where the ground is hardly visible on aerial images but still reached by laser pulses from the LiDAR.

Regarding hydrographic streams, their density depends on the chosen parameter for minimum watershed size. The smallest streams are detected with small values like 100 pixels but the resulting map might be overdetailed. Bigger values for this parameter lead to the representation on the map of only the largest streams. The appropriate value of this parameter should be chosen depending on the dominant map topography. The structured format of the generated layer can be more easily included in a GIS database.

5 Conclusions

LDAR data could play an important role in accelerating the process of topographic map updating especially for features like contour lines and hydrographic streams, either using free software like QGIS or commercial one like ARCGIS, provided that the point cloud is well classified in urban and forest areas and proper parameters for minimum watershed size are specified.

References

ANCFCC Homepage. https://www.ancfcc.gov.ma/nos-métiers/cartographie/processus-production/. Last accessed 17 August 2022.

ARCGIS Homepage. https://desktop.arcgis.com/fr/arcmap/10.3/tools/spatial-analyst-toolbox/how-flow-accumulation-works.htm. Last accessed 15 August 2022.

Awrangjeb, M., Fraser, C., & Lu, G. (2015). Building change detection from LIDAR point cloud data based on connected component analysis. *ISPRS Annals of the Photogrammetry, Remote Sensing and Spatial Information Sciences* (II-3/W5), pp. 393–400.

Cantemir, A., Visan, A. N., Parvulescu, N., & Dogaru, M. (2016). The use of multiple data sources in the process of topographic maps updating. *The International Archives of the Photogrammetry, Remote Sensing and Spatial Information Sciences* (XLI-B4), pp. 19–24.

Petrovič, D., & Podobnikar, T. (2011). Use of airborne laser scanning data for updating topographic maps in hilly and mountain areas. *Proceedings of the 25th International Cartographic Conference, Paris, 3–8 July 2011*, International Cartographic Association, pp. 1–7.

Local Geodatabase as Tool for Monitoring the Landslide "Thracian Cliffs"

Hristo Nikolov and Mila Atanasova

Abstract

The goal of the presented research was to establish sound basis for studying the current geodynamic processes at single landslide area located between the villages of Topola and Bozhurets at the Bulgarian Black Sea coast. The investigated zone has been under high anthropogenic pressure in the last decade by constructing several touristic and holiday properties some of them already collapsed because of lack of preliminary monitoring of the surface motions in it. In order to register past and ongoing surface motions, we used data from satellite-based synthetic aperture radar (SAR), which is the main instrument of the Sentinel-1 (S-1) mission. To this end, created was local geodatabase (LGDB) into which collected all available information concerning this specific zone including, geological and techtonic maps, hydrology inventories, geodetic data from local geodynamic GNSS networks, information for land use/land cover in the said coastal zone, optical imagery from unmanned aerial systems and satellites, to name a few, besides the SAR data. Currently, such detailed and well-organized information for the studied region is not available in centralized repository for public use. Other objective of this study was to create and keep updated the mentioned LGDB using free and open source processing tools, such as QGIS, ingesting publicly available data from different sources e.g. Copernicus data and products. Key element of this research was to harmonize the already available data and to create methodology for their integration into the LGDB as well as to establish a protocol for future data acquisition from the mentioned sources and their integration into the geodatabase. The established LGDB allowed synergetic interpretation of the data thus providing better modelling of the studied geodynamic processes by large number of experts in different application domains. The authors are confident that the resulting information product is of high quality and allows seamless integration the produced LGDB into regional datacubes that are gaining popularity for machine learning applications in Earth sciences and even are used to offer products for the wide public.

Keywords

Geodynamic processes • "Thracian Cliffs" landslide • SAR data • Local geodatabase

1 Introduction

The Northeast part of the Black Sea coast of Bulgaria is known with large number of landslides registered there by the national authorities (Annual Report 2021 of Geozastita–Varna Ltd., https://tinyurl.com/3yutscme) and well studied by the scientific community. In this research, the specific object of study is the active landslide located near the "Thracian Cliffs" golf club (Nankin et al., 2020). From the terrain observations made in the zone of this landslide is estimated that its area is 11 ha with irregular shape, the landslide retreat scarp having height between 12 and 15 m. It was suggested by the experts that the sliding plane is located at depths of 10–15 m. The surface motions in its area are caused by two different, but complementary processes—the first one being the overwetting the earth masses by rains and snow melting as well as other water sources having unknown origin and the second one is the marine erosion. In the study of Nankin et al. (2020), it is stated that main material of the earth masses are aragonite clays, which exhibit considerable sensitivity and are prone to form landslides when wet.

H. Nikolov (✉)
Space Research and Technology Institute (SRTI), Bulgarian Academy of Sciences, 1113 Sofia, Bulgaria
e-mail: hristo@stil.bas.bg

M. Atanasova
National Institute of Geophysics, Geodesy and Geography (NIGGG), Bulgarian Academy of Sciences, 1113 Sofia, Bulgaria

M. Bezzeghoud et al. (eds.), *Recent Research on Geotechnical Engineering, Remote Sensing, Geophysics and Earthquake Seismology*, Advances in Science, Technology & Innovation,
https://doi.org/10.1007/978-3-031-48715-6_27

Fig. 1 Location and approximate contour of the studied landslide area

The location of the investigated zone and a recent picture of the landslide area are provided on Fig. 1a. On it also seen is the extremely steep slope at the lower part of the landslide, which makes the terrain observations impossible. The approximate area of landslide is delineated by the red line on Fig. 1b (Annual Report 2021 of Geozastita–Varna Ltd., https://tinyurl.com/3yutscme). The five buildings visible inside this area are severely damaged and unhabituated. On the road seen between them and the sea seen are a lot of cracks some of them reaching 2–3 cm. With regard to the sea erosion, it needs to be mentioned that it contributes to the activations of the landslide since no measures have been taken to stop it so far.

2 Methods and Data

To meet the objectives of this research to monitor the surface motions in the zone of the studied landslide implemented was an approach that integrated data and other information from several sources. The main piece of information is from interferometric processing of SAR products from S-1 mission, which is made available at regular basis. The initial SAR products were downloaded from one of the S-1 data archive (e.g., https://scihub.copernicus.eu/dhus/#/home). This information from the interferometric processing represents the surface displacements occurred in the line-of sight (LOS) of the satellite between its two consecutive passes over the studied area. To increase the quality of that information only data from descending orbits of the satellite and from seasons with little vegetation were used. To elaborate and to verify the SAR results data from referent stations of national GNSS networks (Kenyeres et al., 2019) as well data from local GNSS networks were used too. In the process of creation of the LGDB the mentioned sources were supplemented by other available ones such as geological (e.g. national geological map scale 1:100 000) and lithological maps, soil maps, seismological information, land cover/land use maps (Copernucus Coastal Zones Datasets, 2022), digital elevation models (DEM) at different scale (EU Dem, https://land.copernicus.eu/imagery-in-situ/eu-dem/eu-dem-v1.1), optical imagery from multispectral instruments with high and middle spatial resolution.

From methodological point of view the structure of the LGDB is flexible and shall accommodate, after some preprocessing, other emerging sources of spatial information concerning the studied object too. Examples of such sources are continuously enhanced spatial resolution of the data regarding precipitation and soil moisture (Soil Water Index, https://land.copernicus.eu/global/products/swi).

3 Results

The results produced are from SAR data processed by SBAS (Berardino et al., 2002) approach. An example of the interferometric images resulting of this research is provided on Fig. 1a. It was produced from S-1 SAR data 29 November 2018 and 28 March 2019 and proves that the registered movements in LOS are located on the east part of the landslide. Other important results are the displacements produced from interferometric images that include the inaccessible parts of the landslide due to steep terrain.

The next important source of information came from the purposely established local geodetic network. The points of this geodynamic network are presented on Fig. 2b and were used to provide GNSS measurements with high accuracy for verification of the SAR results.

On the Fig. 3 below shown are the currently included into the LGDB layers as described in the Sect. 2. The main purpose of LGDB is to serve as tool for experts to better understand the currently ongoing processes at the studied area and thus to produce models that could predict their behavior under different scenarios.

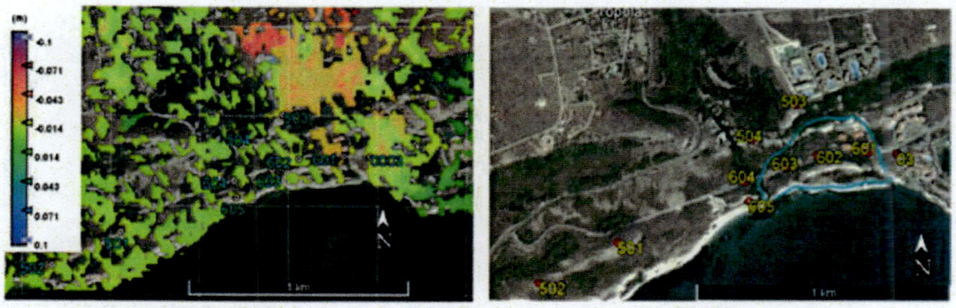

(a) (b)

Fig. 2 Uwrapped interferometric image and local geodetic network for the "Thracian Cliffs" landslide area

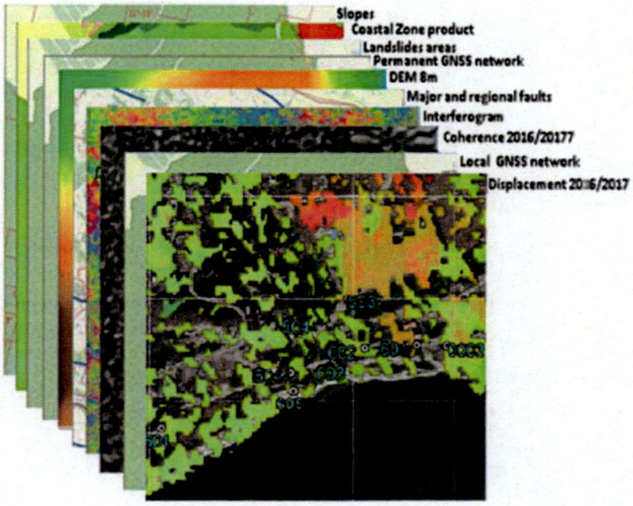

Fig. 3 Structure of the LGDB for "Thracian Cliffs" landslide

4 Discussion

To mitigate the unfavorable effects of the landslide activations the authors created specific LGDB in GIS, which shall lead to better understanding of the reasons and factors that cause them. The set of SAR results was produced from regular processing of the purposely established local SAR archive, which currently consists of more than 250 interferometric images and is constantly growing. The open source based LGDB expands the number of potential users since no licenses are needed.

5 Conclusions

This paper justified the purpose of the produced LGDB as valuable tool to monitor the "Thracian Cliffs" landslide and is considered as work in progress since more data could be added seamlessly to it. This study was supported by Bulgarian National Science Fund under contract KP-06-OPR 06/1.14.12.2018.

References

Annual Report 2021 of Geozastita–Varna Ltd (in Bulgarian). https://tinyurl.com/3yutscme. Last accessed July 2022.

Berardino, P., Fornaro, G., Lanari, R., & Sansosti, E. (2002). A new algorithm for surface deformation monitoring based on small baseline differential SAR interferograms. *IEEE Transactions on Geoscience and Remote Sensing, 40*, 2375–2383.

Copernucus Coastal Zones Datasets. https://land.copernicus.eu/local/coastal-zones. Accessed June 2022.

EU DEM. https://land.copernicus.eu/imagery-in-situ/eu-dem/eu-dem-v1.1. Accessed June 2022.

Kenyeres, A., et al. (2019). Regional integration of long-term national dense GNSS network solutions. *GPS Solutions, 23*, 122. https://doi.org/10.1007/s10291-019-0902-7

Nankin, R., Ivanov, P., & Krastanov, M. (2020). Thracian Cliffs landslide, Northern Bulgarian Black Sea Coast. *Review of the Bulgarian Geological Society, 81*, part 3, pp. 215–217, National Conference with international participation "Geosciences 2020".

Soil Water Index. https://land.copernicus.eu/global/products/swi. Accessed June 2022.

Extraction of the Network of Fractures by the Technique of Remote Sensing Integrating the Methods of Multivariate Analysis and Evaluation of Its Impact on the Layout of the Hydrographic Network (Case of the Coastal Basin of Tarfaya, South-West of Morocco)

Fatima Jira, My Hachem Aouragh, Mohammed Jelbi, Abdellah Lakhouili, and Asma Belasri

Abstract

This study focuses on the recognition and analysis of fracturing in the coastal basin of Tarfaya located in southwestern Morocco, through the extraction of lineaments using the remote sensing method. This extraction was carried out by processing and interpretation of the Landsat-8 OLI satellite image. For a more in-depth analysis of the spatial distribution of the lineaments, we opted for a comparison of the average lengths and directions of the lineaments in the four zones of the basin using Kruskal–Wallis test. The Ascending Hierarchical Classification (AHC) was designed, in our study, to group individuals (lineament directions) that represent the same directions. The lineament analysis carried allowed to highlight a network of 1603 lineaments oriented in three directions: NNE-SSW, NE-SW and WNW-ESE. These directions suggest that the bedrock fracturing of the Tarfaya coastal basin results from several tectonic phases. The application of the tests (ANOVA) proved the absence of an interaction between the length and the orientation of the lineaments. The Ascending Hierarchical Classification (AHC) of the directions of the lineaments allowed the study of their spatial distribution in the basin and subsequently the identification of the fracturing which impacted the layout of the hydrographic network of the basin.

Keywords

Tarfaya coastal basin • Lineaments • ANOVA • AHC • Hydrographic network • Remote sensing

1 Introduction

The Tarfaya coastal basin is located between 27° and 29° North latitude and 11° and 13° West longitude. It belongs to the Atlantic margin of Morocco which was formed following the Triassic rifting during the opening of the central Atlantic (Klingelhefer et al., 2016). Several studies have been carried out mainly to assess the oil potential of the area (Michard et al., 2017). These studies were able to identify structural elements related to the rifting and the opening of the central Atlantic Ocean (Fig. 1).

Secondary and tertiary series were piled up in this pitch then modeled in stepped trays: the upper trays correspond to hamadas; and the lower tray corresponds to the coastal platform inclined 1% at least towards the NE. The morphostructural study carried out by Jira (1993) in the coastal basin of Tarfaya highlighted the existence of an important folding materialized by anticlines and synclines and significant fracturing revealed by discontinuities along the Draa and Chebeika wadis, which made it possible to subdivide the basin into four structural zones. The land outcropping the basin is mainly made-up Cretaceous formations with a monoclinal structure which have recorded the effects of two different tectonics: a distensive synsedimentary tectonic of a Lower Cretaceous age and a compressive tectonic of an Upper Cretaceous to Tertiary age (Abou Ali et al., 2005).

The methodology of our study is designed to achieve the following objectives:

F. Jira (✉) · A. Lakhouili · A. Belasri
Faculty of Sciences and Technology Settat, Hassan First University of Settat, B.P. 577, Settat, Morocco
e-mail: fatima.jira@uhp.ac.ma

M. H. Aouragh
Meknes Faculty of Sciences, Moulay Ismaïl University, B.P 11201, Zitoune-Meknes, Morocco

M. Jelbi
Kenitra Faculty of Sciences, Ibn Tofail University, B.P. 133, Kenitra, Morocco

M. Bezzeghoud et al. (eds.), *Recent Research on Geotechnical Engineering, Remote Sensing, Geophysics and Earthquake Seismology*, Advances in Science, Technology & Innovation,
https://doi.org/10.1007/978-3-031-48715-6_28

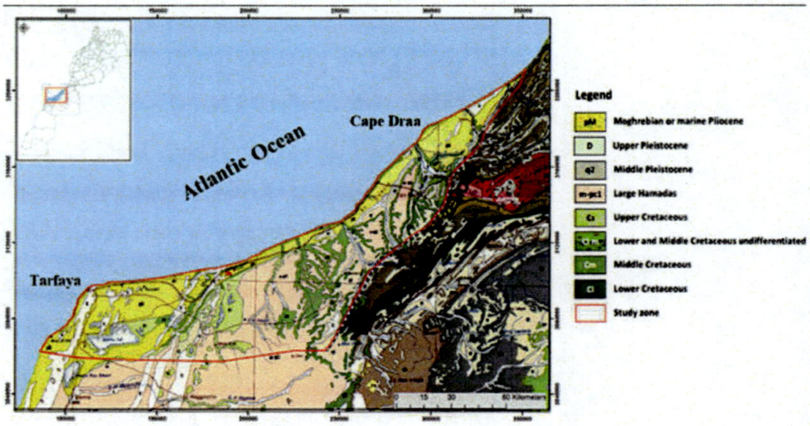

Fig. 1 Geological map of Tarfaya coastal basin and location of the study area (Bensaid, 1985)

- Contribute to a better knowledge of the state of fracturing of the basement of the Tarfaya coastal basin through the characterization and identification of lineaments;
- Highlight the structural details of the surface of the Tarfaya coastal basin characterized by its tabular morphology.
- Detect the impact of the structuring of the basin on the architecture of the hydrographic network that drains it.

The Ascending Hierarchical Classification (AHC) is a multivariate statistical analysis, used to measure (dissimilarity/similarity) between individuals, using Ward's method and Euclidean distance (Ward, 1963). In our case, the utility of the method (CAH) was designed to group individuals (lineament directions) that have the same orientations, distinguish between them, and compare them with the hydrographic network.

2 Methods

As part of this work, the hydrographic network was digitized from topographic maps at 1/100,000 that cover the study area and from Digital Terrain Model.

To highlight the fracturing of the coastal basin of Tarfaya, an extraction of the lineaments usually used by the techniques of remote sensing was carried out by Three Landsat-8 OLI/TIR scenes; The manual lineament extraction is performed by visual interpretation in the GIS tool using the composite color enhancement of the different PCA band combinations. The automated extraction is carried out using the Line algorithm associated with the PCI Geomatica software. The two methods were merged and analyzed statically to produce the directions, lengths, densities of the lineaments.

A statistical study of the orientations and the lengths of the lineaments was adopted, in particular an analysis of variances (ANOVA) to determine the existence or the absence of significant difference between the means of the variables. In our study, the Kruskal–Wallis test was selected since the hypothesis of the normality of the data distribution was rejected. The rank-based Kruskal–Wallis test (Kruskal & Wallis, 1952) was applied based on the p-value (probability), extracted from the Kruskal–Wallis test, and its comparison with the probability corresponding to the significance level of $\alpha = 1\%$ (Kakaï & Lykke, 2016).

3 Results

By examining the lineaments map, we notice that the coastal basin of Tarfaya is fractured by a network of 1603 lineaments with a length varying between 0.27 km and 29.35 km and a density of 1.02 km/km^2 (Fig. 2).

All the lineaments are distributed according to an organization which subdivides the basin into four zones. High-density values are recorded especially in zone 3, zone 1 and also toward the south of zone 4. The lineaments are oriented in three directions: NNE-SSW, NE-SW and WNW-ESE (Fig. 3). A check on the geological map confirms that the lineaments are often consistent with the discontinuities and the lithological contacts. The directional rosettes of the lineaments clearly show a difference between the four zones. However, in the area located in the NE of the Wadi Draa, the direction of the lineaments is NE-SW with a frequency of 8%. In zone 2, between Wadi Draa and Wadi Chebeika, a main ENE-WSW direction has been highlighted with frequency of 6% and a secondary NE-SW direction with more than a 4% frequency. The lineaments in zone 3, which is located between Wadi Chebeika and the cliff of Ed Dzeroua, is marked by an NNE-SSW direction with a frequency of 6%. The lineaments in zone 4, which is located SW of the cliff of Ed Dzeroua, follow a main NE-SW to NNE-SSW direction with a frequency of 6% and a secondary ENE-WSW direction with more than a 4% frequency.

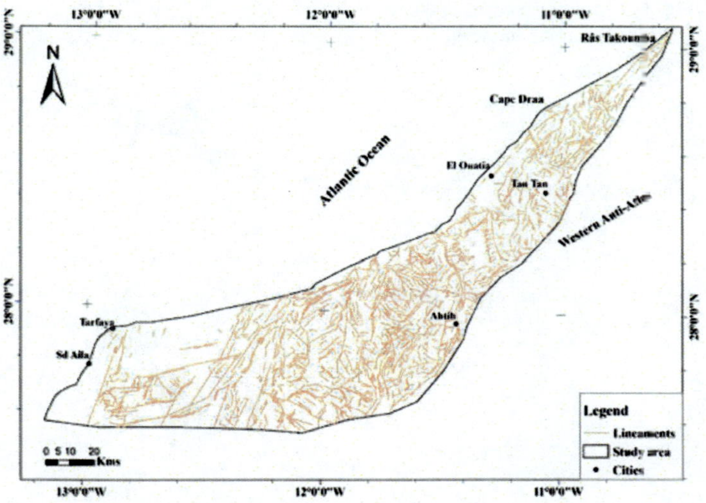

Fig. 2 Results obtained after extracting lineaments from Landsat 8 OLI

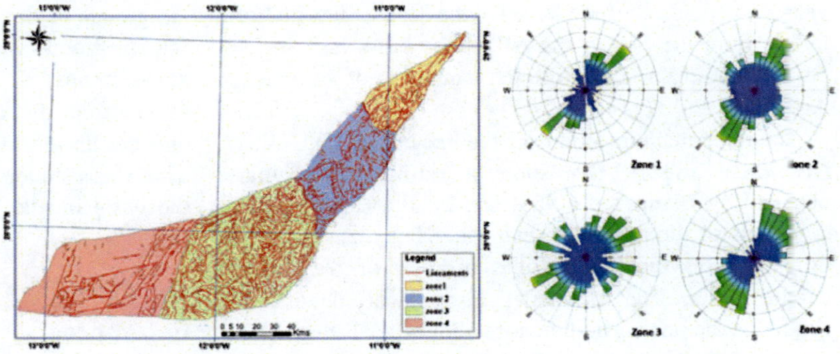

Fig. 3 Zonation of the Tarfaya coastal basin and the directional rosettes

The comparison of the means of the lengths and directions of the lineaments of the four zones according to the Kruskal–Wallis test show that for the directions, the value of the probability (p-value < 0.0001) is lower than the significance level of $\alpha = 0.01$, which allows us to conclude that there is a significant difference between the averages of the lineament directions in the four zones. However, we notice that the average lineament lengths don't show any significant difference with a p-value of 0.967, which is well above 0.01 (Table 1).

Table 1 Kruskal–Wallis test

Kruskal–Wallis test	Variables	
	Orientation	Length
K (Observed value)	52.0665	0.2635
K (Critical value)	7.8147	7.8147
DDL	3	3
p-value	< 0.0001	0.9667
α	0.01	0.01

The application of the ascending hierarchical classification (AHC) to the directions of the lineaments, allowed us to distinguish six groups from the dendrogram. The directional rosettes, shown in Fig. 4 show that the high frequency directions of the six groups are respectively: ENE-WSW, WNW-ESE, NE-SW, NNE-SSW, ENE-WSW and NNW-SSE.

4 Discussion

In our study, the validation of the lineaments is performed by comparing the preferential directions of the automatic lineaments with those of previous works. Also, various works have shown the potential of the LINE algorithm to detect lineaments for hydrogeological studies (Mallast et al., 2011). The network of lineaments shows a high density south of zone 4, on the site of the sebkha Tah where two synclinal furrows-oriented WSW-ENE emerge (Glangeaud, 1953). Glangeaud L. also reported anticlinal and syncline undulations in the Cretaceous bedrock in zone 3 of the basin where high lineament density values were recorded. The high

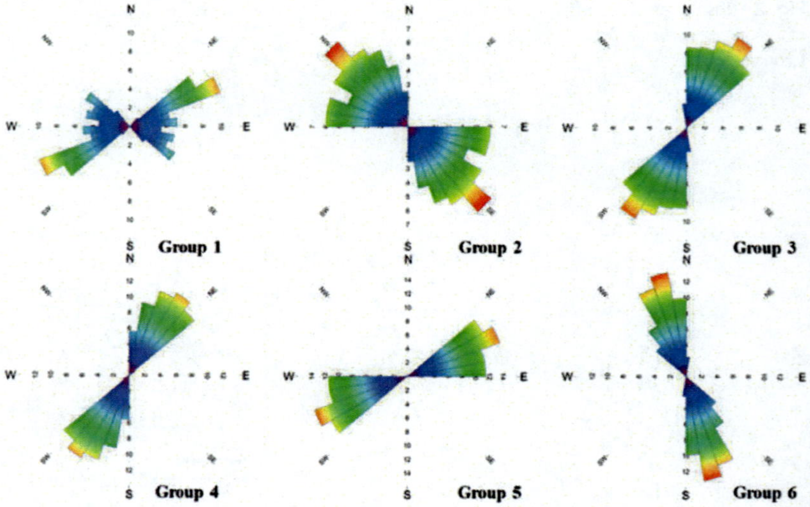

Fig. 4 Directional rosettes of the CAH groups of the lineaments

values noted in zone 1 are linked to the Moghrebian platform falling from the SW to the NE of the basin and the rapprochement of this zone with the mountains of the Western Anti Atlas.

The three main directions of the lineaments ENE-WSW, NNE-SSW and NE-SW show a similarity with those obtained by Glangeaud (1953) and by Jira (1993) and the directions of the tectonic events (the Hercynian and Atlas orogeny's) having affected the bedrock of the Tarfaya coastal basin. The ascending hierarchical classification applied to the lineament directions allowed to distinguish six groups of lineament directions. By confronting these groups of directions with the hydrographic network which drains the basin, it was found that only groups 2 and 6 of direction respectively WNW-ESE, E-W and NNW-SSE, N-S influenced the architecture of the network hydrographic.

5 Conclusions

The extraction technique applied in our study allowed us to detect the maximum of lineaments in the coastal basin of Tarfaya. The arrangement of these lineaments made it possible to subdivide the basin into four zones. The ANOVA test applied to the means of the lineament directions clearly showed a significant difference between the four zones of the basin. The directional rosettes resulting from these lineaments present three main directions; ENE-WSW, NNE-SSW and NE-SW which are in conformity with previous work and the Hercynian and Atlasic orogeny's which affected the basement of our study area. The ascending hierarchy classification applied to the lineaments allowed us to distinguish six groups of directions. The superposition of the maps of the six groups with the map of the hydrographic network made it possible to illustrate the spatial and directional relationships that exist between the elements of the network and the lineaments of the 2nd and 6th directional groups. This classification reflected the effect on the structuring of the basin on the layout of the hydrographic network that drains it.

References

Abou Ali, N., Hafid, M., Chellaï, E. H., Nahim, M., & Zizi, M. (2005). Structure de socle, sismostratigraphie et héritage structural au cours du rifting au niveau de la marge d'Ifni/Tan-Tan (Maroc sud-occidental): Comptes Rendus Géoscience, v. 337, no. 14, pp. 1267–1276.

Bensaid, M. (1985). Carte géologique du Maroc, Ministère d'Energie et des Mines, Direction de la Géologie, Notes et Mémoire n° 260.

Glangeaud, L. (1953). Les Formations Plio-Quaternaires de l'Afrique du Nord. Mémoire du Service Géologique (Maroc) t.1, n°175, pp. 181–195.

Glele Kakaï and Lykke: Aperçu sur les méthodes statistiques univariées utilisée dans les études de végétation. Annales des Sciences Agronomiques20-Spécial Projet Undesert-UE: 113-138 (2016).

Jira, F. (1993). Etude Morphostructurale et Sédimentologique des Formatopns Plio-Pleistocènes du Bassin Côtier de Tarfaya (Maroc Méridionale). Thèse DES, Univ.Cadi Ayyad, Marrakech, 165p.

Klingelhefer, F., Biari, Y., Sahabi, M., Aslanian, D., Schnabel, M., Matias, L., Benabdellouahed, M., Funck, T., Gutscer, M., Reichert, C., & Austin, J. A. (2016). Crustal structure variations along the NW-African continental margin: A comparison of new and existing models from wide-angle and reflection seismic data. *Tectonophysics, 674*, 227–252.

Kruskal, H., & Wallis, A. (1952). Use of ranks in one-criterion variance analysis. *Journal of the American Statistical Association, 47*(260), 583–621.

Mallast, U., Gloaguen, R., Geyer, S., Rodiger, T., & Siebert, C. (2011). Semi-automatic extraction of lineaments from remote sensing data and the derivation of groundwater flow-paths. *Hydrology & Earth System Science Discussions, 8*, 1399–1431.

Michard, A., Saddiqi, O., Missenard, Y., Oukassou, M., & Barbarand, J. (2017). Les drandes régions géologiques du Maroc: diversité et soulèvement d'ensemble. *Revue Géologues, 194*, 4–12.

Ward, J. H., Jr. (1963). Hierarchical grouping to optimize an objective function. *Journal of the American Statistical Association, 58*, 236–244.

A Composite Approach to Assessing Similarity in the Risk Level of Agricultural Drought: An Example of the Tensift and Moulouya Watershed in Morocco

Ismaguil Hanadé Houmma, Sébastien Gadal, Loubna El Mansouri, Rachid Hadria, and Paul Gérard Gbetkom

Abstract

In this chapter, a per-pixel trend analysis approach and the temporal variance of drought risk were used to compare the seasonal dynamics of the multivariate risk level of agricultural drought in two hydrological systems (Tensift and Moulouya watersheds in Morocco). The multivariate risk level is calculated from the time series of the enhanced composite model for agricultural drought monitoring constructed from anomalies in precipitation, evapotranspiration, soil moisture, NDVI and land surface temperature (LST). These five factors of agricultural drought are obtained from multi-sensor remote sensing data (MODIS, CHIRPS) for the period 2004 to 2022. In both watersheds, the results show a significant upward trend in the level of extreme and moderate risk and a very contrasting seasonal variance. At the same time, over the past two decades, risk levels (normal and low) have shown a marked downward trend. The multivariate frequency of generalized extreme risk is identical in both hydrological systems. Over the data period, the seasonal occurrence of multivariate risk is 6/19 (extreme risk), 7/19 (moderate risk) and 6/19 (low to normal risk). The recent spatio-temporal dynamics of the level of extreme risk have significant negative correlations (−0.6 at Tensift and −0.62 at Moulouya) with cereal yield anomalies. Multivariate risk is positively correlated with the SPI index at the seven-month scale of the growing season. The maximum correlation is 0.86 with a p-value of 0.0000 at Tensift and 0.74 with a p-value of 0.0004 at Moulouya. Depending on the year, the multivariate extreme risk can vary from 0 to 100% in terms of exposed areas in both the Tensift and the Moulouya watersheds. It is generalized over all the spatial extents of these watersheds for the years 2022, 2020, and 2016 and non-existent for the years 2004, 2006, 2009, 2010, 2013, and 2015.

Keywords

Drought risk • Index • Multivariate mapping • Remote sensing • Watershed

1 Introduction

In traditional approaches, weather station data are used in drought risk assessments based on hydroclimatic indices such as the Standardized Precipitation Index (SPI), the Palmer Drought Severity Index (PDSI). However, in most underdeveloped countries, weather station networks do not reflect local variability related to topography or latitude, and therefore, the spatial diversity of agricultural and climatic landscapes are poorly considered (Nam et al., 2018). To address this limitation, several studies (Enenkel et al., 2016; Hanadé et al., 2022; Nam et al., 2018) agree on the importance of multivariate mapping based on the integration of multiple climate and biophysical variables. Multivariate drought risk mapping is an essential component of prospective drought modelling in highly changeable environments. By aggregating the historical variance of drought co-variables into a single indicator, multivariate risk-level

I. H. Houmma (✉) · S. Gadal
Aix Marseille Université, Université Côte d'Azur, Avignon Université, CNRS, ESPACE, UMR 7300, 84000 Avignon, France
e-mail: ismaguil.hanade-houmma@etu.univ-amu.fr

I. H. Houmma · L. E. Mansouri
Department of Geodesy and Topography, Geomatics Science and Engineering, Hassan II Institute of Agronomy and Veterinary, Rabat, Morocco

R. Hadria
National Institute of Agricultural Research (INRA), Rabat, Morocco

P. G. Gbetkom
LEGOS, Université de Toulouse, IRD, CNES, CNRS, UPS, 31400 Toulouse, France

M. Bezzeghoud et al. (eds.), *Recent Research on Geotechnical Engineering, Remote Sensing, Geophysics and Earthquake Seismology*, Advances in Science, Technology & Innovation,
https://doi.org/10.1007/978-3-031-48715-6_29

mapping is useful for assessing the magnitude of a drought during its development. In semiarid regions due to high local climate variability, it is commonly observed that even when precipitation is around normal, the extreme variability of other stress factors most often leads to more severely stressed conditions than expected. This situation is frequently observed in many Mediterranean watersheds that have been affected by intense and severe drought episodes, because of very intense heat waves. As a direct consequence, the question of water resources rationalization in the case of prolonged severe droughts arises. Here, reliable indicators are necessary to establish the diagnosis and provide early warning indications of drought onset. In this respect, multivariate risk mapping is a support to decision making because a zonal classification of multivariate risk level allows for the development of emergency measures to priority areas. Thus, this study proposes a comparative approach of multivariate mapping of the level of risk of agricultural drought in two watersheds in the Mediterranean climate of Morocco.

2 Study Area

A comparative analysis of the multivariate risk of drought concerns the Tensift watershed with an area of 20,450 km^2 in west-central Morocco and the Moulouya watershed with an area of 55,500 km^2 of eastern Morocco (Fig. 1). Both areas are characterized by arid and semiarid conditions of a Mediterranean climate. They are particularly vulnerable to the increased frequency of extreme events that often occur simultaneously and significantly impact the agricultural sector.

Moulouya is characterized by an annual rainfall varying between 230 and 380 mm and temperatures 5 °C to 31 °C,

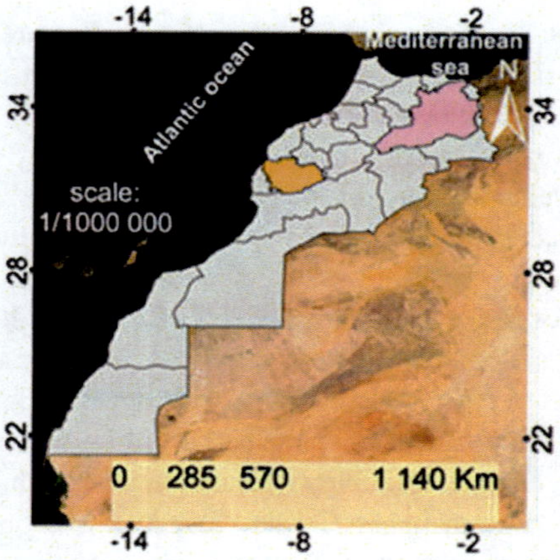

Fig. 1 Location of the study area

strong winds reaching a maximum speed of 50 m/s and its main wadi takes its source in the High Atlas at 2000 m altitude. It has a major socio-economic importance, especially for irrigated crops as it hosts the third largest citrus region of Morocco on an area of 21,200 ha in 2018. Tensift is one of the most important watersheds due to its extent, the heterogeneity of its relief and its strong mobilization of water resources. It extends from 4 to 4141 m altitude and its climate is subject to influence of the Atlantic Ocean. Annual precipitation varies between 300 and 700 mm in the mountainous upstream part and the temperatures are very high in summer (37.7 °C maximum average) and low in winter (4.9 °C minimum average). More than 60% of its land use is used for cereal crops. It is characterized by poor vegetation which varies according to the altitude and the nature of the land and the irrigated area of Haouz Haouz Center.

3 Data and Methods

The remote sensing data used are the 10-day NDVI, the 8-day LST product of MODIS MOD11A2, the monthly evapotranspiration anomaly of the MODIS MOD16A product, the cumulative precipitation over 10 days of the CHIRPS with their spatial resolutions of 250 m, 1000 m, 500 m and 5 km respectively. These data series covering the period 2003–2021 were resampled at a spatial resolution of 250 m. Soil moisture was estimated by considering the difference between the LST day and night values. Each of the variables was transformed into a biophysical indicator through normalization based on the maximum and minimum values of each pixel over the study period. In addition to these remote sensing data, performance anomaly data and data from six weather stations were used as auxiliary data for statistical validation of the approach. Rainfall data from stations were acquired at https://www.infoclimat.fr/climatologie/ and yield anomalies data were based on a World Bank database of major cereal crops (barley, durum, and soft wheat).

After data processing, the improved composite model for drought monitoring (CDMIa_RF) developed by Houmma et al. (2022) was used to evaluate the dynamics of drought conditions in the two watersheds (Eq. 1). After assessment of the state of the historical dynamics of drought given by CDMIa_RF, to assess the multivariate drought risk level, the approach of Enenkel et al. (2016) was used to assess the drought risk level using the composite index (ECDI) in Ethiopia. Thus, the drought risk alert level based on the CDMIa_RF was calculated as shown in Eq. 2. Multivariate risk was classified into four intensity levels for comparative mapping between the two watersheds and statistical analyses. After this step, an R code was used to assess the magnitude of the trend per pixel over the last two decades (2004–2022). The pixels of significant trends were classified as follows: from (−2.5 to −1.8) very significant trends; (−1.8 to 1.6) significant trends and (−1.6 to −1.4) slightly significant trends.

$$\mathrm{CDMIa_{RF}} = 0.36 \times \mathrm{PCI} + 0.16 \times \mathrm{TCI} + 0.12 \times \mathrm{VCI} + 0.17 \times \mathrm{ETCI} + 0.2 \times \mathrm{ISMA} \quad (1)$$

$$\mathrm{MRL} = \frac{\text{Seasonal values of the CDMIa\_RF for the year (i)} - \text{Average values of the series}}{\text{Standard deviation values}} \quad (2)$$

The standardized precipitation index (SPI) at the seven-month scale of the agricultural season (October to April) was used to validate the reliability of the developed approach. Precipitation data (2003 to 2021) from the three stations per watershed were used to calculate the SPI (Eq. 3). The stations of Marrackech, Safi and Essaouria were considered for the Tensift watershed and for the Moulouya watershed the stations of Taza, Oujda and Nador were used.

$$\mathrm{SPI} = \frac{P(i) - \mathrm{Pm}}{\sigma P} \quad (3)$$

where p: Total precipitation of a period (mm); Pm: Historical mean precipitation of the period (mm) and σP: Historical standard deviation of precipitation of the period (mm).

4 Results

4.1 Spatio-Temporal Analysis of Multivariate Risk Similarities

The comparative mapping of five levels of multivariate risk of agricultural drought (Fig. 2) shows high accordance in the spatio-temporal evolution of drought risk in the two watersheds. This accordance is observed in its temporal occurrence, the intensity of the multivariate risk level, and its spatial magnitude. The level of extreme to moderate risk is dominant at the expense of normal conditions (zero risk). Over the period 2004 to 2022, the exceptional and generalized normal conditions of 2009 remained atypical.

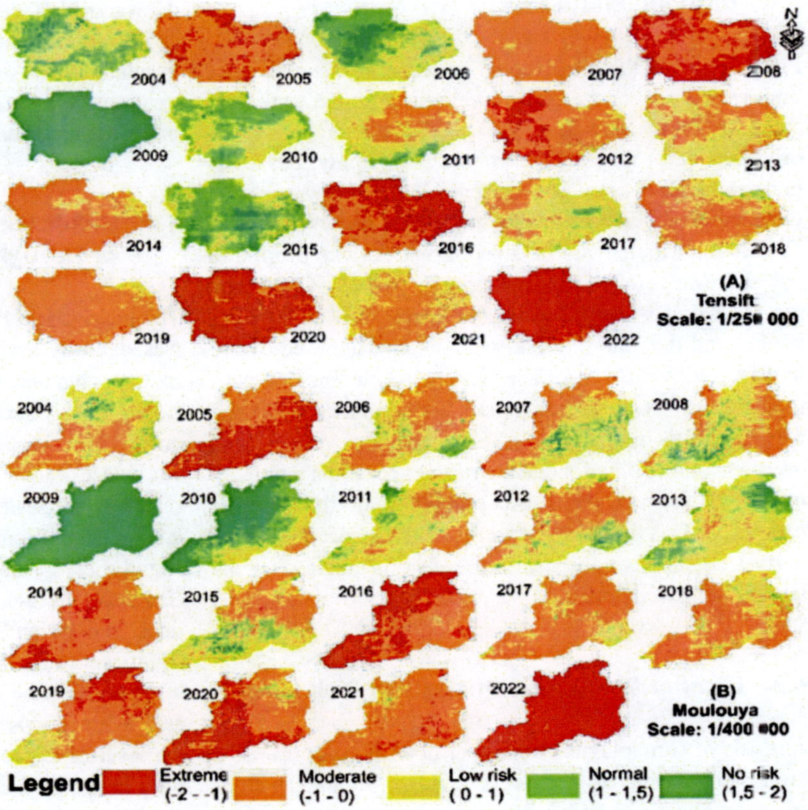

Fig. 2 Comparative mapping of the level of multivariate risk from 2004 to 2022

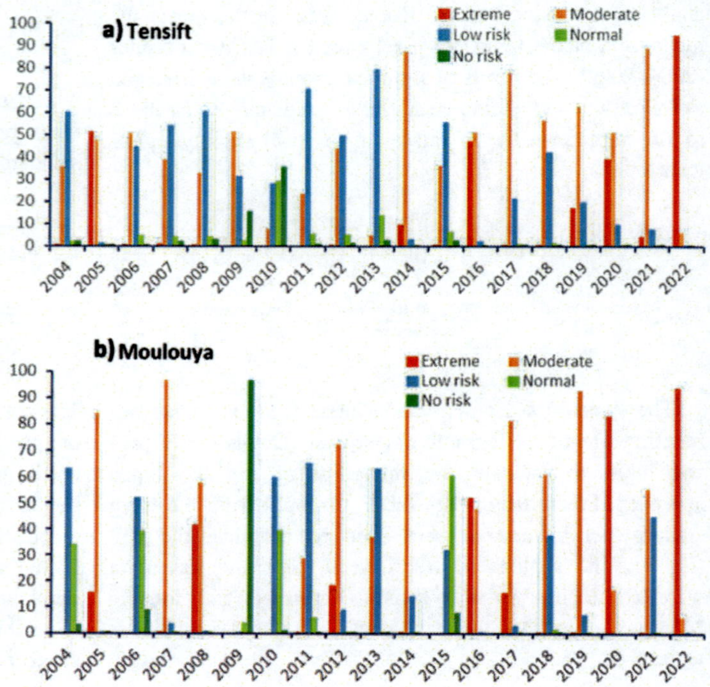

Fig. 3 Proportion in percentage of the distribution of the chronological occurrence of five multivariate risk level classes from 2004 to 2022, **a** Tensift and **b** Moulouya

4.2 Comparative Histograms of the Level of Multivariate Risk

Figure 3 shows the proportional variation of five classes of multivariate risk of the agricultural season at the watershed scale of Tensift and Moulouya. The multivariate risk of drought reflects very high interannual variability, but it has great similarity I terms of occurrence in the two watersheds. High annual differences in extreme risk are observed in both Tensift and Moulouya. The level of risk can vary from 0 to 100%. It was zero for the exceptionally wet year (2009) and generalized at 94% in Tensift and 93.6% in Moulouya for the hot and dry year of 2022 which is distinguished by the highest multivariate risk. The seasonal occurrence of multivariate risk is 6/19 (extreme risk), 7/19 (moderate risk) and 6/19 (low to normal risk). However, the proportional occurrence of low and moderate risk is very dominant at 12 out of 19 years. Overall, the proportion of the level of extreme risk is increasing sharply and the occurrence of normal conditions translates into a clear decrease over the last decade of the series.

4.3 Comparative Trend Analysis of Risk Level

Analysis of chronological trends per pixel (Fig. 4) reveals a highly significant magnitude of changes in the level of multivariate risk in both watersheds. In Tensift and Moulouya, the intensity of the change in multivariate risk is mainly driven by seasonal vegetation cover. In a large part of the watersheds, the trend of change in multivariate risk is not significant. But the most significant trend is mainly observed in rainfed areas such as the province of Taza. The temporal trend in the proportion of pixels of extreme risk level illustrates the magnitude of the change.

4.4 Validation of Multivariate Modeling of Drought Risk Level

The multivariate risk of drought derived from five variables was statically evaluated using the standardized precipitation index recommended by the World Meteorological Organization as a reference index. The results in Fig. 5 show positive and very high correlations between the Min, Max and Mean values of multivariate risk and the SPI at the seven-month scale of the agricultural season. These correlations range from 0.78 to 0.86 in the Tensif watershed and slightly less in the Moulouya watershed from 0.68 to 0.74. However, the p-value test suggests that the statistical relationship between SPI and multivariate risk level is statically significant (0.0017 to 0.0000) in both watersheds. Similarly, when considering cereal yield anomalies on a national scale as a reference variable, recent spatio-temporal dynamics of the level of extreme risk show significant negative correlations (− 0.60 in Tensift and − 0.62 in Moulouya) with cereal yield anomalies. These correlations may be higher by considering only the agricultural yield data at the scale of the

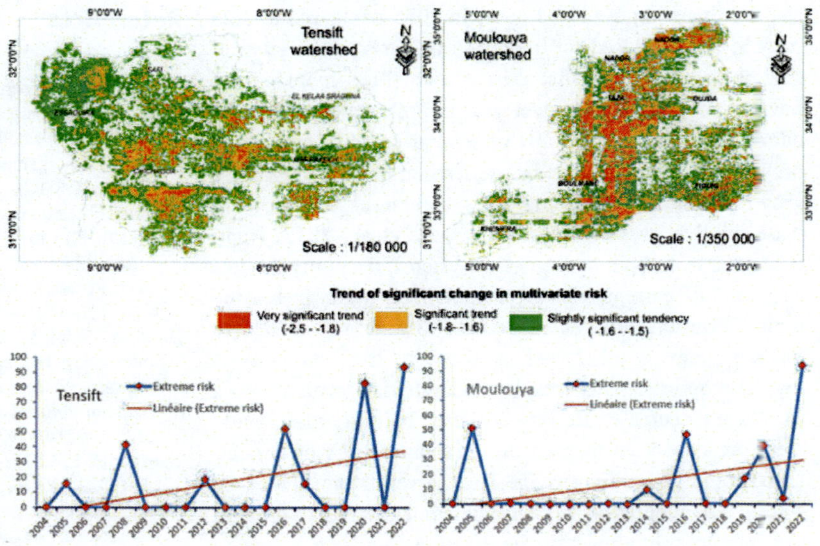

Fig. 4 Temporal variability of extreme risk level and trend magnitude per pixel

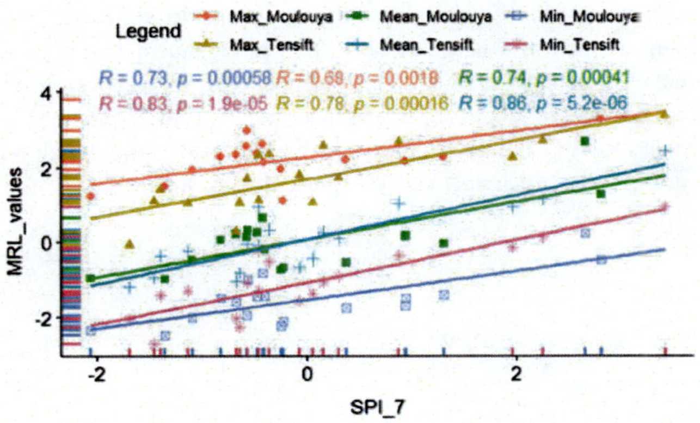

Fig. 5 Statistical relationship between SPI (October to April) and multivariate drought risk level

two watersheds, as this would reduce the imprecision that may affect the yield data as a carrier of the impact of climate stress.

5 Discussion and Conclusions

In this chapter, agricultural drought risk was assessed by adopting a multifactorial approach based on open-source remote sensing data. Given the multifactorial dimension of drought, the approach would be more suitable for spatialized drought risk assessment than using hydroclimatic indices based on in situ measurements. Unlike approaches based on in situ measurements, it allows a spatially continuous assessment of agricultural drought risk at different intensity levels. The drought risk warning level is provided in five multivariate risk levels and was statistically assessed by the SPI index for the growing season. In both watersheds, it was found to be strongly correlated with the SPI index with a maximum correlation of 0.86 on the Tensift scale and a very significant p-value ($0.0000 < 0.05$). Spatio-temporal analysis of the similarities of the multivariate risk established reveals a very large interannual variability in the level of risk of agricultural drought but a high accordance is highlighted between the two watersheds. Three categories of years can be distinguished according to the spatial magnitude of the level of risk: years when extreme multivariate risk is generalized, intermediate years of moderate risk, and years when extreme risk was almost zero. In terms of occurrence, we can also distinguish an alternation of years of extreme peaks as well as a succession of years of normal conditions. The differences between years are such that the level of multivariate risk can increase from 0 to 94% from one year to the next. In the Tensift area, the year 2009, which was distinguished by the lowest level of risk, the flow of rivers was about four times higher compared to 2008. Similarly, in 2015, which was characterized by a low risk, the annual rainfall total (256 mm) was three times greater than the rainfall of 2016 (73 mm) which clearly has a higher risk level than in 2008. The difference in terms of river flow was also 10 times larger (11.04 m^3/s) in 2015 compared to 1.25 m^3/s in 2016). Based on frequency analysis of the level of risk, the study shows that there is two times more risk of having a dry year than a wet year in both watersheds. This corroborates with the results of several other studies based on the analysis of climate data series that have highlighted a growing trend of increasingly dry conditions and a downward trend in wet conditions (Fniguire et al., 2017; Hadri et al., 2021). In both hydrological systems, the chronological evolution of extreme risk is characterized by an increasing trend. Occurrence of generalized multivariate risk is 9/19 or on average every other year. At the same time, the chronological proportion of normal to moderate risk levels shows a significant decrease. Based on these findings, one should therefore expect an increase in multivariate risk of agricultural drought in coming years. Thus, given the results of the study, the integration of multivariate agricultural drought risk into drought impact prediction modeling is an avenue to explore in future research. Nevertheless, in this analysis, lagged effects of variables were not considered. Recent studies such as Khatun et al. (2022) have shown that the delayed impacts of rainfall and temperature on vegetation condition are about four months and 5.2 to 3 months respectively. Future studies on multivariate agricultural drought risk assessment may explore this avenue.

References

Enenkel, M., Steiner, C., Mistelbauer, T., Dorigo, W., Wagner, W., See, L., Atzberger, C., Schneider, S., & Rogenhofer, E. (2016). A combined satellite-derived drought indicator to support humanitarian aid organizations. *Remote Sensing, 8*, 340. https://doi.org/10.3390/rs8040340

Fniguire, F., Laftouhi, N. E., Saidi, M. E., Zamrane, Z., El Himer, H., & Khalil, N. (2017). Spatial and temporal analysis of the drought vulnerability and risks over eight decades in a semi-arid region (Tensift basin: Morocco). *Theoretical and Applied Climatology, 130*, 321–330.

Hadri, A., Saidi, M. E. M., & Boudhar, A. (2021). Multiscale drought monitoring and comparison using remote sensing in a Mediterranean arid region: A case study from west-central Morocco. *Arabian Journal of Geoscience, 14*, 118. https://doi.org/10.1007/s12517-021-06493-w

Hanadé, H. I., El Mansouri, L., Hadria, R., Emran, A., & Chehbouni, A. (2022). Retrospective analysis and version improvement of the satellite-based drought composite index: A semi-arid Tensift-Morocco application. *Geocarto International, 37*, 3069–3090.

Khatun, M. M., Chakraborty, D., & Ifterkharul, A. L. A. M. (2022). Clarifying the impact of climatic parameters on vegetation in Moulvibazar district. *Turkish Journal of Engineering, 6*, 211–222.

Nam, W., Tadesse, T., Wardlow, B. D., Hayes, M. J., Svoboda, M. D., Hong, E., Pachepsky, Y. A., & Min-Jang, W. (2018). Developing the vegetation drought response index for South Korea (VegDRI-SKorea) to assess the vegetation condition during drought events. *International Journal of Remote Sensing, 39*, 1548–1574.

Topography-Based Estimation of Evapotranspiration at High Altitudes in Semi-arid Regions

Badreddine Sebbar, Olivier Merlin, Saïd Khabba, Vincent Simonneaux, Marine Bouchet, and Abdelghani Chehbouni

Abstract

Assessing the surface water balance of mountains is a real challenge given notably the extreme variability of meteorological conditions and the sparsity of in-situ monitoring. While mountains are recognized as water towers feeding the surrounding plains, there is only unconsolidated knowledge about the individual water balance components especially the evapotranspiration (ET). Satellites land surface temperature (LST) along with air temperature (Ta) and incoming solar radiation (Rg) can be used to assess the energy budget and provide a reasonable estimation of instantaneous ET. Nevertheless, over mountains, the Ta and Rg, respectively, undergo strong topographical changes due to elevation and sun exposure effects. Moreover, upscaling the instantaneous ET to its daily value is expected to be uncertain in mountains as the evaporative fraction (EF, defined as the ratio of ET to available energy ratio) of a given pixel can no longer be considered constant during daytime until proven otherwise. In this context, this contribution focuses on a topography-based estimation of ET using the two-source energy balance (TSEB) model. We also examine the variability of hourly and daily EF estimates through both satellite and in-situ monitoring. An eddy covariance tower was installed at 3850 m.a.s.l over the High Atlas Mountains in central Morocco and has been operating since September 2020 to present. The 30 m resolution LST is derived from thermal data collected by Landsat-7, 8, and 9 on clear sky days. Rg is estimated at the Landsat (30 m) resolution from the SRTM's digital elevation model (DEM) and two different topography-based approaches: a physically based model (DART) and a simplified semi-empirical model. The 9 km resolution ERA5-Land's air temperature product is spatialized at the same (30 m) resolution by applying the environmental lapse rate (ELR) retrieved at the Landsat overpass time over a 9 km^2 area including the eddy covariance tower. Satellite-derived estimates of ET and EF are compared to instantaneous station measurements for three and nine dates in 2020 and 2021, respectively. The variability during daytime of the in-situ EF is also assessed to evaluate the potential for upscaling instantaneous remotely sensed ET to a daily scale.

Keywords

Evapotranspiration • Topography • Evaporative fraction • Remote sensing • Energy balance

B. Sebbar (✉) · S. Khabba · A. Chehbouni
Center for Remote Sensing Applications, Mohammed VI Polytechnic University, Benguerir, Morocco
e-mail: Badreddine.SEBBAR@um6p.ma
Badreddine.SEBBAR@univ-tlse3.fr

B. Sebbar · O. Merlin · V. Simonneaux · M. Bouchet · A. Chehbouni
CESBIO, Université de Toulouse, CNES/CNRS/INRAE/IRD/UT3-Paul Sabatier, 18, Avenue Edouard Belin, 31401, Toulouse, France

S. Khabba
LMFE, Physics Department, Faculty of Sciences Semlalia, Cadi Ayyad University, Marrakech, Morocco

A. Chehbouni
International Water Research Institute (IWRI), Mohammed VI Polytechnic University (UM6P), Benguerir, Morocco

1 Introduction

The estimation of evapotranspiration (ET) in mountainous areas using temperature-based models that rely on remotely sensed land surface temperature (LST) is challenging due to the influence of topographical effects. The elevation-dependent air temperature (Ta) and incoming shortwave radiation (Rg) (including from adjacent surfaces) make it difficult to spatially distribute the necessary inputs. To overcome this challenge, a new methodological approach is proposed in this study. This approach considers these effects prior to estimating ET in mountainous regions.

M. Bezzeghoud et al. (eds.), *Recent Research on Geotechnical Engineering, Remote Sensing, Geophysics and Earthquake Seismology*, Advances in Science, Technology & Innovation,
https://doi.org/10.1007/978-3-031-48715-6_30

2 Materials and Methods

The study area is the Rheraya sub-basin located in the High Atlas Mountains in central Morocco. The area is marked by a semi-arid climate, and its elevation varies from about 1000 to 4127 m. The high-altitude regions are characterized by low temperatures, rugged terrain, and sparse vegetation cover.

Malbeteau et al. (Malbéteau et al., 2017) normalized LST for topographical effects. Dynamic environmental lapse rate (ELR) was physically inverted through an energy balance model, whereas Rg was spatialized based on the DART model (Gastellu-Etchegorry et al., 1996). Ta was then spatialized following the expression:

$$T_{\text{pixel}} = T_{\text{station}} + \text{ELR}\left(E_{\text{pixel}} - E_{\text{station}}\right) \quad (1)$$

We adapted a similar methodological approach through replacing in-situ measurements with ERA5-Land meteorological data. Rg was estimated using an alternative approach based on direct radiation (Samani et al., 2007) that was calibrated on smaller sub-zone using the DART model (which is potentially very precise but difficult to apply over large areas).

Using these meteorological forcing data that has been corrected for topography, along with LST observations, ET is estimated as a residual of the two-source energy balance (TSEB) model. The latter calculates the energy balance of the soil-canopy-atmosphere continuum, where transpiration is initially determined by the Priestley–Taylor equation (Kustas & Norman, 2000).

3 Results

3.1 Ta and Rg Spatialization

Figure 1 depicts scatterplots of simulated Rg and Ta compared to in-situ observations using the Samani et al. (2007) and ELR methods, respectively. Resulting net radiation (Rn as a function of Rg, Ta, and LST) is also compared. The RMSE values for Rg, Ta, and Rn are 49.8 W/m², 1.9°C, and 43.3 W/m² respectively. Corresponding R^2 values are: 0.80 for Rg, 0.85 for Ta, and 0.76 for Rn.

3.2 EF Daily Variation

Figure 2 compares the model-based instantaneous EF estimate at the satellite overpass to hourly and daily measurements. The satellite's midday overpass gives an interesting configuration; the modelled/measured instantaneous EF values are fairly close to the daily observed EF. The highest recorded bias is approximately 5%, which corresponds to the August 7, 2021, when the observed EF around sunset rises abruptly due to instrumentation faults.

3.3 ET Estimation

Taking topographic effects into account (Fig. 3) results in an RMSE of 27.08 W/m^2 and an R2 of 0.81 for the simulated ET indicating a strong model performance in capturing ET dynamics. For Rn, these values are 42.73 W/m^2 and 0.67; and for the sensible heat flux (*H*)56.43 W/m^2 and 0.79 demonstrating a reasonably high degree of model fidelity. Although the conductive flux (G), approximated as a percentage of Rn, shows a lower performance with an RMSE of 73.26 W/m² and an R^2 of 0.12, its small magnitude suggests that the overall impact on the model's accuracy is still acceptable.

4 Discussion

The results of our study indicate that the topographically corrected Rg, Ta, and Rn used in the TSEB-PT model are highly effective in predicting energy fluxes, with values that closely align with observations. Our analysis also revealed that the instantaneous EF could potentially be used to extrapolate instantaneous to daily ET values, particularly when the satellite overpass occurs around midday. These findings are in extention consistent with previous studies that have demonstrated the utility of the EF in estimating daily energy fluxes over flat regions (e.g., Crago, 1996; Hoedjes et al., 2008). Furthermore, the present study highlights the significant influence of topography on modeled fluxes, as evidenced by strong correlations between modeled and measured fluxes when meteorological forcing are topographically

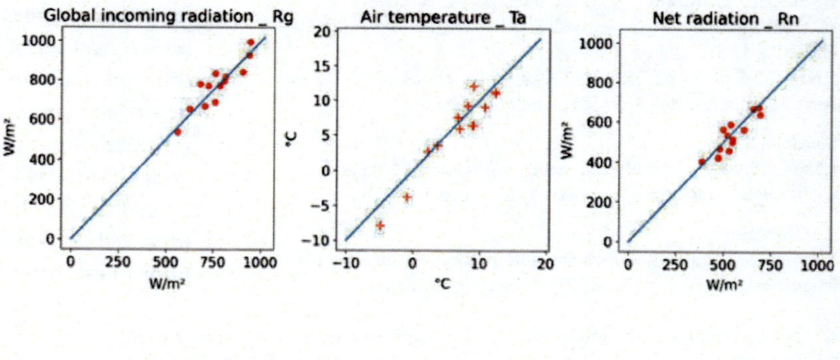

Fig. 1 Scatterplots of measured versus modeled Rg (left), Ta (middle), and Rn (right).

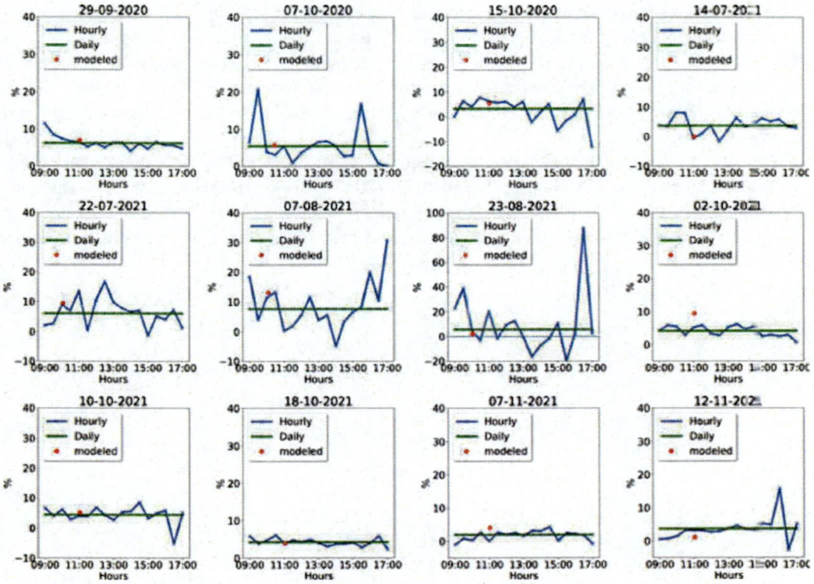

Fig. 2 Instantaneous estimated EF(red dot) versus daily/hourly (green line/blue line) in-situ measured EF

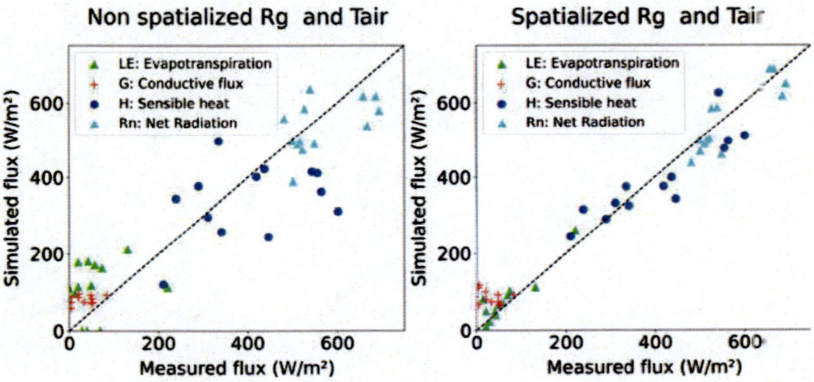

Fig. 3 Comparison of topographically corrected and non-corrected estimates of energy balance fluxes with eddy-covariance observations

corrected. These results are consistent with previous studies that have shown the importance of incorporating topographic information in energy flux estimation models (e.g., Hao et al., 2021; Rana et al., 2007). Overall, this study contributes to the growing body of literature on energy flux estimation in terrestrial ecosystems and supports the use of the TSEB-PT model in accurately predicting energy fluxes in various meteorological and topographical conditions.

5 Conclusions

The estimation of ET over mountainous terrain is the primary emphasis of this work. The removal of topographic influences is essential for achieving an accurate estimate of the spatial variation in ET values. When topographic effects are considered, the TSEB-PT model exhibits satisfactory performance. As a perspective, our aim is to estimate ET by applying a more straightforward contextual model, such as the LST-VI model while considering topographic normalization of LST.

References

Crago, R. D. (1996). Conservation and variability of the evaporative fraction during the day-time. *Journal of Hydrology, 180*(1–4), 173–194.

Gastellu-Etchegorry, J. P., Demarez, V., Pinel, V., & Zagolski, F. (1996). Modeling radiative transfer in heterogeneous 3-D vegetation canopies. *Remote Sensing of Environment, 58*(2), 131–156.

Hao, D., Bisht, G., Gu, Y., Lee, W. L., Liou, K. N., & Leung, L. R. (2021). A parameterization of sub-grid topographical effects on solar radiation in the E3SM Land Model (version 1.0): Implementation and evaluation over the Tibetan Plateau. *Geoscientific Model Development, 14*(10), 6273–6289.

Hoedjes, J. C. B., Chehbouni, A., Jacob, F., Ezzahar, J., & Boulet, G. (2008). Deriving daily evapotranspiration from remotely sensed instantaneous evaporative fraction over olive or-chard in semi-arid Morocco. *Journal of Hydrology, 354*(1–4), 53–64.

Kustas, W. P., & Norman, J. M. (2000). A two-source energy balance approach using directional radiometric temperature observations for sparse canopy covered surfaces. *Agronomy Journal, 92*(5), 847–854.

Malbéteau, Y., Merlin, O., Gascoin, S., Gastellu, J. P., Mattar, C., Olivera-Guerra, L., Khabba, S., & Jarlan, L. (2017). Normalizing land surface temperature data for elevation and illumination effects in mountainous areas: A case study using ASTER data over a steep-sided valley in Morocco. *Remote Sensing of Environment, 189*, 25–39.

Rana, G., Ferrara, R. M., Martinelli, N., Personnic, P., & Cellier, P. (2007). Estimating energy fluxes from sloping crops using standard agrometeorological measurements and topography. *Agricultural and Forest Meteorology, 146*(3–4), 116–133.

Samani, Z., Bawazir, A. S., Bleiweiss, M., Skaggs, R., & Tran, V. D. (2007). Estimating daily net radiation over vegetation canopy through remote sensing and climatic data. *Journal of Irrigation and Drainage Engineering, 133*(4), 291–297.

Evaluating the Accuracy of iPhone Lidar Sensor for Building Façades Conservation

Sahar F. Abbas and Fanar M. Abed

Abstract

Apple company recently started to include lidar sensor in iPhone 12, 13 Pro, and iPad Pro system devices. Lidar was introduced by Apple to improve camera focusing and support augmented reality (AR) applications. However, Apple lidar attracted app developers and users to start scan indoor and outdoor environments within the specified range configurations. Although Apple has not revealed the type of the 3D scanner used, some scholars believe Apple's lidar is based on direct time of flight (DToF) lidar. It is a mobile handheld scanner that provides greater flexibility and control when scanning close objects, and therefore, more affordable and much cheaper than TLS devices in very close ranges. This study aims to assess the positional accuracy of the new lidar sensor in the iPhone 12 Pro Max in outdoor environment, particularly for scanning building façades. Sitescape app was used for scanning objects using this device where it is primarily concerned with extracting 3D point clouds following specific scanning pattern. The study is carried on the façade of a selected building, where a number of artificial targets are affixed and observed using three different laser-based remote sensing techniques (total station, iPhone lidar, and TLS). The research study is divided into two stages where total station measurements used as reference ground-truth dataset. The first stage compares mobile DToF to static TLS sensor in terms of precision. The data are collected in the same range and environmental conditions to analyze the level of accuracy. The result showed that the accuracy of the iPhone lidar and TLS are comparable. The RMSE of TLS data was 3.44 mm, while the iPhone lidar delivered 4.89 mm. In the second stage, the sensor accuracy was tested over various ranges starting from 0.25 to 5 m to show range dependency on the positional accuracy. The accuracy shows convergent results; however, errors increased slightly as range increased. The data acquisition time was generally short and directly proportional to the scene size. Results are later analyzed and studied for future conservation applications.

Keywords

iPhone lidar • DToF • Mobile laser scanning • Accuracy assessment • Augmented reality • Façades conservation

1 Introduction

At the end of 2020, Apple included a lidar sensor in the iPad Pro and iPhone 12 Pro, which is mainly utilized for augmented reality (AR) applications as well as photographic purposes such as improving low-light auto-focusing shots. It is a more affordable sensor and has flexible and speed usage when scanning close objects. Apple's lidar considers to be a DTOF sensor based on a single-photon avalanche diode (SPAD) and a pulsed laser source (Murtiyoso et al., 2021; Spreafico et al., 2021; Tontini et al., 2020).The potential to integrate a lidar sensor in a tablet or smartphone appears promising for rapid surveying as the sensor can quickly capture reliable 3D point clouds sufficiently for 1:200 scale architectural mapping (Spreafico et al., 2021). The Apple lidar sensors are verified to be identical for close-range applications (Luetzenburg et al., 2021). It can scan the topography of small and medium-sized landmasses quickly by delivering fine spatial resolution, and replacing traditional topographic land surveying methods like TLS and SfM-MVS photogrammetry (Luetzenburg et al., 2021). The objective of this paper is to validate the accuracy of the new

S. F. Abbas (✉) · F. M. Abed
College of Engineering, University of Baghdad, Baghdad, Iraq
e-mail: sahar.abbas1512m@coeng.uobaghdad.edu.iq

F. M. Abed
e-mail: fanar.mansour@coeng.uobaghdad.edu.iq

F. M. Abed
University of Exeter, Exeter, UK

M. Bezzeghoud et al. (eds.), *Recent Research on Geotechnical Engineering, Remote Sensing, Geophysics and Earthquake Seismology*, Advances in Science, Technology & Innovation,
https://doi.org/10.1007/978-3-031-48715-6_31

Apple lidar sensor in iPhone 12 Pro Max, as well as assessing the range dependency on the spatial accuracy between the minimum and maximum scan distances when scanning a typical building facade.

2 Materials and Methods

In this research, the accuracy of the iPhone 12 Pro Max lidar sensor was evaluated through two main tests, which were carried out on a wall designed for this experiment. A bunch of artificial targets with (8 * 8 cm) size was pinned to the wall (Fig. 1a) and observed with the total station device (Topcon GM-50) to use its data as a reference to analyze mobile laser scanner (MLS) accuracy in two individual tests. Both tests were carried out using the Sitescape app on the iPhone, which generates a 3D point cloud of the scanned object (https://www.sitescape.ai/) for further processing, by using high-quality scanning mode (Max detail and high point density). In the first test, the iPhone 12 Pro Max lidar was compared to terrestrial laser scanning (Stonex X300) to assess the accuracy of the MLS in comparison with the TLS (Fig. 1b), which is accurate and widely used.

The adopted scanning pattern of MLS in both tests was walking in a straight line from right to left while holding the iPhone on a mini portable tripod to maintain the same height and angle during the scanning. This is important to avoid these two parameters influencing the scan quality. For the initial test, observation with MLS and TLS was accomplished in the same range, which was 4.60 m away from the wall to be scanned, and under the same weather conditions. However, the objective of the second test was to verify the accuracy and the level of details (LOD) of the MLS sensor in different ranges (0.25, 0.5, 1.25, 2, 3, 4 and 5 m), where 5 m is the maximum range of the lidar sensor.

3 Result and Discussion

Following cloud compare data analysis, TLS data found to deliver 3.44 mm RMSE, whereas MLS delivers 4.89 mm in the first test. Figure 2 illustrates a comparison of the scanning results using TLS and MLS sensors.

The study's findings reveal that the iPhone MLS sensor scans the building's facade with excellent coverage in comparison with TLS; however, the TRMSE for individual sensors is comparable. This indicates that in close range, the precision of this innovative MLS sensor is nearly equal to TLS. In contrary, the accuracy results of range dependency test (second test) analysis are revealed in Table 1.

It can be shown from Table 1 that the errors in MLS measures increased as the distance increases, and as illustrated in Fig. 3.

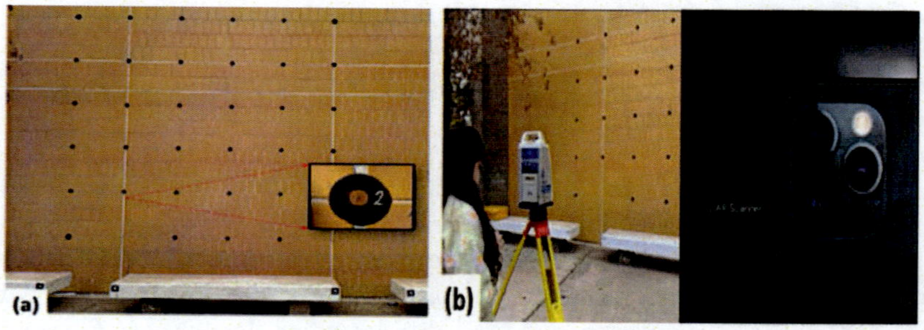

Fig. 1 Scanning the artificial targets of the selected building façade. **a** Artificial targets and the selected building façade. **b** TLS device (Stonex X300) and iPhone 12 Pro Max lidar

Fig. 2 Scan pattern results: **a** MLS **b** TLS

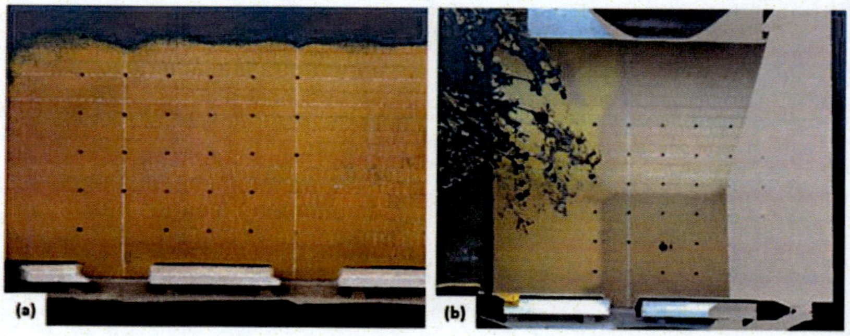

Table 1 TRMSE of the iPhone MLS in various ranges

Range	0.25 m	0.5 m	1.25 m	2 m	3 m	4 m	5 m
Number of clouds	7,390,336	9,149,659	11,569,899	8,389,714	9,096,701	8,107,558	5,372,031
TRMSE (mm)	2.02922	2.66485	2.96118	3.74224	4.06383	4.34152	5.22048

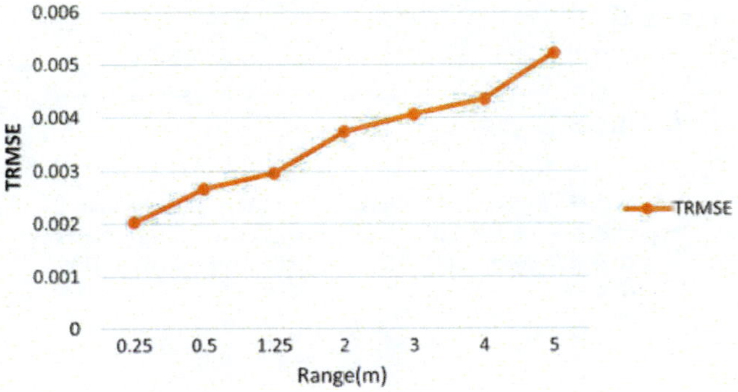

Fig. 3 RMSE in iPhone Lidar (MLS) sensor measures in multiple ranges

An additional scan was done at a distance of 6.16 m to verify if the tracking of this sensor can extend over 5 m. It only scanned the concrete benches and a portion of the floor, as shown in Fig. 4, and the total number of points has decreased to 1,305,712 points.

The scans were analyzed and also found to be subjected to roughness level within the range of 0.25 m where the STD was 0.382 mm in 0.25 m range compared to 0.962 mm in 5 m range. The findings of this analysis show that the amount of noise grows with distance, and the level of detail is higher in the very close range; however, the coverage of the building facade in the 5 m range is greater than 0.25 m, see Fig. 5a. Automatic noise filtering was applied and shown improvements in STD values to be 0.16 and 0.103 mm for 0.25 and 5 m range, respectively. Some noticeable changes in the LOD were also targeted after noise filtering, see Fig. 5b.

4 Conclusions

The primary objective of this study is to evaluate the performance of the new lidar iPhone12 MLS sensor for building facades conservation applications. To do so, it is better to use a proper scanning pattern with a steady motion. In this

Fig. 4 Tracking of the iPhone lidar sensor at range of 6.16 m

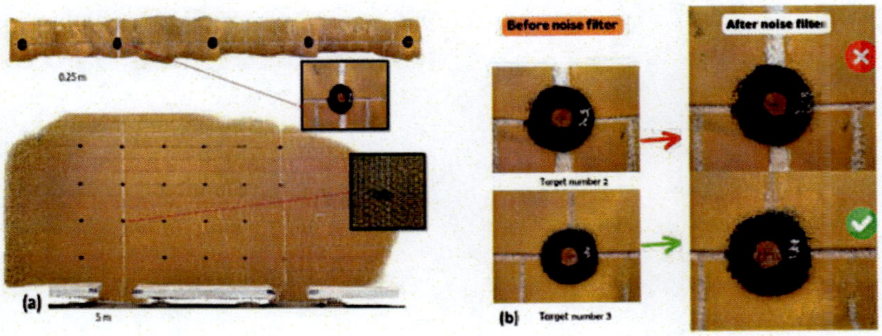

Fig. 5 LOD Analysis after noise removal: **a** LOD @ 0.25 and 5 m ranges. **b** The noise filter impact on targets

study, the iPhone sensor provided millimeter accuracy. The scanning time with iPhone lidar was generally short and only takes a few minutes. Therefore, the study findings revealed the promised results of this sensor in conserving building's facades in close-range environment. It is vital to note that all findings are based on using the Sitescape app with the highest quality setting; other apps may produce different outcomes.

References

Luetzenburg, G., Kroon, A., & Bjørk, A. A. (2021). Evaluation of the Apple iPhone 12 Pro LiDAR for an application in geosciences. *Science and Reports, 11*(1), 22221. https://doi.org/10.1038/s41598-021-01763-9

Murtiyoso, A., Grussenmeyer, P., Landes, T., & Macher, H. (2021). First assessments into the use of commercial-grade solid state lidar for low cost heritage documentation. In *XXIV ISPRS Congress (2021 Edition), 5–9 Juillet 2021, Nice (En Ligne)* (p. 43).

Spreafico, A., Chiabrando, F., Teppati Losè, L., & Giulio Tonolo, F. (2021). The IPAD pro built-in Lidar sensor: 3D rapid mapping tests and quality assessment. *The International Archives of the Photogrammetry, Remote Sensing and Spatial Information Sciences, XLIII-B1-2021*, 63–69. https://doi.org/10.5194/isprs-archives-xliii-b1-2021-63-2021

Sitescape Homepage. https://www.sitescape.ai/. Last accessed 16 May 2022.

Tontini, A., Gasparini, L., & Perenzoni, M. (2020). Numerical model of spad-based direct time-of-flight flash lidar CMOS image sensors. *Sensors (Switzerland), 20*(18), 1–19. https://doi.org/10.3390/s20185203

Flood Inundation Mapping Using Earth Observation Data in the Po River (North of Italy)

Meriam Lahsaini and Meriame Mohajane

Abstract

Floods have caused severe consequences on the environment in different areas of the world and also in Italy. Therefore, this study explores the opportunity of combining the Earth Observation data (both radar and optical images) for flood mapping in the Po River, located in the north part of Italy. Based on a multi-temporal approach, synthetic aperture radar Sentinel 1 images between August and November 2016 were used in the period. The methodology adopted is semi-automatic method using GEE to detect a flooded area. We selected images before and after flood event using VH band flood detection. It was concluded that the use of synthetic aperture radar (SAR) data produces results with low accuracy especially in building areas due to the spectral signatures of urban features. To deal with this, we used an optical image (Landsat 8 OLI) based on automatic threshold using Normalized Difference Vegetation Index (NDVI) and Normalized Difference Water Index (NDWI) for change detection to accentuate changes likely caused by flooding. The results were calibrated by hydraulic model using HEC-RAS software. Thus, the approach developed in this study could be a crucial step toward adequate flood risk mitigation and management strategies.

Keywords

Earth observation (EO) • Flood • SAR image • Landsat 8 OLI • HEC-RAS

M. Lahsaini (✉)
CNR-IMAA, Potenza, Italy
e-mail: meriam.lahsaini@usmba.ac.ma

M. Mohajane
CNR-ITC, Napoli, Italy

1 Introduction

The flood event caused lot of losses in life and economic scale, and it is difficult to manage the flood risk with missing field data. However, this task could be done based on monitoring and prediction results generated by the use of spatial data collected from Earth Observation (EO) data (Çubukçu et al., 2023; Gaagai et al., 2022). In addition, an indirect assessment of the interval flood events in a poorly gauged basin was tried by combining hydraulic modeling and remote sensing. Based on the synthetic aperture radar (SAR) image classification, it is possible to provide valuable information for the monitoring flood events that represent the hazard. Unfortunately, it does not represent well the flood details in the urban area, because the SAR data produces a result with low accuracy in densely populated areas due to the increased amount of backscattering from houses and other objects (Franceschetti et al., 2002; Hennig et al., 2007). For this reason, the methodology developed here is based on the combination with an optical image (Landsat 8 OLI) based on automatic threshold using Normalized Difference Vegetation Index (NDVI) and Normalized Difference Water Index (NDWI) for change detection to accentuate changes likely caused by flooding. Then, the HEC-RAS model was used for simulating flooded area and calibrate the results.

2 Data Collection and Methods

The methodology flowed in this work based on the approach of thresholding for water extraction coupling the flood detection method using two types of images, SAR images (Franceschetti et al., 2002; Hennig et al., 2007) useful for flood mapping since it covering the study area in descending orbit have been obtained. Level-1 ground-range detected (GRD) products have been acquired and Landsat 8 OLI. In order to evaluate this approach, the results were compared to

M. Bezzeghoud et al. (eds.), *Recent Research on Geotechnical Engineering, Remote Sensing, Geophysics and Earthquake Seismology*, Advances in Science, Technology & Innovation,
https://doi.org/10.1007/978-3-031-48715-6_32

Table 1 Location of the in situ data along the Po River (AIPO, Agenzia Interregionale Fiume Po)

Name	Lat	Long	Elevation	Average width	Average flow velocity	Average discharge
	(°)	(°)	(m)	(km)	(m/s)	(m3/s)
Piacenza	45.06	9.70	42.37	106	0.61	933
Cremona	45.13	9.99	29.03	157	0.82	1075
Borgoforte	45.05	10.75	14.05	164	0.84	1313
Sermide	45.02	11.29	9.50	329	0.44	1378
PonteLagoscuro	44.89	11.61	3.48	175	3.01	1477

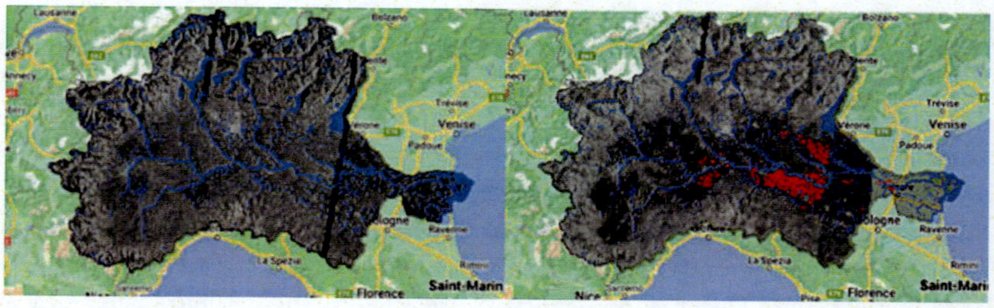

Fig. 1 Extracted flood-prone areas using synthetic aperture radar (SAR) images

those of hydraulic-based flood mapping using HEC-RAS (version 6) software. We used annual maximum flows (AMFs) recorded by COSs for calculating discharge (Table 1).

3 Results

All SAR images downloaded were georeferenced and stacked to perform thresholding process in Google earth engine (Cloud platform). To calibrate the model, we adopted an automatic algorithm in SAR based on flood mapping; we have used multi-temporal Sentinel 1 images to monitor severe and dynamic flood event in the north of Italy, in November 2016. The flooded area was detected by thresholding and classification algorithm using Google Earth Engine (GEE) platform. The objective of this method is to evaluate the affected zone by flood risk and compared the results with hydraulic modeling to validate our simulation.

4 Discussion

Based on the thresholding method, detecting water area using Sentinel 1 is to obtain accurate results. Figure 1 shows the post-flooded area and pre-flooded area. In SAR images, the low backscattering value shown in red color means that the pixel is covered by water, where we have a lower VH intensities due to specular reflection typical of smooth surfaces.

Permanent water surfaces like the lakes in the northern part of the study area constantly have low backscattering values, even in the pre-flood image. Their backscattering values are similar to the flooded pixels. These areas were masked out to minimize the challenges caused by the backscattering similarities of these areas and flooded areas.

We validated the accuracy of thresholding method applied in Po basin, and we calculated the feature statistical agreement F1 and F2 values of backscattering threshold values, using validated pixels selected on the high-resolution (Bates & Roo, 2000; Lim & Brandt, 2019; Mason et al., 2009).

5 Conclusion

SAR images from Sentinel 1 GRD and Landsat OLI 8 are used and applied thresholding algorithm for deriving flood extent using GEE, to calibrate and validate the HEC-RAS model. The overall accuracies were higher than 90%, confirming the applicability of the approach adopted. Thus, the approach developed in this study could be a crucial step toward adequate flood risk mitigation and management strategies.

References

Bates, P. D., & De Roo, A. P. J. (2000). A Simple raster-based model for flood inundation simulation. *Journal of Hydrology, 236*, 54–77. https://doi.org/10.1016/S0022-1694(00)00278-X

Çubukçu, E. A., Demir, V., & Sevimli, M. F. (2023). Modeling of annual maximum flows with geographic data components and artificial neural networks. *International Journal of Engineering and Geosciences, 8*(2), 200–211. https://doi.org/10.26833/ijeg.1125412

Franceschetti, G., Iodice, A., & Riccio, D. (2002). A canonical problem in electromagnetic backscattering from buildings. *IEEE Transactions on Geoscience and Remote Sensing, 40*, 1787–1801. https://doi.org/10.1109/TGRS.2002.802459

Gaagai, A., Aouissi, H. A., Krauklis, A. E., et al. (2022). Modeling and risk analysis of dam-break flooding in a semi-arid montane watershed: A case study of the Yabous Dam, northeastern Algeria. *Water, 14*(5), 767. https://doi.org/10.3390/w14050767

Hennig, T. A., Kretsch, J. L., Pessagno, C. J., Salamonowicz, P. H., & Stein, W. L. (2007). The shuttle radar topography mission. *Lecture Notes in Computer Science, 2181*, 65–77. https://doi.org/10.1007/3-540-44818-7_11

Lim, N. J., & Brandt, S. A. (2019). Are feature agreement statistics alone sufficient to validate modelled flood extent quality? A study on three Swedish Rivers using different digital elevation model resolutions. *Mathematical Problems in Engineering, 2019*, 9816098. https://doi.org/10.1155/2019/9816098

Mason, D. C., Bates, P. D., & Dall' Amico, J. T. (2009). Calibration of uncertain flood inundation models using remotely sensed water levels. *Journal of Hydrology, 368*, 224–236. https://doi.org/10.1016/j.jhydrol.2009.02.034

Effects of Urbanization on Urban Ecosystem Services (UESS)—A Framework

Deeksha Nayak, Anoop Kumar Shukla, and Nandineni Rama Devi

Abstract

Ecosystem services are a complex relationship between humans and nature. Interactions among these ecosystems are self-sustainable if they are unaltered. But, due to rapid urban expansion, changes in land use land cover (LULC), and anthropogenic activities, these ecosystem services (ES) are exerted under tremendous pressure. In the Mangaluru urban agglomeration, a coastal region of Karnataka, this study examined changes in land use and land cover (LULC). In order to categorize the spatiotemporal changes in LULC over the course of five decades, 1980, 1990, 2000, 2010, and 2022, the study used the object-based image analysis (OBIA) technique. Deriving knowledge about various LULC classifications is the primary goal of this work. The study area is divided into five categories using the OBIA technique: built-up area, water body, forest area, agricultural land, and barren land. The categorization accuracy is assessed using images from Google Earth, SoI Topo maps, and on-the-ground confirmation. We discover that between 1980 and 2022, the area used for agriculture has decreased, the area used for forests has increased, the area used for buildings has dramatically increased, while the area used for water bodies and arid land has primarily remained the same. The results of the accuracy evaluation demonstrate that the LULC variations discussed in this work are legitimately correct and applicable to subsequent uses. In order to expand the city in the future, decision-makers may find the study's findings helpful in determining the best course of action. This study also demonstrates GIS and remote sensing (RS) in LULC applications, particularly in the coastal regions.

Keywords

Urban ecosystem services • LULC • Urbanization • GIS • Remote sensing

1 Introduction

Due to rapid urbanization, ecosystem services are deteriorating because anthropogenic interventions impact biochemical cycles and earth's energy equilibrium, causing global warming and climate changes (MEA, 2005; Shukla et al., 2020). However, studying the state of the ecosystem at the regional level is a critical issue in corroborating sustainable development. There is an insufficient knowledge repository on the factors that impact ecosystem services caused by rapid urbanization. Establishing the relationship among the ecosystem services and to understand the trade-offs and synergies among these services is a tedious challenge. Other problems associated with the insufficient knowledge of ecosystem services at the regional/local level are seen in the decision-making of land use/cover (LULC). The regional processes are interlinked with global events contributing to sustainable global parameters; therefore, they must be assessed and maintained.

Urbanization has occurred in the present due to population growth and the country's economic development. As a result, the LULC change has a variety of effects on ecosystem services and intensifies natural hazards due to hydrology. One of the main benefits of using Earth Observation satellite data is the ability to detect these spatiotemporal changes in LULC (Apostolopoulos & Nikolakopoulos, 2021; Dhanaraj & Angadi, 2022; Mihi et al., 2019; Shukla et al., 2018). Due to societal factors like politics, economics, culture, and other factors, the land cover is constantly changing. The analysis of landscape elements, the detection of development patterns, the prevention of destructive processes, and the measurement of natural and man-made

D. Nayak · A. K. Shukla (✉) · N. R. Devi
Manipal School of Architecture and Planning,
Manipal Academy of Higher Education, Manipal, India
e-mail: anoop.shukla@manipal.edu

M. Bezzeghoud et al. (eds.), *Recent Research on Geotechnical Engineering, Remote Sensing, Geophysics and Earthquake Seismology*, Advances in Science, Technology & Innovation,
https://doi.org/10.1007/978-3-031-48715-6_33

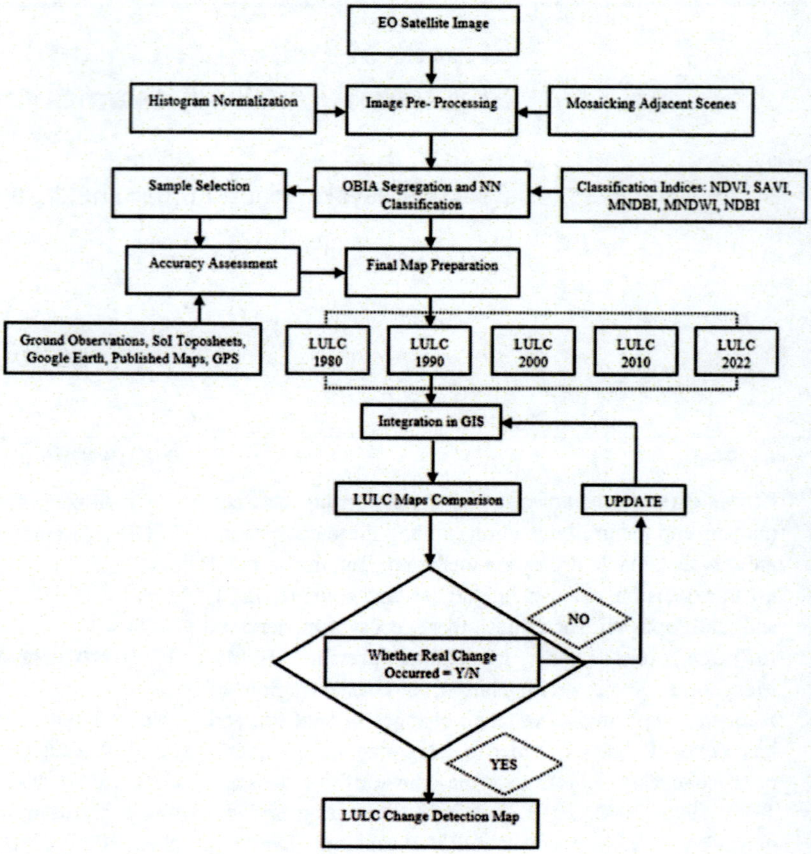

Fig. 1 Framework the dynamics of land use land cover changes

damage to the landscape can all be done by keeping track of long-term land cover changes. The tier II city of Mangaluru is located on the western coast of India. The city spans 569 km^2 in total. Geographic limitations constrain urban growth on the north and south sides. Understanding the pattern of LULC change at a regional scale is critical due to the current rapid urbanization (Dhanaraj & Angadi, 2021; Mohajane et al., 2018). The innovative aspect of this study is the method suggested for evaluating changes to the LULC for Mangaluru city. This study aims to use multitemporal satellite images using OBIA classification approach (Mahmoud et al., 2022; Tabib Mahmoudi & Hosseini, 2021) to examine the LULC changes in Mangaluru city, Karnataka, India.

2 Methodology

Remote sensing and GIS technologies are used to extract the images of LULC. The maps were created using satellite imageries from United States Geological Survey Earth Explorer. Image of 1980 was collected from Landsat 5 (MSS), 1990 from Landsat 5 (TM), 2000 from Landsat 7 (ETM+), 2010 from Landsat 7 (ETM+), and 2022 from Landsat 9 (OLI/TRS), respectively. The methodology implemented for LULC classification and change detection is showed in Fig. 1. The initial step involved in procuring the decadal images from satellite form 1980–2022 and processed. Initial image processing was done with the help of ERDAS Imagine image software, where image pre-processing mosaicking is performed, followed by mapping done using ArcGIS. Finally, the LULC classification was created with the help of eCOgnition software.

3 Results and Discussion

Multi-temporal remotely sensed images were used to classify LULC accurately. The classified LULC maps of the study area are shown in Fig. 2. The study shows how the LULC classifications (agricultural land, forest area, water body, built-up area, and barren land) have changed geographically between 1980, 1990, 2000, 2010, and 2022. According to the LULC percentage changes of these classes over the timeline given above, agricultural land decreased by 38%, while forest areas increased by 27%, water bodies decreased by 1%, and there was a significant increase in built-up area by 23% wasteland also reduced by 11%. In

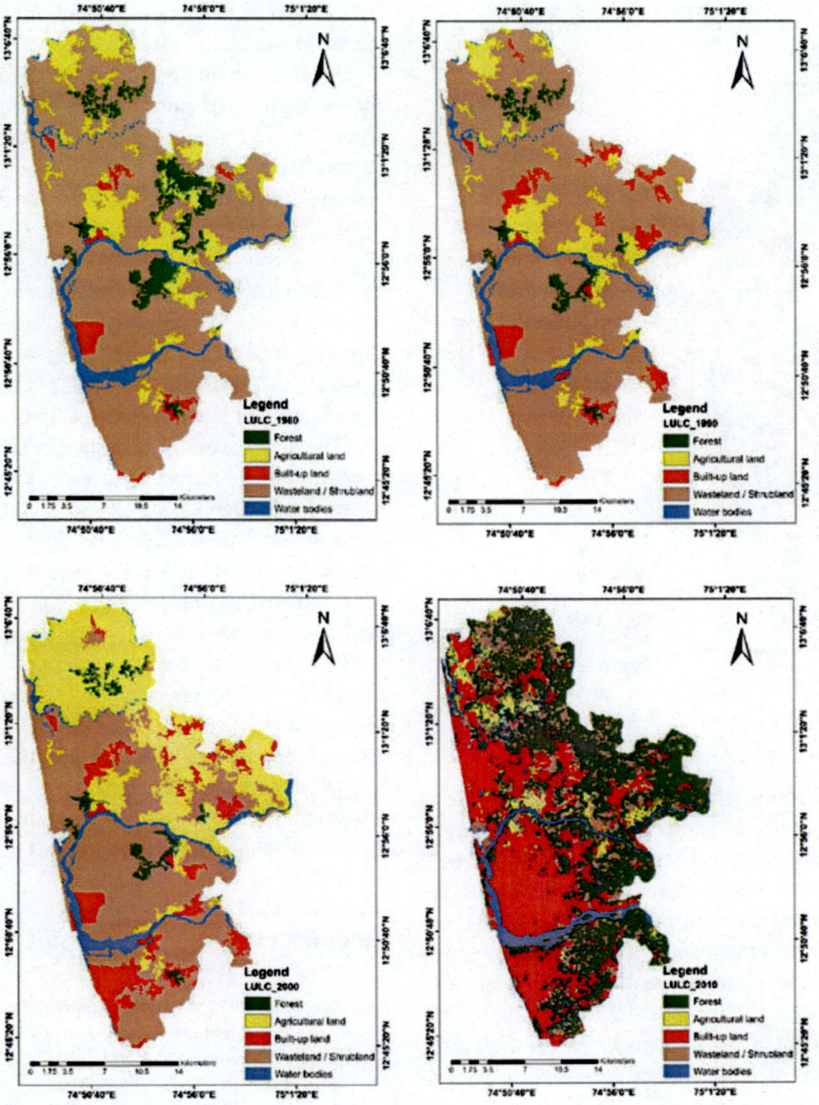

Fig. 2 Classified LULC map of the years 1980, 1990, 2000, 2010, and 2022

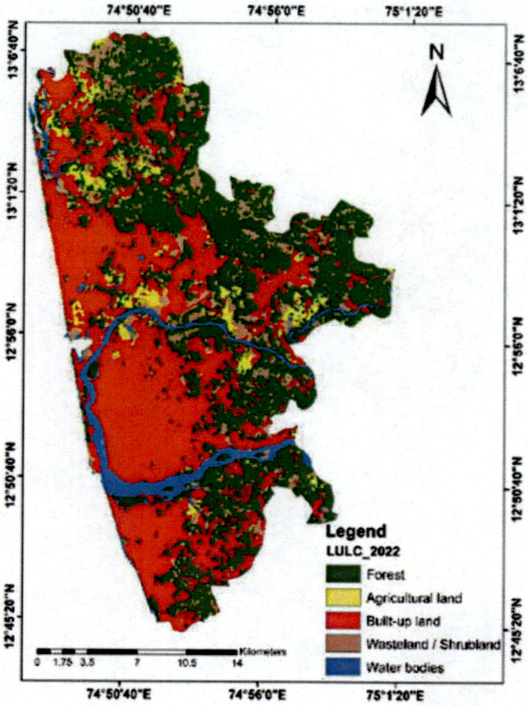

Fig. 2 (continued)

thematic mapping from remotely sensed data, the term accuracy is used typically to express the degree of correctness of a map of classification (Arar & Chenchouni, 2012; Foody, 2002). The confusion matrix is currently at the mostly used accuracy assessment method, which includes a simple cross-tabulation of the mapped class label against that observed in the ground or reference data for a sample of cases at specified locations (Congalton & Green, 2008). It is difficult to carry out accuracy assessment for all the LULC maps because of lack of ground truth data. Hence, we utilized the 2002 image for ground truth verification and accuracy assessment of the LULC analysis. Confusion matrix for 2022 LULC map was prepared by ground truthing at several location on the field as well as from secondary sources. The images from Google Earth tool, which has temporal information is used as secondary data source for carrying out accuracy assessment. For the 2022 LULC, a total of 517 known location points corresponding to image objects were randomly generated and verified with reference data. The results showed an overall accuracy of 92.65% and kappa index of agreement of 0.91. In terms of producer's accuracy, all classes were over 90%, while in terms of user's accuracy, all classes were over 90% except forest class. The minimum threshold of the overall classification accuracy for a LULC map should be 85% (Anderson, 1976). A good overall classification accuracy of 92.65% and kappa coefficient (K) of 0.91 is obtained for LULC map of the year 2022, which is in accordance with the Anderson's minimum threshold for LULC classification.

The study is limited to Mangaluru urban agglomeration, addressing the gap in the relative effect of LULC changes on urban ES. The result points toward the emphasis on conducting a regional study on ES changes, which could give a better picture of damages that occurred than the national scale.

4 Conclusions

Utilizing data from Landsat 5, 7, 8, and 9 from the years 1980, 1990, 2000, 2010, and 2022, a thorough OBIA LULC classification of the Mangaluru City agglomeration is carried out. The land coverings identified in the study region include agricultural land, forest area, water body, built-up area, and barren land. The LULC classes show significant changes between 1980 and 2022. The built-up area has increased between 1980 and 2022. The results of the accuracy evaluation demonstrate the validity and applicability of the LULC variations described in this study. The outcome of this study can help the urban designers/planners and policymakers in undertaking adequate steps to enhance and conserve the ecological relationships of the coastal city of Mangaluru and take appropriate decisions in future LULC planning. It also helps us understand the percentage of contribution by specific ecosystem services to global warming and climatic changes that can be considered critical issues and address the same.

References

Anderson, J. R. (1976). *A land use and land cover classification system for use with remote sensor data* (Vol. 964). US Government Printing Office. https://doi.org/10.3133/pp964

Apostolopoulos, D., & Nikolakopoulos, K. (2021). A review and meta-analysis of remote sensing data, GIS methods, materials and indices used for monitoring the coastline evolution over the last twenty years. *European Journal of Remote Sensing, 54*(1), 240–265. https://doi.org/10.1080/22797254.2021.1904293

Arar, A., & Chenchouni, H. (2012). How could geomatics promote our knowledge for environmental management in Eastern Algeria? *Journal of Environmental Science and Technology, 5*(5), 291–305. https://doi.org/10.3923/jest.2012.291.305

Congalton, R. G., & Green, K. (2008). *Assessing the accuracy of remotely sensed data: Principles and practices*. CRC Press.

Foody, G. M. (2002). Status of land cover classification accuracy assessment. *Remote Sensing of Environment, 80*(1), 185–201. https://doi.org/10.1016/S0034-4257(01)00295-4

Dhanaraj, K., & Angadi, D. P. (2021). Urban expansion quantification from remote sensing data for sustainable land-use planning in Mangaluru, India. *Remote Sensing Applications: Society and Environment, 23*. https://doi.org/10.1016/j.rsase.2021.100602

Dhanaraj, K., & Angadi, D. P. (2022). Land use land cover mapping and monitoring urban growth using remote sensing and GIS techniques in Mangaluru, India. *Geojournal, 87*(2), 1133–1159. https://doi.org/10.1007/s10708-020-10302-4

Mahmoud, A. S., Mezaal, M. R., Hameed, M. R., & Naje, A. S. (2022). A framework for improving urban land cover using object and pixel-based techniques via remotely sensed data. *Nature Environment and Pollution Technology, 21*(5), 2189–2200. https://doi.org/10.46488/NEPT.2022.v21i05.013

MEA. (2005). *Ecosystems and human well-being: Synthesis*, Island Press.

Mihi, A., Nacer, T., & Chenchouni, H. (2019). Monitoring dynamics of date palm plantations from 1984 to 2013 using landsat time-series in sahara desert oases of Algeria. In El-Askary, H. M., et al. (Eds.), *Advances in remote sensing and geo informatics applications* (pp. 225–228). Springer. https://doi.org/10.1007/978-3-030-01440-7_52

Mohajane, M., Essahlaoui, A., Oudija, F., el Hafyani, M., el Hmaidi, A., el Ouali, A., Randazzo, G., & Teodoro, A. C. (2018). Land use/land cover (LULC) using landsat data series (MSS, TM, ETM+ and OLI in azrou forest, in the central middle atlas of Morocco. *Environments, 5*(12), 1–16. https://doi.org/10.3390/environments5120131

Shukla, A. K., Ojha, C. S. P., Garg, R. D., Shukla, S., & Pal, L. (2020). Influence of spatial urbanization on hydrological components of the Upper Ganga River Basin, India. *Journal of Hazardous, Toxic, and Radioactive Waste, 24*(4), 04020028. https://doi.org/10.1061/(ASCE)HZ.2153-5515.0000508

Shukla, A. K., Ojha, C. S. P., Mijic, A., Buytaert, W., Pathak, S., Garg, R. D., & Shukla, S. (2018). Population growth, land use and land cover transformations, and water quality nexus in the Upper Ganga River basin. *Hydrology and Earth System Sciences, 22*(9), 4745–4770. https://doi.org/10.5194/hess-22-4745-2018

Tabib Mahmoudi, F., & Hosseini, S. (2021). Three-dimensional building change detection using object-based image analysis (case study: Tehran). *Applied Geomatics, 13*(3), 325–332. https://doi.org/10.1007/s12518-020-00349-w

An Approach for Salinity Recovery Using in Situ and Satellite Observations on the Example of the Sea of Azov

Tatyana Shulga and Vyacheslav Suslin

Abstract

The work proposes a salinity recovery method for the Sea of Azov based on implementation of a general regression compiled from archival in situ data and regional biooptical parameters obtained from standard MODIS L2 products. The observational data from the open Internet services were acquired directly from data providers. The authors' procedures for quality control and merging were implemented for these data. We researched the following biooptical parameters: $a_{ph}(678)$ is the absorption coefficient by phytoplankton at 678 nm, *Tchl* is the sum concentration of chlorophyll-*a* and pheophetin-*a*, $a_{tot}(438)$ is the total absorption coefficient by all optically active components at 438 nm, $a_{CDM}(438)$ is the absorption coefficient by colored detrital matter at 438 nm, and $b_{bp}(438)$ is the particulate backscattering coefficient at 438 nm. We chose these variables because they are the operational satellite ocean color products of the MODIS (NASA). Each measurement of satellite data with 1 km spatial resolution processed to spatial maps of five biooptical parameters on a regular grid of the Sea of Azov. Based on the linear regressions satisfying the condition $R \geq 0.5$, general equations of the following form were compiled $y = (a_{aver} \pm \sigma_1) \cdot x + (b_{aver} \pm \sigma_2)$, where a_{aver} and b_{aver}—averaged linear coefficients *a* и *b*, σ_1 and σ_2—standard deviations, *x*—regional biooptical products, *y*—salinity (‰). The results of the study showed the possibility of using different approaches to building generalized empirical regressions for the spring and summer. The result is merged regressions for spring and summer designed for reconstruction of salinity and obtaining the data sets for spatiotemporal variability analysis of salinity. Average values of salinity recovered indirectly using our proposed method are within the 95% confidence bands for the long-term average seasonal trends for periods 1986–2018 and 2000–2018 from in situ data. The main result is merged regressions for spring and summer seasons designed for reconstruction of salinity and obtaining the data sets for spatiotemporal variability analysis of salinity in the Sea of Azov. In addition, the values of salinity reconstructed from $a_{CDM}(438)$ are found to reflect its changes most realistically within the observed salinity range (1–18‰). These results allow us to use in the future reconstructed salinity datasets in the assimilation procedures of the 3D hydrodynamic model and for retrospective salinity recovery.

Keywords

Sea of Azov • Salinity • Ocean-color data • Biooptical regional products • Regression analysis

1 Introduction

The most continental, shallow, and almost closed Sea of Azov is characterized by significant variability of the main hydrometeorological parameters associated with the seasonal volume of river flow, the inflow of salt water, weather conditions, and anthropogenic pollution. Unlike temperature, obtaining regular information on salinity from remote sensing data is possible only indirectly, for example, by establishing empirical regression relationships with biooptical parameters (Glukhovets & Goldin, 2018). At the same time, comparison of in situ and salinity data obtained from off-the-shelf satellite tools (e.g., SNAP, https://smap.jpl.nasa.gov/) revealed weak data correlation. This is due to the fact that the Sea of Azov is one of the smallest and freshest in the world, which leads to low sensitivity of satellite

T. Shulga (✉) · V. Suslin
Marine Hydrophysical Institute of RAS, Sevastopol, 299011, Russia
e-mail: shulgaty@mail.ru

T. Shulga
State Oceanographic Institute N.N. Zubova, Roshydromet, Sevastopol, 299011, Russia

M. Bezzeghoud et al. (eds.), *Recent Research on Geotechnical Engineering, Remote Sensing, Geophysics and Earthquake Seismology*, Advances in Science, Technology & Innovation,
https://doi.org/10.1007/978-3-031-48715-6_34

measurements. In this work, the approach to the restoration of the salinity fields of the Sea of Azov is based on obtaining generalized regression equations relating in situ archival data with regional biooptical parameters obtained using standard products of the second-level MODIS (oceancolor. gsfc.nasa.gov).

2 Materials and Methods

At the first stage of this study, based on the systemized and analysis of global set of RV's in situ data for 1913–2018 (atlas.ssc-ras.ru/azs/azs-invent.html), conclusions were drawn about the long-term seasonal variability of temperature and salinity (Nacef et al., 2016). For the purpose of use at the next stages of the work, these data were distributed in arrays the average long-term values of temperature and salinity for three sub-regions of the Sea of Azov: sea surface, middle, and deep water. At the second stage, the data cover the recent period of MODIS satellite ocean-color data between 2002 and 2018 which were analyzed and systematized. Salinity data are obtained at the third stage of this study based on the proposed method of salinity recovery from the ocean-color data regional products proposed in Suslin and Churilova (2016). We research the following biooptical parameters: $a_{ph}(678)$, $TChl$, $a_{tot}(438)$, $a_{CDM}(438)$, and $b_{bp}(438)$. The listed variables were chosen as they are the operational satellite ocean-color products of the MODIS: R_{rs} is the remote sensing reflectance. The each set of satellite data with kilometer spatial resolution was converted into spatial maps of all biooptical parameters on a regular grid of the Sea of Azov with a resolution of 1/59° × 1/84°. At the final stage, the results of salinity recovery were verified using the established generalized empirical (regression) dependencies. For this purpose, the average values of the restored salinity were compared with the average long-term trends in the salinity of the Sea of Azov according to in situ data for 1986–2018 and 2000–2018.

3 Results

3.1 Generalized Regression for the Spring–Summer Seasons in the Sea of Azov

For each of the fives biooptical parameters for the Sea of Azov in the spring and summer seasons, regression dependencies were obtained that connect them with salinity in situ data using the OriginLab (originlab.com). Table 1 presents generalized linear regressions equations obtained by averaging the coefficients of single linear regressions in 2000–2018 in the presence of linear relationships with correlation coefficients > 0.5 (Fig. 1).

3.2 Results of Salinity Recovery in the Sea of Azov

The established general (averaged) equations of linear regressions, given in Table 1, were used in the procedure for restoring salinity fields in the Sea of Azov. As an example, Fig. 2 shows the salinity fields reconstructed from the biooptical parameter $a_{CDM}(438)$ for June 24, 2009. As can be seen in Fig. 2, in general, the salinity fields reconstructed by various biooptical parameters are of a common nature, but there are also differences related to areas with higher and lower salinity values.

3.3 Verification of Salinity Recovery in the Sea of Azov

A comparison of the results of the restored salinity values was carried out for the modern periods of 1986–2018 and 2000–2018 (see Fig. 3), the choice of which is due to the availability of sounding data since 2000. The comparison is carried out of the Sea of Azov in the spring and summer seasons. For this analysis, two arrays for the summer and

Table 1 Generalized linear regression equations

	Spring	Summer
$a_{ph}(678)$ (m^{-1})	y = (0.038 ± 0.008)x + (− 0.334 ± 0.132)	y = (0.041 ± 0.024)x + (− 0.355 ± 0.250)
$Tchl$ (mg m^3)	y = (2.442 ± 0.110)x + (− 20.940 ± 2.594)	y = (1.894 ± 1.008)x + (− 16.500 ± 11.004)
$a_{tot}(438)$ (m^{-1})	Nonlinear relations	y = (− 0.192 ± 0.078)x + (3.654 ± 0.515)
$a_{CDM}(438)$ (m^{-1})	y = (− 0.287 ± 0.303)x + (4.710 ± 3.816)	y = (− 0.261 ± 0.071)x + (4.217 ± 0.548)
$b_{bp}(438)$ (m^{-1})	y = (− 0.082 ± 0.029)x + (1.006 ± 0.334)	y = (− 0.043 ± 0.029)x + (0.561 ± 0.296)

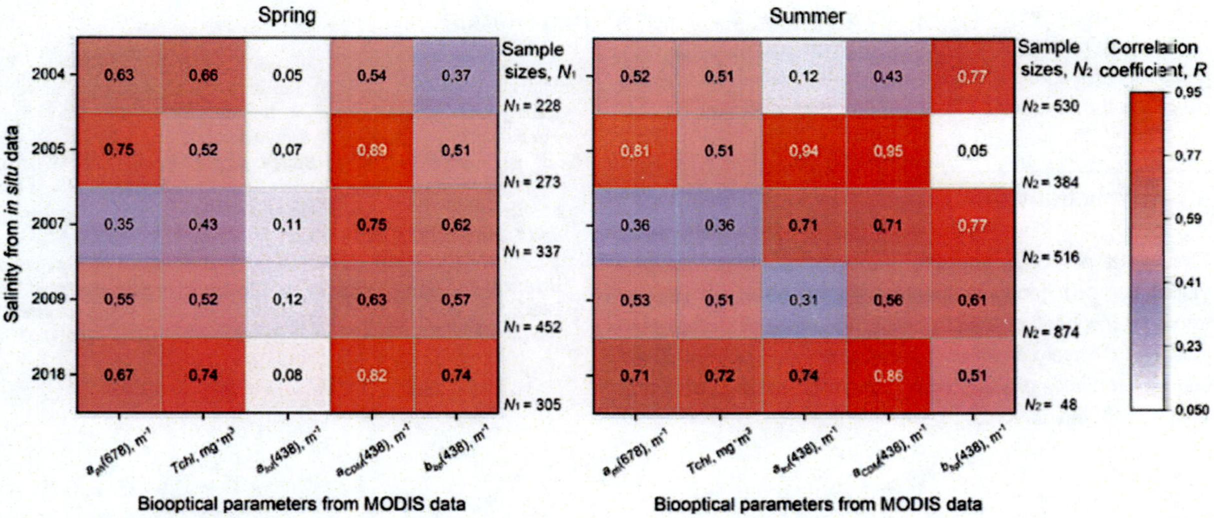

Fig. 1 Correlation coefficients of the linear regressions with corresponding sample sizes

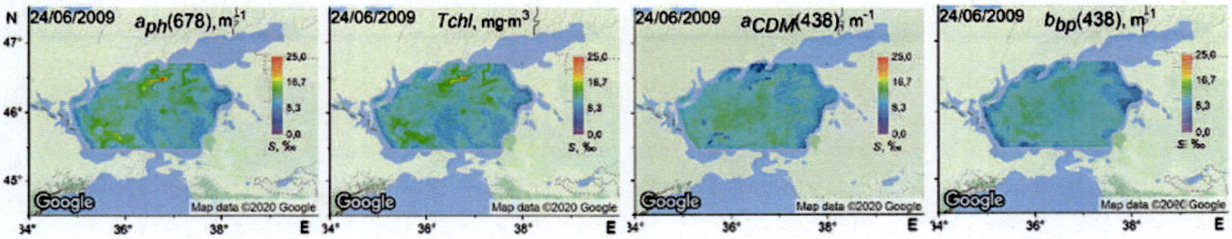

Fig. 2 Results of recovering salinity fields of the Sea of Azov

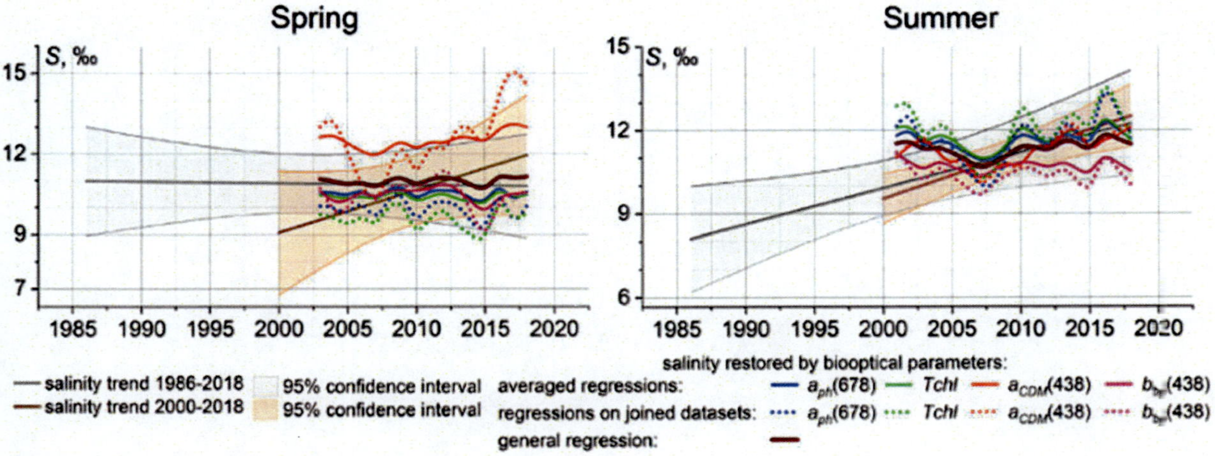

Fig. 3 Summer and spring salinity trends for the periods 1986–2018 and 2000–2018

spring seasons were formed from in situ datasets and regional MODIS products. These data are use in the salinity recovery procedure for each of the biooptical parameters based on the equations (Table 1) and then averaged.

4 Conclusions

The result of comparing the reconstructed mean salinity values showed good agreement with the observed salinity trends. It has been established that the graphs of the average values of the restored salinity are in the region of 95% of the confidence bands of the modern long-term average trends in 1986–2018 and 2000–2018.

References

Glukhovets, D., & Yu, G. (2018). Research of the relationship between salinity and yellow substance fluorescence in the Kara Sea. *Fundamental and Applied Hydrophysics, 11*(3), 34–39.

Nacef, L., Bachari, N. E. İ, Bouda, A., & Boubnia, R. (2016). Variability and decadal evolution of temperature and salinity in the Mediterranean Sea surface. *International Journal of Engineering and Geosciences, 1*(1), 24–33.

Suslin, V., & Churilova, T. (2016). A regional algorithm for separating light absorption by chlorophyll-a and coloured detrital matter in the Black Sea, using 480–560 nm bands from ocean colour scanners. *International Journal of Remote Sensing, 37*(18), 4380–4400.

A New Approach for Sand and Dust Storm Monitoring Using Satellite Data

Mohsin Jamil Butt, Mazen Ebraheem Assiri, and Essam Mohammed Alghamdi

Abstract

The aim of the current study is to monitor sand and dust storm events, which are one of the key environmental apprehensions over the Kingdom of Saudi Arabia, using MODIS satellite data. A new approach, Saudi Dust Detection Index (SDDI), is presented based on spectral characteristics of sand and dust features to monitor SDS events. The index-based results are validated by using MODIS combined dark target (DT) and deep blue (DB) aerosol optical depth (AOD) product and AOD product from AERONET station data. The accuracy of the SDDI is determined by using probability of false-positive detection (POFD), probability of correct positive detection (POCD), and algorithm accuracy approach. Results of the study show that SDDI can identify SDS events based on various threshold values (between 0 and 0.5) with high accuracy. This research is at a preliminary stage, and the future objectives are further improvements and validation of the developed index.

Keywords

SDDI • MODIS • SDS • POFD • POCD

1 Introduction

Sand and dust storm (SDS) plays an important role in the environment of the earth through the loading of aerosols in the atmosphere. Sand and dust uplifted in the air can form SDS events and thereby increasing the concentration of aerosols in the atmosphere which have severe climatological affects (Choobari et al., 2014). Many studies have been conducted related with SDS events in the past in different parts of the world (Butt & Mashat, 2018; Butt et al., 2017). However, there is a need to strengthen the SDS monitoring network system by using modern techniques and technologies. The data collection from ground-based stations are time consuming and spatially limited which gave the pave to use satellite-based data due to many advantages. Researchers have been using satellite data for the monitoring of SDS events in different parts of the world (Albugami et al., 2018; Bibi et al., 2015; Huang et al., 2007; Kaskaoutis, 2012; Kaskaoutis et al., 2007; Labban & Butt, 2021; Qu et al., 2006). At present, the most commonly used sources of satellite data for SDS monitoring are from Spinning Enhanced Visible and Infrared Imager onboard Meteosat Second Generation, from NOAA's Advance Very High Resolution Radiometer, and from Moderate Resolution Imaging Spectroradiometer onboard Terra and Aqua satellite systems. MODIS due to high spectral resolution and moderate temporal and spatial resolution is the main source of data in the current study. Studies have also been conducted to develop indices for the monitoring of SDS events using MODIS data in different parts of the world (Karimi et al., 2012; Qu et al., 2006; Samadi et al., 2014).

In the KSA, severe SDS events are very frequent and devastating (human and animal diseases, traffic accident and disruption in flights due to reduced visibility, and major infrastructure and economic losses), and therefore, it is very important to monitor SDS events in a near real time.

2 Materials and Methods

A near real-time detection approach with reasonable accuracy is the need of the hour to mitigate hazards associated with SDS events. In the current research, Saudi Dust Detection Index (SDDI) based on spectral signatures of sand and dust in MODIS bands 3, 4, and 7 (Eq. 1) is applied.

M. J. Butt (✉) · M. E. Assiri · E. M. Alghamdi
Department of Meteorology, King Abdulaziz University, Jeddah 21589, Saudi Arabia
e-mail: mbutt@kau.edu.sa

M. Bezzeghoud et al. (eds.), *Recent Research on Geotechnical Engineering, Remote Sensing, Geophysics and Earthquake Seismology*, Advances in Science, Technology & Innovation,
https://doi.org/10.1007/978-3-031-48715-6_35

$$\text{SDDI} = \left(\rho_{2.13\mu m} - \left(\left(\rho_{0.55\mu m} + \rho_{0.469\mu m}\right)/2\right)\right)/\left(\rho_{2.13\mu m} + \left(\left(\rho_{0.55\mu m} + \rho_{0.469\mu m}\right)/2\right)\right) \quad (1)$$

where $\rho_{0.468\ \mu m}$, $\rho_{0.55\ \mu m}$, and $\rho_{2.13\ \mu m}$ are reflectance in MODIS bands 3, 4, and 7, respectively. We have selected eight SDS events that were reported during the year 2003 in the current study (Table 1) while the event of March 26, 2003, is discussed in detail. MODIS and five days (two days before and after the SDS event) AERONET AOD data located at solar village (SV) are retrieved from NASA while MODIS DT-DB AOD product is retrieved form Giovanni. The accuracy of the index is tested by using Eqs. 2, 3, and 4.

$$\text{POFD} = (b/(a+b)) \times 100 \quad (2)$$

$$\text{POCD} = (a/(a+c)) \times 100 \quad (3)$$

$$\text{Accuracy} = (a+d/(a+b+c+d)) \times 100 \quad (4)$$

where a is the true positives (number of time both NCM and index indicate SDS event), b is the false positives (number of time when NCM indicates no SDS but index indicates SDS), c is the false negatives (number of time when NCM indicates SDS and index indicates no SDS), and d is the true negatives (number of time both NCM and index indicate no SDS).

3 Results

The spectral analysis of all the SDS events was performed by applying SDDI index, and various threshold values (between + 1 and − 1) corresponding to different features were detected over the study area. The determination of threshold values is based on a number of factors that includes visual interpretation, image processing, and ground-based data that classify SDS events into different classes, and DT-DB AOD product. For example, negative values are detected for clouds (aqua marine color), greater than 0 but less than 0.5 (0–0.2 is weak, 0.2–0.4 is moderate, and 0.4–0.5 is sever) for SDS events (peach color), and greater than equal to 0.5 for surface features (red color). The results of the SDDI for the year 2003 SDS events are given in Table 1. Similarly, the result of SDDI index on March 26, 2003, event is presented in Fig. 1a, while MODIS DT-DB AOD product and AERONET station based AOD are shown in Fig. 1b, c, respectively.

4 Discussion

It is evident from Table 1 that the SDDI has successfully detected all the SDS events which were studied in the current research over the KSA. This can be further elaborated from the SDDI index based resultant image (Fig. 1a) where sand and dust features are clearly separated from other objects as well as the index successfully differentiated atmospheric sand and dust features (peach color in Fig. 1a) from that located on the earth (red color in Fig. 1a).

In the process of validation, it is evident from Fig. 1b (MODIS DT-DB AOD product) that large amount of aerosols are present in the atmosphere in the north, east, and south. It is also evident from Fig. 1c (AERONET station based AOD) that the region has lowest amount of aerosols two days before SDS event which gradually increases and reaches at its maximum on the day of the event. However, the aerosols started decreasing again once the event day is passed.

Finally, the results of the accuracy analysis of the index indicate that the POFD, POCD, and accuracy of the index are 3%, 100%, and 97%, respectively, for the events of the year 2003 over the KSA. This clearly indicates that the SDDI can successfully detect SDS events over the KSA with high accuracy.

Table 1 SDS events for the year 2003

Date	Stations reported SDS event	SDDI detection
25/02/2003	Dammam, Hfr Batin, Qaisumah	YES
19/03/2003	Hfr Batin, Jouf, Qaisumah	YES
26/03/2003	Ahsa, Arar, Dahran, Dammam, Guraiat, Hfr Batin, Jouf, Qaisumah, Rafha, Sharorah, Turaif	YES
16/05/2003	Hfr Batain, Riyadh, Qaisumah	YES
17/05/2003	Arar, Riyadh, Qaisumah, Rafha, Wadi Dawaser	YES
25/05/2003	Guraiat, Jouf, Turaif	YES
01/06/2003	Gassim, Hfr Batin, Qaisumah	YES
12/08/2003	Baha, Bishah	YES

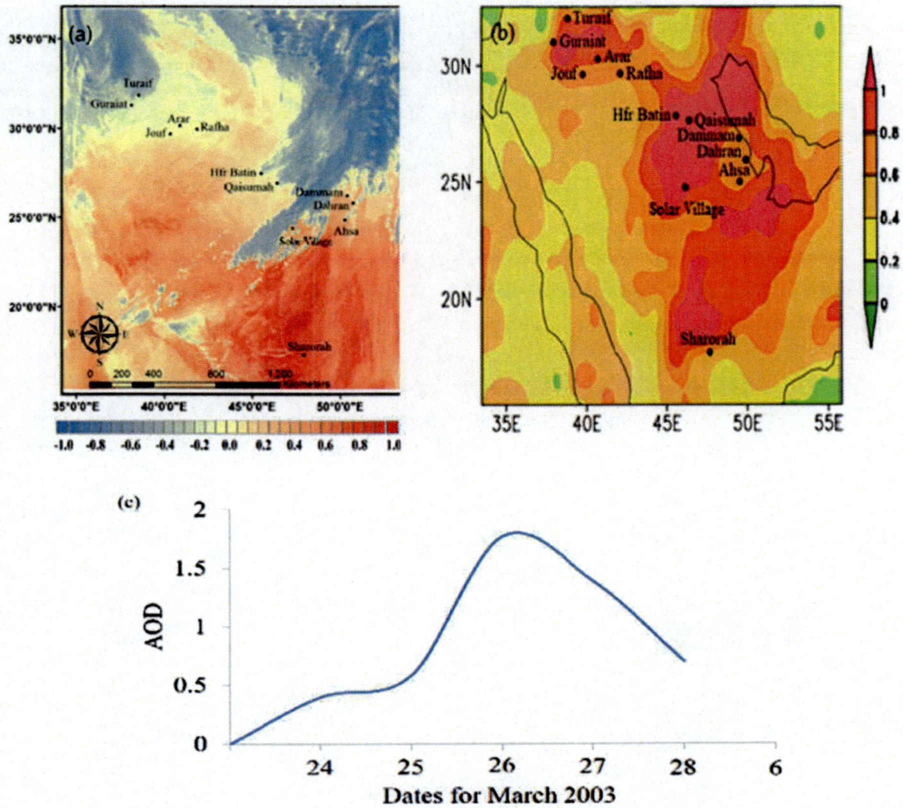

Fig. 1 Application results for the March 26, 2003, SDS event after applying **a** SDDI index, **b** validation by MODIS DT-DB AOD, and **c** AOD at SV AERONET station

5 Conclusions

The SDDI tested in the current research is capable of separating sand and dust features from water and ice clouds in the atmosphere with various threshold values. The inclusion of band 3 and band 4 has enabled the index to further classify SDS events into light, moderate, and intense again based on the threshold values. Though not all factors can be explained that produces errors in the monitoring of SDS, the contributions from the surface features cannot be ignored in the applied indices. This the future goal of our research that can further improve the index results over the KSA.

References

Albugami, S., Palmer, S., Meersmans, J., & Waine, T. (2018). Evaluating MODIS dust-detection indices over the Arabian Peninsula. *Remote Sensing, 10*(12), 1–17. https://doi.org/10.3390/rs10121993

Bibi, H., Alam, K., Chishtie, F., Bibi, S., Shahid, I., & Blaschke, T. (2015). Intercomparison of MODIS, MISR, OMI, and CALIPSO aerosol optical depth retrievals for four locations on the Indo-Gangetic plains and validation against AERONET data. *Atmospheric Environment, 111*, 113–126. https://doi.org/10.1016/j.atmosenv.2015.04.013

Butt, M. J., Assiri, M. E., & Ali, M. A. (2017). Assessment of AOD variability over Saudi Arabia using MODIS Deep Blue products. *Environmental Pollution, 231*. https://doi.org/10.1016/j.envpol.2017.07.104

Butt, M. J., & Mashat, A. S. (2018). MODIS satellite data evaluation for sand and dust storm monitoring in Saudi Arabia. *International Journal of Remote Sensing*. https://doi.org/10.1080/01431161.2018.1488293

Choobari, O. A., Zawar-Reza, P., & Sturman, A. (2014). The global distribution of mineral dust and its impacts on the climate system: A review. *Atmospheric Research, 138*. 152–165. Elsevier Ltd. https://doi.org/10.1016/j.atmosres.2013.11.007

Huang, J., Ge, J., & Weng, F. (2007). Detection of Asia dust storms using multisensor satellite measurements. *Remote Sensing of Environment, 110*(2), 186–191. https://doi.org/10.1016/j.rse.2007.02.022

Karimi, N., Moridnejad, A., Golian, S., Samani, J. M. V., Karimi, D., & Javadi, S. (2012). Comparison of dust source identification techniques over land in the middle east region using MODIS data. *Canadian Journal of Remote Sensing, 38*(5), 586–599. https://doi.org/10.5589/m12-048

Kaskaoutis, D. G., et al. (2012). Synergistic use of remote sensing and modeling for tracing dust storms in the Mediterranean. *Advances in Meteorology*. https://doi.org/10.1155/2012/861026

Kaskaoutis, D. G., Kosmopoulos, P., Kambezidis, H. D., & Nastos, P. T. (2007). Aerosol climatology and discrimination of different types over Athens, Greece, based on MODIS data. *Atmospheric Environment, 41*(34), 7315–7329. https://doi.org/10.1016/j.atmosenv.2007.05.017

Labban, A. H., & Butt, M. J. (2021). Analysis of sand and dust storm events over Saudi Arabia in relation with meteorological parameters and ENSO. *Arabian Journal of Geosciences, 14*(1), 1–12. https://doi.org/10.1007/S12517-020-06291-W/FIGURES/7

Qu, J. J., Hao, X., Kafatos, M., & Wang, L. (2006). Asian dust storm monitoring combining terra and aqua MODIS SRB measurements. *IEEE Geoscience and Remote Sensing Letters, 3*(4), 484–486. https://doi.org/10.1109/LGRS.2006.877752

Samadi, M., Boloorani, A. D., Alavipanah, S. K., Mohamadi, H., & Najafi, M. S. (2014). Global dust Detection Index (GDDI); A new remotely sensed methodology for dust storms detection. *Journal of Environmental Health Science and Engineering, 12*(1), 1–14. https://doi.org/10.1186/2052-336X-12-20

Temporal Issues in 3D Strata Management

Usman Mehmood, Uznir Ujang, Suhaibah Azri, and Tan Liat Choon

Abstract

Due to population growth and urbanization, high-rise residential strata have developed in Malaysia's big cities. People from diverse socioeconomic backgrounds must share a ground parcel and use the same common utilities. Strata tenants must deal with several management issues to maintain an attractive, standardized living environment. Poor strata administration caused several problems for owners. Unlike conventional housing, residential strata require an executable model for proactive management. This research aims to identify the temporal management issues in 3D residential strata and join them with the existing Land Administration Domain Model (LADM) in the form of the management package in the LADM. A conceptual domain model in unified modeling language (UML) is developed in this research. Temporal management issues, actors involved, and management procedures for residential strata are identified by literature review from 2000 to 2021. Furthermore, it is concluded that proactive and preventive management is required to increase the residential strata's market value and life span. The developed generic conceptual domain model for temporal management of residential strata can further be used by the management bodies worldwide to perform proactive management by developing database management systems and software based on this model.

Keywords

UML • Domain model • LADM • Strata management

U. Mehmood (✉) · U. Ujang · S. Azri · T. L. Choon
3D GIS Research Lab, Faculty of Built Environment and Surveying, Universiti Teknologi Malaysia, 81310 Johor, Malaysia
e-mail: usmanmehmood1991@gmail.com

U. Mehmood
No. 8 Field Party, Directorate of Photogrammetry, Survey of Pakistan, Rawalpindi, 46000, Pakistan

1 Introduction

High-rise residential strata (HrRS) are growing more popular worldwide, especially among city dwellers. HrRS is a partitioned building with accessory parcels and provisional blocks (if any). Most individuals choose HrRS for its housing amenities (Che Ani et al., 2010); corridors, pools, lifts, external barriers, parking, open space, mosques, and water tanks are common properties (Mohamad et al., 2015).

HrRS is unique as there are many apartments on a single ground parcel (Linariza & Ashok, 2003). Management corporation (MC) controls every facility in collaboration with residents. Due to the different origins and ethnicities of for-sale property buyers/owners, housing plan management must be systematic and acceptable. This protects most owners while not exploiting the minority (Liias, 1998).

2 Temporal Strata Management Issues

MC of HrRS strata must deal with budget administration, risk management, project, and renovations, building operation and maintenance, licenses and certificates, and legal documentation governing owner rights and obligations (Hui, 2005). Furthermore, (Baldwin, 1994; Chong et al., 2021; Kamarudin, 2014), in their work, highlight the issue of strata disputes. Strata management bodies must frequently resolve tenant disputes.

The annual management fee levied on residents is essentially fixed, and poor collection is also noticed throughout the residential strata (Baldwin, 1994; Hui, 2005). Fewer funds are collected, which leads to poor management facilities (Chong et al., 2021). Many strata apartments with many occupants who live close together have resulted in numerous disputes among strata residents, which always entail by-law violations, unlawful use of common property, repairs and maintenance difficulties, and questions about the strata managing agency (Chong et al., 2021; Kamarudin, 2014; Mohamad, 2015).

M. Bezzeghoud et al. (eds.), *Recent Research on Geotechnical Engineering, Remote Sensing, Geophysics and Earthquake Seismology*, Advances in Science, Technology & Innovation,
https://doi.org/10.1007/978-3-031-48715-6_36

Furthermore, residents are concerned about hazards such as fire, gas leaks, typhoons, flooding, landslides, and power outages (Hui, 2005). Similarly, in the event of an emergency, evacuation of HrRS neighborhoods is a vital concern (Chu et al., 2019). Strata should be properly surveyed to identify the hazards under the MC to avoid mishaps. Lastly, to satisfy the tenants' requirements, improvise security and safety, up-gradation, and meet the latest market requirement, renovation projects of strata must be planned from time to time (Hui, 2005).

In addition, HrRS must be fully maintained and renovated to retain their rental and sale values, especially the common properties like lifts, corridors, etc. (Au-Yong et al., 2018; Mat Noor & Eves, 2011; Yen, 2018). Here in this work, the issues directly related to MC will be joined with the Land Administration Domain Model (LADM) as a management package.

3 LADM and Temporal Strata Management

LADM consists of four packages, i.e., the Party package, which involves people and organizations, basic administration, RRR (rights, restrictions, and responsibilities), and the spatial unit package LADM is a conceptual framework that any country can adopt for land administration (van Oosterom & Lemmen, 2015). Temporal issues can be seen in many, i.e., BIM, and smart city applications (Azri et al., 2018; Basir et al., 2018) Now, in relation to temporal strata management issues with LADM, a conceptual domain model is presented, as shown in Fig. 1.

4 Discussion

HrRS have various challenges over time when individuals from different cultures, religions, mindsets, and professions share similar facilities, including lifts, parking spaces, lobbies, and corridors. The parcel owner usually maintains and renovates the apartments. MC holds and renovates common properties. Poor collection of funds is one of the main reasons for poor strata management since surveys show that many residential strata do not trust MC. Tenants are not satisfied with the quality of maintenance and renovation works performed by the MC.

On the other hand, MC depends upon the funds to perform its management operations. Every residential strata

Fig. 1 Conceptual domain model for temporal strata management issues

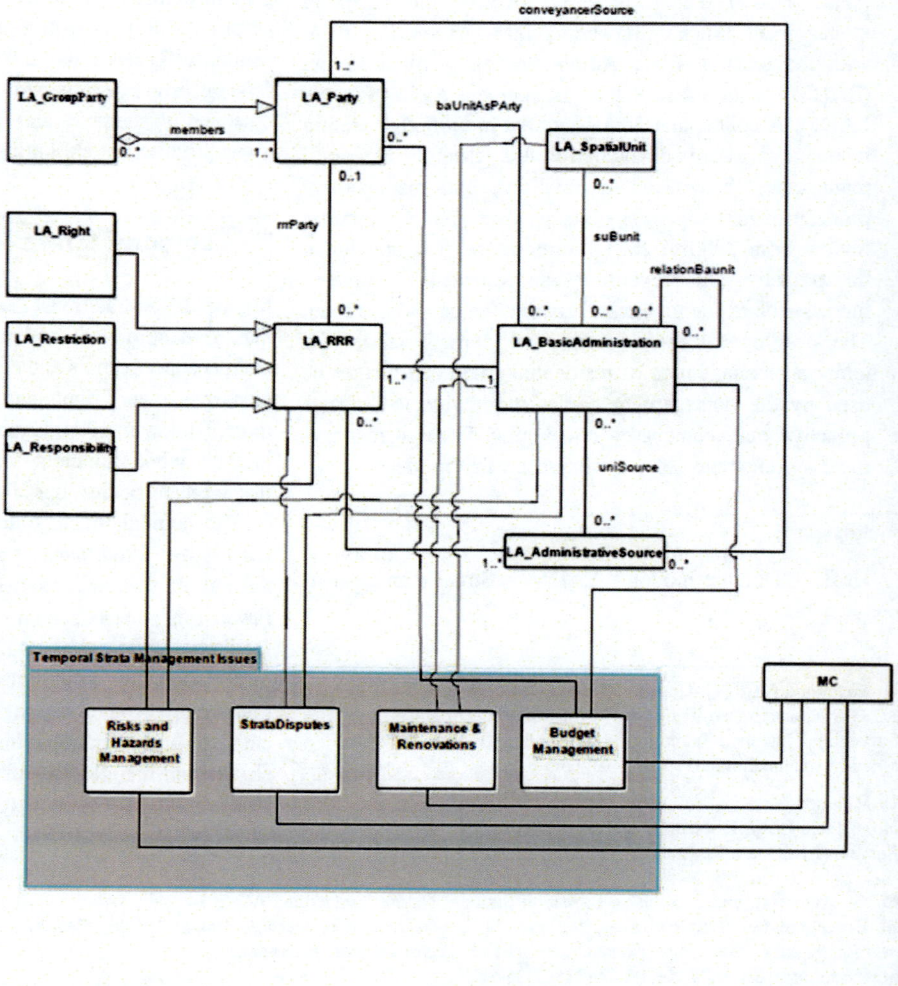

have specific renovation responsibilities. Tenants remodel their residences according to their own will and ideas and do not consider renovation roles and regulations. These tenants' ignorance may induce risks and hazards in the strata. MC usually contracts with maintenance services provider contractors in private residential strata to maintain common property. Delay in contractor response and poor maintenance work quality also leads to the low-quality living standard of the strata.

Neighbors and other tenants of the strata are very important to consider as they all have to share many common properties within the strata. It is found that more are more strata disputes arises when different cultural and ethical background people share the walls. It is suggested that a trustworthy relationship between the tenants and MC of strata should be established by frequently arranging meet-ups and group activities for the tenants of residential strata to tackle budget issues and disputes.

5 Conclusion

In Malaysia, increase in urbanization leads to increase the development of HrRS in the cities. Nowadays, we can observe hundreds of HrRS in the major cities of Malaysia. Management and life style of HrRS are completely different from conventional landed houses due to two reasons, i.e., mixed distribution of ownership and multiple management bodies. Within a same HrRS scheme, there are two or three management bodies. Parcel owners for the upkeep of their owned parcel, MC for the management of common properties and subsidiary management corporation for the upkeep of limited common property. This lead to the requirement of a conceptual unified domain model which will be helpful to manage HrRS all over the world more effectively.

Acknowledgements This research is supported by the Ministry of Education (MOE) through Fundamental Research Grant Scheme (FRGS/1/2021/WAB07/UTM/02/2).

References

Au-Yong, C. P., Ali, A. S., & Chua, S. J. L. (2018). A literature review of routine maintenance in high-rise residential buildings: A theoretical framework and directions for future research. *Journal of Facilities Management*.

Azri, S., Ujang, U., & Rahman, A. A. (2018). Dendrogram clustering for 3D data analytics in smart city. *The International Archives of Photogrammetry, Remote Sensing and Spatial Information Sciences, 42*, 247–253.

Baldwin, G. (1994). Property management in Hong Kong: An overview. *Property Management*.

Basir, W. W. A., Majid, Z., Ujang, U., & Chong, A. (2018). Integration of GIS and BIM techniques in construction project management—A review. *The International Archives of Photogrammetry, Remote Sensing and Spatial Information Sciences, 42*, 307–316.

Che Ani, A., Tawil, N., Sairi, A., Abdullah, N., Tahir, M., & Surat, M. (2010). Facility management indicators for high-rise residential property in Malaysia. *WSEAS Transactions on Environment and Development, 6*(4), 255–264.

Chong, K. G., Eni, S., & Safian, E. E. M. (2021). Impact of having an early briefing among residents/owners in reducing strata disputes in Malaysia. *International Journal of Real Estate Studies, 15*(S1), 18–26.

Chu, H., Yu, J., Wen, J., Yi, M., & Chen, Y. (2019). Emergency evacuation simulation and management optimization in urban residential communities. *Sustainability, 11*(3), 795.

Hui, E. Y. (2005). Key success factors of building management in large and dense residential estates. *Facilities*.

Kamarudin, F. (2014). *The development of an effective and efficient dispute resolution processes for strata scheme disputes in Peninsular Malaysia*. Queensland University of Technology.

Liias, R. (1998). Housing stock: The facilities for future development. *Facilities*.

Linariza, H., & Ashok, V. (2003). Facility management: An introduction. *Journal of the Malaysian Surveyor, 38*, 13–19.

Mat Noor, N., & Eves, C. (2011). *Malaysia high-rise residential property management: 2004–2010 trends and scenario*. Paper presented at the Proceedings of the 17th Annual Pacific Rim Real Estate Society Conference.

Mohamad, M., Sufian, A., & Mohamad, N. (2015). Strata Management in Peninsular Malaysia: An Overview. *Jurnal Pentadbiran Tanah*.

Mohamad, N. A. (2015). Developing a model for pre-action dispute resolution for strata dispute management in Malaysia. *UUM Journal of Legal Studies, 6*, 97–104.

van Oosterom, P., & Lemmen, C. (2015). The Land Administration Domain Model (LADM): Motivation, standardisation, application and further development. *Land Use Policy, 49*, 527–534. https://doi.org/10.1016/j.landusepol.2015.09.032

Yen, L. C. (2018). How to manage your condo without killing anyone. *Property Management, 101*.

Road Traffic Noise Visualization Modeling in Two- and Three-Dimensional Space: A Review

Nevil Wickramathilaka, Uznir Ujang, Suhaibah Azri, and Salfarina Samsuddin

Abstract

Traffic noise causes 90% of urban noise pollution. Visualizing noise is challenging. Three-dimensional (3D) space noise visualization reflects natural movement or acoustic dispersion. However, improving traffic noise visualization is hardly addressed. This review paper aims to fulfill the gap mentioned earlier. Noise models such as the Federal Highway Administration (U.S.), RLS-90 Model (Germany), Road Traffic Noise Prediction (U.K.), Stop and Go model (Thailand), and Henk de Kluijver model are compared and described. Noise is visualized using hotspots, contours, and building facades. The shape and direction of building facades do not affect 3D noise visualization. Traffic volume, composition, speed, road gradient, surface, ground covering, and distance to traffic sources affect noise. Green spaces absorb noise and provide acoustic relief. The 2D and 3D propagation should address noise screening and diffraction. Interpolating noise for visualization uses IDW, Kriging, and TIN. TIN outputs correct noise contours. Flat triangles must be avoided to remove unpredictable oscillation of interpolated noise levels in 3D space. For 2D noise visualization, raster cell size and noise level accuracy are essential. Noise interpolation requires 2 m minimum observation points. Plus, this review identifies the IDW distance weighted factor and the optimal Kriging variogram. Furthermore, this review suggests purple instead of red to denote high noise levels.

Keywords

Road traffic noise model • Two-dimensional (2D) • Three-dimensional (3D) • Spatial interpolations

1 Introduction

Noise pollution is 90% of urban noise pollution (Butler, 2004). A maximum of 55 dB can be exposed without physical and mental disturbances. Road traffic noise combines engine noise, road surface type, traffic, and environmental conditions (Ibili, 2021). The noise emanating from the considered point source is dispersed throughout the environment. Therefore, noise visualization in 3D is essential rather than in 2D (Alam et al., 2020). In the domain of reality, 3D visualization provides a more comprehensive understanding (Azri et al., 2018; Mohd et al., 2018; Ujang et al., 2013). The 2D only displays patterns, trends, and relationships in visualization (Hasara et al., 2020; Alam et al., 2022; Dias et al., 2022; Rupathunga et al., 2022).

2 Methods

But still, there are no conversions regarding improving the quality of traffic noise visualization in 2D and 3D. This review paper aims to fulfill the gap mentioned above. The hundred (100) research papers were collected for review. The fifty (50) best research papers for this review were selected concerning the sub-topics: road traffic noise models, noise propagation, noise interpolations, and noise visualization. Moreover, 11 papers were selected to improve the discussion and conclusion (see Fig. 1).

N. Wickramathilaka (✉) · U. Ujang · S. Azri · S. Samsuddin
3D GIS Research Lab, Faculty of Built Environment and Surveying, Universiti Teknologi Malaysia, Johor Bahru, 81310 Johor, Malaysia
e-mail: nevilvidyamanee@kdu.ac.lk

N. Wickramathilaka
Southern Campus, General Sir John Kotelawala Defence University, Edison Hill, Nugegalayaya, Sewanagala, Sri Lanka

M. Bezzeghoud et al. (eds.), *Recent Research on Geotechnical Engineering, Remote Sensing, Geophysics and Earthquake Seismology*, Advances in Science, Technology & Innovation,
https://doi.org/10.1007/978-3-031-48715-6_37

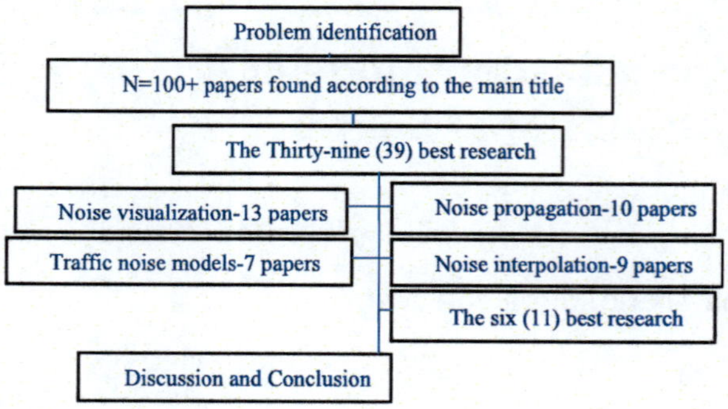

Fig. 1 Flowchart

3 Results

Traffic noise models: The FHWA model (Seong et al., 2011), the RLS-90 (Quartieri et al., 2009), the CoRTN model (Rajakumara and Gowda, 2008), the Stop and Go model (Pamanikabud and Tharasawatpipat, 1999), and the Henk de Klujijver's model (Kurakula, 2007) are widely used in traffic noise calculations. Noise propagation: The traffic propagation path may be changed because of the noise barriers (Lu et al., 2017). Green spaces help to mitigate traffic noise by providing an acoustic barrier (Wickramathilaka et al., 2022). Reducing barriers is vital to visualize traffic noise (Dudiev & Tupov, 2020). When propagation, the noise will reduce from log(r), r: distance from the noise source to receiver (Ranjbar et al., 2012). Therefore, the horizontal and vertical intervals between noise observation points are vital (Kurakula, 2007). The building's orientation and shape influence the horizontal noise effect. Hence, 3D building shapes may not be as valid as 2D. Noise interpolation: The observed sample noise levels are not represented the entire study area, and the noise levels in significant locations of the city area will be considered (Puyana-Romero et al., 2020). Therefore, spatial interpolation is important for noise mapping (Konde and Saran, 2017). The inverse distance weighted (IDW), Kriging, and triangular irregular network (TIN) interpolation techniques have been used to interpolate noise levels in 2D and 3D. Flat triangles have been eliminated to design noise observation points along the vertical axis (Kurakula, 2007). The accuracy comparison is vital for identifying the performance of interpolations (Harman et al., 2016). Noise visualization: Noise visualization uses 2D and 3D hotspots and contours (Farcas, 2020). Most European countries use standard (DIN 18005-2) color schemes for noise visualization. Alberts and Alfe`rez have introduced a new color scheme with better lightness than (DIN 18005-2). The highest dB levels represent by using purple color in the new scheme (Weninger, 2015).

4 Discussion

A valid traffic noise model is vital to calculate the noise levels. Several amendments may be applied to the model according to environmental conditions (Munir et al., 2021).

However, prior knowledge of the accuracy validation for the traffic noise models is essential (Debnath & Singh, 2018). The FHWA noise model's validation results are ± 0.58–1.58, the CoRTN noise model is + 1.4 dB to 50–54.9 dB noise levels, and − 1.3 dB for 80–84.9 dB noise levels, the Stop-and-go noise model is 0.1165 dB (Ranjbar et al., 2012). There are several factors that affect traffic noise. Furthermore, the Henk de Kluijver noise model considers the reflection noise and absorbed noise levels from the air. Thus, several studies have used this model (Kurakula, 2007; Ranjbar et al., 2012). Green space is vital to mitigate traffic noise levels (Dobson & Ryan, 2000). Especially, the trees' leaves absorb the noise (Watanabe & Yamada, 1996). Green areas will absorb noise from 5 dB (A) to 10 dB (A) (Orikpete et al., 2021), and it is from 10% to 24% noise reduction (Li et al., 2021). Observing the noise levels everywhere is not a possible task. Therefore, calculating the noise levels for the noise observation points using a noise model is vital (see Fig. 2). Then, spatial interpolation can be applied to the noise levels (Kurakula, 2007; Harman et al., 2016). When designing noise observation points in 3D space, the flat triangle should be eliminated. It means that the exact *x*, and *y* coordinates must be eliminated for any *z* values (flat triangles) of noise interpolation in 3D space (see Fig. 2) (Ranjbar et al., 2012). Inaccurate distribution of noise receiver locations impacts the accuracy of noise levels (Cai et al., 2015). The uncertainty errors may have occurred in noise visualization; the output raster cell size may not match the accuracy range of noise levels (Escobara, 2021). IDW, Kriging, and TIN spatial interpolations are widely used for

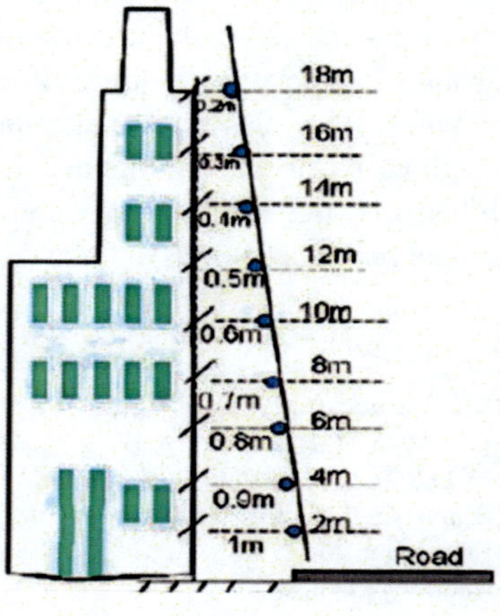

Fig. 2 3D noise points (Kurakula, 2007)

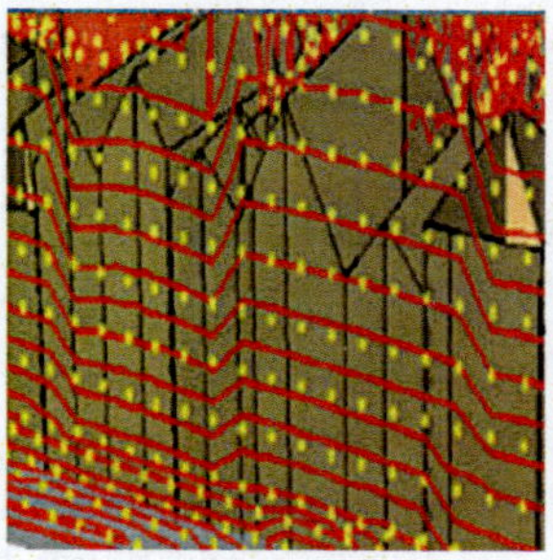

Fig. 3 TIN noise contours (Kurakula, 2007)

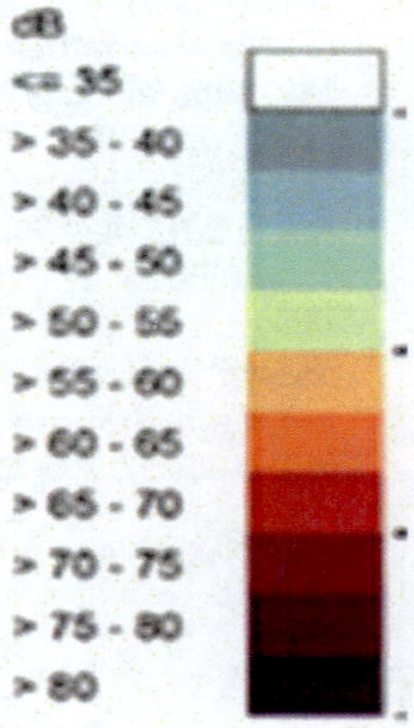

Fig. 4 Color notifications (Weninger, 2015)

noise mapping. Further, elements of spatial interpolation impact the accuracy. The weighting factor of IDW, the Gaussian variogram of Kriging, and the Delaunay triangulation of TIN are vital elements for interpolation (Kurakula & Kaffer, 2008; Tan & Xu, 2014; Harman et al., 2016). IDW is best for 2D noise interpolation, and TIN is best for 3D. Interpolated IDW, 3D visualization, and Kriging noise contours are irregular. The 3D contours do not match noise calculation points. But TIN contours are extracted for noise computation (see Fig. 3) (Kurakula, 2007). The warm color intensity was reduced, and mostly cool color was used in noise visualization. But to enhance the recognition, the green and warm color red represented 65–70 dB noise levels and high purple noise levels (see Fig. 4) (Weninger, 2015).

5 Conclusion

A proposed noise model must consider the research area's environmental and physical elements. Distance reduces noise. Noise interpolation requires a 2 m minimum between noise observation locations. It is important to choose between IDW and kriging for spatial interpolation. For interpolation, consider the IDW distance weighted factor and Kriging variogram. Noise maps with color alerts are better.

Acknowledgements This research was partially funded by UTM SPACE: UTM SPACE Contract Research Grant, Vot R. J130000.7752.4J550.

References

Alam, P., et al. (2020) Noise monitoring, mapping, and modeling studies—A review *21*(4), 82–93.

Alam, P., Mazhar, M. A., Ahmad, K., Aslam, Z., Afsar, S. S., & Husain, A. (2022) Comparative assessment of road traffic noise through 2D noise mapping: A case study of an urban area. *Frontiers in Sustainability, 3*.

Azri, S., Ujang, U., Abdul Rahman, A. (2018). Dendrogram clustering for 3D data analytics in smart city. *The International Archives of the Photogrammetry Remote Sensing and Spatial Information Sciences, XLII-4/W9*, 247–253.

Butler D. (2004). Noise management: Sound and vision. *Nature 427*(6974), 480–481. https://doi.org/10.1038/427480a. PMID: 14765165

Cai, M., et al. (2015). Road traffic noise mapping in Guangzhou using GIS and GPS. *Applied Acoustics, 87*, 94–102. https://doi.org/10.1016/j.apacoust.2014.06.005

Debnath, A., & Singh, P. K. (2018). Environmental traffic noise modelling of Dhanbad township area—A mathematical based approach. *Applied Acoustics, 129*, 161–172. https://doi.org/10.1016/j.apacoust.2017.07.023

Dias, K., Kalpitha, W. K. H., & Wickramathilaka, N. V. (2022). *Impact of Flood on the Built-up Environment: A Case Study of Baddegama DS Division* (p. 2022).

Dobson, M., & Ryan, J. (2000). Trees and shrubs for noise control. In Arboricultural practice notes (pp. 1–8).

Dudiev, T., & Tupov, V. (2020). Method of acoustic calculation of traffic noise barriers. *MATEC Web of Conferences, 320*, 00034. https://doi.org/10.1051/matecconf/202032000034

Escobara, G. G. (2021). *Uncertainty Evaluation of Road Traffic Noise Models in Two Ibero-American Cities*.

Farcas, F. (2020). GIS Tools for Noise Mapping and a Case Study for Skåne Region

Harman, B. I., Koseoglu, H., & Yigit, C. O. (2016). Performance evaluation of IDW, Kriging and multiquadric interpolation methods in producing noise mapping: A case study at the city of Isparta, Turkey. *Applied Acoustics, 112*, 147–157. https://doi.org/10.1016/j.apacoust.2016.05.024

Hasara, K., Singhawansha, J., & Wickramathilaka, N. (2020). Detecting Urban Expansion Trends in Weligama Urban Council Using Remote Sensing and GIS (pp. 23–29).

Ibili, F. (2021) Traffic noise models and noise guidelines: A review. **53** (1–2), 65–79. Available at: https://doi.org/10.1177/09574565211052693

Konde, A., & Saran, S. (2017). Web-enabled spatiotemporal semantic analysis of traffic noise using CityGML. **11**(2), 248–259.

Kurakula, V. (2007). A GIS-based approach for 3D noise Modelling using 3D city models. *International Institute for Geo-Information Science and Earth, 319*(5864), 766–769.

Li, B., Qiu, Z., & Zheng, J. (2021). Impacts of noise barriers on near-viaduct air quality in a city : A case study in Xi' an. *Building and Environment, 196*, 107751. https://doi.org/10.1016/j.buildenv.2021.107751

Lu, L., Becker, T., & Löwner, M. O. (2017). 3D complete traffic noise analysis based on CityGML. *Lecture Notes in Geoinformation and Cartography, 0*(9783319256894), 265–283. https://doi.org/10.1007/978-3-319-25691-7_15

Mohd, Z. H., Ujang, U., Choon, T. L., (2017). Heritage house maintenance using 3D city model application domain extension approach. *International Archives of the Photogrammetry, Remote Sensing & Spatial Information Sciences*.

Mishra, R. K., et al. (2021). Dynamic noise mapping of road traffic in an urban city. *Arabian Journal of Geosciences, 14*(2). https://doi.org/10.1007/s12517-020-06373-9

Munir, S., et al. (2021). Temporal and seasonal variations of noise pollution in urban zones: A case study in Pakistan. *Environmental Science and Pollution Research, 28*(23), 29581–29589. https://doi.org/10.1007/s11356-021-12738-8

Orikpete, O., et al. (2021). Appraisal of industrial and environmental noise regulation in Nigeria and its impact on sustainable national development. Researchgate.Net *10*(09).

Puyana-Romero, V., Cueto, J. L., & Gey, R. (2020). A 3D GIS tool for the detection of noise hot-spots from major roads. *Transportation Research Part d: Transport and Environment, 84*, 102376. https://doi.org/10.1016/j.trd.2020.102376

Quartieri, J., et al. (2009). A review of traffic noise predictive noise models. In Recent advances in applied and theoretical mechanics (pp. 72–80).

Rajakumara, H. N., & Gowda, R. M. M. (2008). Road traffic noise prediction models: A review. 3(3), 257–271. https://doi.org/10.2495/SDP-V3-N3-257-271

Ranjbar, H. R., et al. (2012). *3D analysis and investigation of traffic noise impact from Hemmat Highway located in Tehran on buildings and surrounding areas* (pp. 322–334).

Rupathunga, P., Wickramathilaka, N. V., & Hansamal, P. A. T. (2022). *Spatial pattern of urban expansion and green spaces: A case study of Tangalle area, Sri Lanka* (p. 2022).

Seong, J. C., et al. (2011). Modeling of road traffic noise and estimated human exposure in Fulton County, Georgia, USA. *Environment International, 37*(8), 1336–1341. https://doi.org/10.1016/j.envint.2011.05.019

Tan, Q., & Xu, X. (2014). Comparative analysis of spatial interpolation methods: An experimental study. *Sensors and Transducers, 165*(2), 155–163.

Ujang, U., Anton., F., Azri, S., Rahman, A. A., Mioc, D. (2013). Improving 3D spatial queries search: Newfangled technique of space-filling curves in 3D City Modeling. *ISPRS Annals of the Photogrammetry, Remote Sensing and Spatial Information Sciences, II-2/W1*, 319–327.

Wang, H., Chen, H., & Cai, M. (2018). Evaluation of an urban traffic noise-exposed population based on points of interest and noise maps: The case of Guangzhou. *Environmental Pollution, 239*, 741–750. https://doi.org/10.1016/j.envpol.2017.11.036

Watanabe, T., & Yamada, S. (1996). Sound attenuation through absorption by vegetation. *Journal of the Acoustical Society of Japan (E) (English translation of Nippon Onkyo Gakkaishi), 17*(4), 175–182. https://doi.org/10.1250/ast.17.175

Weninger, B. (2015). A colour scheme for the presentation of sound emission in maps: Requirements and principles for design. *Euro Noise, 2015*, 439–444.

Wickramathilaka, N., Ujang, U., Azri, S., & Choon, T. L. (2022). Influence of urban green spaces on road traffic noise levels: A review. *International Archives of the Photogrammetry, Remote Sensing and Spatial Information Sciences - ISPRS Archives, 48* (4/W3-2022), 195–201.

Applied and Theoretical Geophysics

Muography Applied in Underground Geological Surveys: Ongoing Work at the Lousal Mine (Iberian Pyrite Belt, Portugal)

Pedro Teixeira, Alberto Blanco, Bento Caldeira, Bernardo Tomé, Isabel Alexandre, João Matos, Jorge Silva, José Borges, Lorenzo Cazon, Luís Afonso, Luís Lopes, Magda Duarte, Mário Pimenta, Mourad Bezzeghoud, Paolo Dobrilla, Pedro Assis, Raul Sarmento, Rui Oliveira, and Sofia Andringa

Abstract

The LouMu Project is an ongoing collaboration between the Laboratory of Instrumentation and Experimental Particle Physics (LIP), the Institute of Earth Sciences—University of Évora and the Lousal Ciência Viva Center which is evaluating the muography potential in the Lousal Mine, with the general aim to create the conditions to use muography as a novel method for geophysical surveys in Portugal. The National Laboratory of Energy and Geology is also supporting the project. The aim is to do a first geological survey of the area, mapping already known structures and ore lenses and measuring their densities. In the end, newfound data will be added to the existing information to improve it, but the whole process is also serving to test the performance of the muon telescope and the muographic analysis tools. The muon telescope, developed by LIP, uses RPC detectors to observe the crossing muons in real time. Available geological and geophysical information and new measurements done with seismic refraction and ground penetrating radar are being used to create a 3D model, which provides a reference against which to compare the muography results. From the muography data, an equivalent 3D map of densities will be reconstructed, and the geological model will be used to cross-check and improve the muography results. This paper provides an overview of this ongoing project, focused on the muography work done so far. Its implementation is described, and the first image obtained with the telescope is presented, showing the local geology and in particular the Corona N–S direction strike-slip fault visible in the mine gallery. The planned work progression is shared as a final consideration.

Keywords

Underground muography • Geophysical survey • Lousal Mine • Iberian Pyrite Belt

1 Introduction

Muography is an imaging technique that uses muons, a natural background radiation generated in the atmosphere that can cross large columns of matter enabling non-invasive and remote observations to create average density

P. Teixeira (✉) · B. Caldeira · J. Borges · M. Bezzeghoud · R. Oliveira
Physics Department (ECT), Institute of Earth Sciences (ICT), Earth Remote Sensing Laboratory (EaRSLab), University of Évora, Rua Romão Ramalho nº59, 7000-671 Évora, Portugal
e-mail: pmmt@uevora.pt

A. Blanco · B. Tomé · L. Afonso · L. Lopes · M. Pimenta · P. Dobrilla · P. Assis · R. Sarmento · S. Andringa
Laboratory of Instrumentation and Experimental Particle Physics (LIP), Av. Prof. Gama Pinto, 2, 1649-003 Lisboa, Portugal

I. Alexandre · M. Pimenta · P. Assis
Instituto Superior Técnico (IST), Av. Rovisco Pais, 1, 1049-001 Lisboa, Portugal

J. Matos
Laboratório Nacional de Energia E Geologia (LNEG), Campus de Aljustrel, Bairro da Vale d'Oca, Apartado 14, 7601-909 Aljustrel, Portugal

J. Silva
Physics Department (DF), University of Coimbra, Rua Larga, 3004-516 Coimbra, Portugal

L. Cazon
Instituto Galego de Física de Altas Enerxías (IGFAE) - Universidade de Santiago de Compostela, Rúa de Xoaquín Díaz de Rábago, 15705 Santiago de Compostela, Spain

M. Duarte
University of Minho, Gualtar Campus, CP3, 3.02, 4710-057 Braga, Portugal

M. Bezzeghoud et al. (eds.), *Recent Research on Geotechnical Engineering, Remote Sensing, Geophysics and Earthquake Seismology*, Advances in Science, Technology & Innovation,
https://doi.org/10.1007/978-3-031-48715-6_38

distribution maps. The use of muons withing the geosciences field has been proved successful in numerous applications around the world (Bonechi et al., 2020; Zhang et al., 2020).

This paper aims to share the work done so far, and for the first time in Portugal, of muography in geophysical applications. A collaboration has been established for that purpose between the Laboratory of Instrumentation and Experimental Particle Physics (LIP), the Institute of Earth Sciences of the University of Évora (ICT-UE), the Lousal Ciência Viva Center (science museum), and with the support from National Laboratory of Energy and Geology (LNEG). The use of muography for underground geophysical surveys has been the aim of the Lousal Muography Project (LouMu) (https://pages.lip.pt/loumu/en/loumu/; Teixeira et al., 2022), that is now applying it to the Lousal volcanogenic massive sulfide deposit (VMS), located in South Portugal (Iberian Pyrite Belt mineral province). The mine was exploited for pyrite until 1988 and is currently run as a science museum, standing as an excellent European example of environmental rehabilitation (Relvas et al., 2012).

The project comprises the development of muon telescopes and their use to create density maps and a 3D model. The process involves building and running muography simulations and the development of muography analysis tools, using geological and geophysical standard methods to study the terrain above the mine and the muography implementation in the Lousal Mine. Outreach activities are also elaborated, which aimed to the academic and general publics. The goal of this work is to map the densities of the terrain between the surface and the mine gallery, to evaluate the performance of the detectors and the muography analysis tools and to add the obtained data to the geological knowledge of the area.

2 Methods and Implementation

To make a survey, a telescope measures the muons transmitted through the column of matter in each direction, which is called transmission muography. The muon detection in an underground environment is protected from the background radiation measured on the surface, but the number of muons reaching further in depth is much lower. For the development of the muon detectors, the geological and underground conditions which define the required exposure time were considered.

The muon telescope, built by LIP, use robust resistive plate chambers (RPCs) that detect the crossing muons in real time. RPCs are a type of gaseous particle detectors, used to track charged particles. The passage of muons ionizes the gas and causes an avalanche of electrons that produce an electrical signal. The trajectory of the muons is given by simultaneous detections in the different parallel RPCs. The gas is sent to the RPCs with a low flux, from a container located in a storage house outside the mine, without leakage to the environment. The telescope, nicknamed CorePix, is equipped with four square RPC detectors measuring 1 m 1 m and is installed in an empty storeroom of the Waldemar gallery (the uppermost mine gallery), 18 m below the surface.

In 2019, a smaller scale prototype was installed inside the Waldemar gallery near the Waldemar shaft (Fig. 1, position A, Lousal Mine southern sector), to test the equipment response to the mine environment. After that, the CorePix was assembled and stayed in the laboratory for about a year for calibration and testing while muographing the structure of the building above it. In April of 2022, it was placed in the Waldemar gallery storeroom 4 (Fig. 1, position B) where

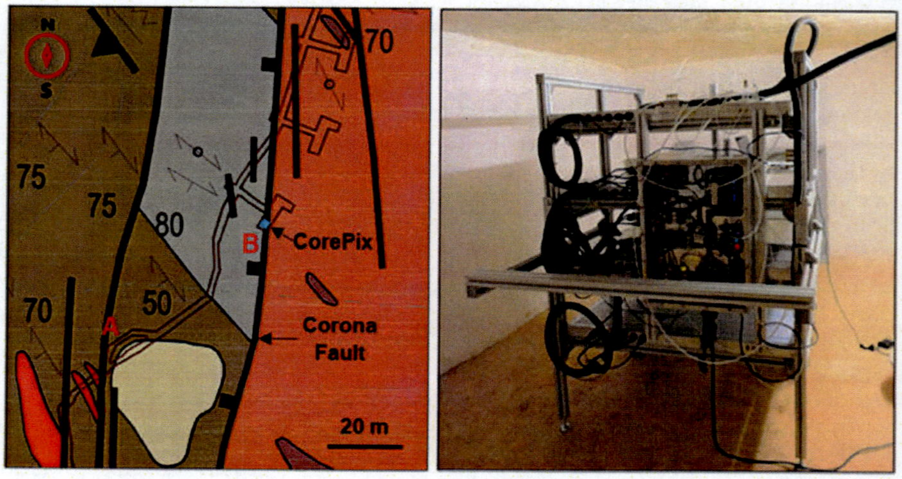

Fig. 1 Left: Excerpt of the Lousal Mine Waldemar gallery geological map showing the CorePix (blue square) location in the storeroom 4 (position B) and the Corona Fault (adapted from Matos et al. 2022); Right: CorePix telescope positioned inside the storeroom 4 of the Waldemar gallery

it has been collecting data. The current observation position (B) was chosen based not only on space availability but also on the local geology setting defined by rock fracking related with the regional late Variscan strike-slip Corona Fault (Matos, 2021). This subvertical N5E direction fault, which crosses the area being surveyed, is, at the moment, the main survey target. Using Geant4 software (Agostinelli et al., 2003), reference simulations, based on the telescope location area, are also being used in the muography analysis.

In this approach, available geological and geophysical information and new measurements taken with ground penetrating radar (GPR) and seismic refraction are being used to build a reference 3D geological model to compare with the muography results.

3 Preliminary Results

The left picture of Fig. 1 is an excerpt of the Waldemar gallery map that shows the different lithologies and geological faults known on the site and the outline of the mine gallery with the indicated location of the CorePix telescope (blue square), inside the storeroom 4. Depicted as orange are phyllites and quartzites from the phyllite-quartzite group, and in color gray are gray shales from the volcano-sedimentary complex. The right picture of Fig. 1 is the CorePix telescope collecting data in its observation position, with the detectors aligned parallel to the vertical.

The image shown in Fig. 2 (left) is a treated muon detection map that is called a muograph and shows the ratio between the detected muon flux (transmission map) and the open-air muon flux (OAF) measured at the surface. The axis values are tan(x) and tan(y) and are calculated as:

$$\tan(x) = px/pz \quad \tan(y) = py/pz \tag{1}$$

where px, py, and pz are the components of the direction vector of the muons trajectory. The color bar indicates the fraction of muons that are transmitted across the ground and reach the detector (e.g., green corresponds to a 10% survival). Yellow corresponds to pixels with more muon detections and blue to pixels with less muon detections. The middle square corresponds to an area of 17 m × 17 m at surface. It is obtained with the 33 cm × 33 cm center pads (the telescope current working resolution) of two RPCs vertically separated by 67 cm. Using other combinations of two RPCs, like at half that vertical separation, the field of view is extended, although with less angular resolution (outer square of 30 m × 30 m).

In this first two-months muograph, the contrast created by the Corona Fault geological zone in the terrain is already visible as the yellow area. The fault zone corridor is characterized by rock fracking and weathering due to the fault movement. Clay fault gauge is also common in a brittle tectonic setting. This geological constrains are reflected into a lower average rock density than the side rock geology, translating into a higher muon transmission rate (yellow). Because of the geological movement happening across the fault, two different geological settings exist on each side of it on the studied area, creating a density contrast that is also measurable. A framing of the muograph with the corresponding surveyed area is shown in Fig. 3. A digital model of the surface, obtained with photogrammetry, and a 3D scanning of the mine gallery, obtained with a ground lidar, were georeferenced using control points acquired by differential GNSS measurements. Figure 3 has a projection of those two clouds seen from above, aligned, and cropped to the studied area, and a white rectangle representing the 5 m width Corona Fault zone, approximately.

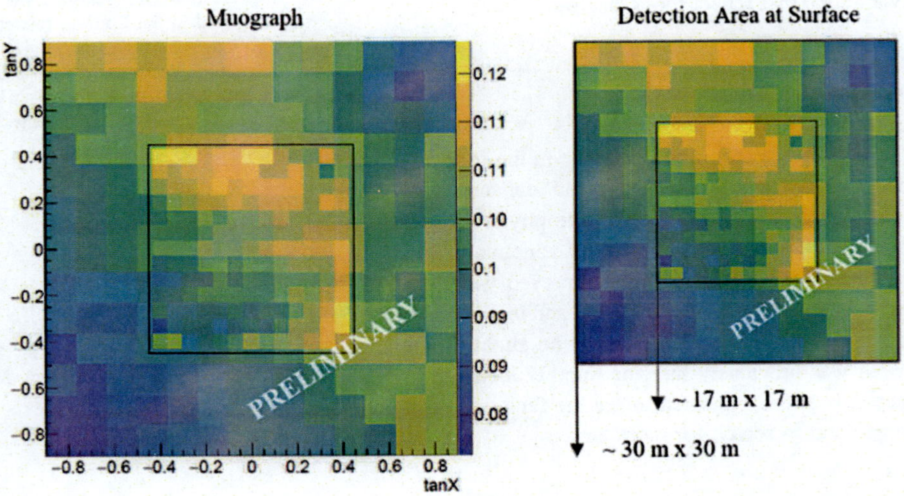

Fig. 2 Left: Muograph from the survey of the telescope, created with two months of data; Right: Corresponding sizes of the detection area at the surface

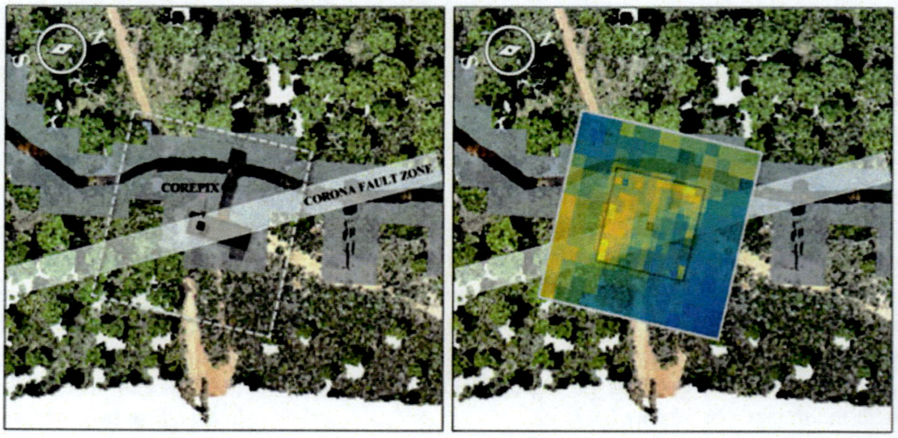

Fig. 3 Framing of the muographic survey map superimposed on the surface area above the telescope and the mine gallery

Referring to Fig. 3 orientation, the geology on the lower right half (blue in the muograph, orange in the geological map) has an average density higher than the fault zone (yellow), which gives a lower muon transmission rate. The differences in the measured muon transmission in the surveyed area correlate with the average density values measured on site, and in most of the area, the topographic variation is less than 1 m. The orientation of the fault zone that can be seen on the muograph seems the same as what is known from the geological map (almost N-S), but the location does not fit perfectly. It is something to check on further analysis.

Disclaimer: the muograph shown on Fig. 2 confirms that in this configuration, the detectors are able to discern the survey target, i.e., the Corona Fault, that is known to exist at the site, and this is a relevant step in the current application of muography. However, this result is obtained with an almost online analysis and is very preliminary, as the data selection and combination are still being optimized as well as the final open-air-flux normalization. Some corrections will still be made, and subsequent results may contain variations that reflect these changes.

4 Final Considerations

The muons detection will continue in the following months, increasing the muon statistic and enhancing the density contrasts in the observed site. The physical changes in the rocks within and outside the fault zone are gradual and not always very clear, even when surveyed with standard geophysical methods. The orientation and inclination of the fault zone above the storeroom might be slightly different than what was previously known. Another location for the telescope is already planned to see the fault zone from another angle and to search for more targets.

The simulations are being improved to match the results and for better understanding them. The geophysical data is being obtained with standard and independent geophysical methods, and the 3D geological reference model will be built as soon as all data processing and its analysis is finished. The end-result will be the reconstruction of a 3D map of densities from the muography data that will use the reference model to help solve the problem of the non-unicity of solutions that appear in this type of analysis.

The inclusion of the telescopes in the mine public tours organized by the science museum offer a great opportunity for education and outreach activities that represent a major aspect of the project. The successful muography survey in the Lousal Mine is the first stage of this collaboration which has the goal to make muography available in Portugal for applications in geophysics, developing stand-alone telescopes for the geophysical community and generalized muography methods to apply in other scenarios.

Acknowledgements This R&D project is financed by National Funds through the FCT—Foundation for Science and Technology, reference EXPL/FIS-OUT/1185/2021. The FCT also funds the Ph.D. scholarship, integrated in the LouMu, reference PD/BD/150490/2019, and within the scope of ICT, the project with reference UIDB/04683/2020. A grateful thanks to the Lousal Ciência Viva team for their support to the project, particularly Vanessa Pais and João Costa, who organize the outreach activities in the museum and help in the muon telescopes maintenance.

References

Agostinelli, S., et al. (2003). Geant4—A simulation toolkit. *Nucl. Inst. and Methods A, 506*(3), 250–303.

Bonechi, L., D'Alessandro, R., Giammanco, A. (2020). Atmospheric muons as an imaging tool. *Reviews in Physics, 5*, 100038. ISSN 2405-4283.

LouMu Webpage: https://pages.lip.pt/loumu/en/loumu/. Last accessed 2022/08/07.

Matos, J. X. (2021). Alteração hidrotermal ácido-sulfato associada aos jazigos de sulfuretos maciços de Lagoa Salgada, Caveira, Lousal, Aljustrel e São Domingos (Faixa Piritosa Ibérica). PhD Thesis, Geology (Metallogeny), University of Lisbon, Faculty of Science.

Matos, J. X., et al. (2022). *Lousal mine southern sector geological map, 1/5.000 scale*. LNEG, LouMu Project.

Relvas, J., Pinto, A., Matos, J. X. (2012). Lousal, Portugal: A successful example of rehabilitation of a closed mine in the Iberian Pyrite Belt. SGA News 31, June 2, 1–16.

Teixeira, P., et al. (2022). *Muography for underground geological surveys: Ongoing application at the Lousal Mine (Iberian Pyrite Belt, Portugal)*. JAIS, vol. 2022.

Zhang, Z. X., et al. (2020). Muography and Its potential applications to mining and rock engineering. *Rock Mechanics and Rock Engineering, 53*, 4893–4907.

Estimating Water Infiltration Rate Using Time-Lapse GPR Technique

Hamdan Hamdan, Daniel Moraetis, Nikos Economou, A. Mahmoud Elsamani, and Mohamed Abdel-Fattah

Abstract

Estimating soil infiltration rate is of utmost importance in agricultural applications. The infiltration rate determines the amount of water to enter the soil, which in turn controls the load of dissolved chemicals (nutrients or pollutants) that will interact with the soil. The soil infiltration rate is commonly measured using the cylinder infiltrometer method, which is tedious and time consuming. Geophysical methods on the other hand have proven to be a powerful tool in providing valuable information about water content in the soil, but rarely used to give a quantitative estimation of the infiltration rate. Ground-penetrating radar (GPR) in particular has the potential to offer a suitable, non-invasive, and cost-effective method for studying the wetting front at shallow depths. A time-lapse GPR experiment was conducted in order to study the feasibility of utilizing GPR method in estimating the soil infiltration rate. A GPR survey was conducted along a 20 m traverse before and immediately after pouring water at two specific locations over the traverse. The GPR measurements were repeated over the same traverse with 5 min time interval, to monitor the downward movement of the wetting front. The time-lapse GPR sections were successful in monitoring the changes of the location of the wetting front with time, which can be used to estimate the vertical soil infiltration rate. The preliminary results are quite encouraging and show that GPR can be efficiently used in monitoring the wetting front downwards movement.

Keywords

Time-lapse GPR • Infiltration rate • GPR migration • Wetting front

1 Introduction

Geophysical methods have been increasingly applied in agriculture over the last few decades. They can provide a fast, reliable, and non-destructive description of the shallow soil, which can be used for assessment of the cultivated area, soil nutrient monitoring, and hydrologic monitoring of agricultural sites (Toushmalani, 2010). The GPR is among the most commonly used geophysical methods in agricultural applications, due to its high resolution and speed of data acquisition and processing. In addition, the electromagnetic waves used in this method are highly sensitive to changes in soil water content, making it a powerful tool in soil hydrologic characterization (Bano, 2006).

Measuring and monitoring the efficiency of an installed irrigation system is critical to the success of any agricultural site. Irrigation water is stored in the soil for some time, therefore allowing it to be available for root uptake, growth of plants, and organism habitation in soil. The rate of the downward movement of the water, represented by the wetting front, dictates the infiltration rate. Fast infiltration rate may decrease water availability for the plant roots, while slow infiltration may rot the plant roots. The soil infiltration rate can be estimated using different versions of the ring infiltrometer method, or more advance methods like the neutron probe method. These methods are time consuming, invasive, and expensive over large areas as the tests have to be carried out at one point at a time.

Many researchers have proved the efficiency of the GPR method in estimating different parameters of the soil wetted body (SWB) which represents the moist area of soil under irrigation systems (Peng et al., 2020; Wang et al., 2022).

H. Hamdan (✉) · D. Moraetis · A. M. Elsamani · M. Abdel-Fattah
Department of Applied Physics and Astronomy,
University of Sharjah, P.O. Box 272272 Sharjah, UAE
e-mail: hhamdan@sharjah.ac.ae

N. Economou
Technical University of Crete, 73100 Chania, Greece

Sultan Qaboos University, 123 Muscat, Oman

M. Bezzeghoud et al. (eds.), *Recent Research on Geotechnical Engineering, Remote Sensing, Geophysics and Earthquake Seismology*, Advances in Science, Technology & Innovation,
https://doi.org/10.1007/978-3-031-48715-6_39

Most of their work was concentrating on measuring the subsurface water content or the geometrical parameters of the SWB like their shape, size, and thickness. They have proved that the depth of SWP can be located using GPR radargrams, which lead us to the conclusion that GPR might be used to estimate the infiltration rate. Here, we will present preliminary results from a time-lapse GPR survey to monitor the changes of the location of the SWP wetting front with time. If these changes are successfully recorded, we can use them to estimate the vertical soil infiltration rate.

2 Methodology and Results

A time-lapse GPR experiment was conducted at a selected location in order to study the feasibility of utilizing GPR method in estimating the soil infiltration rate. The study area was covered with 5–10 cm of gravel, laying over dry sand, which increases in density with depth. Initially, the GPR data were collected using 400 MHz antenna, along a 20 m traverse. Later, sufficient amount of water (around 5 L) was poured at around 12 m from the GPR traverse start point. The GPR data acquisition was repeated on the same traverse every 5 min. Approximately 15 min later, around 7.5 L of water were poured at 6 m of the GPR traverse start point. The time interval was between 5 and 10 min.

Three infiltration tests using the conventional ring infiltrometer were performed close to the GPR traverse. The modeled equilibrium water velocity from the three infiltration tests is 0.12, 2.00, and 1.54 cm/min for Test 1, 2, and 3, respectively. In Fig. 1, we demonstrate the results of the two infiltration tests (Test 3—close to the start of the GPR traverse, Test 2—middle of the GPR traverse).

The great difference in the "dielectric permittivity" between the external and internal zones of the SWB, due to the moisture content, gives clear reflection patterns even in the raw data GPR sections. Reflections of the SWB are shown as high amplitude hyperbolas, which are related to changes of the dielectric constant of the saturated sand compared to the surrounding dry sand. Different schemes for the processing of the GPR data were employed in order to locate the SWB front, including the scheme proposed by Peng et al. (2020) and Wang et al. (2022). For the specific data set, the processing was implemented using the MatGPR software (Tzanis, 2010) involving time zero shifting, removal of DC component, and removal of global background noise. Figure 2 presents a sample of the results (focused at 11–13 m and at 0–3.5 ns) at different times using the wiggle trace display. The velocity of electromagnetic waves was estimated around 0.13 m/s using hyperbola fitting. Since the absolute depth of the SWB is not crucial in estimating the SWB movement rate, we picked another point within the SWB that could be located easily in the different GPR sections (Fig. 2). The depth and time of the same point were located at different time intervals and were used to calculate the velocity of SWB down movement.

3 Discussion

The different locations of the SWB in the time-slice GPR sections were used to estimate the relation of the SWB velocity with time as shown in Fig. 3. The curves look very similar to that estimated using the infiltration test (Fig. 1), where the velocity starts with high values close to 4 cm/min, but stabilizes after the first 3–4 min around 1.5 cm/min. The GPR estimation shows similar results at the first 4–6 min but with fewer readings. However, the GPR results show that the SWB velocity decreases to much smaller values after the first hour. This could be related to the changes in lithology, especially after the first 30 cm, where a denser sand layer is observed. The Horton equation is usually used in hydrogeology to utilize the observed velocities of the water infiltration in estimating the modeled equilibrium water velocity. More analysis is needed if this equation can also be used to utilize the time-lapse GPR readings in calculating the infiltration rate.

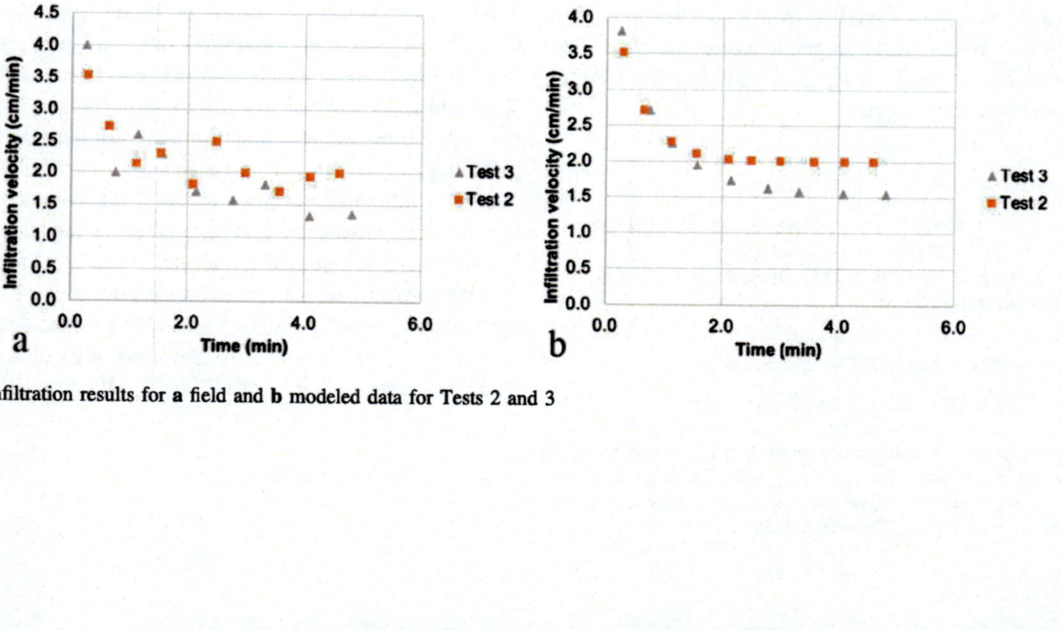

Fig. 1 Infiltration results for **a** field and **b** modeled data for Tests 2 and 3

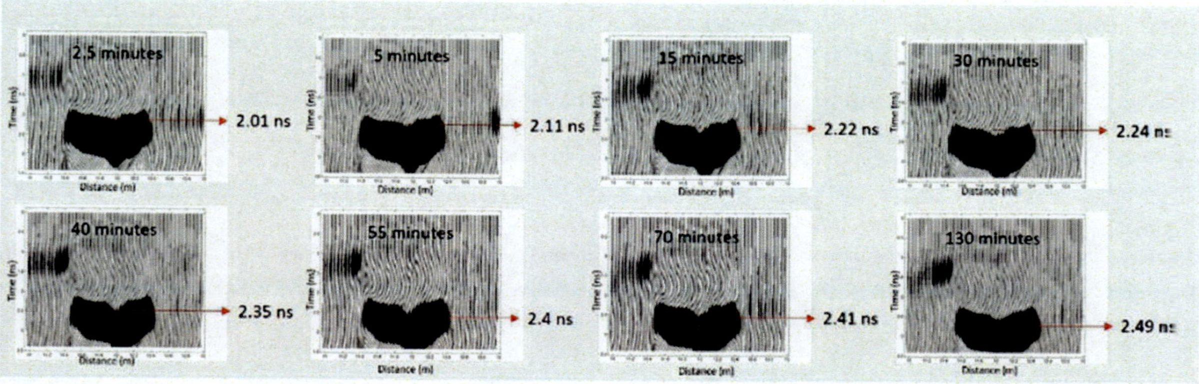

Fig. 2 GPR time-lapse results zoomed between 11 and 13 m of the GPR traverse to show the differences of the SWB depths at different times

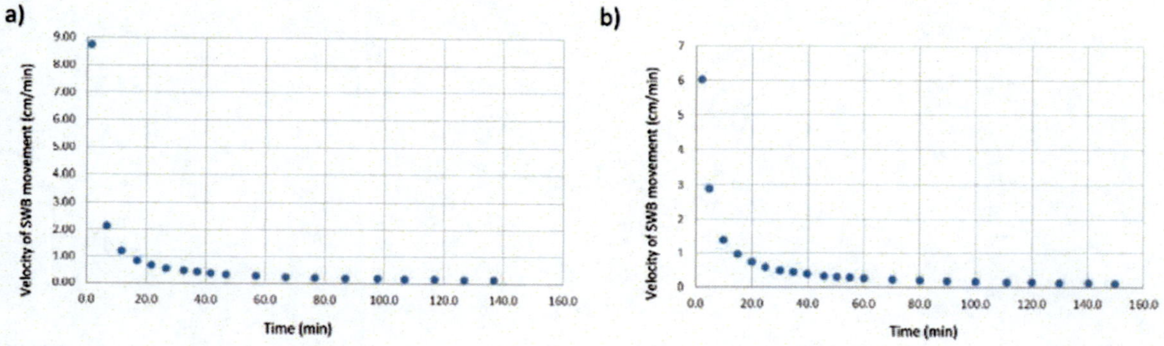

Fig. 3 Relationship between the velocities of the SWB movement and time at **a** 6 m and **b** 12 m of the GOR traverse

4 Conclusions

The use of time-lapse GPR survey showed promising results in estimating the infiltration rate. Although the velocity values calculated using the proposed methodology are different from the results of the conventional ring infiltrometer method, the relationship between time and velocity has similar pattern in both methodologies. Since the infiltration test uses the velocity of the downward movement of water at the surface, while the GPR method estimates the SWB velocity within the subsurface, further analysis is needed to examine if the Horton equation can be replaced by another experimental equation to calculate equilibrium water velocity using the time-lapse GPR data. Future work includes also the use of higher frequency antenna with smaller time interval to improve both the spatial and temporal resolution of the proposed methodology.

References

Bano, M. (2006). Effects of the transition zone above a water table on the reflection of GPR waves. *Geophysical Research Letters, 33*(13)

Peng, G., Ruiyan, W., Gengxing, Z., & Yuhuan, L. (2020). The application of GPR to the detection of soil wetted bodies formed by drip irrigation. *PLoS ONE, 15*(7).

Toushmalani, R. (2010). Application of geophysical methods in agriculture. *Australian Journal of Basic and Applied Sciences, 4*(12).

Tzanis, A. (2010). MatGPR release 2: A freeware MATLAB® package for the analysis & interpretation of common and single offset GPR data. *FastTimes, 15*(1), 17–43.

Wang, R., Yin, T., Zhou, E., & Qi, B. (2022). What indicative information of a subsurface wetted body can be detected by a ground-penetrating radar (GPR)? A laboratory study and numerical simulation. *Remote Sensing, 14*(18).

Advanced Geophysical Processing of Ground-Penetrating Radar and Magnetic Datasets

Rui Jorge Oliveira, Bento Caldeira, Teresa Teixidó, and José Fernando Borges

Abstract

Due to the presence of an excessive amount of noise in the data, archeological geophysics frequently produces results that cannot be used to evaluate the content that may exist in the subsurface. If it is impossible to differentiate between signal and noise, excessive noise will result. Its genesis, when it is attributable to heterogeneities in the ground (overthrows, corners) that produce as many reflections as structures that may exist, prevents a fair assessment of the subsoil composition. Low perceptibility circumstances happen when there is no contrast between buried structures and the surrounding environment. It could be due to elements that are harmful to the method (metals and ceramics in the magnetic method; clay and water in the electromagnetic method) or when the buried objects are formed of the same material as the surrounding medium. The problem of the identification and selection of useful signals in geophysical data is one of the research topics of effective methodologies of archeological geophysics, with the aim of producing more accurate models that allow a more precise interpretation of structures buried beneath the earth. In this paper, three approaches conceived by the team are described, which allow: to increase the sharpness of the GPR models by reducing the background noise through factoring techniques applied in the 2-D spectral domain; to increase the resolution of the models by increasing the density of the profiles with Fourier interpolation; and to increase model information by combining maps from the two geophysical methods, using data fusion techniques that combine mathematical transformations and statistical analysis.

Keywords

Archeological geophysics • Advanced processing • Digital signal processing • Geophysical data enhancement

1 Introduction

This paper describes three geophysical methods for enhancing geophysical datasets. Even under ideal conditions for the use of geophysical technologies, such as a broad region to be surveyed with few obstructions, the data produced may have issues with excessive noise, density/sampling, or lack of perceptibility (Oliveira, 2020). In the sections that follow, some of these challenges are addressed and a proposed solution is offered.

2 Advanced Geophysical Processing to Enhance Geophysical Data

2.1 Removal of Clutter Noise in GPR Datasets

Due to the reflections caused by the collapse of structures, GPR datasets taken from archeological sites are frequently influenced by clutter noise. If this noise predominates, the important data may be obscured, and the user will be unable to draw conclusions regarding hidden structures on the site. The elimination of clutter noise in the spectral domain (Oliveira et al., 2021) is an efficient method for filtering GPR datasets with this issue.

Clutter reflection noise filtering for GPR data (Oliveira, 2020) is a method based on singular value decomposition (SVD) (Chen et al., 2019). Considering the 2-D spectral domain of GPR data, this simplifies the parametrization of filters due to their circular symmetry and constant processing effort, regardless of the amount of data. Using SVD to

R. J. Oliveira (✉) · B. Caldeira · J. F. Borges
Physics Department, Earth and Remote Sensing Laboratory and Institute of Earth Sciences, University of Évora, Évora, Portugal
e-mail: ruio@uevora.pt

T. Teixidó
Andalusian Institute of Geophysics, University of Granada, Granada, Spain

M. Bezzeghoud et al. (eds.), *Recent Research on Geotechnical Engineering, Remote Sensing, Geophysics and Earthquake Seismology*, Advances in Science, Technology & Innovation,
https://doi.org/10.1007/978-3-031-48715-6_40

calculate the modified data, the first principal component corresponds to clutter noise (Oliveira et al., 2021). By reconstructing the data considering all the principal components except the first, followed by the application of the inverse Fourier transform to restore the time–space domain of the GPR data, the filtered dataset contains only the useful information of the B-scan while automatically removing the clutter noise.

The case study used to demonstrate the success of this method displays two B-scans (Fig. 1), input data taken in the laboratory with a 1600-MHz antenna (GSSI SIR-3000) and filtered data containing only the information linked to hidden items.

2.2 Fourier Interpolation of GPR Datasets

The GPR has a subsampling issue as a result of the data acquisition setting, which is caused by the method itself. With a sample rate of 40 to 50 traces per meter, which offers a spacing between traces of 0.025 m and 0.020 m, respectively, the data are dense in the B-scan direction. In an archeological survey, each one is normally spaced between 0.25 and 0.50 m perpendicular to the B-scan direction. There is always a data imbalance in both directions, regardless of how close the B-scans are to one another. This inquiry implies that data density may be a factor to consider. During the acquisition step, for instance, a regular square mesh might be utilized to avoid this transit impact of the antennas on the ground. On the other hand, the proposed method allows for the expansion of data and information from the initial dataset.

An approach for lateral data densification is offered as a supplement to regular processing. Interpolating GPR B-scans in the spectral domain after applying the Fourier transform (Oliveira et al., 2022) enables the creation of B-scans between each other. To calculate spatial derivatives, this method employs automatic event recognition techniques. This enables the identification of spatial discontinuities by detecting very tiny signal changes. This mitigates the subsampling issue of 3-D GPR models and enables a more efficient interpolation devoid of artifacts or signal degradation. Widely utilized in seismic processing, the approach permits the estimation of missing traces from extant traces (Spitz, 1991).

Two 3-D GPR datasets obtained in the Roman villa of Horta da Torre (Fronteira, Portugal) using a 400-MHz antenna and GSSI SIR-3000 equipment serve as the case study for testing this technology (Fig. 2).

The 2013 dataset has a profile spacing of 0.25 m. The space was reduced to 0.125 m after the density increase. Space dropped from 0.50 m to 0.25 m for the 2015 dataset. The results demonstrate an increase in the clarity and graphical information pertaining to subterranean structures of the wall and pavement variety.

2.3 Geophysical Data Fusion of GPR and Magnetic Datasets

Due to the frequent perceptibility issues associated with geophysical data, the concept of data fusion using GPR and vertical magnetic gradient (MAG) datasets was conceived. The fusion method is useful in case it is impossible to interpret the results regarding the existence of subterranean structures due to a significant lack of perceptibility.

The proposed method to combine GPR and MAG data (Oliveira et al., 2022) was based on the work of Gautam and Kumar (2015), who describes a method for fusing images acquired by separate equipment by combining the 2-D wavelet transform; multiresolution singular value decomposition; and gradient computation. The fusion method involves applying the 2-D wavelet transform to each input dataset, dividing them into sub-bands, then calculating the multiresolution singular value decomposition and the gradient. All sub-band pairs are fused according to predetermined fusion rules. We obtain the fused dataset by using the inverse of the wavelet transform to restore the spatial

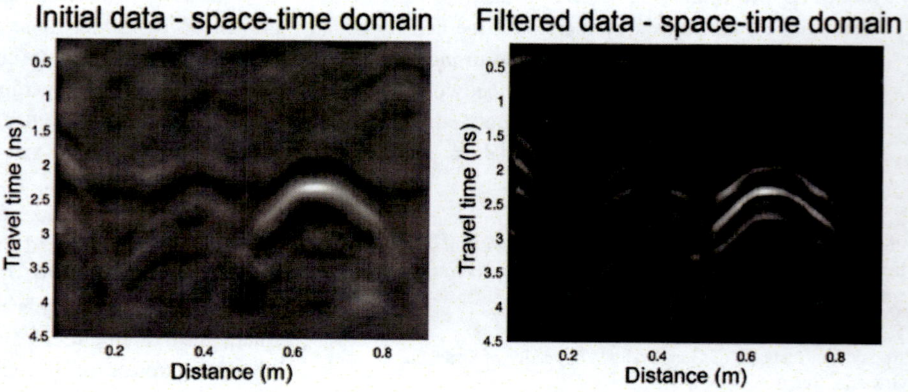

Fig. 1 On the input data, the effect of the clutter reflection noise filtering. Data that have been filtered expose information about objects buried underneath

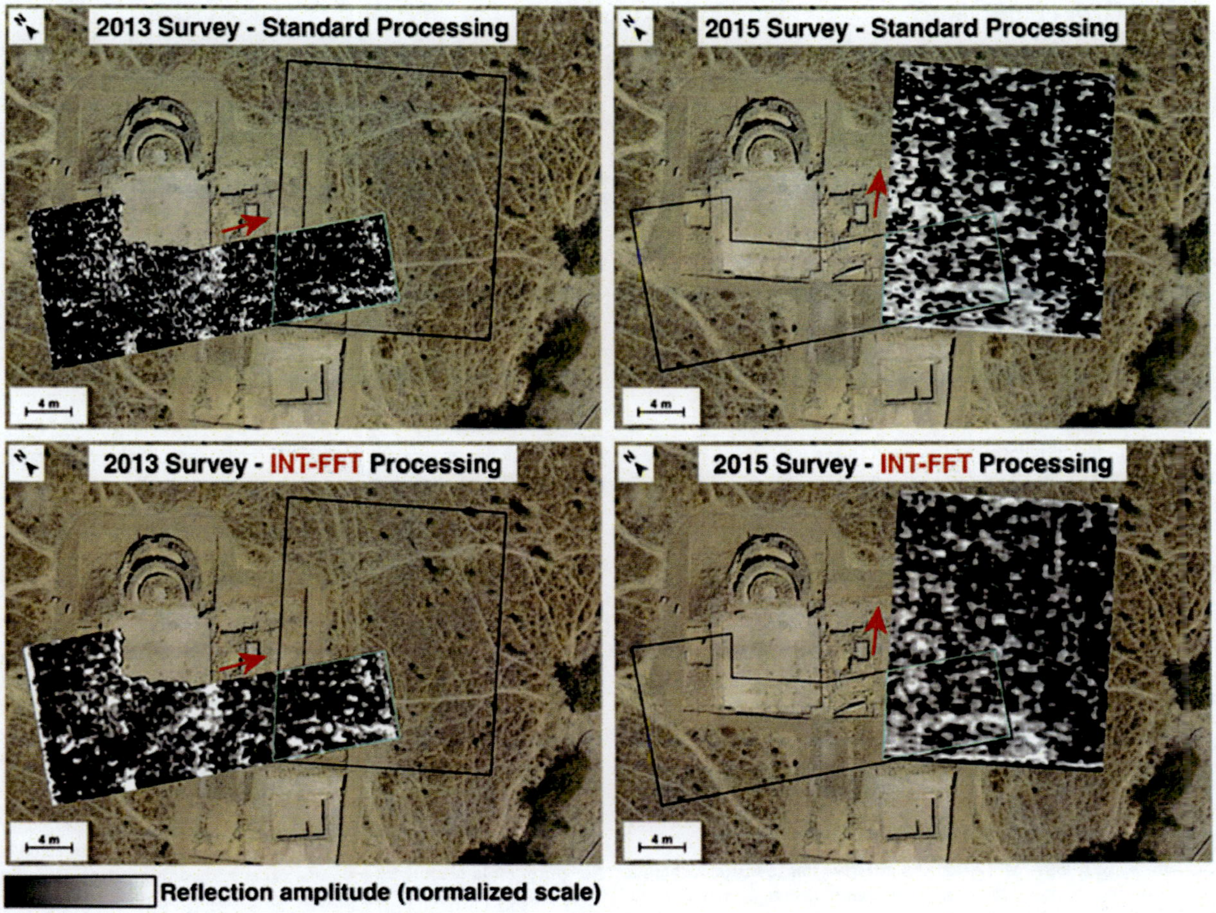

Fig. 2 The impact of Fourier interpolation on 3-D GPR datasets. The densified datasets (INT-FFT) are more precise and contain more information regarding subsurface wall and pavement features. The red arrows represent the acquisition profile direction

domain. Case study used GPR and MAG data from the Roman site of Cortijo de Quintos (Córdoba, Spain) to test data fusion approach (Peña et al., 2008).

The resulting image is of higher quality and sharpness than the individual inputs (Fig. 3). It is possible to observe spatial alignments that match to circular-shaped wall remnants. According to magnetic evidence, these artifacts are associated with ancient ovens. This site was excavated, and Roman ovens were discovered.

3 Conclusion

An important characteristic of archeological geophysics is that it is noninvasive and nondestructive, and each technique has a unique ability to discover a buried item of study. The approaches can provide a model of the subterranean that represents the spatial distribution of a particular property as a representation of reality. These models are not identical by nature, but because they refer to the same space with structures, they must have some characteristics. The greater the resolution of the measured or estimated physical parameters, the simpler it is to comprehend the resemblance between characteristics. A problem for geophysicists is to match geophysical models to glean the best and most meaningful archeological information from the object under investigation.

This work exemplifies this paradigm, as the offered innovative approaches are focused on enhancing the models arising from routine acquisition and processing. The objective of the three proposed methods is to improve the signal-to-noise ratio of GPR and magnetic data under conditions of poor detectability. This study illustrates that the implementation of this processing can improve the resolution of GPR subsurface images by lowering background noise, the resolution of models by increasing the profile density, and the information of models by combining the maps of the two geophysical methods. In addition to the standard geophysical data processing routines, these methods can be applied singly or collectively.

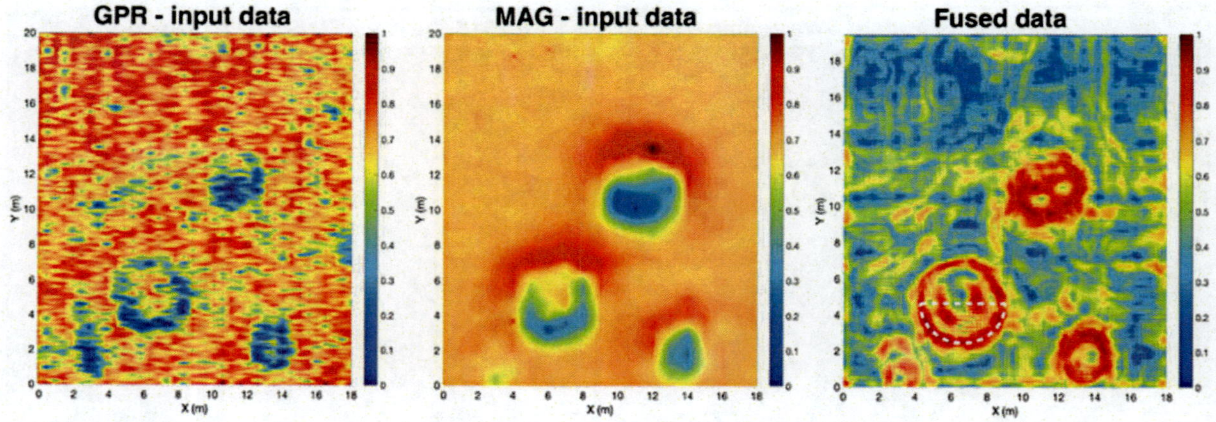

Fig. 3 Data fusion outcomes. The fused image is more informative than each of the inputs separately. Sharper circular forms emerged. The site was excavated (blank dashed line on the fused map), and it was determined to be Roman ovens

Acknowledgements The work was supported by the Portuguese Foundation for Science and Technology (FCT) by the project UIDB/04683/2020—ICT.

References

Chen, G., Fu, L., Chen, K., Boateng, C., & Ge, S. (2019). Adaptative ground clutter reduction in ground-penetrating radar data based on principal component analysis. *IEEE Transactions on Geoscience and Remote Sensing, 2019*(57), 3271–3282.

Gautam, S., & Kumar, M. (2015). An effective image fusion technique based on multiresolution singular value decomposition. *INFO-COMP, 14*(2), 31–43.

Oliveira, R. J. (2020). *Prospeção Geofísica Aplicada à Arqueologia* (Ph.D. Thesis). Institute of Research and Advanced Training, University of Évora, Évora, Portugal.

Oliveira, R. J., Caldeira, B., Teixidó, T., & Borges, J. F. (2021). GPR clutter reflection noise-filtering through singular value decomposition in the bidimensional spectral domain. *Remote Sensing, 13*, 2005.

Oliveira, R. J., Caldeira, B., Teixidó, T., Borges, J. F., & Carneiro, A. (2022). Increasing the lateral resolution of 3D-GPR datasets through 2D-FFT interpolation with application to a case study of the Roman Villa of Horta da Torre (Fronteira, Portugal). *Remote Sensing, 14*, 4069.

Oliveira, R. J., Caldeira, B., Teixidó, T., Borges, J. F., & Bezzeghoud, M. (2022). Geophysical data fusion of ground-penetrating radar and magnetic datasets using 2D Wavelet transform and singular value decomposition. *Frontiers in Earth Science*, 2089.

Peña, J. A., Teixidó, T., Carmona, E., & Sierra, M. (2008). Prospección magnética y radar 3D como métodos para obtener información a priori en la planificación de una excavación arqueológica. Caso de estudio: Yacimiento del Cortijo de Quintos (Córdoba, España). In proceeding of the 6ª Asemblea Hispano Portuguesa de Geodesia y Geofisica, Tomar (Portugal), 11–14.

Spitz, S. (1991). Seismic trace interpolation in the FX domain. *Geophysics, 56*, 785–794.

Ground-Penetrating Radar Application for 'Water in İstanbul' Project

Çiğdem Özkan-Aygün, Caner İmren, Beril Karadöller,
Lutgarde Vandeput, Jim Crow, Stefano Bordoni, Martin Crapper,
and Maria Monteleone

Abstract

'Water in Istanbul: Rising to the Challenge?' is the title of a new 24-month project which brings archaeologists, historians, engineers, and urban scientists together to explore the historical water management infrastructure and evolution of water technology in the city of Istanbul through history. One of the outcomes of this research will be producing a hydraulic model that will help us to understand how the past system functioned and was managed. This in-depth research also hopes to bring a new perspective for the contemporary water-related challenges. The archaeological field work geographically focuses on the I. Hill of the Byzantine city where Ottoman imperial palace called Topkapı gets located over. This is the most challenging topographical area regarding the historical water distribution which functioned according to the principle of gravity. Non-destructive survey methods are crucial in such fragile archaeological areas. Also, the bureaucratical difficulties related to the legal permissions make it a necessity to apply archaeogeophysical methods. In the first field work phase, we preferred ground-penetrating radar (GPR) in order to identify the remains, including supply lines providing freshwater to the Topkapı area. Our survey based on previous research by Hülya Tezcan and the archaeological survey results of Çiğdem Özkan Aygün in order to identify the areas for investigation with GPR. Thanks to GPR, the location, geometry, and depth of the buried historical structures would be possible to determine. GPR, which provides high-resolution information from shallow areas, is a geophysical method frequently used, especially in archaeogeophysical studies. The method is based on recording the travel times of reflected and scattered electromagnetic waves, which are sent to the subsurface with high-frequency antennas, with a receiver. In this study, results of the GPR study, which was carried out using a 350 MHz centre antenna frequency to detect the historical water channels beneath the area surrounding the Topkapı Palace are mentioned. After processing the data sensitively, GPR sections were interpreted and all possible water channels were detected. Depths of the detected channels are between 2 and 5 m below the surface.

Keywords

Archaeology • Archaeogeophysics • Ground-penetrating radar • Hydraulics • Water channels • Cisterns • Drainage • Topkapı Palace • Hagia Sophia • İstanbul

1 Introduction

'Water in Istanbul' project is a multidisciplinary and multinational project bringing together the academicians from İstanbul Technical University, University of Edinburgh and Northumbria University. This work is supported by Scientific Research Projects Department of Istanbul Technical University. Project Number: 43072 and The British Academy, 'Knowledge Frontiers: International Interdisciplinary Research 2021' Scheme.

İstanbul as a city with a history of more than 1700 years and challenging for freshwater since its foundation makes it a perfect laboratory for learning from the past and generating sustainable solutions for the future. Topkapı Palace area is a micro-scale representation of the Eastern Roman and Ottoman historical water technology in Istanbul. That's why we

Ç. Özkan-Aygün (✉) · C. İmren · B. Karadöller
Istanbul Technical University, İstanbul, Turkey
e-mail: ozkanci@itu.edu.tr

L. Vandeput
The British Institute at Ankara, London, SW1Y 5AH, UK

J. Crow · S. Bordoni
The University of Edinburgh, Edinburgh, EH1 1HT, UK

M. Crapper · M. Monteleone
Northumbria University, Newcastle upon Tyne, NE1 8ST, UK

M. Bezzeghoud et al. (eds.), *Recent Research on Geotechnical Engineering, Remote Sensing, Geophysics and Earthquake Seismology*, Advances in Science, Technology & Innovation,
https://doi.org/10.1007/978-3-031-48715-6_41

focus on the Topkapi for our multidisciplinary fieldwork and archival research which will provide data to model the Ottoman water management system.

Dr. Özkan Aygün of Istanbul Technical University (İTÜ) whose work on Istanbul's past water systems has provided new evidences for the subterranean hydraulic infrastructure of Hagia Sophia has also conducted an archaeological survey for documenting the network of cisterns, channels, and wells of Topkapı and surroundings (Özkan Aygün, 2010; Özkan Aygün & Güney, 2014; Özkan Aygün & Kaçan, 2014). Together with her findings, the data coming from Hülya Tezcan (1989) and Kazım Çeçen's (Çeçen & Kolay, 1997) publications construct the basis of the literal sources of this project.

Application of GPR methods has a crucial role for the new data to be obtained as a part of Dr. Aygün's survey for this new project. Those results are being shared with hydraulic engineers and modelled using GIS and engineering software to enable research on the functionality of the system.

2 Application of Ground-Penetrating Radar Method

Ground-penetrating radar is an electromagnetic method that sends high-frequency signals into the ground, especially in shallow geophysical surveys. Ground-penetrating radar was first used in glacier studies by the German geophysicist Walter Stern (1929). The method is based on the principle of recording electromagnetic waves sent into the ground by a transmitting antenna, reflected and scattered from subsurface layers, and recorded by a receiving antenna. Depending on the travel time of the reflected and scattered waves recorded by the receiving antenna, the presence, position, and geometry of structures in the ground can be determined. The electromagnetic signal, which has a harmonic structure, contains a centre frequency. The centre frequency determines the penetration depth, the amount of absorption, the vertical resolution, and the degree of scattering. As the frequency increases, the temporal resolution increases; however, the penetration depth decreases as absorption increases. Conversely, information is collected with a lower resolution from a higher penetration depth Telford et al. (1990).

In this study, approximately 5 km of data was collected using shielded antenna with a central frequency of 350 MHz (Fig. 1, yellow lines). Since the data collection was carried out in an urban environment, the external disturbances that may affect the electro wave were identified individually to consider during the analysis. Due to the presence of narrow areas in the study area, the data were collected in 4, in some cases 5, parallel profiles spaced 0.5 m apart. The data collection parameters were a sampling interval of 0.512 ns, trig-interval of 0.02 m, and recording length of 250 ns.

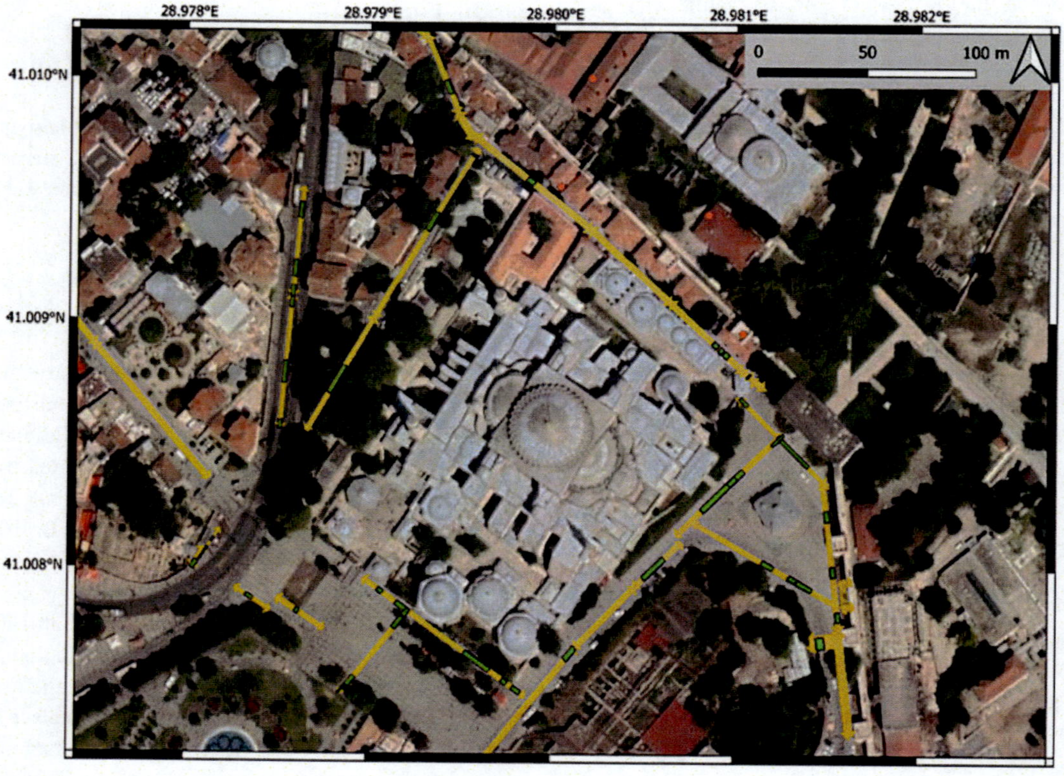

Fig. 1 Possible water channels (green polygons) shown in site location map

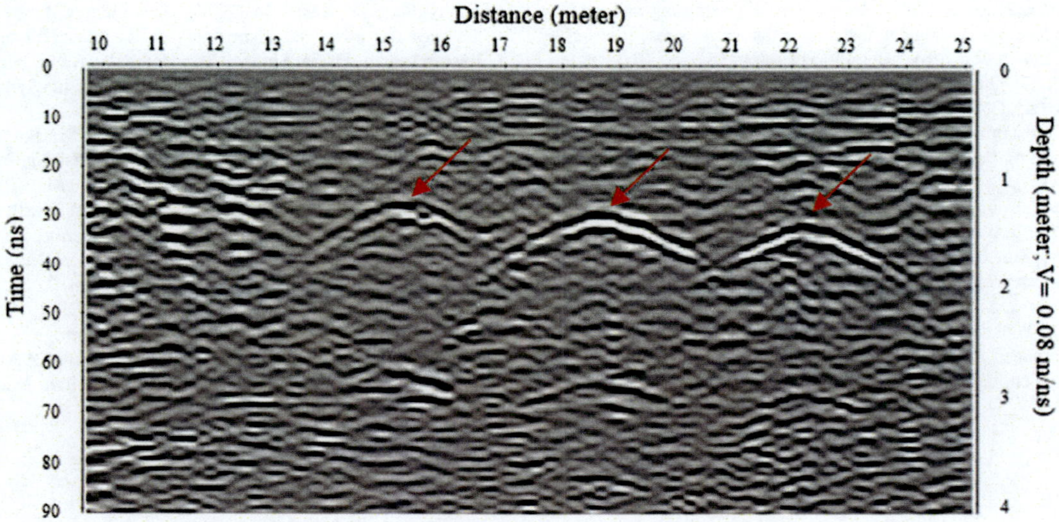

Fig. 2 Possible water channel structures (indicated with red arrows) shown in the GPR Sect. (350 MHz)

The collected raw ground-penetrating radar data were exposed to a sensitive data processing process. Different mathematical approaches are used to improve the quality of the signal, allowing an accurate understanding of the geophysical response from the ground. Ground-penetrating radar sections are analysed by applying various data processing steps to obtain the most realistic image of the subsurface. The data were subjected to different data processing steps such as 1D filtering (e.g. de-wow, bandpass, notch, deconvolution), 2D filtering (e.g. running average, background removal), start time correction, static correction, gain correction, frequency–wavenumber filter, velocity analysis, and migration.

3 Results

The analysis and interpretation of the ground-penetrating radar sections revealed geophysical signatures that are considered to be water channels and associated structures. In the GPR section shown in Fig. 2, which belongs to one of the profiles near the Topkapı Palace Bab-ı Hümayun Gate, hyperbolic features with a depth of approximately 1.5 m and a width of about 1.5 m are interpreted as three adjacent water channel structures in this site.

4 Discussion

When the penetration depth of the GPR sections is analysed, it is concluded that the structure signatures obtained could be lost in some places. Since the characteristics of the topsoil cover are different in the study area and the dielectric properties change due to varying water-holding properties, it is concluded that the 350-MHz antenna is insufficient in some areas. Therefore, it has been planned to investigate the suspected areas in detail using antennas with a larger wavelength of the source signal. With the 250-MHz antenna, the second phase of the study has started and approximately 3 km of data has been collected. The 250 MHz antenna data strongly follow the findings of the 350 MHz antenna data and reveal new information about deeper structures. The new findings suggest that some water channels may go even further down more than 5 m. Thus, lower-frequency data are planned to be collected in the third phase of the study.

5 Conclusions

It has been proven by many studies that GPR surveys are highly effective for archaeological sites in urban areas (Basile et al., 2000; Leucci & Negri, 2006; Leucci et al., 2016; Rabbel et al., 2015). Through this study, the subterranean historical water structures beneath the I. Hill of İstanbul are being studied by the help of GPR for the first time, with very promising results.

References

Basile, V., Carrozzo, M. T., Negri, S., Nuzzo, L., Quarta, T., & Villani, A. V. (2000). A ground penetrating radar survey for archaeological investigations in urban area (Lecce, Italy). *Journal of Applied Geophysics, 44*, 15–32.

Çeçen, K., & Kolay, C. (1997). Topkapı Sarayı'na Su Sağlayan İsale Hatları. İSKİ Yayınları.

Leucci, G., & Negri, S. (2006). Use of ground penetrating radar to map subsurface archaeological features in an urban area. *Journal of Archaeological Science, 33*, 502–512.

Leucci, G., De Giorgi, L., Di Giacomo, G., Ditaranto, I., Miccoli, I., & Scardozzi, G. (2016). 3D GPR survey for the archaeological characterization of the ancient Messapian necropolis in Lecce, South Italy. *Journal of Archaeological Science: Reports, 7*, 302–390.

Özkan Aygün, Ç. (2010). New findings on Hagia Sophia Subterranean and its Surroundings. *Bizantinistica: Fondazione Centro Italiano di studi sull'alto Medioevo di Spoleto,12*, 57–108.

Özkan Aygün, Ç., & Güney, C. (2014). Application of ROV, subbottom profiler and gis at archaeological surveys cases of Hazar Lake Sunken Settlement and I. Hill of Constantinople, Hagia Sophia, Topkapı Palace and Hippodrome, 20th Annual Meeting of the European Association of Archaeologists, Abstracts, pp. 10–14. Istanbul/Turkey.

Özkan Aygün, Ç., & Kaçan, M. (2014). New findings of cisterns, wells, galleries and other water structures on the First Hill of İstanbul: Researches at Hagia Sophia, Topkapı Palace and Hippodrome. ISTYAM (International Istanbul Historical Peninsula Symposium—2013), pp. 258–279. İstanbul.

Rabbel, W., Erkul, E., Stümpel, H., Wunderlich, T., Pašteka, R., Papco, J., Niewöhner, P., Bariş, Ş, Çakin, O., & Pekşen, E. (2015). Discovery of a Byzantine Church in Iznik/Nicaea, Turkey: An educational case history of geophysical prospecting with combined methods in urban areas. *Archaeological Prospection, 22*, 1–20.

Stern, W. (1929). Versuch einer elektrodynamischen Dickenmessung von Gletschereis. *Gerlands Beitrage Zur Geophysik, 23*, 292–333.

Telford., W. M., Geldart, L. P., & Sheriff, R. E. (1990). *Applied Geophysics*. Cambridge University Press. Cambridge.

Tezcan, H. (1989). Topkapı Sarayı Çevresinin Bizans Devri Arkeolojisi. Türkiye Turing ve Otomobil Kurumu Yayınları. İstanbul.

Exploration and Evaluation of Groundwater Potential Using Geoelectrical Method in Arid Zones: A Case Study of Majel Bel Abbes Aquifer (West-Central Tunisia)

Mouez Gouasmia, Ferid Dhahri, Abdelkader Mhamdi, Moumni Lahmadi, and Mohamed Soussi

Abstract

The region of Majel Bel Abbes, west-central Tunisia, is characterized by a semi-arid to arid climate. Groundwater is the main source of water for domestic, agricultural and industrial purposes in the region. In order to better recognize the architecture and the lithology of the aquifer system of Majel Bel Abbes and evaluate its hydrogeological potentialities, a geophysical study based on electrical prospection was carried out. The latter allowed the delimitation of two potential aquifers including (i) a Plio-Quaternary hydrogeological unit exhibiting 30 to 50 Ω.m resistivity values and variable lithologies that thicken toward the central part of Majel Bel Abbes basin and (ii) a Miocene sandy hydrogeological unit exhibiting ∼20 Ω.m resistivity values and acknowledged as Beglia Formation. This unit rests on the Campanian–Maastrichtian limestones of the Abiod Formation that overlays the Cenomanian limestones to the east of the basin and disappears toward their borders. The new results concerning the Majel Bel Abbes basin architecture and their hydrogeological unit lithologies, thickness and potentialities should be considered by the local authorities in the planning and design of new drilling hydraulics wells to ensure safe management of water supplies.

M. Gouasmia (✉) · F. Dhahri · A. Mhamdi
Faculty of Sciences of Gafsa, University of Gafsa, Sidi Ahmed Zarroug, 2112 Gafsa, Tunisia
e-mail: mouez.gouasmia@yahoo.fr

M. Gouasmia · A. Mhamdi · M. Soussi
Faculty of Sciences of Tunis, LB18ES07: Sedimentary Basins and Petroleum Geology, University of Tunis El Manar, El Manar II, 2092 Tunis, Tunisia

F. Dhahri
Faculty of Sciences of Tunis, Laboratory LR18ES37: Geodynamics, Geomaterials and Geo-Digital, University of Tunis El Manar, El Manar II, 2092 Tunis, Tunisia

M. Lahmadi
Water Resources Division, 2100 Gafsa, Tunisia

Keywords

Geoelectrical method • Groundwater • Majel Bel Abbes

1 Introduction

The region of Majel Bel Abbes located in west-central Tunisia is characterized by a continental arid to semi-arid climate with irregular and scarce rainfall (170 mm/year). Their watersheds are characterized by limited hydrographic networks with seasonal and often occasional surface flows making groundwater the main source of water supply. In order to better recognize the hydrogeological units in this region and to evaluate their potentialities, a geophysical study using vertical electrical sounding was carried out. The geoelectrical methods have been successfully used in groundwater exploration worldwide. They use the electrical properties of rocks to determine the nature of the geological layers and their fluids content. Also, the thickness and depth of the aquifer formations and the chemical quality of the water are well assessed (Gouasmia et al., 2018, 2021).

2 Site Description

The Majel Bel Abbes basin exemplifies a synclinal trough in the so-called "central Tunisian atlas." The region exhibits several NE–SW folded structures affected by multidirectional fault network. The study area is limited to the north and to the east by the Gour Souane, Hoguef, Toual, Nadhour and Sidi Aich Cretaceous cored anticlines, to the west by the Algerian border (Fig. 1), and to the south, by the Gafsa North plain drained by Kbir and Sidi Aich wadis. It constitutes a basin filled by continental detrital sediments that thicken toward its subsiding central part. Faults are especially mappable within the competent

M. Bezzeghoud et al. (eds.), *Recent Research on Geotechnical Engineering, Remote Sensing, Geophysics and Earthquake Seismology*, Advances in Science, Technology & Innovation,
https://doi.org/10.1007/978-3-031-48715-6_42

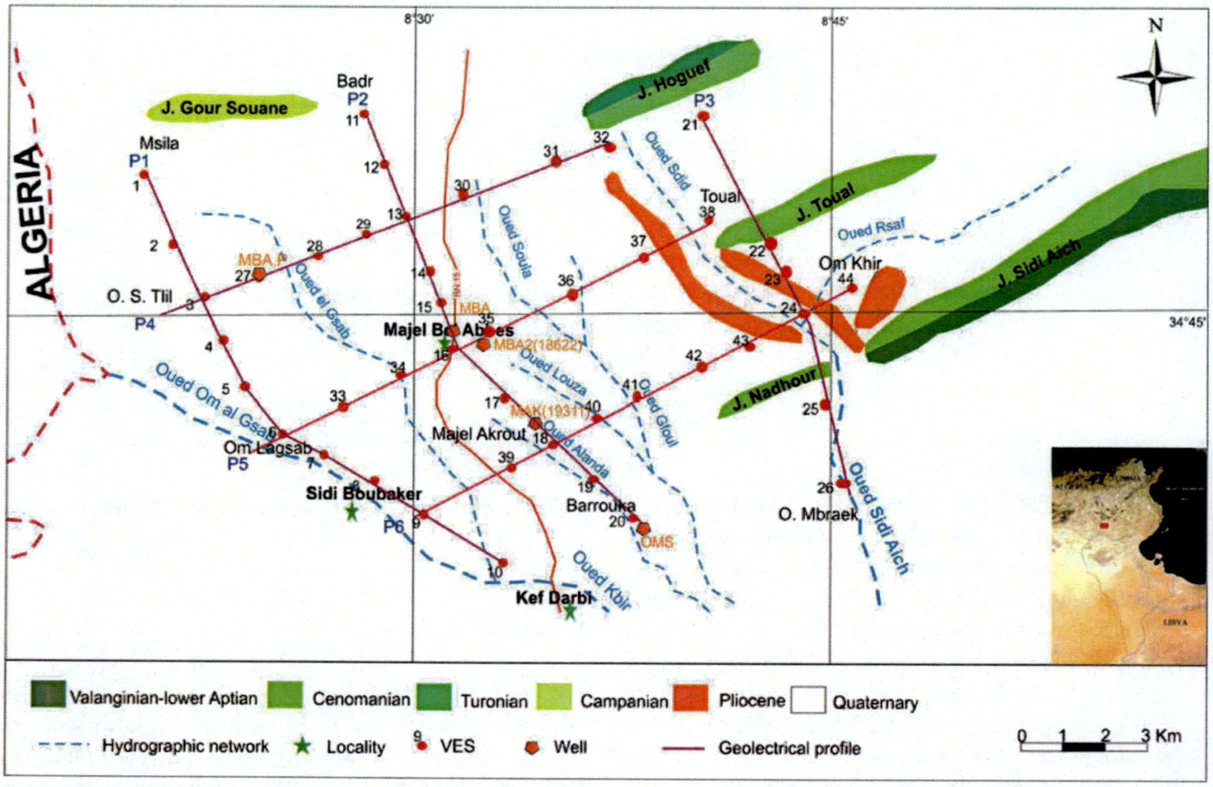

Fig. 1 Simplified geological map of the study area with location of the vertical electrical sounding and the geoelectrical profiles

Cretaceous rocks that core the anticlinal structures; however; near Majel Bel Abbes plain, they are sealed by the detrital Mio-Plio-Quaternary fill. The most important aquifer of this region is the Plio-Quaternary, and indeed, it is intensively exploited. The Miocene sandstone aquifer characterized by a high porosity and a significant thickness is however exploited by a unique well (Gouasmia et al., 2022).

3 Results and Discussion

To achieve the main objectives of this study, we have implemented 46 vertical electrical sounding (VES) using Schlumberger field arrangements with 1000–2000 m length AB. These VES allowed to establish three ENE–WSW profiles (P1–P3) and three NNW–SSE profiles (P4–P6) (Fig. 1). Only P2 and P5 are illustrated in Fig. 2. The geoelectrical profiles were established after the interpretation of VES to determine the geoelectric parameters of the different layers, namely the real resistivity and thickness. The inversion of the VES data must be calibrated by wells data that provide real information on the lithology and hydrogeological nature of the different investigated levels. Lithological data from MBA2 (18622) and MAK (19311) wells (Fig. 1) were used to validate the geophysical results and to estimate the resistivities of the different lithological levels. The VES model was used to elaborate the geoelectrical profiles that allowed to distinguish the following six different geoelectrical layers, from top to bottom:

- The first level includes the uppermost detrital materials with contrasted resistivity attributed to the heterogeneous Quaternary facies;
- The second geoelectrical unit displays 30 to 50 Ω.m resistivities varying corresponding to Plio-Quaternary sandy, clayey and clayey–sandy facies. It constitutes a main hydrogeological unit that thickens toward the center of the Majel Bel Abbes basin and exceeds 400 m thick to the SW of Majel Bel Abbes city.
- The third level is moderately conductive with 15–27 Ω.m resistivities. It corresponds to the Miocene sandy Beglia Formation drilled by MAK19311 well in Majel Akrout (Fig. 1). The top of this Formation deepens to about 580 m under Majel Bel Abbes city. It exhibits ~100 m thick near Majel Bel Abbes city; however, it thins progressively toward Majel Akrout, in the SE, due to the deformation of the subsurface geologic layers responsible

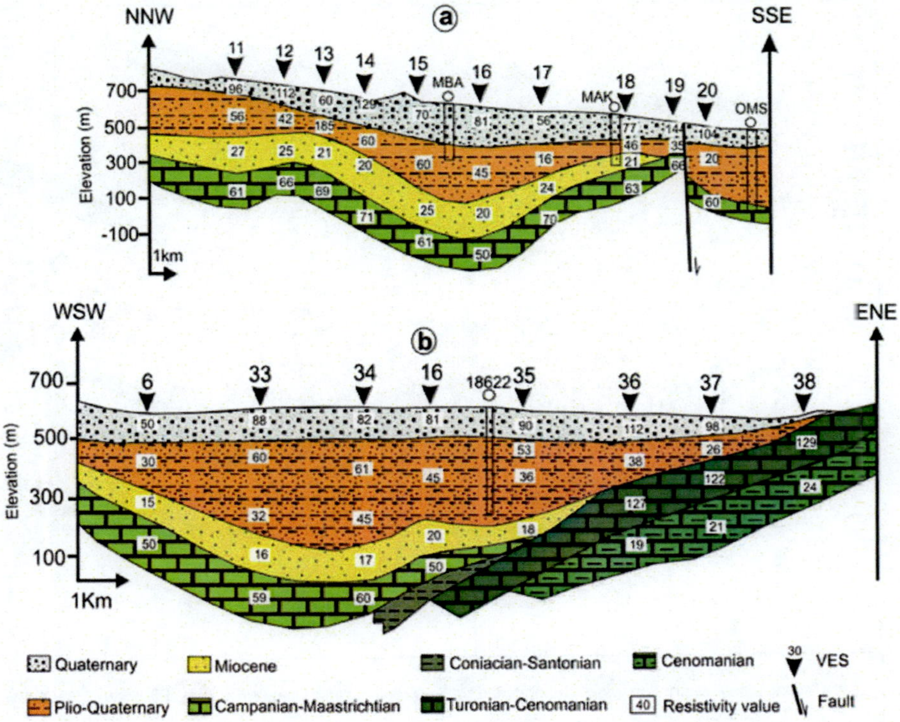

Fig. 2 Geoelectrical profiles (**a** P2 and **b** P5)

also of the gap of this Formation in the eastern part of the study area where Plio-Quaternary series rest unconformably over Upper Cretaceous series (Fig. 2).

- The fourth and fifth levels are both resistant and exhibit, respectively, 50–72 and >100 Ω.m resistivity. These levels are interpreted as carbonated facies and are attributed to the Campanian–Maastrichtian (Abiod Formation) and the Turonian–Cenomanian, respectively.
- The last level exhibits low resistivities of 19–24 Ω.m and corresponds plausibly to the Cenomanian marls and limestones alternations.

The lateral and vertical distribution of VES data and their possible association in many ranges of resistivity intervals allowed to distinguish six geoelectrical layers. The Upper Cretaceous and Miocene layers appear to be deformed; however, the post-Miocene series come to fill the generated trough. The geoelectrical interpretation seems accurately done for a depth of investigation near 700 m; however, in the deepest part of the geoelectrical profiles, some uncertainties can occur due to tectonic deformations and similar resistance of rocks from different formations. Therefore, the clayey facies that appears between the fourth and fifth geoelectrical levels in Fig. 2b is not easy to decern at a significant depth, and it is represented as a logical interpretation of the Coniacian–Santonian facies expected to be intercalated there.

From a hydrogeological point of view, the geoelectric results have well characterized the two aquifers and elucidated the basin architecture. Indeed, the Plio-Quaternary aquifer seems to be unconfined and occupies wide area. Its recharge is ensured by the direct infiltration of meteoric waters and surface flows from wadis. The unexploited Miocene sandy aquifer represents a potential strategic reserve for future water supply. Thus, the results allowed us to underline the lateral (and so vertical) extension of these sands, and Fig. 3 delimits the area of possible exploitation of this aquifer. As recommendation, a piezometric well can be drilled to the southwest of Majel Bel Abbes to monitor groundwater levels, determine the Miocene aquifer parameters and provide groundwater sample for chemical analysis. It will help also in deciphering possible hydrodynamic relations between the aquifers systems of Majel Bel Abbes in the north and Gafsa North in the south.

4 Conclusions

The geoelectrical study of the aquifer system of Majel Bel Abbes basin allowed to recognize the basin architecture, the different lithologies of its sedimentary fill and their hydrogeological characterization at about 700 m depth. The results showed that Majel Bel Abbes basin exemplifies a structural trough in which Mio-Plio-Quaternary series

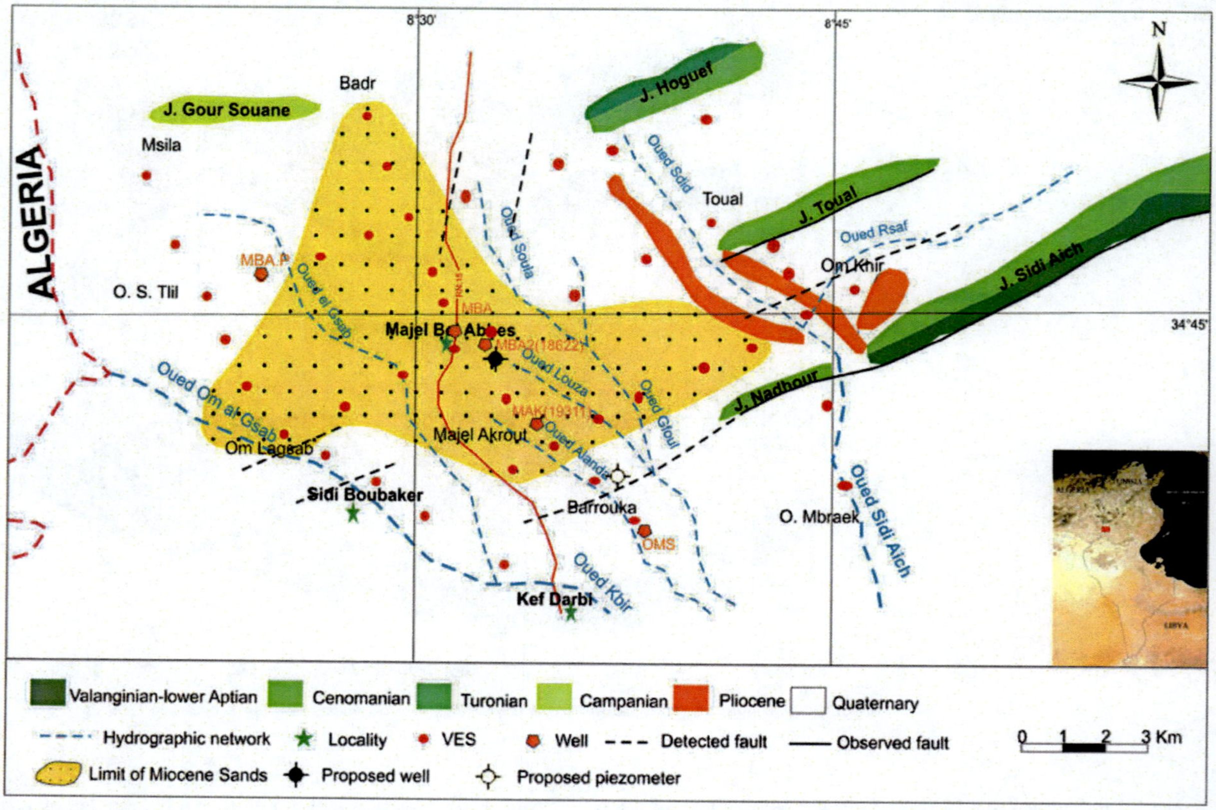

Fig. 3 Synthesis map showing the favorable area for the exploitation of the Miocene aquifer in Majel Bel Abbes region

thicken toward the central part. These series are mainly made of heterogeneous detrital sediments with important hydrogeological potentialities. The Miocene sands drilled in Majel Akrout and the Plio-Quaternary aquifer units constitute together an unconfined aquifer system. The deep Miocene aquifer in the region is not exploited yet and is considered a potential strategic reserve for future water supply. The results of this study will be useful for the local authorities in the planning and design of new monitoring and drilling wells to ensure safe management of groundwater resources in the region.

References

Gouasmia, M., Dhahri, F., Abdelkader, M., Salhi, I., Gabtni, H., & Soussi, M. (2021). Integrated hydrogeological study of a tectonically controlled aquifer system: The Rohia-Sbiba graben, Central Tunisia. *Journal Earth System Sciences, 130*, 221. https://doi.org/10.1007/s12040-021-01718-8

Gouasmia, M., Khorchani, H., Mhamdi, A., Dhahri, F., & Soussi, M. (2018). Hydrogeological characterization of a carbonate aquifer using geophysical and geochemical approach: Case of the Krachoua Formation in Tataouine area. *Southern Tunisia. Arabian Journal of Geosciences, 11*(24), 786. https://doi.org/10.1007/s12517-018-4150-x

Gouasmia, M., Mhamdi, A., Dhahri, F., & Soussi, M. (2022). Hydrogeological and hydrochemical characterization of Mejel Bel Abbes aquifer (West-Central Tunisia). In Chenchouni, H. et al. (Eds.), *New prospects in environmental geosciences and hydrogeosciences. CAJG 2019. Advances in science, technology & innovation (IEREK interdisciplinary series for sustainable development)*. Springer. https://doi.org/10.1007/978-3-030-72543-3_123.

On the Origin of Sand Injectites in Bahrah Area, Northern Kuwait Bay: A Fluid Dynamic Numerical Modeling

Filippo Marchelli, Renzo Di Felice, Mubarak Al-Hajeri, and Aimen Amer

Abstract

This work aims at describing from a fluid dynamic point of view the occurrence and the origin of sand injectites outcrops observed in two sites within the Bahrah coastal plain area, northern Kuwait Bay, through analogies with the knowledge on spouted beds. For non-cohesive systems, the channel formation can be easily reproduced through Eulerian–Eulerian computational fluid dynamics (CFD) simulations. When a certain fluid inlet velocity is set, the fluid breaks through the bed of particles creating the central channel. Cohesiveness, most likely present and relevant in the field of interest, was introduced by simulating the fluid–sand multiphase system with the CFD software adapted to satisfactorily describe previously reported field observation. To simulate such systems, the volume-of-fluid (VOF) method yielded the best results. In fact, cohesiveness can be taken into account through the Bingham model, which introduces additional parameters to estimate, and these parameters determine whether the fluid can create a central vertical channel or diagonal cracks. The CFD simulations proved to be a valid tool to reproduce laboratory-scale observations and may further explain the mechanisms behind these enigmatic formations, confirming the role of fluid–solid drag in the creation of the injectites. Future works will address further development of this approach and the application on larger scales.

Keywords

Sand injectites • Computational fluid dynamics • Multiphase systems • Kuwait Bay • Fluid–solid momentum exchange

1 Introduction

The origin and properties of sand injectites have been investigated by scientists for centuries, with many aspects still being unclear (Hurst et al., 2011). Two recent works have focused on those that can be observed in the Bahrah coastal plains, northern Kuwait Bay, pointing out their features and differences through various analyses (Al-Hajeri et al., 2020, 2021).

Several geological phenomena, including sand injectites, are caused by fluid–solid transport phenomena. The underlying physical mechanisms are similar to those that can be found in different fields, such as fluidization in chemical engineering. Thus, similar modeling strategies may be useful to obtain numerical predictions that would not be possible to achieve experimentally. Fluidization has often been reproduced through computational fluid dynamics (CFD) simulations, based on the numerical solution of local balance equations. Different techniques permit reproducing the solid phase, with advantages and disadvantages (Guo & Yu, 2017; Moliner et al., 2019a; Pan et al., 2016), also for the case of soils (Zhou et al., 2019).

Experimental observations from the literature can be relevant in the context of sand injectites formation: Nermoen (Nermoen, 2010) shows a comparison of the different behaviors of non-cohesive and cohesive solid particles when subjected to a fluid flow. This is comparable to the discrepancies between Site 1 and Site 2 reported in the previous papers about the Bahrah injectites (Al-Hajeri et al., 2020, 2021). Cohesive forces are common in soils, complicating the simulation setup. Several approaches have been

F. Marchelli · R. Di Felice (✉)
University of Genova, 16145 Genova, Italy
e-mail: renzo.difelice@unige.it

M. Al-Hajeri
Exploration Group, Kuwait Oil Company, Al Ahmadi, Kuwait

A. Amer
Schlumberger, Al Khobar, Saudi Arabia

M. Bezzeghoud et al. (eds.), *Recent Research on Geotechnical Engineering, Remote Sensing, Geophysics and Earthquake Seismology*, Advances in Science, Technology & Innovation,
https://doi.org/10.1007/978-3-031-48715-6_43

proposed, whose validity depends on the origin and intensity of the forces (Xu et al., 2022). This preliminary work investigates the application of such approaches to a physical configuration that may be useful in understanding the origin of the Bahrah injectites, reinforcing the hypotheses on their origin.

2 Methods

We performed the simulations with the well-known commercial program Ansys Fluent. For the sake of brevity, the equations of the various models are not included here but can be found in the program's guide or in previous publications. The expected fluidization pattern resembles that of spouted beds (Moliner et al., 2017), which are often simulated via CFD, with both discrete- (Marchelli et al., 2019) and continuum-based (Moliner et al., 2019) methods for solids. Since the involved particles are very fine, the latter are more appropriate due to the lower computational requirements (Moliner et al., 2019). Non-cohesive particles are often simulated through the reliable and established kinetic theory of granular flows (KTGF) (Moliner et al., 2019). This approach can be modified to include cohesive forces (Zhong et al., 2020), but depending on the mechanism, the volume-of-fluid (VOF) approach may be more suitable (Chen et al., 2014; Wang & Song, 2019), especially when the solid medium is so compact that the fluid cannot pass through it. In this case, the two phases can be considered as immiscible. Cohesiveness can thus be included in the viscosity, often through the Bingham approach (Nogami & Yagi, 2004), which was employed here.

The simulations focused on a 2D case, with a 7 m width and a 12 m height, meshed with squared cells with dimensions of 1 cm and a fluid (air or water depending on the case) entering the volume vertically upward from a 10-cm nozzle at the center of the base. The simulations began with a bed of sand (density 2500 kg/m^3 and diameter 0.25 mm) settled at the bottom, with an initial height of 6 m, and were run until a steady state was reached (typically a few seconds).

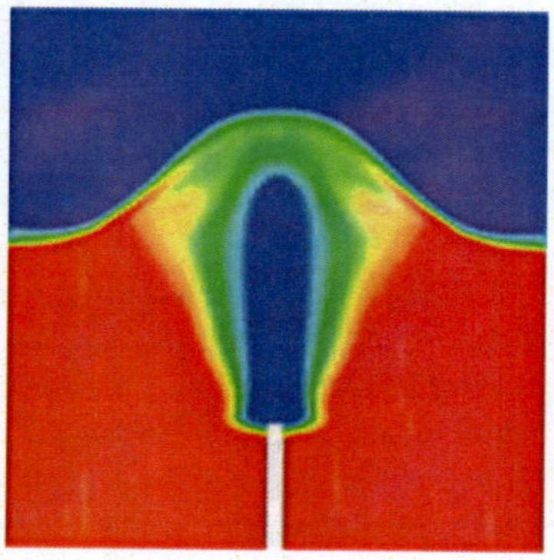

Fig. 1 Volume fraction of sand and air in a non-cohesive case; red is where sand reaches its maximum volume fraction (0.6), whereas blue is 100% water

3 Results and Discussion

For the case of non-cohesive particles, Eulerian–Eulerian simulations employing the KTGF seem adequate to reproduce the expected behavior of the solid bed. As an example, Fig. 1 shows the results of a simulation of this kind, which is similar to the fluid dynamic regime of a spouted bed (Moliner et al., 2019). The mathematical treatment for this case is straightforward as the literature is abundant. The values of the fluid superficial velocity or its pressure, combined with the solids properties, are usually enough to verify whether it is sufficient to break through the bed of solids and create a channel, which can be verified through either empirical correlations (Moliner et al., 2017) or simulations (Marchelli et al., 2017).

For the cohesive case, various approaches were tested; Fig. 2 shows a selection. The VOF approach seems more adequate to represent the less ideal cases, in which cohesive forces are so strong that hinder the upward movement of the flow. Conversely, with the KTGF, the fluid is always able to pass through the bed of particles, which always has a non-null porosity. Cohesive particles are more challenging

Fig. 2 Volume fraction of sand and air in cohesive cases, which from left to right are: KTGF air–sand simulation with the approach of Zhong et al. (2020); KTGF water–sand simulation with a null restitution coefficient; VOF water–sand simulation with low cohesive forces; VOF water–sand simulation with high cohesive forces

but more interesting from a modeling point of view, as it is a much more unexplored topic. The combination of the VOF method and the Bingham approach seems to be the key to reproduce some peculiar formations that involve asymmetrical shapes. On the other hand, the large number of input parameters may be detrimental for a field in which experimental observations and measurement are limited, so it is fundamental to limit the modeling complexity as much as possible.

4 Conclusions

CFD simulations are able to confirm our initial speculations about the analogies between the geological phenomena that lead to the formation of the Bahrah sand injectites and spouted beds. The theoretical and empirical knowledge on spouted beds can be fruitfully applied to the formation these injectites, whose formation process can be simulated by different CFD techniques. The flexibility and detail of these simulations may prove to be useful in future studies to show the influence of several phenomena and operating variables, helping to shed more light on these enigmatic formations.

References

Al-Hajeri, M. M., Green, D., Amer, A., Najem, A., Al-Refaei, Y., Naqi, M., & Al-Qattan, F. (2021). The diagenetic evolution of sand injectite outcrops in the mixing groundwater–seawater zone of Bahrah coastal plains, northern Kuwait Bay. *Arabian Journal of Geoscience, 14*, 19.

Al-Hajeri, M. M., Amer, A., Djawair, D., Green, D., & Al-Naqi, M. (2020). Origin of enigmatic sand injectite outcrops associated with non-tectonic forced-folding structure in Bahrah area, northern Kuwait Bay. *Marine and Petroleum Geology, 115*, 104268.

Chen, T. L., Zhang, L. Y., & Zhang, D. L. (2014). An FEM/VOF hybrid formulation for fracture grouting modelling. *Computers and Geotechnics, 58*, 14–27.

Guo, Y., & Yu, X. (2017). Comparison of the implementation of three common types of coupled CFD-DEM model for simulating soil surface erosion. *International Journal of Multiphase Flow, 91*, 89–100.

Hurst, A., Scott, A., & Vigorito, M. (2011). Physical characteristics of sand injectites. *Earth-Science Review, 106*, 215–246.

Marchelli, F., Bove, D., Moliner, C., Bosio, B., & Arato, E. (2017). Discrete element method for the prediction of the onset velocity in a spouted bed. *Powder Technology, 321*, 119–131.

Marchelli, F., Moliner, C., Bosio, B., & Arato, E. (2019). A CFD–DEM study of the behaviour of single-solid and binary mixtures in a pyramidal spouted bed. *Particuology, 42*, 79–91.

Moliner, C., Marchelli, F., Ong, L., Martinez-Felipe, A., Van der A, D., & Arato, E. (2019). Sensitivity analysis and validation of a Two Fluid Method (TFM) model for a spouted bed. *Chemical Engineering Science, 207*, 39–53.

Moliner, C., Marchelli, F., Bosio, B., & Arato, E. (2017). Modelling of spouted and spout-fluid beds: Key for their successful scale up. *Energies, 10*, 38.

Moliner, C., Marchelli, F., Curti, M., Bosio, B., Rovero, G., & Arato, E. (2019). Spouting behaviour of binary mixtures in square-based spouted beds. *Particuology, 43*, 193–201.

Moliner, C., Marchelli, F., Spanachi, N., Martinez-Felipe, A., Bosio, B., & Arato, E. (2019). CFD simulation of a spouted bed: Comparison between the Discrete Element Method (DEM) and the Two Fluid Model (TFM). *Chemical Engineering Journal, 377*, 120466.

Nermoen, A. (2010). *Some effects of gas-induced fluidization in dry granular media*. University of Oslo.

Nogami, H., & Yagi, J. (2004). An application of Bingham model to viscous fluid modeling of solid flow in moving bed. *ISIJ International, 44*, 1826–1834.

Pan, H., Chen, X. Z., Liang, X. F., Zhu, L. T., & Luo, Z. H. (2016). CFD simulations of gas-liquid-solid flow in fluidized bed reactors—A review. *Powder Technology, 299*, 235–258.

Wang, T., & Song, B. (2019). Study on deepwater conductor jet excavation mechanism in cohesive soil. *Applied Ocean Research, 82*, 225–235.

Xu, H., Wang, W., Ma, C., Zhong, W., & Yu, A. (2022). Recent advances in studies of wet particle fluidization characteristics. *Powder Technology, 409*, 117805.

Zhong, H., Zhang, Y., Xiong, Q., Zhang, J., Zhu, Y., Liang, S., Niu, B., & Zhang, X. (2020). Two-fluid modeling of a wet spouted fluidized bed with wet restitution coefficient model. *Powder Technology, 364*, 363–372.

Zhou, Z. Q., Ranjith, P. G., Yang, W. M., Shi, S. S., Wei, C. C., & Li, Z. H. (2019). A new set of scaling relationships for DEM-CFD simulations of fluid–solid coupling problems in saturated and cohesiveless granular soils. *Computational Particle Mechanics, 6*, 657–669.

Fluid Identification in Carbonate Stringer Reservoirs: A Modeling Study from Oman

Hilal Al-Obaidani and Mohammed Farfour

Abstract

Carbonate stringers are self-charging reservoirs residing inside salt structures and known for their challenging characterization and fluid identification. We build a geological model that conceptualizes stringers in some fields from south Oman Salt Basin. We use the model to produce P-wave and S-wave velocity, and density models. The models are used to create pre-stack gathers from which AVO attributes are extracted. The gathers are inverted back to impedances and density models using pseudo-well logs that are created from the input models. The AVO analysis demonstrated that AVO attributes may not be able to detect the carbonate reservoirs because of the increase of Vp/Vs from the Ara salt to the reservoir rock. However, the post-stack and pre-stack inversion results suggested that the reservoirs can be detected using different elastic products, namely Vp/Vs, Poisson ratio, bulk modulus, and Lambda Rho among other impedance-based fluid indicator. The result from the study will be used for the interpretation of a new seismic data which is acquired in the area.

Keywords

Carbonate reservoir • Seismic • Fluid • AVO • Stringers • Oman

1 Introduction

AVO analysis attempts to study the behavior of amplitude with respect to offset (or angles) and relate amplitude variations to lithology and fluids (Farfour, 2020). Generally, fluid expressions can be addressed using fluid indicators computed from primary AVO attributes, namely the intercept and gradient or P-wave and S-wave reflectivities. Over the years, a number of successful uses of AVO for hydrocarbon detection and reservoir characterization have been published. Examples include: Farfour (2020) and Foster et al. (2010).

Seismic inversion is the process that allows interpreter to retrieve impedances of geological layers from processed seismic traces. Different inversion types and workflows are available in industry. Some of the inversions convert stacked seismic traces to impedances, while others convert pre-stack NMO-corrected gathers to P-wave impedance, S-wave impedance, and density models. Geophysicists use Zoeppritz equations and their approximations to model or invert seismic data. These equations help compute impedances and density from traces or compute traces from the impedance and density models.

Carbonate stringers are known for their complicated characterization due to their heterogeneity, deposition, and diagenetic properties. Carbonate stringers in southern part of the Sultanate of Oman are a typical example of such challenging reservoirs. Different challenges have been encountered in their imaging and characterization (Al-Barawani & McClay, 2005; Al-Siyabi, 2005). The aim of this work is to improve the understanding of these reservoirs by testing different approaches used in reservoirs delineation and characterization. This includes AVO modeling and analysis, post-stack and pre-stack inversion.

2 Methods

We create P-wave velocity, S-wave velocity, and density model. The models are inspired from Al-Jahdhami and Healey (2013) and from Al-Siyabi (2005). The elastic models are used to create synthetic gathers which are then implemented to compute AVO attributes. The AVO

H. Al-Obaidani (✉) · M. Farfour
Sultan Qaboos University, Muscat, Oman
e-mail: hilal.saleh@tethysoil.com

M. Bezzeghoud et al. (eds.), *Recent Research on Geotechnical Engineering, Remote Sensing, Geophysics and Earthquake Seismology*, Advances in Science, Technology & Innovation,
https://doi.org/10.1007/978-3-031-48715-6_44

attributes are used to perform AVO analysis by computing different fluid factors. Next, the gathers are used for AVO inversion to derive impedance and density models within seismic resolution. From P-wave and S-wave impedances, we compute different impedance-based attributes such as Vp/Vs, Poisson ratio, bulk modulus, lame parameters, and impedance fluid indicators introduced by Russell (2014).

3 Results and Discussion

Figure 1 shows P-wave model that we built for the field of the study. The model shows that the carbonate stringers are characterized by their high P-wave velocity, which differentiates them from their surrounding salt bodies. Due partly to their high compaction and to the high attenuation of seismic energy caused by the host salt, fluid presence may not cause large drop in velocity. This makes identifying fluid-saturated interval challenging. However, Vp/Vs model in Fig. 2 showed a better capability of fluid detection.

We used values for Vp, Vs, and density from published papers and from well logs in a nearby field to build the models and to conduct AVO and fluid modeling. The analysis inferred that fluid expressions are not easy to observe in seismic data in general and in pre-stack gathers in particular. The difficulty of detecting fluid using AVO attributes is because of the increasing Vp/Vs at the top of the reservoir unlike the case of common AVO classes in clastic reservoirs. However, we found also that in case the fluid presence will cause a considerable drop in Vp/Vs of the reservoir to values lower than that of the salt, we can start see fluid expressions.

Figure 3 illustrates modeled gathers using Zoeppritz equations at the present conditions and in case Vp/Vs drops. We observe that we will have clear drop in amplitude at far offset which will reduce the amplitude in stack data.

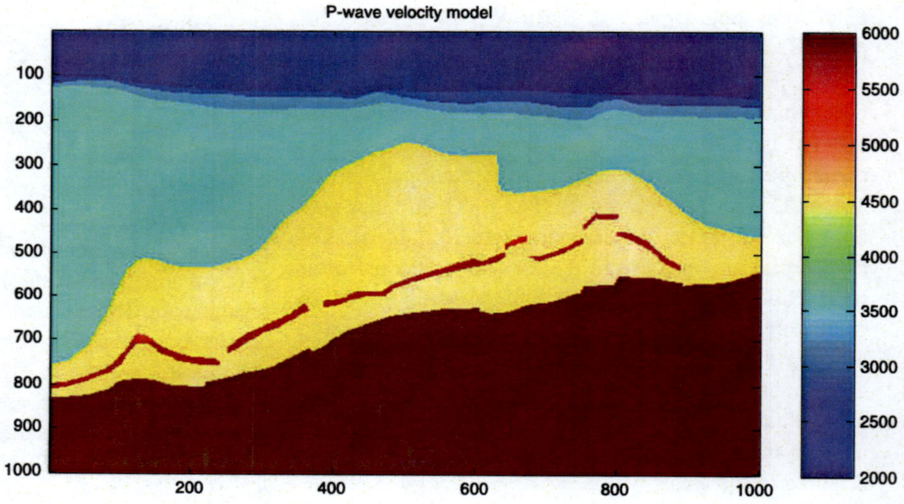

Fig. 1 P-wave velocity model showing carbonate stringers in Ara salt

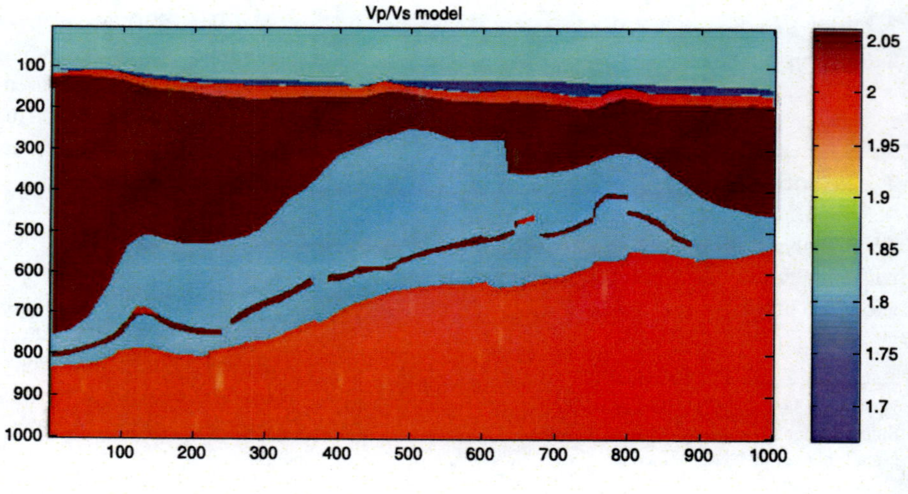

Fig. 2 Vp/Vs model showing carbonate stringers in Ara salt. Notice that fluid-saturated zones are observable

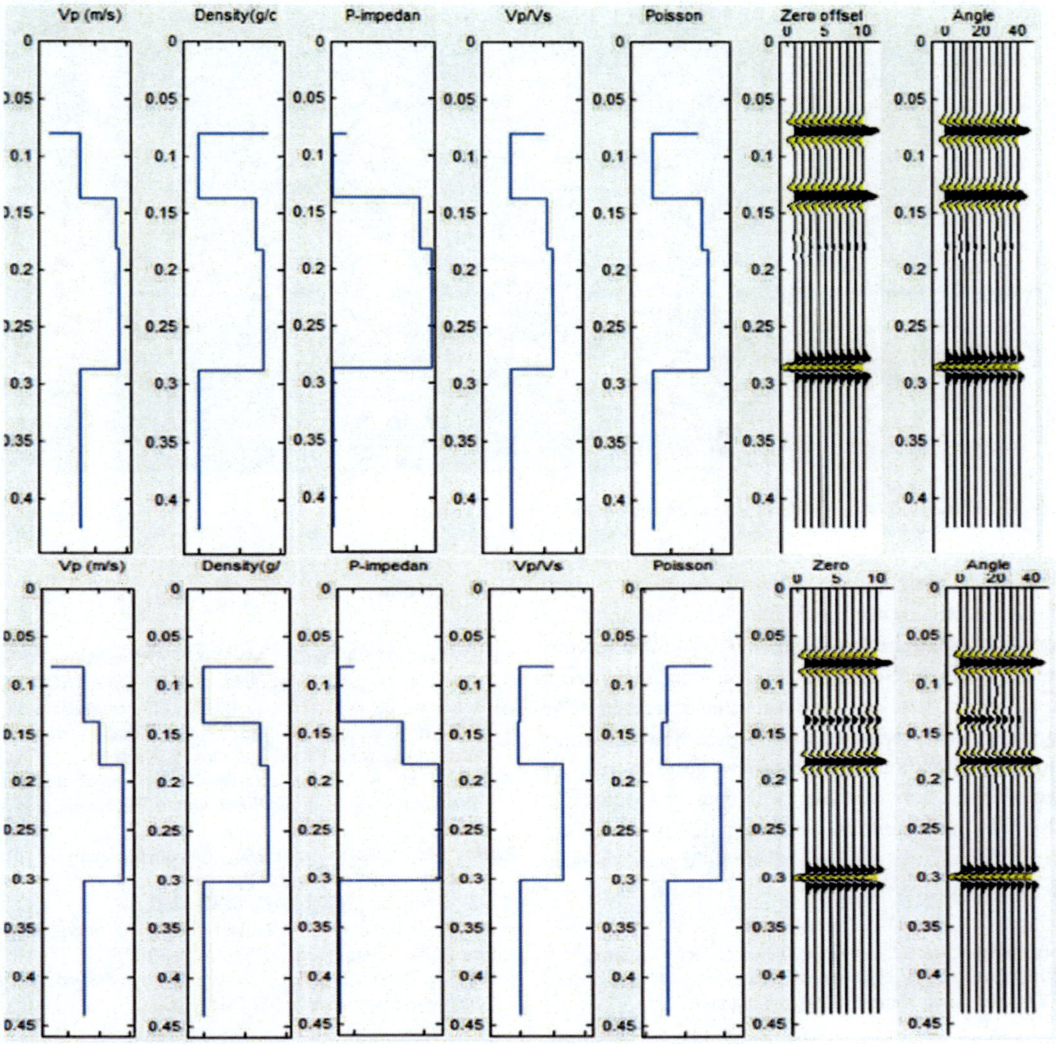

Fig. 3 AVO modeling results with present Vp/Vs (top) and after Vp/Vs drops (bottom). Note the drop in amplitude of the reservoir at far offset (middle reflection)

Next, to go beyond reflectivity-based AVO, we carried out pre-stack seismic inversion to produce P-impedance, S-impedance, and density models from the gathers. We created a pseudo-well passing by the fluid-saturated interval. The data in the well include: Vp, Vs, and density. A number of inversion workflows were implemented and evaluated. The inversion tests demonstrated the importance and the impact of the initial model on the inversion results. We first used only horizons picked at the top and bottom of the Ara salt. We found that the salt trend will influence the trend of the reservoir and will even mask it in some areas. Next, we included a horizon picked at the top of the carbonate reservoir to better constrain the inversion process. Editing the latter horizon has improved significantly the inversion models and their imaging ability and accuracy. From the inversion, we extracted a number of impedance-based fluid indicators such as Vp/Vs, Poisson ratio, bulk modulus, and Lambda Rho, and others. The latter attributes demonstrated a better capability in detecting the fluid in the reservoir (Fig. 4).

4 Conclusions

In this study, we conducted a geophysical modeling which involves elastic properties (Vp, Vs, and density) and seismic properties (synthetic gathers and stacks) to examine the ability of AVO reflectivity and impedance attribute for detecting hydrocarbon-saturated carbonate reservoirs from

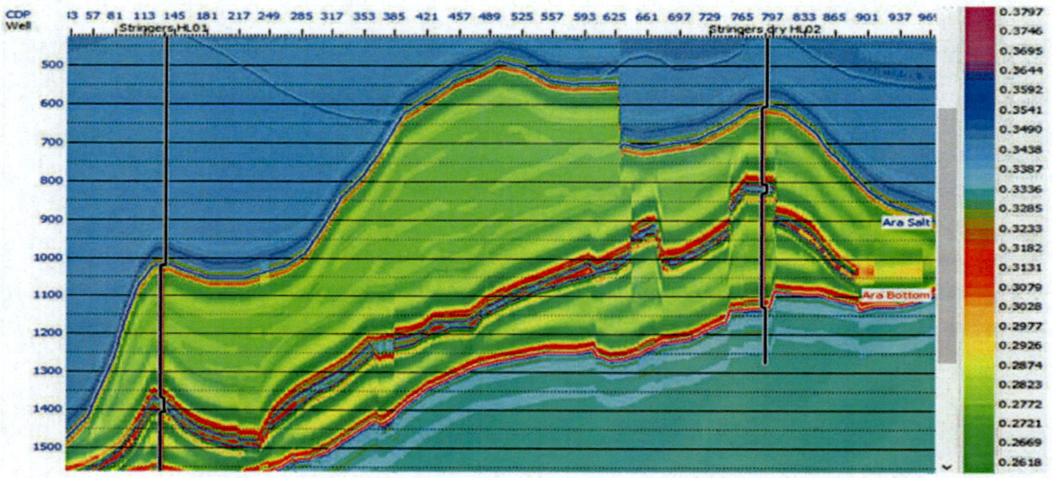

Fig. 4 Poisson ratio derived from Vp/Vs which is derived from pre-stack inversion

South Oman Salt Basin. The AVO attributes derived from the synthetic gathers showed a difficulty in detecting fluid expressions because of the increasing Vp/Vs of the reservoir. However, attributes derived from AVO inversion showed a better capability in detecting fluid-saturated reservoirs. The study suggests that to achieve a good detection and characterization of the carbonate reservoirs, a good number of interpreted seismic horizons and well data is required. It shows also that the reservoirs will be visible in AVO and inversion attributes if the presence of the fluid causes a drop in Vp/Vs of the reservoir.

Acknowledgements We acknowledge Tethys Oil for supporting this research project and for the PhD scholarship. We would like also to thank CGG for providing Hampson-Russell software.

References

Al Jahdhami, N., & Healey, M. (2013). *SPE middle east oil and gas show and conference, Manama, Bahrain* (SPE-164227-MS).

Al-Barawani, B., & McClay, K. (2008). Salt tectonics in the Thumrait area, in the southern part of the South Oman Salt Basin: Implications for mini-basin evolution. *GeoArabia, 13*(4), 77–108.

Al-Siyabi, H. A. (2005). Exploration history of the Ara intrasalt carbonate stringers in the South Oman Salt Basin. *GeoArabia, 10*, 39–72.

Farfour, M. (2020). Amplitude components analysis: Theory and methods. *The Leading Edge Journal, Society of Exploration Geophysicists (SEG). 39*(1), 62a1–62a6.

Foster, D. J., Keys, R. G., & Lane, F. D. (2010). Interpretation of AVO anomalies. *Geophysics, 75*(5), 75A3–75A13.

Russell, B. (2014). Prestack seismic amplitude analysis: An integrated overview. *Interpretation, 2*, SC19–SC36.

Structural Interpretation of the Essaouira Basin and Its Coastal Zone in Central-Western Morocco from Gravity Data: Hydrogeological Implications

Abdelah Khouz, Mohammed Jaffal, Blaid Bougadir, Fatima El Bchari, Jorge Trindade, Siham Afraou, Azzouz Kchikach, Mustapha El Ghorfi, Mourad Jadoud, Ahmed Manar, and Jean-Louis Bodinier

Abstract

Located in the central-western part of Morocco, where the climate is arid and semi-arid, the Essaouira basin is an area at high risk of water shortage due to the decrease in precipitation as a consequence of climate change on the one hand, and on the other hand to the incessant increase in the demand for this resource necessary for life. That is why, more and more underground water resources are used to meet the needs of the population, whether for domestic use or for irrigation. However, the mobilization of these resources requires a good knowledge of the aquifers of the Essaouira basin. The present study was undertaken with this perspective. Its main purpose is to achieve a better understanding of the deep structure of this basin. This study is based on the analysis of gravity data. The methodological approach involves, in addition to the qualitative interpretation of these data, the use of various filtering processes that help map hidden geological structures. The obtained results reveal the existence of a network of faults that complement the structural map of the study area. Some negative anomalies are clearly associated with evaporate deposits. However, most of them are in all probability related to structural lows and subsided depressions that represent favorable areas for groundwater accumulation and, consequently, favorable zones for drilling water exploitation boreholes.

Keywords

Gravity • Filtering • Contact analysis • Structure • Hydrogeology • Essaouira basin • Morocco

1 Introduction

During the last decades, Morocco has been strongly affected by climate change, which was responsible for the random advent of several periods of drought, especially in arid and semi-arid regions. Located in the central-western part of Morocco, where this kind of climate prevails, the Essaouira basin is not spared from the effects of these unforeseen hazards. It constitutes an area of high risk of water scarcity because of the rainfall decline and the increasing demand for this necessary resource. Consequently, the groundwater of the Essaouira basin is subjected to intense exploitation in order to meet the needs of water supply for irrigation and domestic use. However, the mobilization of these resources requires a good knowledge of the hydrogeology of this

A. Khouz (✉) · B. Bougadir
Laboratory of Applied Sciences for the Environment and Sustainable Development (SAEDD), Higher School of Technology Essaouira, Cadi Ayyad University, Marrakech, Morocco
e-mail: abdellah.khouz@gmail.com; 1902243@estudante.uab.pt

A. Khouz · J. Trindade
Centre for Geographical Studies, IGOT, Universidade de Lisboa, Lisbon, Portugal

Universidade Aberta, Lisbon, Portugal

M. Jaffal · B. Bougadir · S. Afraou · A. Kchikach · M. E. Ghorfi
Georessources, Geoenvironment and Civil Engineering Laboratory, Faculty of Sciences and Techniques, Cadi Ayyad University, 40000 Marrakech, Morocco

F. E. Bchari
Department of Earth Sciences, Polydisciplinary Faculty, Cadi Ayyad University, Safi, Morocco

M. Jaffal · A. Kchikach · M. E. Ghorfi · J.-L. Bodinier
Geology and Sustainable Mining Institute, Mohammed VI Polytechnic University, 43150 Benguerir, Morocco

M. Jadoud
Geosciences and Environmental Techniques Laboratory, Faculty of Sciences El Jadida, Chouaïb Doukkali University, 24000 El Jadida, Morocco

A. Manar
Ministry of Energy Transition and Sustainable Development, Rabat, Morocco

J.-L. Bodinier
Geosciences Montpellier, University of Montpellier and CNRS, 34095 Montpellier, France

M. Bezzeghoud et al. (eds.), *Recent Research on Geotechnical Engineering, Remote Sensing, Geophysics and Earthquake Seismology*, Advances in Science, Technology & Innovation,
https://doi.org/10.1007/978-3-031-48715-6_45

region, especially the structure of its aquifers. The present study was undertaken with this perspective. It aims to investigate the deep geological structure of the Essaouira basin using a geophysical approach based on the interpretation of gravity data. The methodology involves, besides the qualitative analysis of these data, the use of different filtering techniques to delineate the major structural features of the study area, which includes the Essaouira basin and its coastal zone. This area represents a vast domain that extends over 3000 square kilometers and is circumscribed by a rectangle defined by the latitudes 31.22°N and 32.01°N and by the longitudes 9.77°W and 9.1°W. The study area belongs to the Atlantic plains zone of central Morocco, which extends from El Jadida to Agadir. In the absence of a perennial stream network in this area, groundwater is the main source of water supply for domestic needs and irrigation.

2 Methodology

The gravity data used in this study were kindly provided by the Moroccan Ministry of Energy and Mines. They are available as a Bouguer anomaly map established with a reduction density of 2.3 g/cm3 after a measurement campaign conducted by the General Geophysics Company in 1975 on behalf of the Bureau of Research and Mining Participations (BRPM, Morocco). The map provided as a scanned image was digitized in order to have a digital database necessary for the subsequent processing and mapping steps. Such processing particularly includes the contact analysis approach that allows the detection of geological structures and the determination of their dip. This approach is based on the analysis of the gravity field gradient using horizontal gradient and upward continuation techniques. It constitutes a useful tool for gravity data interpretation which is particularly effective in mapping tectonic faults delineated by density contrasts (Archibald et al., 1999; Jaffal et al., 2010, 2022).

3 Results and Discussion

The residual gravity map shows several positive and negative anomalies designated P1 to P10 and N1 to N13, respectively. The superimposition of these anomalies on the geological map of the study area provides information on their origin. Indeed, some negative anomalies are clearly associated with salt deposits. This is the case of the anomalies that coincide with the anticlinal folds with Triassic saline core of Jebel Hadid (N1) and Tidzi (N8 and N9). The N10 anomaly is also related to Triassic evaporates recognized in the Smimou area. In addition, the circular shape of the N11 anomaly would similarly indicate the presence of a salt dome in the south of Essaouira. The other negative anomalies and the positive anomalies would, respectively, reflect structural lows and highs due to the synclinorium structure of the basin, as evidenced by anomalies N5, N6 and N7, which coincide with synclinal troughs whose core is occupied by Eocene formations. Therefore, except for salt-core anticlines, we can consider that the negative anomalies would represent depressions that act as receptacles for groundwater accumulation.

Furthermore, the filtering of the gravity data using the horizontal gradient (HG) and upward continuation (UC) techniques helps detect various geological structures and determine their dip direction. The maxima of the HG that underline the highlighted contacts were superimposed on the residual gravity map. This clearly shows that these contacts mark the boundary between the positive and negative anomalies. They have different shapes, lengths, directions and dips. In fact, most of the contacts represent linear structures. Some extend over several tens of kilometers in different directions: N–S, NE–SW or NW–SE. Others have circular shapes that mark the boundary of salt domes (N8, N9, N10 and N11) or relatively closed synclinal troughs (N7).

At the end of this study, a map summarizing all the results was established. This document shows that the study area is affected by a faults network of varying length and direction. The corresponding rose diagram indicates that these faults are mainly oriented NNE–SSW, i.e., parallel to the Atlantic margin. This means that the structuring of the Essaouira basin is mainly marked by the distensive phase that initiated the Triassic rifting of the central Atlantic (Hafid, 2000). The detected faults delineate structural depressions where groundwater would converge, particularly at the approach to Jebel Hadid. The latter corresponds to an anticlinal structure with a saliferous core (Hafid, 2000) that generates a strong negative gravity anomaly. The existence of this anticline has major consequences for the hydrogeology of the study area. Indeed, on the one hand, it constitutes a barrier that separates the groundwater, which circulates in the Cenomanian–Turonian aquifer of the Essaouira basin (Jalal et al., 2001), from that contained in the coastal Plio-Quaternary aquifer (Fig. 1).

4 Conclusions

In conclusion, the interpretation of the gravity data based on the contact analysis technique has revealed a network of geological structures that complement the structural map of the Essaouira basin. The presence of evaporates deposits in

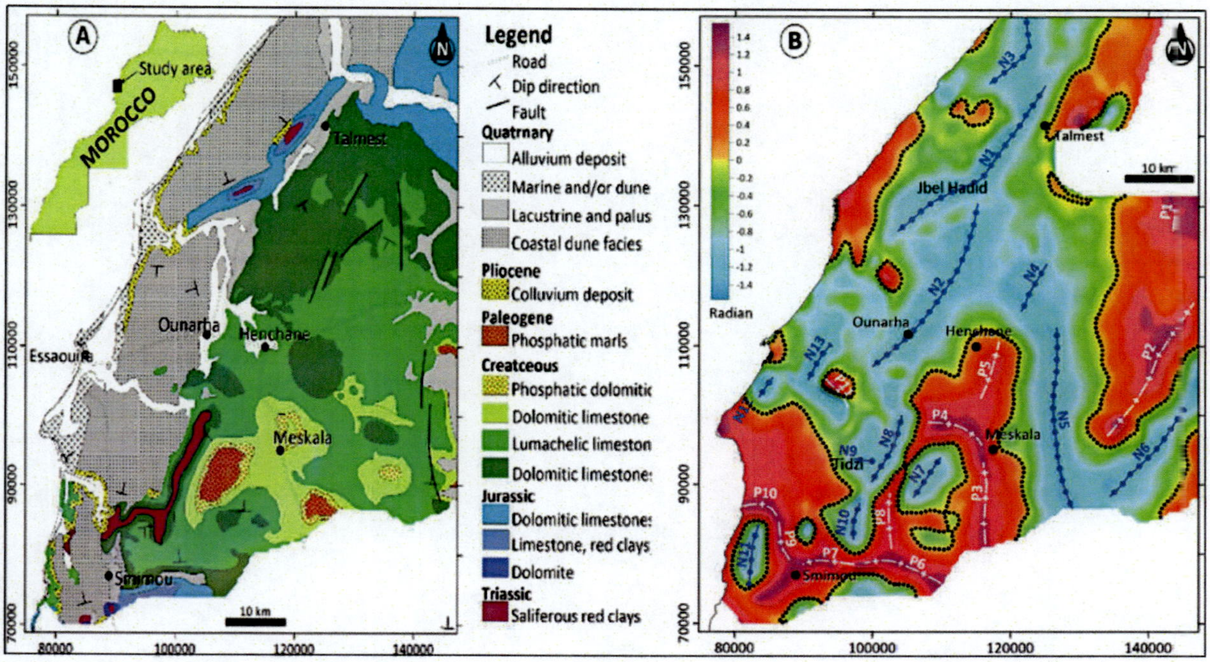

Fig. 1 a Geological map of the study area, b tilt derivative of the residual gravity anomaly

the Essaouira basin explains some negative gravity anomalies, but most of them are most likely related to structural lows that represent favorable areas for groundwater accumulation and, therefore, potentially productive zones.

On the other hand, the evaporate deposits that form the core of the anticline may affect the salinity of the groundwater of the two aquifers depending on the structural interconnection that the latter may have with the core of the anticline, which was confirmed with some wells sampling.

References

Archibald, N., Gow, P., & Bochetti, F. (1999). Multiscale edge analysis of potential field data. *Exploration Geophysics, 30*, 38–44. https://doi.org/10.1071/EG999038

Hafid, M. (2000). Triassic–early Liassic extensional systems and their Tertiary inversion, Essaouira Basin (Morocco). *Marine and Petroleum Geology, 17*, 409–427.

Jaffal, M., Charbaoui, A., Kchikach, A., El Ghorfi, M., Khaldoun, A., Safhi, A. E., Bodinier, J.-L., Khadiri Yazami, O., & Jourani, E. (2022). Gravity study of the Western Bahira Basin and the Gantour Phosphatic Plateau, central Morocco: Interpretation and hydrogeological implication. *Journal of African Earth Sciences, 193*, 104581.

Jaffal, M., El Goumi, N., Kchikach, A., Aïfa, T., Khattach, D., & Manar, A. (2010). Gravity and magnetic investigations in the Haouz basin, Morocco. Interpretation and mining implications. *Journal of African Earth Science, 58*(2), 331–340.

Jalal, M., Blavoux, B., Bahir, M., Bellion, Y., Laftouhi, N. E., Puig, J. M., Mennani, A., & Daniel, M. (2001). Etude du fonctionnement du système aquifère karstique Cénomano-Turonien de l'oued Igrounzar (Bassin d'Essaouira, Maroc). *Journal of African Earth Sciences, 32*(4), 803–817.

Geophysical Responses of Paleoproterozoic Rocks and Structures in Western Mali: Magnetic and Electromagnetics Data Analysis

Adama Youssouf Koné, Imen Hamdi Nasr, Adnen Amiri, Mohamed Hedi Inoubli, Wajdi Belkhiria, Aboubacar Denon, Souleymane Sangaré, and Saïdou Ly

Abstract

The western part of Mali belongs to the West African Craton (WAC) which abounds in significant potential for gold resources hosted in the Birimian Paleoproterozoic formations. These formations and associated structures are most often masked by the thickness sedimentary and lateritic covers. It is why need to use of geophysical methods. The geophysical signatures of these geological formations are little studied specially in the Inlier of Kedougou Kenieba. Previous studies have proposed a magnetic signature model of geological structures in western Mali, allowing to characterize some gold-bearing structures. However, many auriferous structures have a low magnetic response requiring the use of other geophysical methods such as electromagnetism. This allows understanding the resistant or conductor character of structures. The electromagnetic and magnetic data will be analyzed out in order to characterize the geophysical signatures of the structures related to gold mineralization. These results will be calibrated to geological field data in order to consolidate the geophysical interpretation. This study helped to assess the degrees of magnetization and conductivity of rocks and linear structures within western Mali. Structures related to gold mineralization have low-to-moderate magnetism with a high conductivity. This geophysical signature model is well known in volcano-sedimentary rocks along the main shear zone as well as NE–SW trend faults. These results support previous work in western Mali and are also corroborated by surface geological data. The geology shows the presence of gold in quartz veins and facies bearing sulfide minerals. Generally, the massif sulfide minerals are known by their high influence on the conductivity of rocks.

Keywords

Geophysical signature • Western Mali • Gold mineralization • Kedougou-Kenieba Inlier • Birimian

A. Y. Koné (✉) · S. Ly
Faculté des Sciences et Techniques (FST), Université des Sciences, des Techniques et des Technologies de Bamako (USTTB), Bamako, BP: E3206, Mali
e-mail: koneadama2005@gmail.com

I. Hamdi Nasr
Faculté des Sciences de Bizerte (FSB), Université de Carthage, 7021 Bizerte, Tunisie

A. Amiri · M. H. Inoubli
Faculté des Sciences de Tunis, Université de Tunis El Manar, 2092 Tunis, Tunisie

W. Belkhiria
Office National des Mines (ONM), Tunis, Tunisie

A. Denon · S. Sangaré
EurekaGeo SARL, Geoscientist Consultants, Hamdallaye ACI 2000, Bamako, Mali

1 Introduction

In western Mali, gold-bearing structures are less expressed on the surface, which gives an important place to geophysical studies in this region. Previous studies (Diallo et al., 2020; Koné et al., 2018, 2019, 2021) have characterized geological structures mostly linked to the gold showings. This was useful for gold exploration in this region. However, that there are less information on the geophysical signatures of these structures, proving in this study the reinterpretation of available geophysical data of Kossanto area. This study aims to propose geophysical responses to these structures.

2 Geological Setting

The study area is located in the Kedougou-Kenieba Inlier (KKI) within the West African Craton (WAC), which is affected by many orogenic events (e.g., eburnean orogeny) (Feybesse et al., 2006; Koné et al., 2021). WAC is composed with the plutonic, volcano-sedimentary, metamorphic and

M. Bezzeghoud et al. (eds.), *Recent Research on Geotechnical Engineering, Remote Sensing, Geophysics and Earthquake Seismology*, Advances in Science, Technology & Innovation,
https://doi.org/10.1007/978-3-031-48715-6_46

migmatites rocks. These formations are known in the Reguibat shield (in north) and that of Leo (in south) as well as the Precambrian base inlier of Kayes and Kedougou-Kenieba. The Kossanto area is located in the western within magmatic and sedimentary rocks (Fig. 1).

3 Data and Methods (Magnetic and Electromagnetic)

Aerodat collected magnetic data in 1996 with flight line spacing of 200 m, flight height-120 m and line direction of N65-245°. Quality control was completed by Sysmin project (2001). Flight characteristics are: line spacing of 3000 m, height of 120 m, line direction of N155-335° (Konate & Reid, 2010). The total magnetic intensity (TMI) map has been done and interpreted, in order to characterize magnetic signature of geological structures. The TMI allows understanding the variations of magnetic contrast linked the presence of ferromagnesian rocks. Also, it allows highlighting the magmatic sources (Blakely, 1996; Koné et al., 2018; Diallo et al., 2020).

Electromagnetic (EM) data have been collected by Aerodat in 2002 with the GEOTEM System and have the follow characteristics: line spacing of 200 m, flight height of 120 m and direction of N245° during Sysmin project (Konate & Reid, 2010). For other parameters, we have: basic frequency of 75 Hz, impulsion of 3 ms, emitter buckle of three turns, dipolar moment of 4.65 × 105 Am^2, receptor sensor with three induction coils (Konate & Reid, 2010). The conductance has been calculated using the dipolar moment. The conductivity anomalies can be due to the presence of metal-bearing material such as gold and associated minerals (sulfide).

4 Interpretation

Analysis of the TMI map shows many lineaments oriented N–S to NNE–SSW, NE–SW, NW–SE and circular anomalies (Fig. 2a). The circular anomalies constitute the magnetic response of crystalline rocks. The high magnetic signals coincide to the basic formations zones (Fig. 2a). The N–S to NNE–SSW direction is central lineament and marked by a moderate disturbance of the magnetic signal along its direction. NE–SW lineaments have low magnetic signal. Some NW–SE lineaments are marked by low magnetic response while others are marked by an abrupt passage between two poles of opposite magnetization (Fig. 2a).

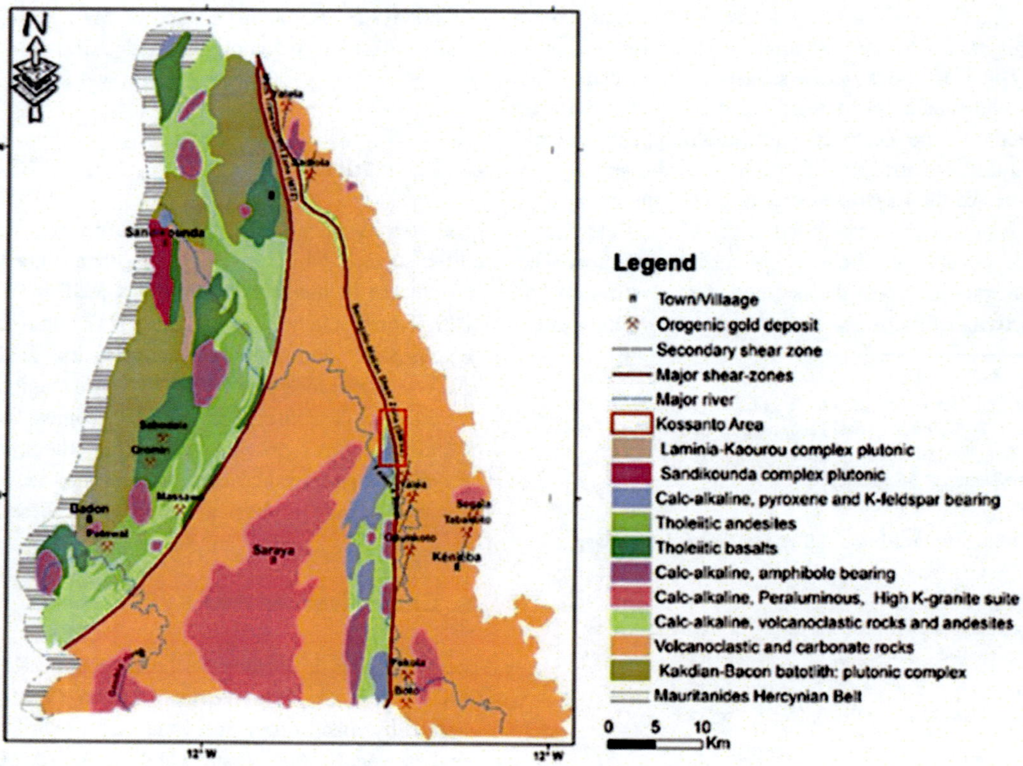

Fig. 1 Geological map of Kedougou-Kenieba Inlier showing Kossanto area (Masurel et al., 2016)

The conductivity values range from 931.66 to 11,227.33 microsiemens. The conductivity map highlights a central conductive lineament (N–S to NNE–SSW) with high conductivity (Fig. 2b). It coincides with the central magnetic lineament. It is noted NE–SW and NW–SE trends which have also a high conductivity signal. In the western part, the conductive NE–SW and NW–SE directions intersect. It is also noted that many circular anomalies coincide with resistive anomalies (Fig. 2b).

5 Results and Discussions

The central lineament corresponds to a major accident, namely senegalo-malian shear zone (SMSZ). Characterized NE–SW and NW–SE lineaments are spatially tied to this SMSZ (Fig. 2). They can be qualified as secondary structures of the SMSZ. Most of lineaments present a high conductivity response (Fig. 2b). High circular magnetic signals are mainly associated to crystalline rocks that contain a high magnetic state minerals such as ferromagnesian and/or others basic minerals. These rocks are distributed in two anomaly areas: conductive and resistive (Fig. 2b). Anomalies of Saboussiré, Kéniébandi and Koussili coincide with intermediate to basic rock zones while the Linguekoto anomaly is related to tourmaline sandstones. Those of Kolomba and Sekoto correspond to carbonates/andesites intercalated within volcano-sedimentary.

SMSZ is defined as the largest shear zone of the region with a moderately disturbed magnetic signature along its direction and a high conductivity (Koné et al., 2018, 2019, 2021). NE–SW directions have a low magnetic anomaly. They are mostly conductive and coincide with quartz vein zones and fault (Koné et al., 2018, 2019, 2021). NE–SW directions often cross those NNE–SSW, implying the presence of a shear corridor. NW–SE trends are expressed by the abrupt passage between the high and low magnetic values (Koné et al., 2018). These directions present a low conductivity, except for a few west of SMSZ. NW–SE directions mostly coincide with contact zones between magmatic and metamorphic rocks. These structures have significant geophysical responses facilitating their identification (Fig. 2).

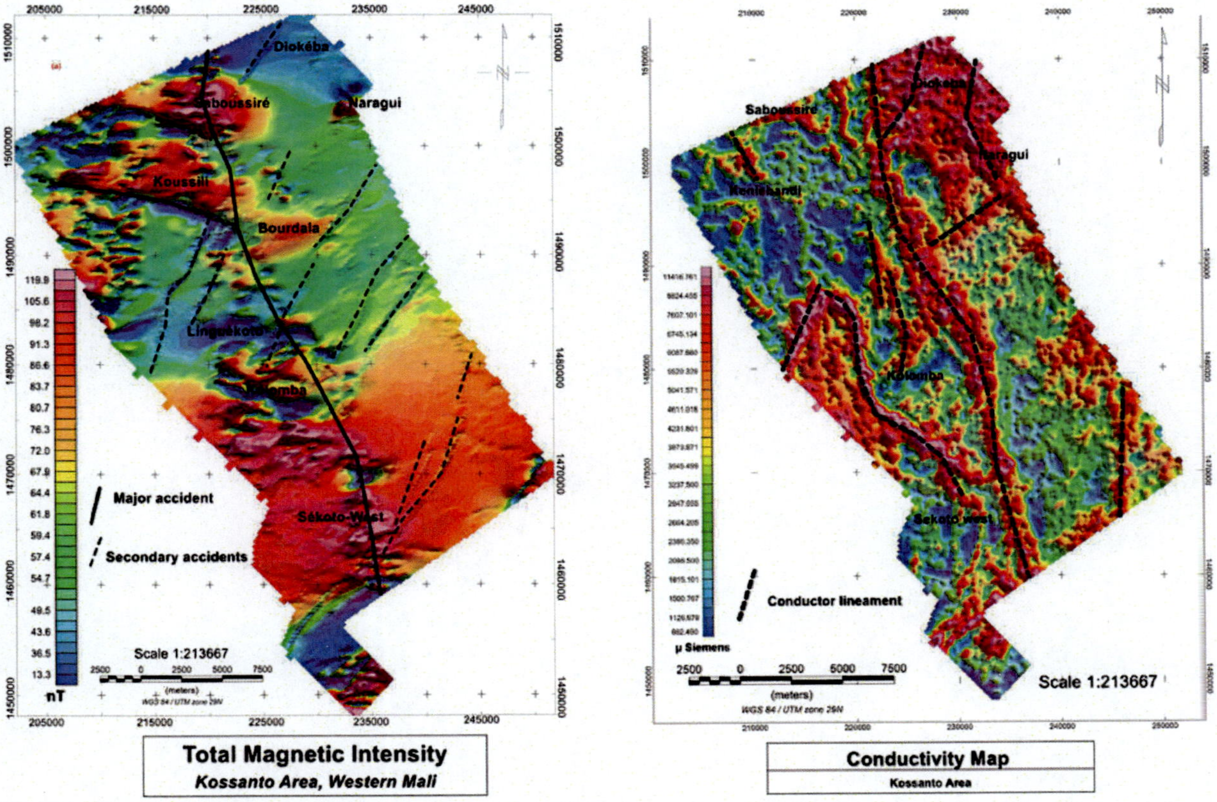

Fig. 2 **a**) Total magnetic intensity (Koné et al., 2018) and **b**) conductivity map of Kossanto area

6 Conclusions

Geophysical interpretation helped identifying circular anomalies comparable to rock areas, which have a high magnetic signature and low conductivity response. In geology, they correspond to outcrop rock areas. Also, this study allows the characterization of structural directions that have significant responses magnetic and EM. These directions correspond to SMSZ and its ramifications. Integration of magnetic and EM data permits to define geophysical signatures of geological structures in western Mali that will be contributed to the structures characterization in KKI or even the WAC.

References

Blakely, R. J. (1996). *Potential theory in gravity and magnetic applications*. Cambridge University Press.

Diallo, M., Baratoux, L., Dufréchou, G., Jessell, M. W., Vanderhaeghe, O., Ly, S., & Baratoux, D. (2020). Structures of Paleoproterozoic Kédougou-Kéniéba Inlier (Senegal-Mali) deduced from gravity and aeromagnetic data. *Journal of African Earth Science, 162*, 103732.

Feybesse, J. M., Sidibé, Y. T., Konaté, C. M., Lacomme, A,. Miehe, J. M., Lambert, A., & Zammit, C. (2006). BRGM, CPG, DNGM: Synthèse géologique du Birimien malien-Bamako (Mali): Ministère des Mines, de l'Énergie et de l'Eau, 32p.

Konate, O., & Reid, A. (2010). SYSMIN airborne geophysical surveys in Mali. *SEG Technical Expanded Abstracts*, 1117–1121.

Koné, A. Y., Nasr, I. H., Belkheria, W., Inoubli, M. H., Amiri, A., & Ly, S. (2018). Structural setting of Western Mali insights from magnetic data analysis. In Sundararajan, N., Eshagh, M., Saibi, H., Meghraoui, M., Al Garni, M., & Giroux, B. (Eds.), *On significant applications of geophysical methods. CAJG 2018. Advances in science, technology and innovation (IEREK Interdisciplinary Series for Sustainable Development)*. Springer.

Koné, A. Y., Nasr, I. H., Belkheria, W., Inoubli, M. H., Sangaré, S., & Traoré, B. (2019). Structural controls of gold mineralization in western Mali: Insights from Electromagnetic data analysis. *SEG Global Meeting Abstracts*, 383–385.

Koné, A. Y., Nasr, I. H., Traoré, B., Amiri, A., Inoubli, M. H., Sangaré, S., Qaysi, S. (2021). Geophysical contributions to gold exploration in Western Mali according to airborne electromagnetic data interpretations. *Minerals, 11*(2), 126.

Masurel, Q., Miller, J., Hein, A. A. K., Hanssen, E., Thébaud, N., Ulrich, S., Kaisin, J., & Tessougue, S. (2016). The Yatela gold deposit in Mali, West Africa: The final product of a long-lived history of hydrothermal alteration and Weathering. *Journal of Africa Earth Sciences, 113*, 73–87.

Contribution of Geophysical and Borehole Data in the Study of the Geological Structures of the Gantour Plateau and Bahira Basin (Morocco)

Anas Charbaoui, Azzouz Kchikach, Mohammed Jaffal, Jean Louis Bodinier, Bouazzaoui Eljabbar, Oussama Khadiri Yazami, Mourad Guernouche, and Es-Saïd Jourani

Abstract

Geophysical methods and geological investigations are efficient techniques for studying the geological structures of the sedimentary basins (Jaffal et al. Journal of African Earth Sciences 193, 2022). By integrating subsurface and surface geological information, we can improve geological modeling and resource exploration. To gather this information, we use an indirect approach that relies on geophysical methods and direct observation through boreholes and surface geological investigations. These investigations help acquire a variety of geoscientific data that can be used in resources exploration. This study focuses on the Bahira basin that hosts, with the Gantour plateau, an important part of Morocco's phosphate reserves. The main objective is to provide a better understanding of the deep structure of the study area. The methodological approach adopted in this study concerned (i) the combined analysis and reinterpretation of existing geophysical and borehole data and (ii) the implementation of new electrical resistivity tomography surveys. The processing and interpretation of gravity data highlight the major geological structures such as faults and geological contacts that are totally or partially covered by plio-quaternary deposits. Additionally, the gravity anomalies perfectly delineated the structural highs in the basement and sedimentary thickening in depressions and grabens. Furthermore, the combined analysis of geoelectrical and borehole data provides more precisions about the deep geological structure of the phosphatic series of the Bahira basin, in terms of depth and lateral changes in thickness.

Keywords

Phosphatic series • Gravity • Electrical resistivity tomography • Borehole • Bahira basin • Morocco

1 Geographical and Geological Settings

The Bahira basin is located in central Morocco, about 30 km north of Marrakech. It covers an area of about 5000 km^2 and is limited by the Rehamna Paleozoic massif to the north, the Mouissate Jurassic plateau to the west, the Jebilet massif to the south, and Oued Tessaout to the east.

The study area is part of the Moroccan Meseta domain. It represents a synclinal basin that separates the Jebilet and the Rehamna Hercynian massifs (Karroum, 2015). From the geological point of view, the Bahira basin presents a transition from Paleozoic to Quaternary deposits (Fig. 1), and includes the Maastrichtian-Eocene phosphatic series, which outcrops in the Gantour plateau where it is currently being extracted in open pit mines (Ihbach et al., 2020).

2 Materials and Methods

The Bahira basin has been the subject of regional gravity and vertical electrical sounding (VES) surveys (Fig. 1). The gravity measurements were made by the Moroccan Ministry of Energy and Mines for mineral exploration purposes, while the VES survey was carried out by the Tensift Basin Hydraulic Agency as a part of a hydrogeological investigation project. The survey consisted of twelve VES profiles

A. Charbaoui (✉) · A. Kchikach · M. Jaffal · J. L. Bodinier · B. Eljabbar · E.-S. Jourani
Mohammed VI Polytechnic University, Geology and Sustainable Mining (GSM), Benguerir, Morocco
e-mail: anas.charbaoui@um6p.ma

A. Kchikach · M. Jaffal
Cadi Ayyad University, Marrakech, Morocco

J. L. Bodinier
University of Montpellier and CNRS, Geosciences Montpellier, Montpellier, France

O. K. Yazami · M. Guernouche
OCP Group, Casablanca, Morocco

M. Bezzeghoud et al. (eds.), *Recent Research on Geotechnical Engineering, Remote Sensing, Geophysics and Earthquake Seismology*, Advances in Science, Technology & Innovation,
https://doi.org/10.1007/978-3-031-48715-6_47

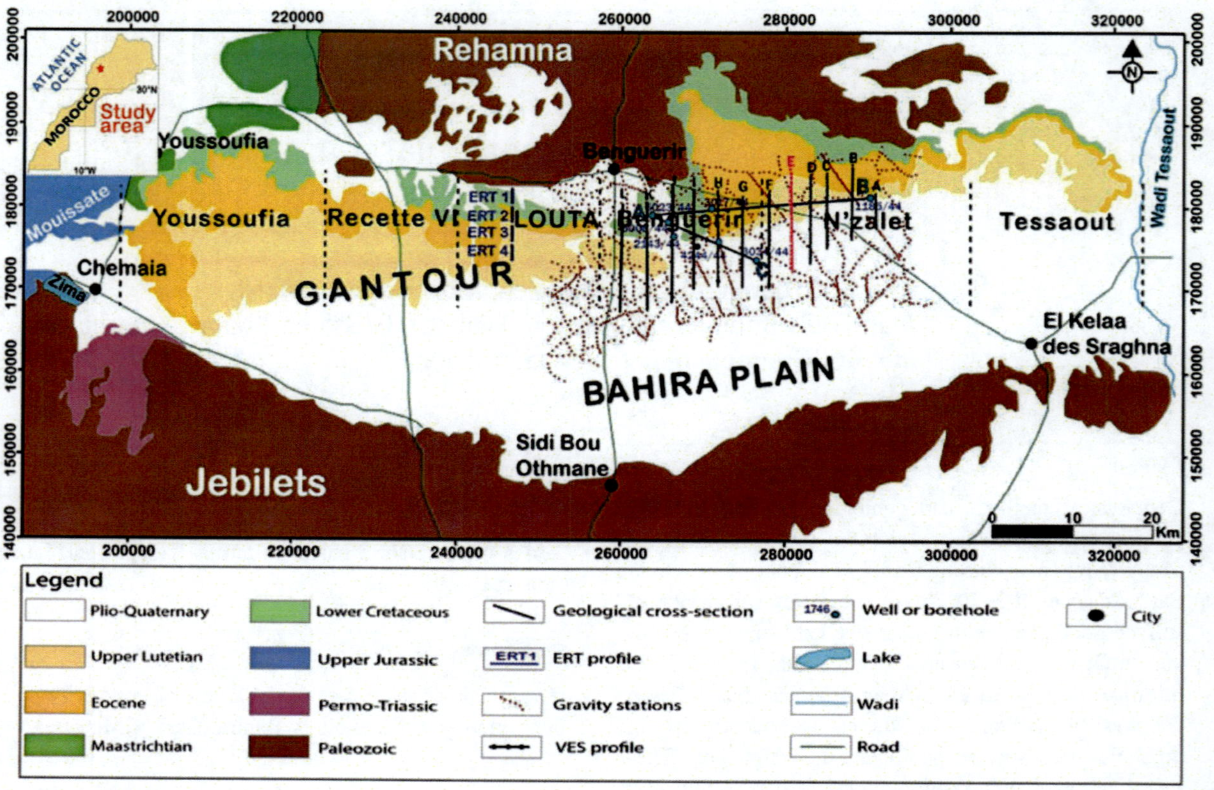

Fig. 1 Location of the geophysical surveys and the boreholes correlation profiles on the geological map of the study area

along the eastern part of the Bahira basin (Fig. 1) (Jaffal et al., 2022).

In this study, data from these two geophysical campaigns were analyzed. As part of the compilation and interpretation of the available data, we established twelve geoelectrical models (denoted A to L) and two borehole correlation sections (AB and AC) (Fig. 2b). These results help to characterize the Bahira basin in terms of overall structure and thickness variation of the sedimentary deposits.

Furthermore, to provide detailed images of these deposits in the subsurface, ERT surveys were carried out along four profiles in the study area. The electrical prospection was performed using a Wenner-Schlumberger array with a unit electrode spacing of 10 m, and the acquired resistivity data was processed using the Res2Dinv software. An example of the elaborated ERT geoelectrical models is represented in Fig. 2c.

3 Results and Discussion

The analysis of the gravity data reveals the existence of several positive and negative anomalies (Fig. 2a). A comparison of gravimetric and geological data shows that, in general, positive anomalies correspond to outcropping areas of the Paleozoic basement, while negative anomalies correspond to the sediment fillings of depressions.

To better visualize the structure of the plio-quaternary cover and the Lutetian limestones, the thickness of these two formations was interpolated through the 12 geoelectrical sections obtained from VES data. The established map for the Eocene thickness, presented in Fig. 2b, clearly shows that the Bahira basin corresponds to an extended east–west depression where the Eocene thickness exceeds 80 m (red–orange colors). Local depressions, where this thickness is above 90 m, are observed at profiles J and G.

The cross-sections established from borehole data illustrate the thickness changes of the Eocene formations which overlay either the Senonian marls or directly the Paleozoic basement.

The analysis of ERT sections reveals the existence of a first high-resistive superficial level (>280 Ω.m) (Fig. 2c). This level represents calcareous encrustations that can be easily seen in the field and play a crucial role against the erosion of the phosphatic series. The resistivity increases progressively with depth, especially in the central part of the profiles. In this area, the ERT model shows moderate resistivity, represented by yellow and green colors. Such

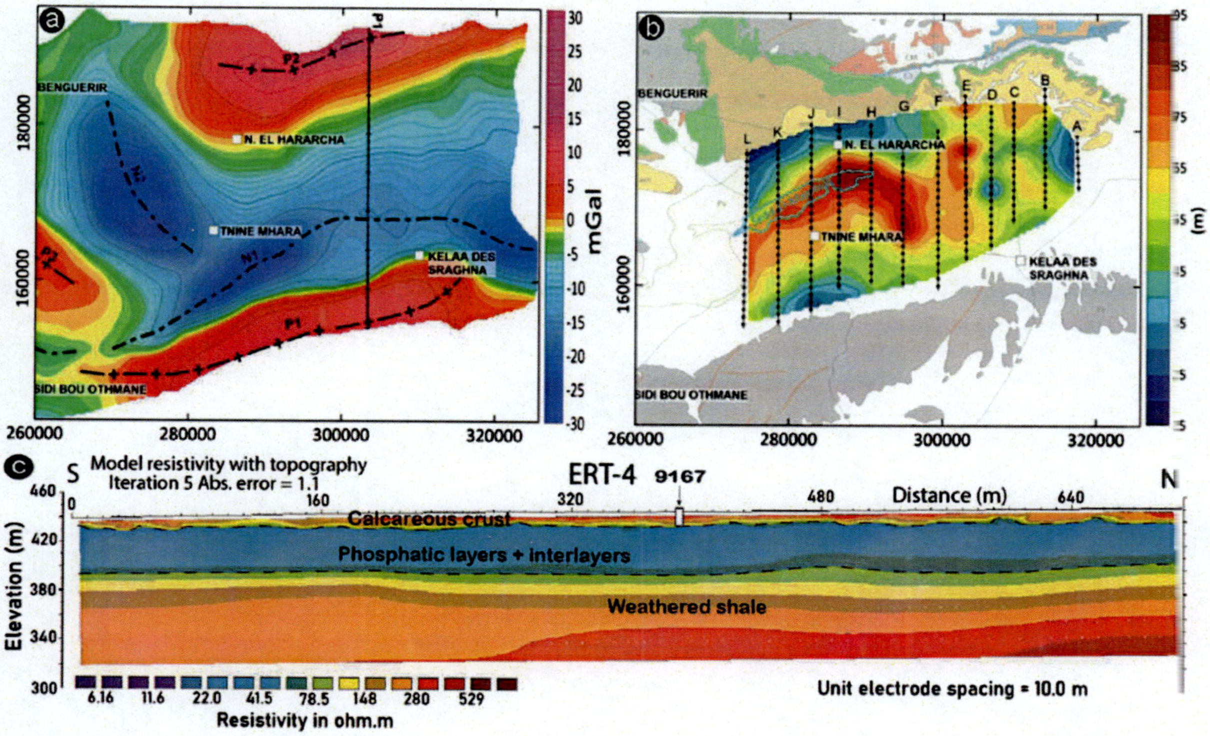

Fig. 2 **a** Residual gravity map (interval of 5 mGal). **b** Electrical map showing the Eocene thickness. **c** Geoelectrical model of ERT 4 profile

values may correspond to the weathered upper part of the basement. Further in depth, resistivity values are higher (>200 Ω.m) and correspond to the electrical signature of the Hercynian basement.

4 Conclusions

The aim of this study was to investigate the deep structure of the Bahira basin and the Gantour plateau which hosts significant phosphate reserves of Morocco. A multidisciplinary approach was conducted to provide a more accurate understanding of the sedimentary fill of the Bahira basin and its lateral variations. The combined analysis of geophysical and borehole data was able to reconstruct the deep geological structure of this basin. These findings have important implications for future mining and hydrogeological investigations in the studied region.

References

Ihbach, F.-Z., Kchikach, A., Jaffal, M., El Azzab, D., Khadiri Yazami, O., Jourani, E.-S., Olaiz, O. A., & Dávila, L. V. (2020). Geophysical prospecting for groundwater resources in phosphate deposits (Morocco). *Minerals, 10*(10), 842. https://doi.org/10.3390/min10100842

Jaffal, M., Charbaoui, A., Kchikach, A., El Ghorfi, M., Khaldoun, A., Safhi, A. E., Bodinier, J. L., Yazami, O. K., Jourani, E., & Manar, A. (2022). Gravity study of the Western Bahira Basin and the Gantour Phosphatic Plateau, central Morocco: Interpretation and hydrogeological implications. *Journal of African Earth Sciences, 193*, 104581. https://doi.org/10.1016/j.jafrearsci.2022.104581

Karroum, M. (2015). *L'apport de la géophysique et de la géochimie dans l'identification hydrogéologiques et la qualité des eaux de la plaine de la Bahira (Maroc centrale)* (p. 210). Université Cadi Ayyad, Marrakech.

Information System-Articulated Geophysical Engineering and Survey Designs and Plans—Managing Field Operations and Environments

Shastri Nimmagadda, Andrew Ochan, Mahtab Rashidifard, and Ausama Giwelli

Abstract

Surveyors, explorers and managers investigate for improvements in geophysical operations in hassle-free environments, including planning, survey designs and budgetary regulations. The critical tasks are field crew induction, wildlife and domestic animal protection, oil spill prevention, land management, built-up areas and shot hole drilling. Geophysical field-data acquisition tasks include survey planning, landowner and community consultations, ecological awareness, operations, road and truck movements and campsite establishment. Besides, field and cable layouts, survey type, arrays, machinery and source usage and latitude and longitude of survey areas need attention while scrutinizing survey operations, including a detailed examination of geological features. Waste management, explosive ordinance areas, abandoning survey areas and campsites are added challenges while managing geophysical field operations. Under survey operations, careful handling of chemicals and hazardous material management, emergency responses, geophysical survey completions and environmental audits need attention in project executions. Manageable crop compensations in the fields can ease settlements with farmers and landowners whose lands are damaged by geophysical surveys. Finally, rehabilitation of field areas is focused on abandoning seismic lines and tracks, including campsites. The research aims to examine the challenges of the geophysical surveys in a diverse range of environments, including land use. We interpret various entities and dimensions of the geophysical survey designs, plans to articulate, and model field operations and the environments for which databases are operationalized. We construe the entities and dimensions for articulating Information Systems (IS) artefacts in geophysical engineering. Several schemas are prepared pertinent to geophysical operations, environment awareness, men, machinery and materials movement and management, including hazards, time and budgetary constraints to integrate data-driven structures in repository systems. We propose a shared common ontology approach with knowledge-based connectivity between attribute dimensions. Multidimensional structures are designed to unify metadata and its management. The IS artefacts are reusable by changing key attributes such as period and geography attribute dimensions. The approach facilitates easy handling of geophysical operations and environmental challenges by explorers, surveyors and managers involved in geophysical surveys, including land procurement and disaster management, to acquire quality geophysical data.

Keywords

Geophysical surveys • Field operations • Environment management • Survey designs

S. Nimmagadda (✉)
School of Management, Curtin University, Perth, WA, Australia
e-mail: shastri.nimmagadda@curtin.edu.au

A. Ochan
National Petroleum Authority, Entebbe, Uganda

M. Rashidifard
University of Western Australia, Perth, WA, Australia

A. Giwelli
Commonwealth Scientific and Industrail Research Organization, Canberra, WA, Australia

1 Introduction

In prospecting areas, acquiring quality data is decisive for interpreting prospective land-based drillable exploratory locations in the upstream businesses (Dentith & Mudge, 2018; Nimmagadda, 2015). Unless we have valuable connectable multidimensional data, it is difficult to interpret the petroleum prospects in the investigating areas. The geophysical survey planning, survey design, survey operations,

M. Bezzeghoud et al. (eds.), *Recent Research on Geotechnical Engineering, Remote Sensing, Geophysics and Earthquake Seismology*, Advances in Science, Technology & Innovation,
https://doi.org/10.1007/978-3-031-48715-6_48

including rehabilitation are vital entities of any data acquisition task (Aleshin et al., 2018; Curtis & Maurer, 2000). Different attribute dimensions are construable from these entities. Several dimensions and types are interpreted in each entity to model the attribute models for exploratory metadata.

2 Aims of the Study

The research purpose is to identify several attributes of geophysical survey planning, survey designs and operations and interconnect them to assess how each affects the other. We take help of the Information System (IS) tools and technologies to resolve the issues and challenges (Nimmagadda, 2015). Objectives of the study is to design various multidimensional data schemas that depict the knowledge-based attribute models. Secondly, develop an integrated framework to manage data acquisition practices in any land-based data acquisition systems.

3 Methodology

To build constructs and models (data models) that integrate multiple dimensions and accommodate in an integrated framework (Nimmagadda, 2015), we have designed five different schemas pertaining to seismic survey planning, survey design, seismic survey operations and rehabilitation, including pandemic entities and dimensions to draw the framework with various tasks (Aleshin et al., 2018; Curtis & Maurer, 2000; Dentith & Mudge, 2018). We have identified several attribute dimensions for mapping and modelling, as summarized in Table 1.

Table 1 Identifying entities and dimensions for mapping and modelling

Entities and dimensions	Connectable attribute dimensions	Primary keys	Secondary keys to connect with primary keys
Survey planning	Planning, risk assessment	Planning ID	Environment ID, survey ID and operation ID
Regulation	Acquisition, contracting	Regulation ID	Planning ID, survey ID and operations ID
Landholder and community involvement	Landowner, community involvement	Landowner ID	Road and tracks ID, planning ID, campsite ID
Environment awareness	Regulatory, rehabilitation, field operations, campsites	Environment ID	Field operation ID, roads and tracks ID, rehabilitation ID
Operating procedure	Operations, designs, roads	Operations ID	Field designs ID, roads and tracks ID
Roads and tracks	Field designs, operations, roads	Roads and tracks ID	Field design ID, operations ID, regulatory ID
Campsites	Campsites, environments, crews, planning	Campsite ID	Environment ID, crew ID, planning ID
Survey design	Survey operations, designs, machinery use, planning	Survey design ID	Field operation ID, planning ID, machinery use ID
Timing	Completion, operations, resources management, landowner, time	Timing ID	Operation ID, survey design ID, resources management ID
Grid layout	Wildlife protection, operations, designs, environment	Grid layout ID	Design ID, operation ID, environment ID, wildlife protection ID
Machinery use	Operations, designs, environments, landowner agreements	Machinery use ID	Operation IS, design ID, environment ID, wildlife protection ID
Survey operations	Crews, environment, wildlife protection, land management, shot holes, waste management	Survey operation ID	Environment ID, crew ID, shot hole ID, land management ID, emergency ID
Crew induction	Safety and health, pandemic management, vaccination, environment, testing, campsite	Crew ID	Campsite ID, environment ID, pandemic ID
Wildlife protection	Environment, operations, survey designs, shot holes, spill prevent	Wildlife protection ID	Environment ID, spill prevention ID, shot hole ID, survey operation ID
Land management	Landowner, regulatory, lease, environment, operations, designs, campsites	Land management ID	Environment ID, operations ID, design ID, planning ID, campsite ID
Shot holes	Environment, surveys, designs, roads and tracks, land management	Shot Hole ID	Operation ID, design ID, environment ID
Waste management	Environment, campsites, operations, designs, rehabilitation	Waste management ID	Environment ID, campsite ID, operation ID, design ID, rehabilitation ID, regulatory ID

(continued)

Table 1 (continued)

Entities and dimensions	Connectable attribute dimensions	Primary keys	Secondary keys to connect with primary keys
Survey area	Survey operations, designs, environment, planning, survey areas	Survey area ID	Operation ID, design ID, environment ID, planning ID
Chemicals management	Regulatory, safety, hazardous material management, operations	Chemical management ID	Regulatory ID, safety and security ID, operations ID
Survey completions—environment audits	Survey completions, land use, landowner, regulatory, planning, operations	Survey completion ID	Operation ID, design ID, rehabilitation ID, environment ID, planning ID, compensation ID
Rehabilitation	Planning, regulatory, timing	Rehabilitation ID	Planning ID, survey ID, operation ID, roads and tracks ID
Geophysical lines	Survey designs, operations, roads and tracks, rehabilitation	Geophysical lines ID	Operation ID, design ID, roads and tracks ID, safety ID, crew ID
Campsites	Rehabilitation, planning, regulatory, roads and tracks	Campsite ID	Rehabilitation ID, planning ID regulatory ID, roads and tracks ID
Pandemic	Crew induction, environment	Pandemic ID	Environment ID, campsite ID

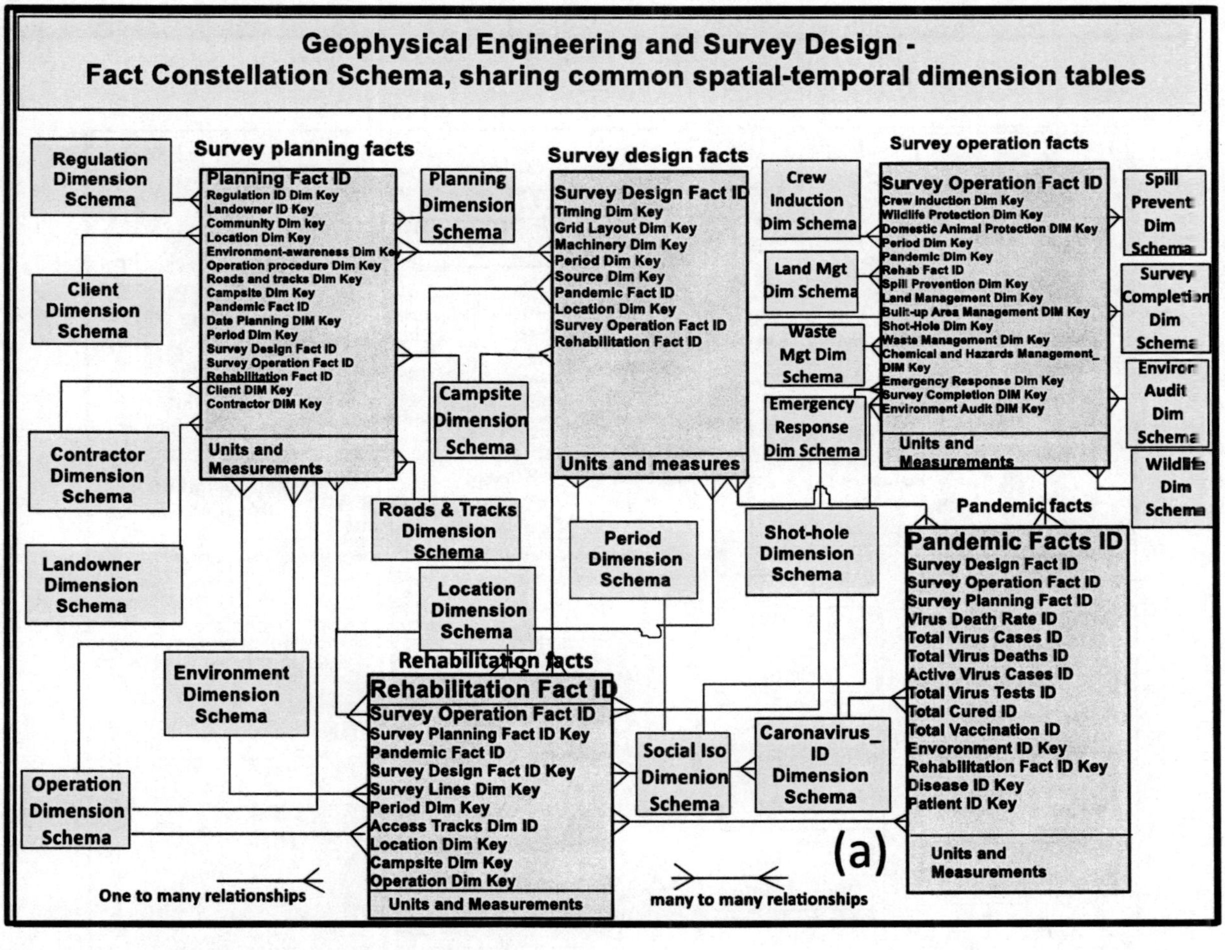

Fig. 1 Fact constellation schema—connecting multiple subschemas

These attribute dimensions are used to build various data schemas. One of such schemas is presented in Fig. 1. Various subschemas relevant to survey planning, design and operations are connectable in a large fact constellation schema.

4 Discussions

An integrated framework is designed to map and model the attributes of geophysical survey planning, survey design and field operations, so that the entities and dimensions are made relatable to each other. Multidimensional warehouse or repository is purported to conduct data mining and visualization of survey data views and their interpretation. The framework can facilitate data acquisition managers to carry out and reiterate survey designs and operations. In addition, data acquisition resources can be optimized through these robust modelling methodologies. Additionally, ontologies can be developed to support methodologies, as detailed in Nimmagadda (2015), with systematic development of data repositories (Fig. 2).

The tasks of the framework are staged in a single canvass to benefit the integration process. Several artefacts in articulating the framework are analyzed as mentioned in different stages (1) to (7). Stages are (1) data acquisition, (2) big data characteristics, (3) types of data structures, (4, 5) mapping and modelling, (6) data warehousing and (7) data mining, visualization and interpretation tasks.

5 Conclusions and Future Vision

The documented entities and dimensions are useful for modelling multidimensional attributes. So far, we have described the precautions taken during geophysical data acquisition and exploration of energy resources. The framework is extendable to various other entities, such as drilling, production and marketing entities. The fact constellation schema can be adaptable and flexible in different geological settings. The framework has immense value in organizing survey designs and operations. Typical beneficiaries of the research are data acquisition managers, data analysis and energy explorers.

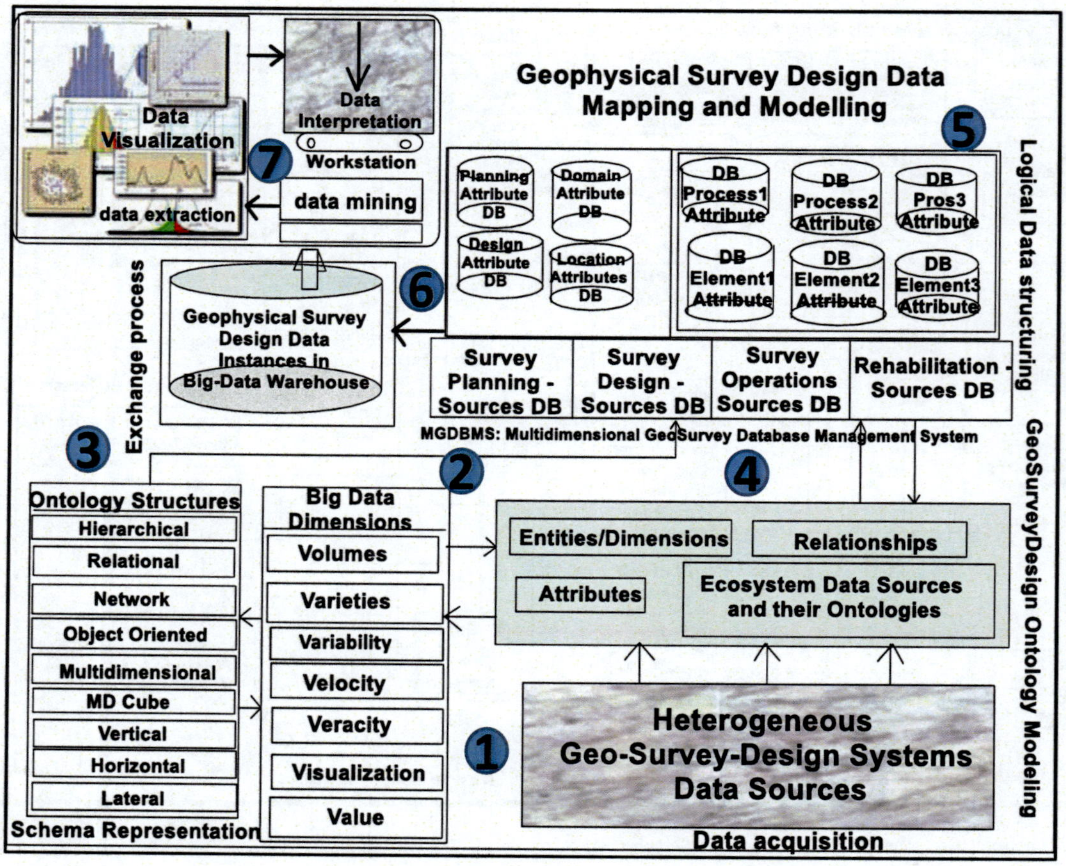

Fig. 2 Framework development—mapping and modelling survey operations

References

Aleshin, I. M., Burguchev, S. S., Perederin, F. V., et al. (2018). Versatile geophysical data acquisition system. *Seismic Instruments, 54*, 562–564. https://doi.org/10.3103/S074792391805002X

Curtis, A., & Maurer, H. (2000). Optimizing the design of geophysical experiments: Is it worthwhile? *EOS Transactions AGU, 81*(20), 224–225. https://doi.org/10.1029/00EO00161

Dentith, M., & Mudge, S. (2018). Geophysical data acquisition, processing and interpretation. https://doi.org/10.1017/cbo9781139024358.004

Nimmagadda, S. L. (2015). *Data warehousing for mining of heterogeneous and multidimensional data sources*. Verlag Publisher

Forward Modeling and Inversion of MRS Relaxation Signal: Application to the Phosphate Series in Youssoufia and Khouribga Deposits (Morocco)

Fatim-Zahra Ihbach, Azzouz Kchikach, Mohammed Jaffal, Konstantinos ChaliKakis, Jean-Louis Bodinier, and Es-Said Jourani

Abstract

The magnetic resonance sounding (MRS) method is one of the most used techniques in hydrogeophysics that directly quantifies the water content distribution from surface measurements. Direct MRS modeling is usually performed as a part of the planning phase of an MRS field survey, with the aim of determining the feasibility and usefulness of this method in a given hydrogeological context, by evaluating the MRS signal amplitude of one or more aquifer layers, using their lithological and piezometric characteristics derived from actual field data. The direct modeling of the MRS signal, of Youssoufia and Khouribga phosphate deposits, was calculated using the lithological and piezometric characteristics derived from boreholes data. The MRS signal was then calculated considering the maximum, average and minimum values of the water content (WMRS) and the relaxation time (T2*) for different electromagnetic noise levels. MRS modeling results show that the Youssoufia mining sites are highly adapted to the use of the MRS method even with significant magnetic noise levels. However, the application of the MRS method in the hydrogeological context of Khouribga mining sites depends strongly on the electromagnetic noise level in the investigated area, but also on the characteristics of the aquifer.

Keywords

Hydrogeophysics • MRS modeling • Phosphate deposits • Youssoufia • Khouribga • Morocco

1 Introduction

Direct MRS modeling makes it possible to calculate a model of the subsoil similar to that obtained by inversion, by evaluating the MRS signal amplitude of one or more aquifer layers, deduced from their real field lithological and piezometric data. This direct calculation in MRS was carried out during the field planning phase (Boucher et al., 2003), in order to determine the applicability of the method in a given geological context, to know the MRS signal magnitude that can be expected while carrying out field measurements, but also to have an idea of the MRS signal measurement conditions (Legchenko, 2013).

2 Methodology

The MRS direct modeling requires prior knowledge of several parameters, the most important are: the lithological units of the sedimentary cover, their thicknesses, their resistivities, the relaxation times T2* corresponding to each type of geological formation and finally the piezometric level necessary to determine the saturated formations and therefore the water content. The modeling process is usually done in several steps (Fig. 1):

F.-Z. Ihbach (✉) · A. Kchikach · M. Jaffal
Georessources, Geoenvironment and Civil Engineering Laboratory, Cadi Ayyad University, Marrakech, Morocco
e-mail: ihbachfatimzahra@gmail.com

A. Kchikach · M. Jaffal · J.-L. Bodinier
Mohammed VI Polytechnic University, Geology and Sustainable Mining Institute (GSMI), Benguerir, Morocco

K. ChaliKakis
UMR 1114 EMMAH - Mediterranean Environment and Agrohydrosystems Modelling, Avignon University, Avignon, France

J.-L. Bodinier
University of Montpellier and CNRS, Geosciences Montpellier, Montpellier, France

E.-S. Jourani
OCP Group, Casablanca, Morocco

M. Bezzeghoud et al. (eds.), *Recent Research on Geotechnical Engineering, Remote Sensing, Geophysics and Earthquake Seismology*, Advances in Science, Technology & Innovation,
https://doi.org/10.1007/978-3-031-48715-6_49

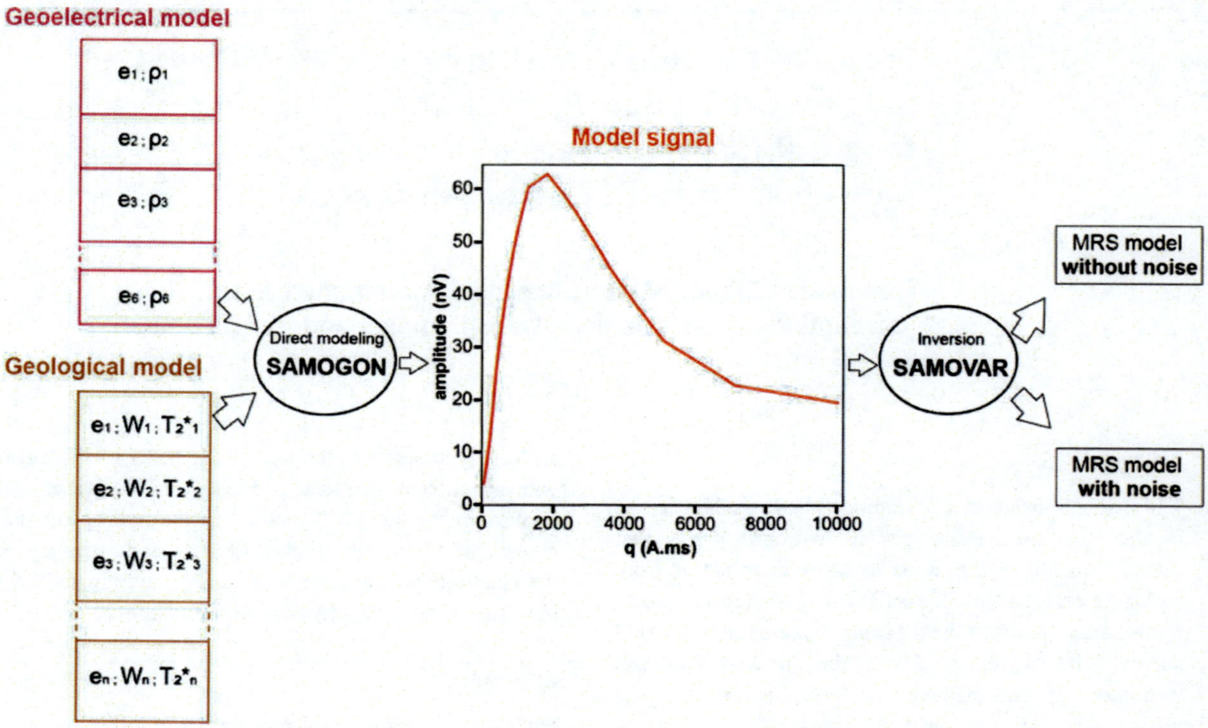

Fig. 1 Summary of the MRS direct modeling process (Ihbach, 2021)

- First, the creation of the geoelectrical model of the studied site with n layers, n resistivity and n−1 thickness. In this study case, the subsurface geoelectrical models were mainly obtained from the resistivity data acquired from the ERT and TDEM methods applied previously on these sites. These models are considered in the calculation of a matrix used in the modeling process.
- Establishment of the geological model: It consists of defining all the lithological layers and assigning to each the water content and the relaxation time according to the geological and lithological nature.
- The last step concerns the modeling of the data with the "SAMOGON" software. It aims to visualize the signal corresponding to the model created. When the latter is saved, all the files necessary for a simulation of an RMP sounding are created automatically and can be reversed by the "SAMOVAR" software, taking into account the level of electromagnetic noise.

Thus, the MRS response of the phosphatic series was analyzed according to the following parameters:

- Water content (W_{MRS})
- Relaxation time (T2*)
- Electromagnetic noise level.

3 Results and Discussion

3.1 Youssoufia

Results obtained from the MRS modeling process (Fig. 2) in the Youssoufia site show that the MRS method is highly applicable in the hydrogeological context of Youssoufia mining sites.

The average conditions of T2* and WMRS at this site allowed us to have significant data even with noise values

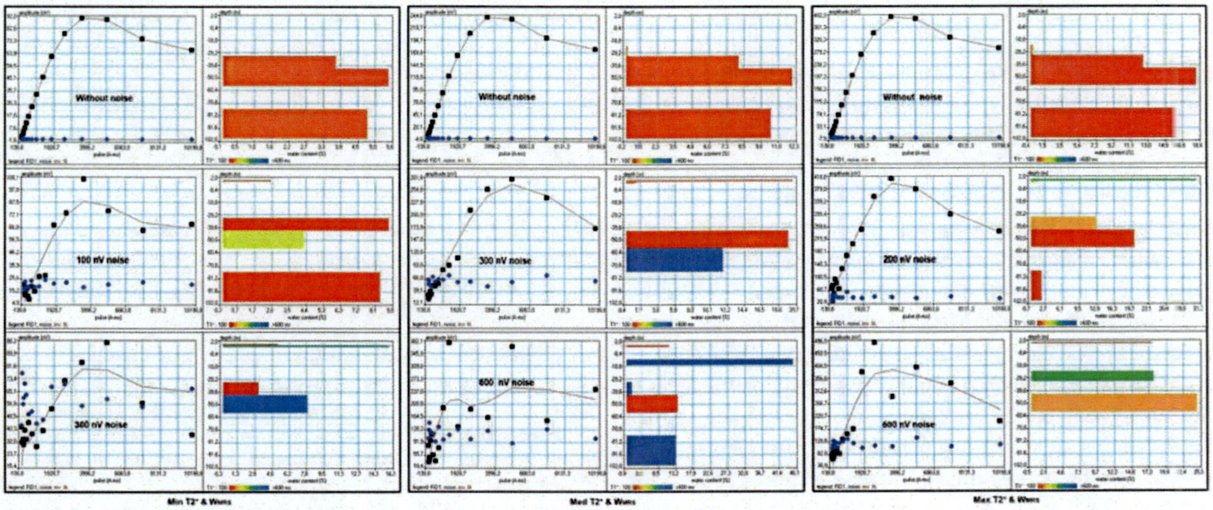

Fig. 2 Results of the MRS inversion model at Youssoufia site with the minimum, average and maximum values of T2* and WMRS at different noise levels

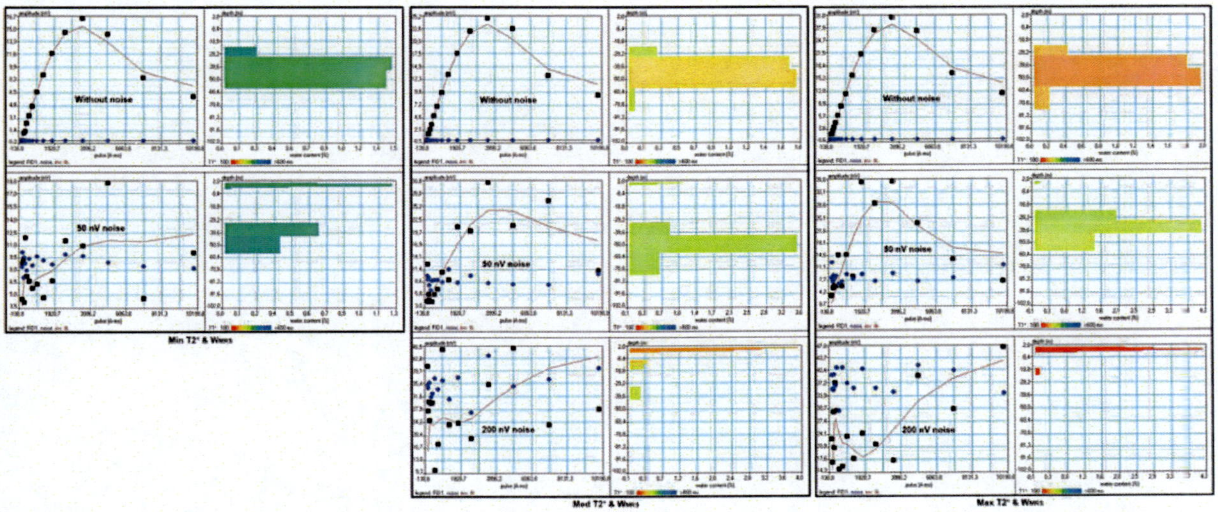

Fig. 3 Results of the MRS inversion model at Khouribga site with the minimum, average and maximum values of T2* and WMRS at different noise levels (Ihbach, 2021)

exceeding 300 nV, and it is only from 600 nV that the MRS signal becomes more or less noisy and insignificant.

3.2 Khouribga

Results obtained from the MRS modeling process (Fig. 3) in the hydrogeological context of Khouribga mining sites show that the application of the MRS method depends strongly on the electromagnetic noise level in the investigated area, but also on the characteristics of the aquifer. They also show that even if we consider an aquifer with good characteristics, its detectability will be very limited if the electromagnetic noise exceeds 50 nV. This is mainly due to the significant depth of the aquifer in this context, but also to the conductive nature of layers above the aquifer that attenuates the response of the MRS signal deeper (Girard et al., 2007; Legchenko, 2013).

4 Conclusions

The direct MRS modeling results of the Youssoufia deposits showed that this site is highly favorable for the use of the MRS as a groundwater exploration technique. It clearly shows that a significant MRS signal can be measured even for an aquifer with mediocre characteristics and moderate noise levels. It also reveals that only a noise level greater than 600 nV can affect the application of the MRS method if we consider an aquifer with medium characteristics.

However, the direct MRS modeling of the phosphatic series of the Khouribga sites shows that the application of the MRS method depends strongly and firstly on the level of electromagnetic noise, and then on the characteristics of the aquifer in this area. It also shows that the detectability of the aquifer can be very limited if the electromagnetic noise exceeds 50 nV, even with an aquifer with good characteristics. This is mainly due to the significant depth of the aquifer in this context, but also to the conductive nature of the layers above the aquifer, which attenuates the MRS signal response deeper.

References

Boucher, M., Legchenko, A., Baltassat, J. M., & Mathieu, F. (2003). Étude des variations temporelles du signal RMP en comparaison avec le niveau piézométrique sur le site de Villanblain (Loiret): BRGM/RP-52310-FR.References. References.

Girard, J.-F., Boucher, M., Legchenko, A., & Baltassat, J.-M. J. J. o. A. g. (2007). 2D magnetic resonance tomography applied to karstic conduit imaging. *Journal of Applied Geophysics, 63*(3–4), 103–116.

Ihbach, F. (2021). Étude Hydrogéphysique des aquifères encaissés dans la série phosphatée: cas des sites miniers de Khouribga, Youssoufia et Benguerir (MAROC). Université Cadi Ayyad, Faculté des sciences et techniques, Marrakech, 224 p.

Legchenko, A. (2013). *Magnetic resonance imaging for groundwater*. Wiley.

Cesium Magnetometry for Characterization and Preservation of an Archeological Site at Argamum, Romania

Sorin Anghel, Andrei Gabriel Dragos, and Gabriel Iordache

Abstract

Argamum-(Orgame) is one of the four oldest Greek settlements established on the shore of the Black Sea alongside Callatis, Tomis, and Histria. Founded around the middle of the seventh century BC, on the northern coast of Dobrudja-Romania, the ancient city occupies the entire area of Cape Dolojman, which dominates an important lagoon complex located south of the Danube delta. The main purpose of the geophysical surveys, which were made in the spring of 2022 by using the cesium vapor magnetometer (G-864), was to identify the fortification elements and adjacent buried structures in the main gate zone specific to this type of settlement and to establish a new archeological research strategy. The delimitation area for the magnetic survey was chosen in accordance with the presence of ceramic material on the surface. By subsequently using a Trimble R4 GPS system, two grids, aligned in a north–south direction inside the fortress near the main gate, were assigned and traced with respect to the geomorphological configuration and vegetation present on the site. The data were bidirectionally collected by surveying a grid characterized by a line spacing of 1 m and a sample rate of 25 readings per meter. This paper presents the interpretation of the large magnetic anomalies observed on the two investigated grids. Archeological research carried out in 2022 confirmed the results obtained by using the geomagnetic survey method and by revealing the existence of fortification walls and adjacent buildings in the main gate zone. In the second phase, we used the Potent Q software to model the dimensions, positions, shapes, and tilt of the buried artifacts. Finally, the geophysical study performed at the ancient city's main gate area demonstrated that magnetometry is capable of providing a rapid overview of the distribution of buried artifacts, which may prove useful for future excavation planning.

Keywords

Cesium magnetometry • Sensor • Archeology • Buried structures • Argamum

1 Introduction

The most important archeological site, Orgame/Argamum (Fig. 1), is located on the Black Sea coast of Romania, providing more than two millennia of inhabited area.

The Orgame/Argamum site, dated to the seventh century BC (Lungu et al., 2017), is located in Tulcea County, Romania, and has proven to be one of the largest archeological sites in Romania. GeoEcoMar conducted several geophysical investigations on this site in the spring of 2022, including magnetic surveys. The magnetic maps obtained through this method revealed many linear buried structures, interpreted as ancient walls.

2 Methodology

The Argamum archeological site was investigated via a magnetic method in the spring of 2022. The survey was carried out on two perimeters located on both sides of the main gate. Data acquisition faced some difficulties caused by the touristic features.

S. Anghel (✉) · A. G. Dragos · G. Iordache
National Institute for Research and Development on Marine Geology and Geo-ecology—GeoEcoMar, 024053 Bucharest, Romania
e-mail: soanghel@geoecomar.ro

M. Bezzeghoud et al. (eds.), *Recent Research on Geotechnical Engineering, Remote Sensing, Geophysics and Earthquake Seismology*, Advances in Science, Technology & Innovation,
https://doi.org/10.1007/978-3-031-48715-6_50

Fig. 1 Argamum archeological site—aerial view (left side)—data acquisition with magnetometer G-864 at Argamum archeological site (right side)

The measurements were performed using a Geometrics G-864 cesium magnetometer (Fig. 1) following longitudinal and transversal profiles with a spacing of 1 m. The diurnal variation was recorded using a G-856 proton magnetometer (Butler, 2005). The magnetic measurements were referenced in real time through the employment of a high-precision Talisman GPS, incorporated in the G-864 magnetometric system, and for improved localization of the perimeter's networks, another RTK GNSS system (Trimble R4) was subsequently used.

For an enhanced interpretation, a digital terrain model was created by processing a set of aerial images acquired by a DJI Phantom Pro V2.0 UAV system.

3 Results and Discussion

Following the summer campaigns of 2022, we have achieved the first data acquisition for the Argamum archeological site. The integrated interpretation of the geological and geophysical information and the archeological excavation made in the field, corroborated with the information from the specialized literature, highlighted enclosure walls, other buried walls, and buried streets.

Cesium magnetometer surveys were conducted to identify buried walls, pithouse boundaries, and activity areas outside of pithouses in support of archeological investigations, and to test this instrument's ability to image these features.

The increased availability of cesium magnetometric investigations provides the possibility to extend the investigation of individual archeological sites into a much wider archeological context, maximizing the information that can be recovered in a relatively short period of fieldwork.

The magnetic field anomaly map for each perimeter was created using the Oasis Montaj software (Figs. 2, 3, and 4). A large positive magnetic anomaly (red color) was detected and associated with walls and streets because many building materials that were used in antiquity retain remnant magnetism induced by ceramic, bricks, and other terracotta components. In conclusion, all discovered walls are composed of a mixture of limestone, ceramic remains, and bricks. Considering the mortar used for the building of ancient walls, it contains certain concentrations of ceramic elements characterized by a high magnetic susceptibility.

The high sensitivity of the cesium magnetometer used in this study helps identify some low amplitude anomalies in

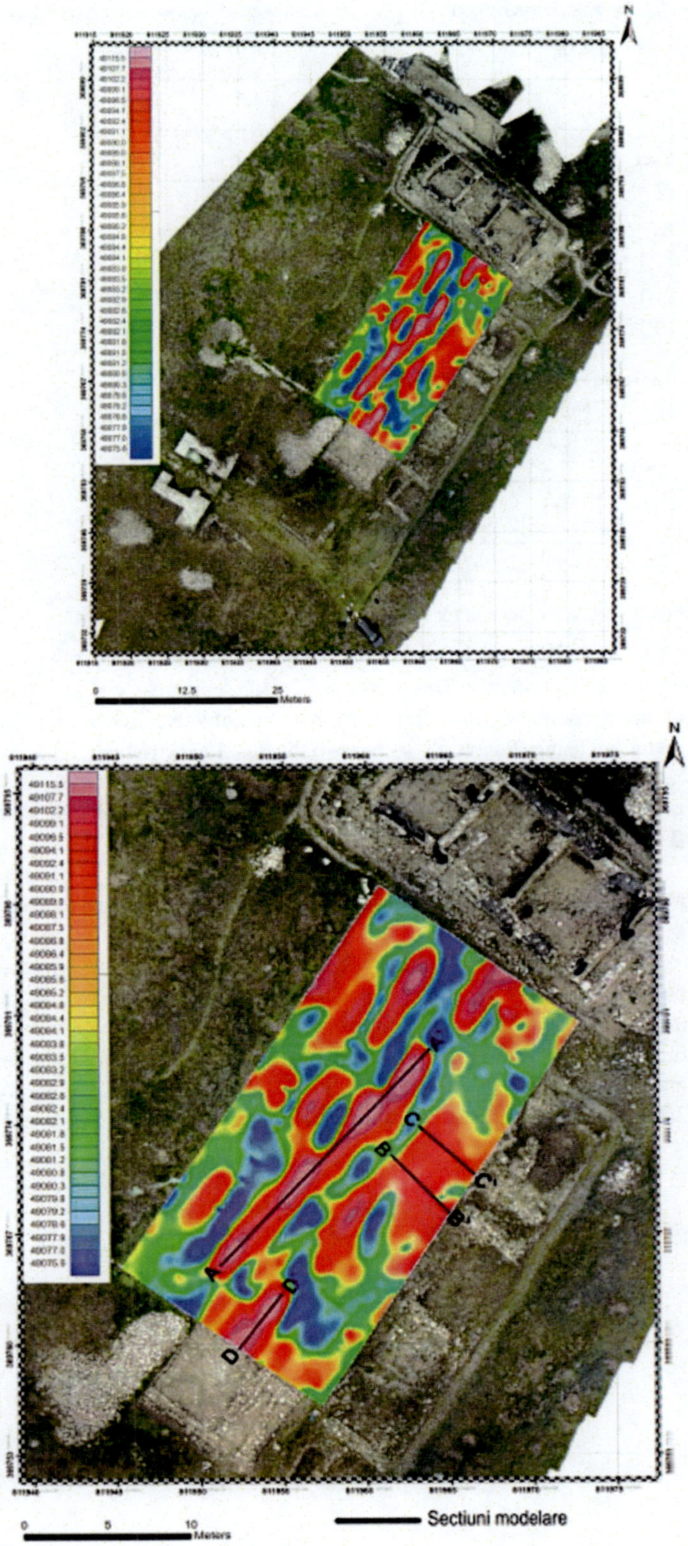

Fig. 2 Magnetic field anomaly map—left side perimeter

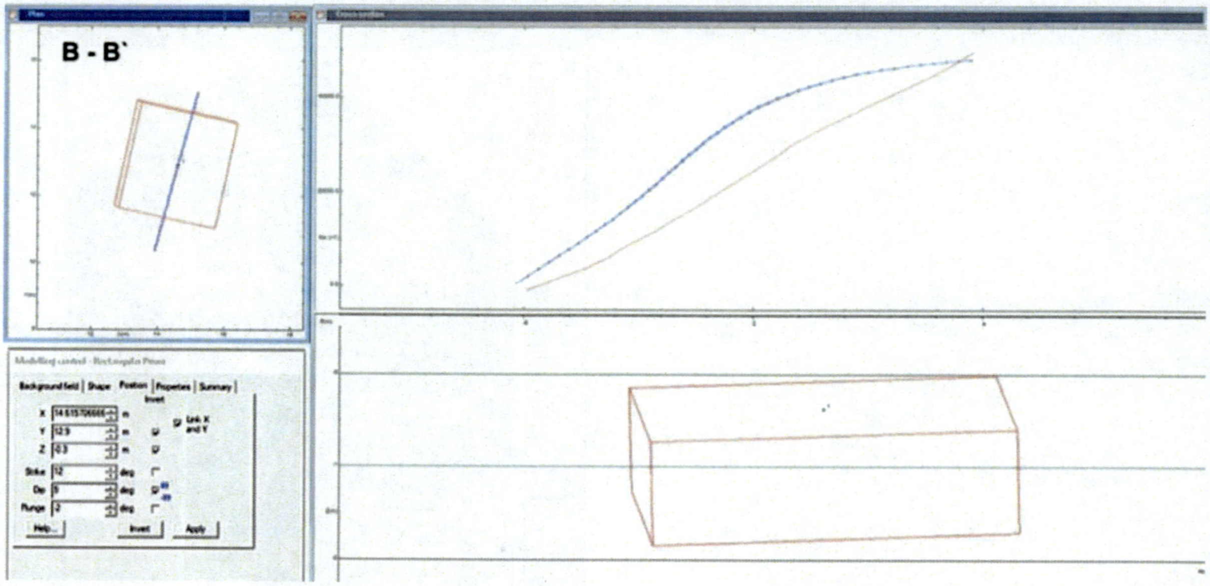

Fig. 3 Modeling process for section B-B'

the left side perimeter. The profile B-B' sampled across one of these anomalies has been subject to quantitative interpretation using Potent Q software (Iliceto, 1971); which helps characterize the causative magnetic source, considered as a buried wall, in terms of form, inclination, and depth (Fig. 3).

We choose the B-B` section because it was validated by archeological excavation executed after the magnetic data acquisition (Fig. 2).

In the final part, all geophysical investigations were confirmed through the archeological survey (Fig. 2). Related to the perimeter from the right side of the main gate, we observe from the magnetic anomaly map the major positive anomaly correlated with buried walls (Campana et al., 2008), ceramic artifacts, and burned area (Fig. 4).

The positive anomalies (red color) have the same direction as the walls discovered in archeological excavations.

4 Conclusions

The employment of interdisciplinary research methods in archeology has, once again, proven to be highly valuable and effective, as exemplified by the magnetometric prospections of the Argamum archeological site. That the use of noninvasive techniques and their correlation with topographical information can be used to pertinently elaborate an excavation strategy for the archeologist constitutes the main idea advanced in this paper. Furthermore, the magnetic method presented above proved to be adequate for

Fig. 4 Magnetic field anomaly map—right side perimeter

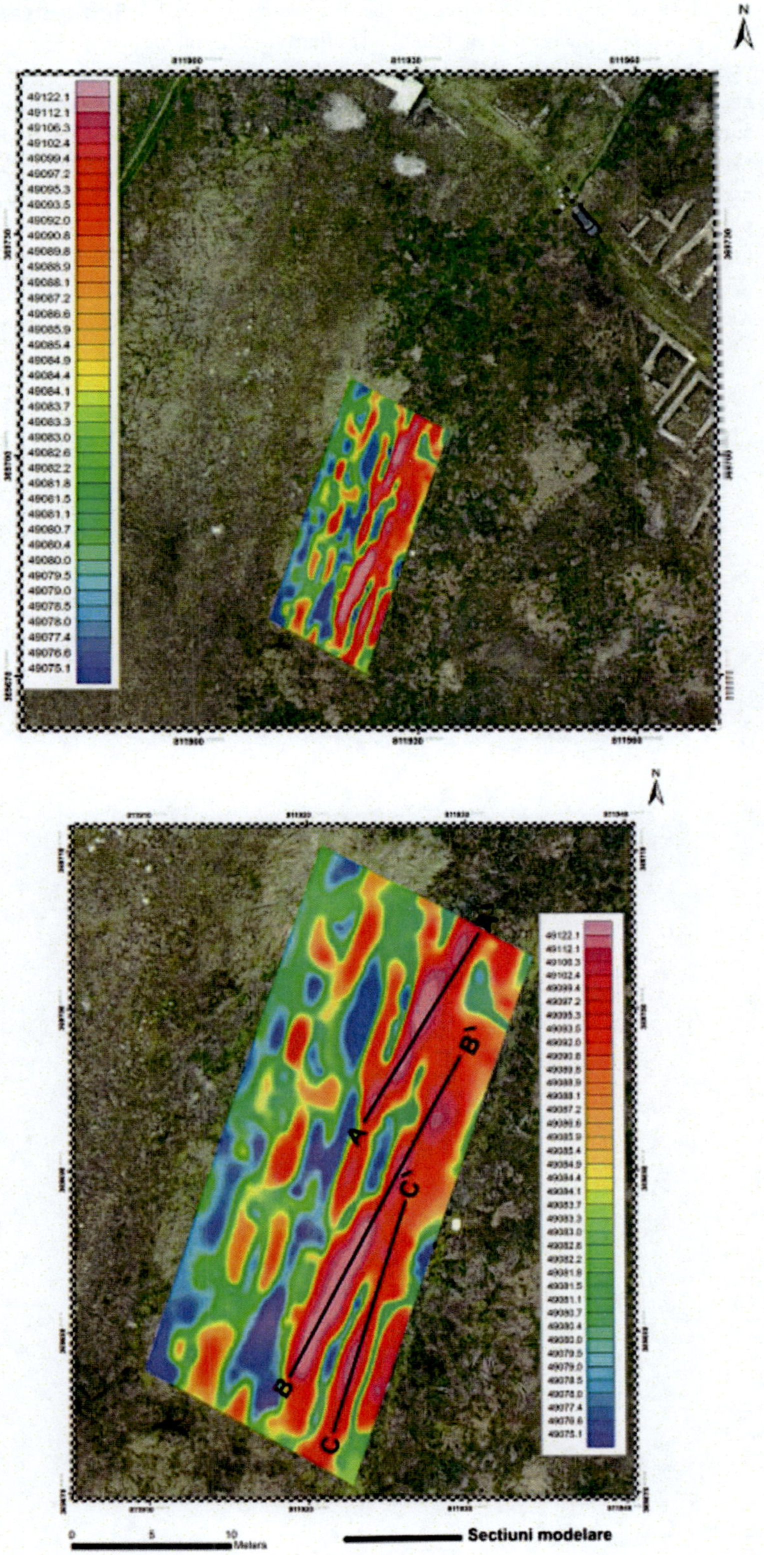

identifying possible fortification. As the data were registered and processed, there was a notable similarity, especially concerning the artifacts discovered using the cesium magnetometer and buried artifacts discovered in archeological surroundings on the land. Magnetometry instigations were also useful in identifying differential destruction of the archeological record in various parts of the site as a result of plowing and erosion. Identifying sources and extent of disturbance has practical importance for archeologists.

Acknowledgements This article was carried out in the frame of the Core Programme PN19200501 research project.

References

Butler, D. K. (2005). *Near–surface geophysics*. Society of Exploration Geophysicists.

Campana, S., & Piro, S. (2008). Seeing the unseen. Geophysics and landscape archaeology.

Iliceto, V. (1971). Nouvelles prospections geophysique du cimetiere merovingien de Galery – Prospezioni archeologiche Fondazione Lerici.

Lungu, V., & Caiete, A. R. A. (2017). Architectural finds and founder cult evidence in the Heroon at Orgame.

Application of Deep Learning for Low-Frequency Extrapolation to Marine Seismic Data in the Sadewa Field, Kutei Basin, Indonesia

Asido Saputra Sigalingging, Sonny Winardhi, Ekkal Dinanto, Wahyu Triyoso, Andri Hendriyana, Sigit Sukmono, Pongga D. Wardaya, Erlangga Septama, and Rusalida Raguwanti

Abstract

The success of full-waveform inversion (FWI) modeling depends on many factors, i.e., initial model, source wavelet, and low-content. Low-frequency seismic data is extremely important for FWI as it allows for accurate sub-surface imaging and increasing the resolution of sub-surface features, while also providing information on overall sub-surface geological features. However, the lack of low-frequency data can lead to cycle-skipping which could negatively affect the results of the inversion process. To resolve this issue, low-frequency data can be reconstructed using artificial intelligence or deep learning algorithms. In this study, we utilized a convolutional neural network (CNN) algorithm to automatically extrapolate the low-frequency data from band-limited Common Shot Gather (CSG) seismic data in time domain without any preliminary processing steps. The CNN model was tested and validated with synthetic seismic data to select the best model. The best model was then applied to marine seismic data in the Kutei Basin, East Kalimantan, Indonesia. The results demonstrate that deep learning has been accurately applied to extrapolate low frequencies in marine seismic data with an RMS error below 1% and that it produces results that are in line with the broadly established in conventional seismic processing methods (i.e., deghosting). Therefore, our study can be a powerful and effective alternative method to estimate the low-frequency content in the seismic reflection data of other areas.

Keywords

Deep learning • Convolutional neural network (CNN) • Low-frequency • Seismic imaging

1 Introduction

Full waveform inversion (FWI) is one of the sub-surface modeling techniques that provides high-resolution values of the physical properties of rocks through the tomography process of seismic data (Pratt, 1990; Tarantola, 1984). The results of the FWI optimization process frequently may not converge to a global minimum and instead get stuck at a local minimum due to several factors, including inaccurate initial models that do not represent the complex geological structure, source wavelets, seismic noise, and the loss of low-frequency content in the seismic data.

The main factor affecting FWI results is the loss of low-frequency content in seismic data (Bunks et al., 1995) which naturally contains information about geological trends. Therefore, the low-frequency content is essential for the FWI process to avoid the velocity model from converging to a local minimum (Hu et al., 2017). In this study we have selected Sun and Demanet's modified convolutional neural network (CNN) model (Sun & Demanet, 2018, 2020) as the regression model, since deep learning has the advantage of being able to directly predict low-frequency content without preprocessing of the raw data, which is rarely practicable by conventional methods. Therefore, we used this model to achieve the objective of extrapolating low-frequency content as an end-to-end process in 2D seismic data (i.e., synthetic and real data).

A. S. Sigalingging (✉)
Geophysical Engineering, Institut Teknologi Sumatera, Jati Agung, Indonesia
e-mail: asido.saputra@gmail.com

A. S. Sigalingging · S. Winardhi · E. Dinanto · W. Triyoso · A. Hendriyana · S. Sukmono
Geophysical Engineering, Faculty of Mining and Petroleum Engineering, Institut Teknologi Bandung, Jalan Ganesha 10, Bandung, 40132, Indonesia

P. D. Wardaya · E. Septama · R. Raguwanti
Pertamina Upstream Research and Technology Innovation, Jalan Mega Kuningan Barat III, Sopo Del Tower 50th 8 Floor, Jakarta, 12950, Indonesia

M. Bezzeghoud et al. (eds.), *Recent Research on Geotechnical Engineering, Remote Sensing, Geophysics and Earthquake Seismology*, Advances in Science, Technology & Innovation,
https://doi.org/10.1007/978-3-031-48715-6_51

2 Theory and Methods

Artificial neural networks (ANNs) are machine learning techniques that mimic biological learning mechanisms. Neurons, cells in the human nervous system, are modeled as perceptrons in ANNs. Each perceptron processes input and performs a linear calculation to generate output. Nonlinear processing is achieved by transforming outputs through an activation function. The perceptron represented mathematically as $y = f(x, w)$, where x is input, y is output, and w is a parameter that sets the function. Convolutional neural networks (CNN) are a type of artificial neural network (ANN) commonly designed to process data with a grid-like topology, such as images. The CNN utilized convolutional layers to scan and analyze image data, by using filters and pooling layers to reduce dimensionality, which allows the CNN to extract features from images. The feature extraction process is executed through the convolution of the input data and a filter kernel, which enhances the informativeness and pattern sensitivity of the input data. In this study, instead of applying CNN for 2D image classification, we used CNN working as regression for 1D signal processing which was successfully experimented by Sun and Demanet (2018) to perform low-frequency prediction of seismic data.

In this study, the low-frequency content of seismic data is extrapolated using the convolutional neural network (CNN) algorithm. The input data for the CNN consists of band-limited seismic data, which is dominated by high-frequency content, and the output generated is the low-frequency content of the seismic data. The architecture of the CNN comprises six blocks consisting of convolutional layers, dropout layers, and fully connected layers, as illustrated in Fig. 1. The first five convolutional layers are equipped with 32 filter kernels, each with a size of 80, while the final layer has only one filter kernel, reducing the number of model parameters. Regularization is achieved through the implementation of a 20% dropout layer, with two dropout layers, to prevent overfitting. The number of units in the fully connected layer is set to twice the number of samples in the input layer. Both the convolutional layer and the fully connected layer utilize PreLU (Hu et al., 2017) as the activation function. The optimization method employed is adaptive moment estimation (Kingma & Ba, 2014) with a learning rate of 5×10^{-4}.

The dataset used in this study comprises of seismic synthetics, which were generated by simulating acoustic wave propagation through the finite difference method, and velocity models obtained from the Marmoussi models depicted in Fig. 1. The simulation process involved using 30 shots and 256 receivers for each model, with a 15 Hz dominant frequency Ricker wavelet serving as the source, and a 6 s time length, with a 10 m grid space in both the horizontal and vertical axes. The open-source Python library, Devito (Luporini et al., 2020), was used to perform the

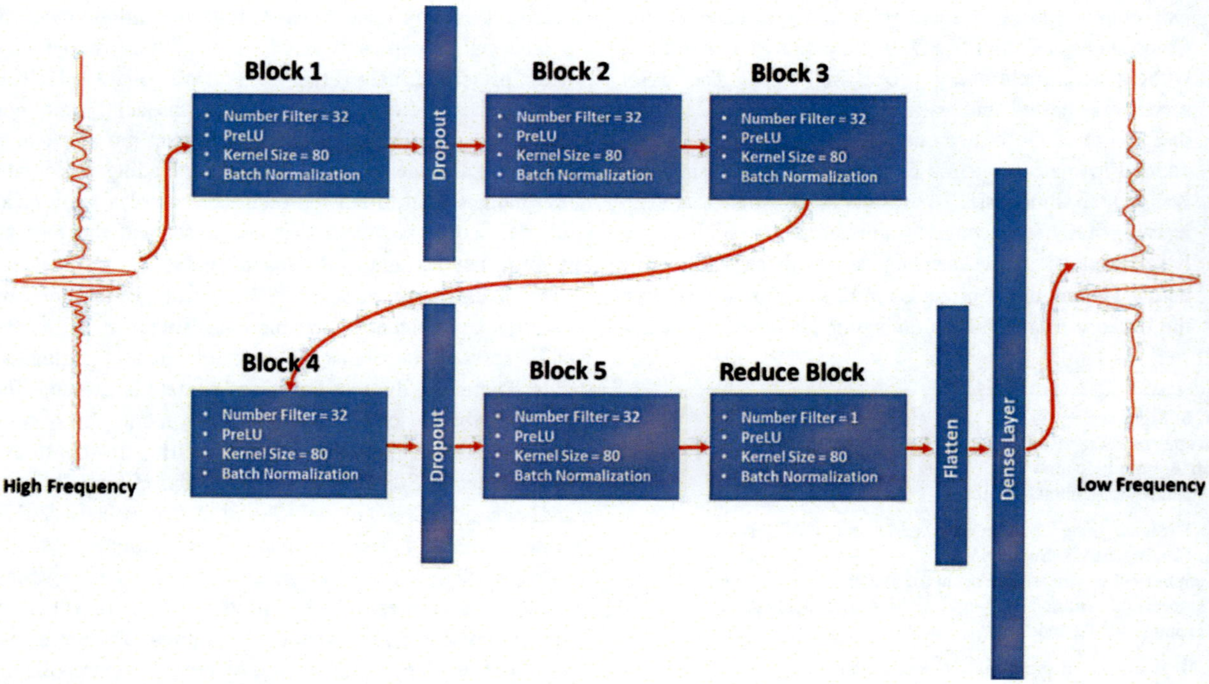

Fig. 1 CNN regression model architecture consists of six convolutional blocks with regularization and a fully-connected layer (Sun and Demanet, 2018)

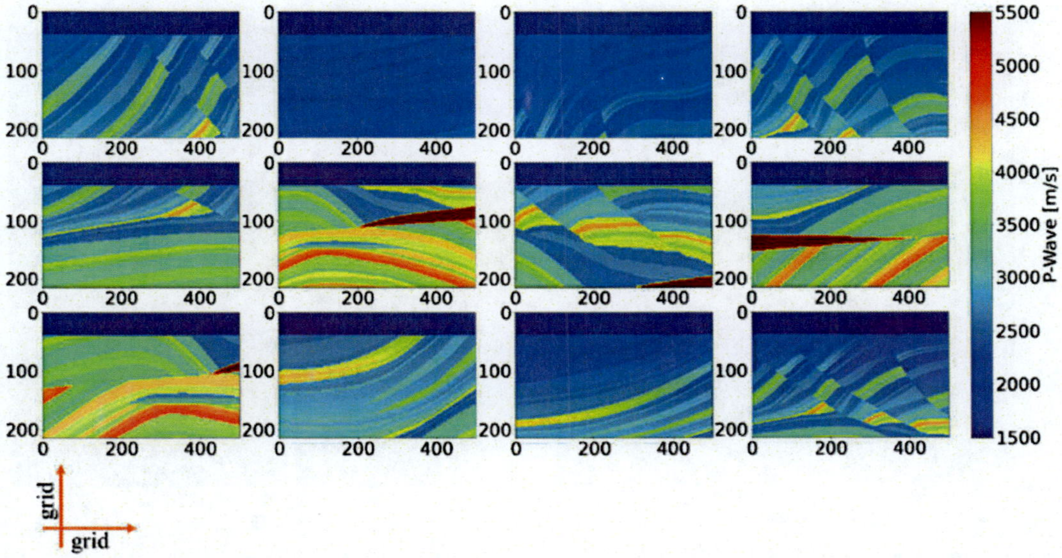

Fig. 2 12 submodels were extracted from the Marmousi model to compile the training dataset. In these submodels, 30 shots per model and 256 receivers per shot were utilized. Approximately 70% of the total traces generated were utilized for training data, while the remainder was used for testing

simulations. After applying a Butterworth filter to divide the synthetics into low- and high-frequency domains, the input and output were then sampled at 8 ms, which helped reduce the parameters and normalize the data. The dataset was ready for training only after all preprocessing steps had been completed (Fig. 2).

The implementation of the model shown in Fig. 1 was carried out using the popular deep learning frameworks in Python, TensorFlow, and Keras. The training process was performed on a GPU Quadro M2000, taking more than two days with 300 epochs and a batch size of 64. As observed in Fig. 3, a faster decrease or convergence of MSE error occurred at smaller batch sizes. However, after testing each combination of models with varying batch sizes, it was determined that a batch size of 32 produced smaller and more stable average prediction results and prediction error.

The convergence of the training process is faster with a smaller batch size since the model learns fewer data patterns. However, this also results in a reduction in the model's ability to predict the data in general, making it important to choose the appropriate batch size. Based on the results of the parameter tuning analysis, the optimal batch size was determined to be 32. The model was trained with a larger number of epochs and a batch size of 32 on real data (Fig. 4).

3 Results

The selected deep learning model, with a batch size of 32 and 300 epochs, was trained and tested on seismic data obtained from the forward modeling of the BP-Tooth

Fig. 3 Training is carried out with the same dataset, but the batch size training parameter is made to varied 16, 32, 64, and 128, respectively, the error value (RMS) is represented by a line

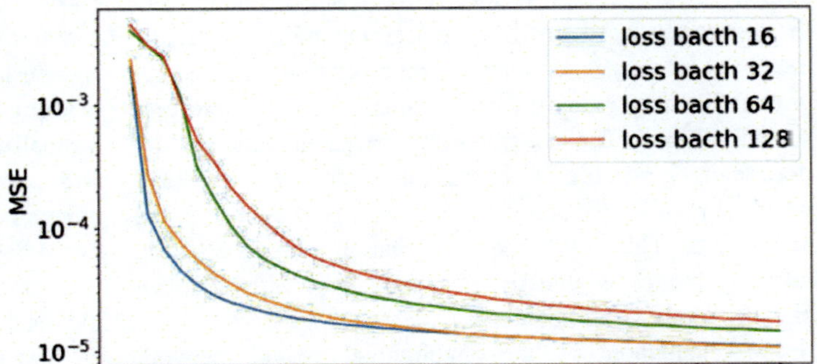

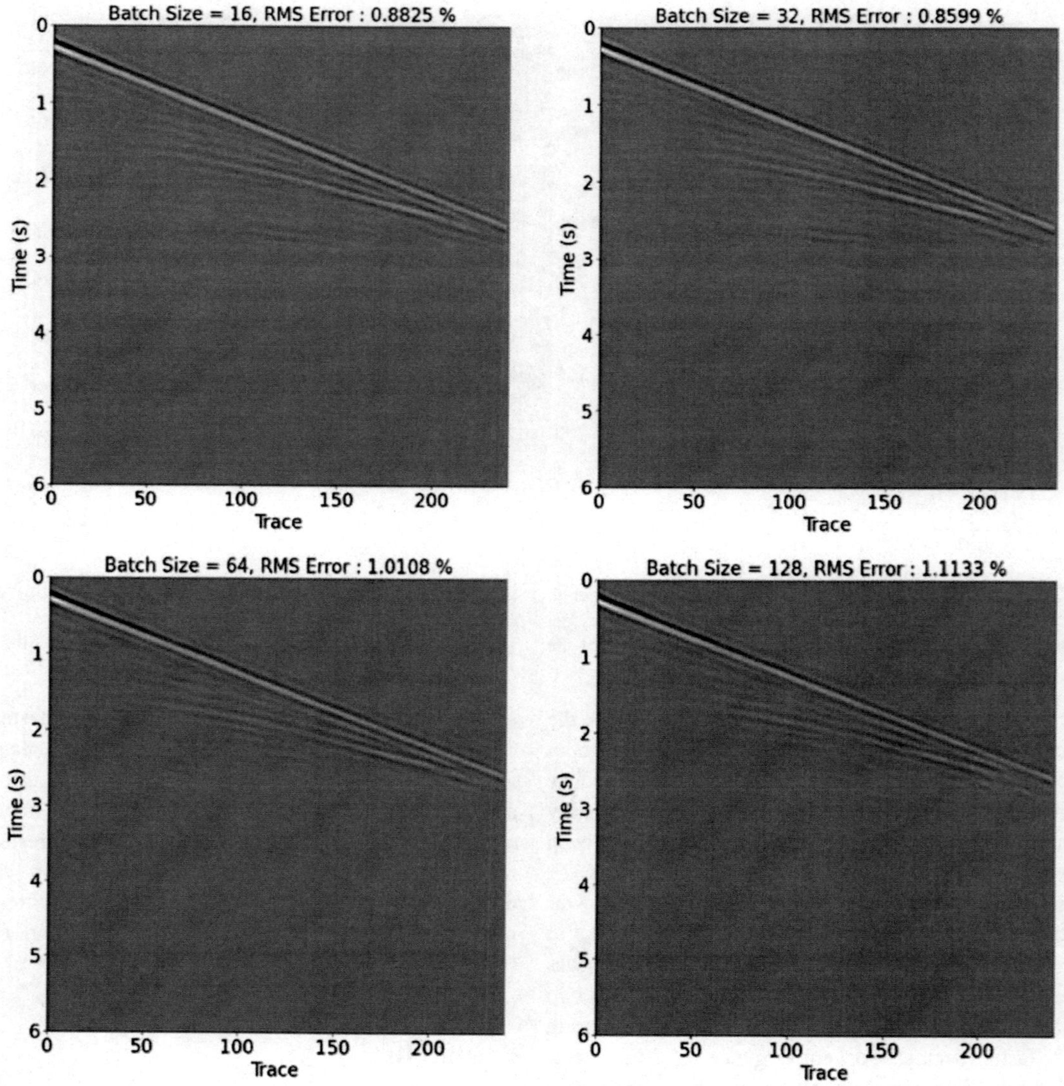

Fig. 4 Prediction of seismic low-frequency content was performed using several models that varied in batch size. The results showed that the model with a batch size of 32 produced a smaller error value compared with other models. Thus, this model was selected as the optimal model for the prediction of real data

velocity model. The BP-Tooth velocity model differs from the submodels used in the deep learning training process, and the wave propagation simulation process was carried out with different geometries. The results of the test are presented in Fig. 5. The low-frequency prediction generated by deep learning was highly accurate, as evidenced by an RMS error value of $< 1\%$, and its trend was highly similar to the actual data. This result indicates that the generated deep learning model is effective in predicting with different velocity models and geometries.

An evaluation of a convolutional neural network (CNN) deep learning model using synthetic data has been carried out, and the model has been applied to predict low-frequency seismic content in the Sadewa Field, Kutei, Indonesia. Prior to its utilization as input for the deep learning model, the raw seismic data was preprocessed by normalizing its amplitude and applying high-pass filtering with a cut-off frequency of $\sim$ 10 Hz. Figure 6 presents a comparison of the CNN model's performance with conventional deghosting methods for low-frequency reconstruction to evaluate the accuracy of the deep learning model's predictions. The comparison results indicate that the deep learning model's predictions are effective on real data, as the predicted low-frequency trend is in line with the actual data,

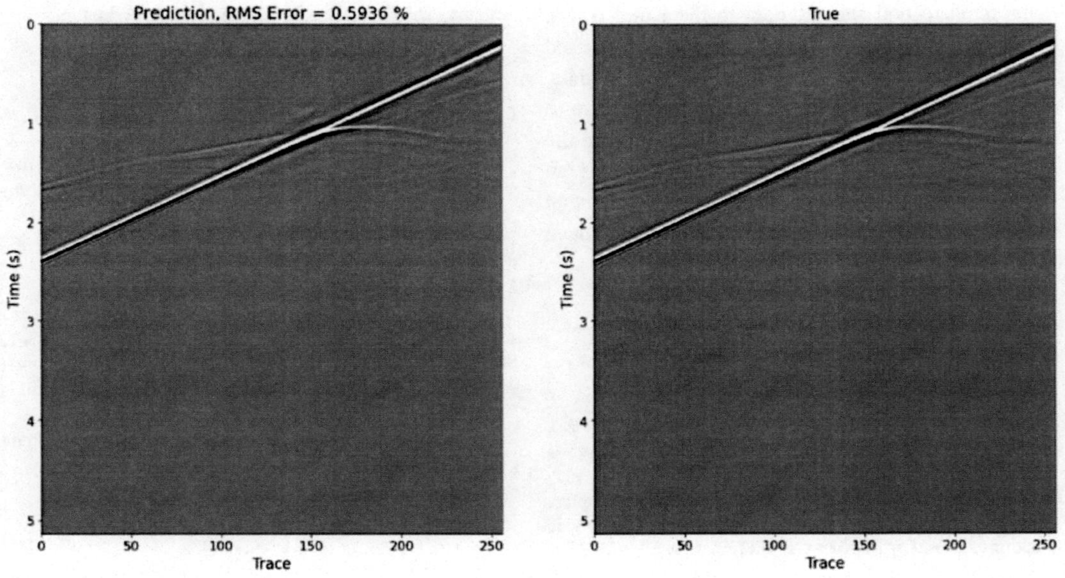

Fig. 5 Results of a low-frequency prediction from a BP-Tooth velocity model; the predicted results are on the left and the actual (synthetic) results are on the right. Comparison of the two results indicates that deep learning is highly effective and accurate in making predictions, even when the data being predicted differs in terms of velocity model and acquisition geometry

although some shallow time seismic events may not be well reconstructed due to strong seismic noise. The successful testing of the deep learning models on both synthetic and real data holds promise for this method to serve as an alternative approach for predicting low-frequency content in other field seismic data, owing to its demonstrated accuracy and effectiveness.

4 Conclusion

The performance of the CNN model was evaluated on synthetic data and was found to have the ability to predict low frequency, despite differences in the velocity and acquisition geometry models used for training. The best

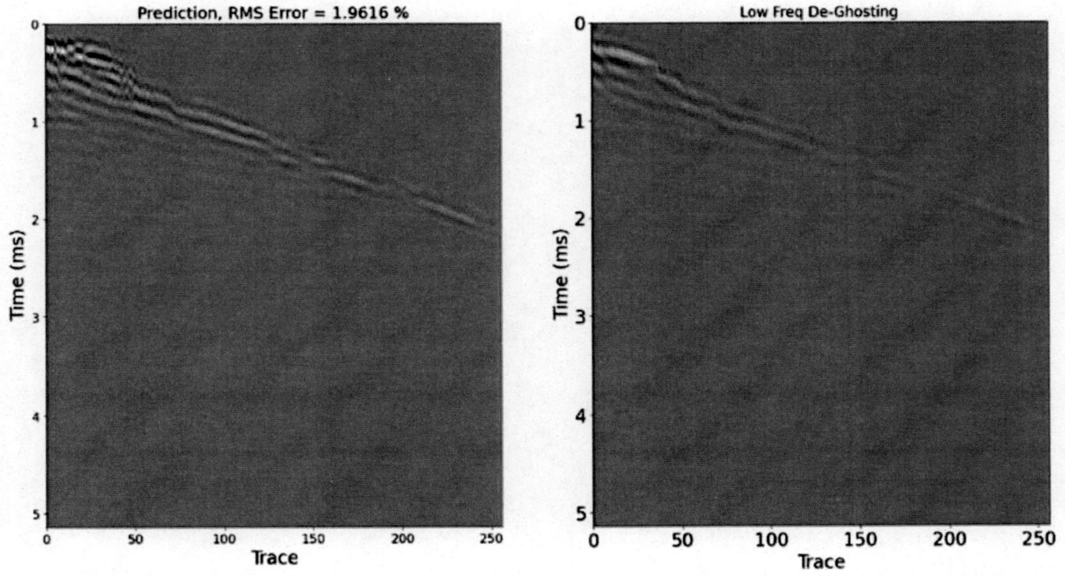

Fig. 6 Illustrates the results of a low-frequency prediction performed on real data from the Sadewa Field in Kutei, Indonesia. The results demonstrate the efficacy of the deep learning model on real data, as the predicted low-frequency trend is in agreement with the actual data. However, it is noted that some shallow time seismic events may not be accurately reconstructed due to the presence of strong seismic noise

model was applied to real seismic data in the Kutei Basin. The results showed that deep learning can effectively be used to extrapolate low-frequency content in marine seismic data, with an RMS error below 2%. However, it is imperative to eliminate any noise present in the real data before using the model. The utilization of deep learning algorithms as an alternative method for predicting low-frequency content in seismic data has shown promise, as preliminary results have demonstrated its effectiveness and relative accuracy.

References

Bunks, C., Saleck F. M., Zaleski, S., & dan Chavent G. (1995). Multiscale seismic waveform inversion. Geophysics, 60, 1457–1473.

Hu, Y., Han, L., Xu, Z., Zhang, F., & Zeng, J. (2017). Adaptive multi-step full waveform inversion based on waveform mode decomposition. *Journal of Applied Geophysics, 139*, 195–210.

Kingma, D. P., & Ba, J. (2014). Adam: A method for stochastic optimization. Article retrieved from Proceedings of the 3rd International Conference on Learning Representations 2015, San Diego, United States

Luporini, F., Lauboutin, M., Lange, M., Kukreja, N., Witte, P., Huckelheim, J., Yount, C., Kelly, P. H., Herrmann, F., & Gorman, G. (2020). Architecture and performance of Devito: A system for automated stencil computation. *Association for Computing Machinery, 46*, 1165–1187.

Pratt, R. G. (1990). Inverse theory applied to multi-source cross-hole tomography. *Geophysical Prospecting, 38*, 311–329.

Sun, H., & Demanet, L. (2018). Low frequency extrapolation with deep learning. In *Proceedings of SEG technical program expanded abstracts 2018, California, Amerika Serikat* (pp. 1267–1274)

Sun, H., & Demanet, L. (2020). Deep learning for low frequency extrapolation of multicomponent data in elastic full waveform inversion

Tarantola, A. (1984). Inversion of seismic reflection data in the acoustic approximation. *Geophysics, 49*, 1259–1266.

New Modified Design of Accelerated Weight Drop (AWD) for Shallow Seismic Survey: A Field Test, Results and Evaluations

Alfian Bahar and Fatkhan Fatkhan

Abstract

The reflection seismic is technique to image subsurface structures to depth of thousand meters. In addition, the method is also applied to shallow targets with depths from 10 to 1000 m, for engineering problems and site investigations. Even though shallow seismic method is only scaling of reflection seismic method, problems encountered in the survey operation are not simple. Problems are cost and environmental issues, particularly related to sources used. The use of explosive seismic sources has implications for costs and environmental issues. One promising source option is an accelerated weight drop (AWD). The use of AWD as seismic source is highly dependent on surface conditions of survey area. Therefore, an AWD design that is suitable for various types of survey areas is needed. A new prototype has been developed as an improvement over previous prototype. The design modification is intended to increase penetration depth and widen frequency band of the source signal. Important criteria have been considered in AWD's design, such as impact energy (penetration and bandwidth), repeatability, portability, economical and environmentally friendly. This paper reports performances test of prototype AWD for generating seismic sources to obtain high-quality data. The survey was conducted to test performances. There is no specific target of geological structure in areas. The survey located in a densely populated area was conducted using 100 units of wireless seismometer (5 m interval) and shot point interval of 15 m. The recorded data has been processed using a standard seismic data processing workflow. The results show that the reflected signal is visible at a depth of 50 ms to 800 ms or about 50–700 m, with a signal frequency of up to 100 Hz with good image quality. The criteria of penetration depth and frequency band have been achieved. Survey cost to obtain 0.5 km is less than 500 USD. This AWD has met other expected design criteria, economical and low environmental impact. Results are better than the old prototype and product in the market.

Keywords

Accelerated weight drop (AWD) • Design of seismic sources • Shallow seismic survey

1 Introduction

Shallow seismic survey is conducted to image subsurface geological structures at a depth of 10 to 1000 m. The method is applied to engineering, site investigation, etc. The shallow seismic method is a scale of conventional seismic methods. However, problems encountered are different, such as costs, permitting and environmental issues. Three issues are directly related to selections of seismic source used.

There are types of seismic sources for land that are explosive (dynamite) and non-explosive. The use of explosive sources is effective in terms of energy and signal bandwidth. However, it often encounters permitting issues and costs, and environmental impacts. Meanwhile, non-explosive sources such as weight drops, vibrators, are alternative choices. Weight drops or accelerated weight drop (AWD) is a promising option, because it is simple compared with others. Even if there are designs and specifications, these AWDs are not necessarily suitable for all area conditions such as residential areas, forests in which access roads are not always available. Therefore, a reliable AWD design is needed by considering all requirements in terms of operational and performances.

A. Bahar · F. Fatkhan (✉)
Bandung Institute of Technology, Bandung, Indonesia
e-mail: fatkhan@gf.itb.ac.id

M. Bezzeghoud et al. (eds.), *Recent Research on Geotechnical Engineering, Remote Sensing, Geophysics and Earthquake Seismology*, Advances in Science, Technology & Innovation,
https://doi.org/10.1007/978-3-031-48715-6_52

2 The Designed AWD

The AWD generates seismic wave using an impact mechanism against the ground surface. Basic principle of the weight drop system is mass weight m is dropped from a height h to the ground. When the mass hits the base plate, the momentum is completely transferred to the ground surface. This transfer of momentum occurs in the impact time interval (Δt). The interval represents pulse width of action force applied to the ground directly related to frequency of source signals.

In market there are several commercial AWD products available. These products use a different designed concept, since they have a different target survey area and penetration depth. There is no single type of AWD that is suitable for all purposes. To deal with those problems, a design concept has been chosen for a new AWD which is simple, flexible and of high performance. Simple means using easy in equipment and operation. Flexibility is related to its use suitable for most of terrains and targets. High performance means that seismic signal has sufficient energy (target depth), and good signal bandwidth (target resolution), repeatability and economical.

The first prototype of AWD has been designed and built by handmade (Bahar et al., 2019). The prototype has a mass of 250 kg and used an elastomer rubber belt as an accelerator system. The lifting system of mass weight uses a chain-gear mechanism driven by gasoline engine. The prototype, named AWD250V1.0, is mounted on a trailer. Even though the AWD technically works well, the criteria for energy and bandwidth are still not met. A new modified AWD design has been built to address these issues.

Goal of the designed this new prototype is that energy effectiveness is increased without changing weight mass. This can be done by reducing the bounce back effect by providing a base plate retaining mass. The frame and base plate mass are balanced with the weight mass. To reduce the bounce back effect of the weight mass, weight is divided into two parts, the solid mass at the bottom and the loose sand in the container at the top. Accelerator is not utilized, however it will be used when energy is still not enough.

3 The Test Survey

Testing performances were conducted in a densely populated area. The survey line is a 500 m long following access roads. These conditions are ideal for the use of a weight drop. The line location is heavy traffic due to population activities.

3.1 Recording Method

Recording data use 100 wireless geophones, with a receiver interval of 5 m. The data are acquired using a fixed receiver position, while the shot point position moves along the survey line with the interval of 15 m. Each shot point location was recorded 5 times, then the records were stacked. During the recording process, traffic activities are not stopped. This survey is not intended to study subsurface structures but just to evaluate the performance of AWD source.

As seen in Fig. 1, the recording is consistent, both in terms of energy and bandwidth of signal. To test the repeatability of the source, it is necessary to study the

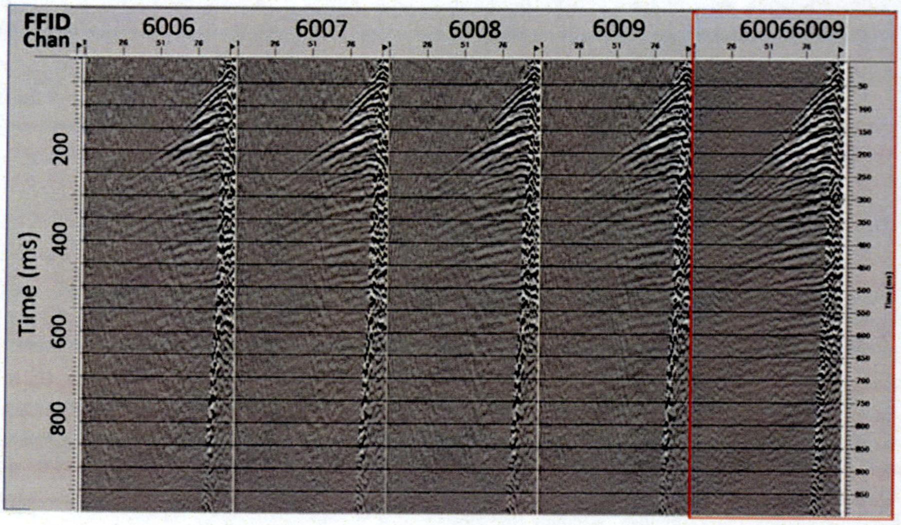

Fig. 1 Single shot records (ffid 6006–6009) and (red square) is stacked shot records

changes in the amplitude and wavelet shape from each recording of the same shot point.

3.2 Processing Data

Seismic processing is applied to field recording data to obtain a stack section. Sequentially, processing steps applied are data input, preconditioning (filter and stacking single shot records), geometry setup, denoising, amplitude correction, deconvolution, velocity analysis and CMP stack. For display purposes, scaling (AGC), filtering and FX decon processes are utilized. The display of semblance velocity is shown in Fig. 2. Figure 3 shows final stack before and after bandpass filter.

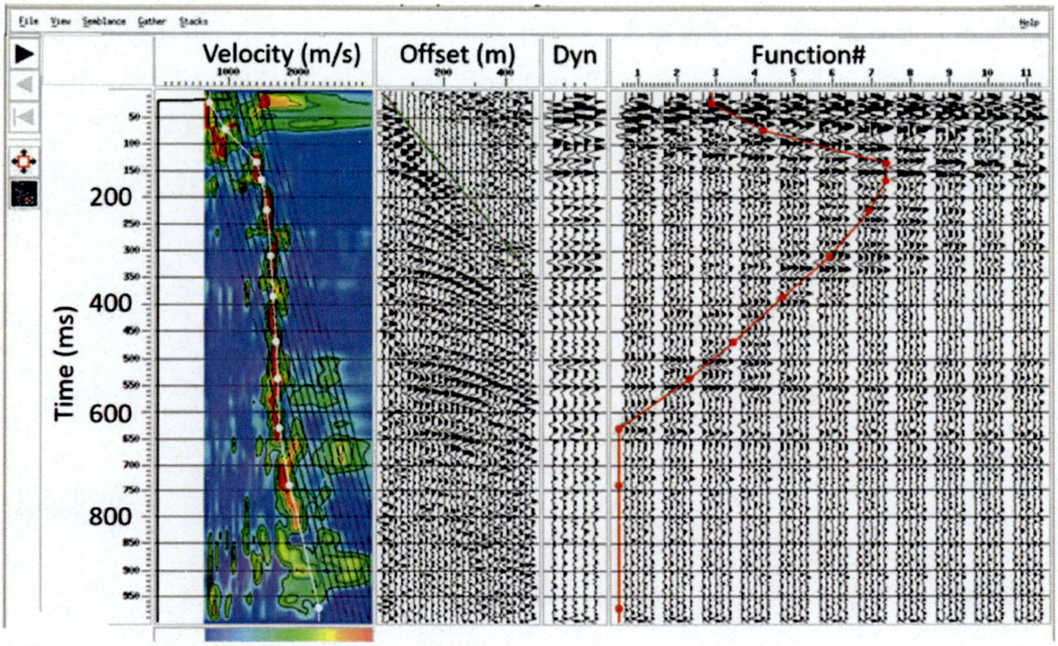

Fig. 2 Semblance velocity analysis shows good reflection signal coherency

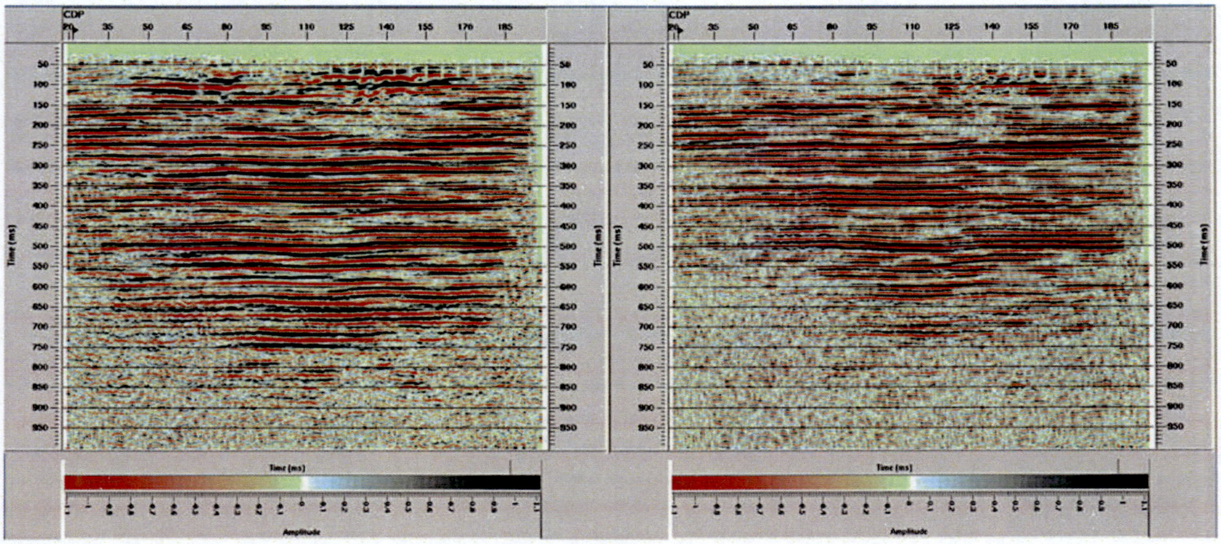

Fig. 3 Final stack section without filter (left) and After bandpass filter 60-18-150-72 (right)

4 Conclusions

- The second prototype of the AWD generates source well. Moreover, it can produce good data quality as seen in its signal responses.
- The AWD source meets requirements needed, such as repeatability, portability and economical, provided surface and subsurface are relatively flat. It may be ideal for area with limited access and sensitive environmental issue.

Reference

Bahar, A., Fatkhan, M., & Sule, R. (2019). A field test and performance evaluations of acceleration weight drop (AWD250 v1.0) for shallow seismic survey.

Break-Out Prediction as a Chip Cutting by Milling Machine for Wellbore Stability in Geomechanics

Mohatsim Mahetaji, Jwngsar Brahma, and Rakesh Kumar Vij

Abstract

Wellbore instability is caused by formation break-out or drilling-induced fracture. Formation break-out and drilling-induced fracture depend on the formation strength and in situ stresses near the borehole. The main objective of this study is to use the concept of milling machine theory to predict the break–out depth of the formation. A new approach is proposed for the prediction of formation break-out dimensions for maximum wellbore stability. The formation break-out is due to the stress concentration around the borehole. This stress concentration acts as an additional shear cutting force to break a formation. The shear cutting force is similar to the milling cutting tool force for the same formation to produce an equal dimension chip. The formation cutting force causes break-out by an imaginary milling tool. This additional shear cutting force plays an important role in the calculation of the break-out dimension. The concept of milling machine theory is applied to develop the mathematical model for the prediction of break-out dimension. Through this study, it has been observed that the depth of the break-out is the function of the break-out angle and yield strength of the formation. Shear cutting force around the wellbore is the maximum at a point near the minimum horizontal stress direction that produces break-out. Break-out angles can be calculated theoretically, empirically (using failure criterion) and actual (using image log). It has been shown that the radial stress acting on the boundary of the wellbore is not responsible for the shear failure of the formation. In this study, a relationship between break-out and stresses around the borehole has been established. The new relation established through this study may help to predict break-out dimension from the shear strength of the formation and stress profile. Till date, no such literature is available for the prediction of break-out dimensions for the wellbore stability in geomechanics. The proposed mathematical model helps us to design a cost-effective and safe well in the geologically complex reservoir for exploitation and exploration of hydrocarbon.

Keywords

Wellbore stability • Geomechanics • Break-out prediction • Mathematical modeling

1 Introduction

Wellbore instability on the geomechanical stress point of view is due to the stress concentration around the borehole. Three perpendicular principal stress acting on the borehole is given as a vertical stress, minimum horizontal stress and maximum horizontal stress. Inclination angle of the bore hole is also impacted on the stress contraction around the borehole (Al-Ajmi, 2006). Prediction of the break-out is not easy as there are lot of other reasons for it like formation strength, drilling and completion practice, tectonic effect and more. In this paper, a new theory is derived for the break-out estimation from stress profile. Hypothesis for this study is "if cutting force acting on the chip is same as a force acting on the formation due to tangential force acting on borehole then dimension of the break-out and chip cut by milling tool is same". This theory is also applied for finding the maximum stress near to the wellbore from break-out dimensions (from image log) (Zhang, 2019).

M. Mahetaji (✉) · R. K. Vij
School of Petroleum Technology, Pandit Deendayal Energy University—PDEU, Gandhinagar, Gujarat, India
e-mail: mohatsimmehtaji@gmail.com

J. Brahma
School of Technology, Pandit Deendayal Energy University—PDEU, Gandhinagar, Gujarat, India

M. Bezzeghoud et al. (eds.), *Recent Research on Geotechnical Engineering, Remote Sensing, Geophysics and Earthquake Seismology*, Advances in Science, Technology & Innovation,
https://doi.org/10.1007/978-3-031-48715-6_53

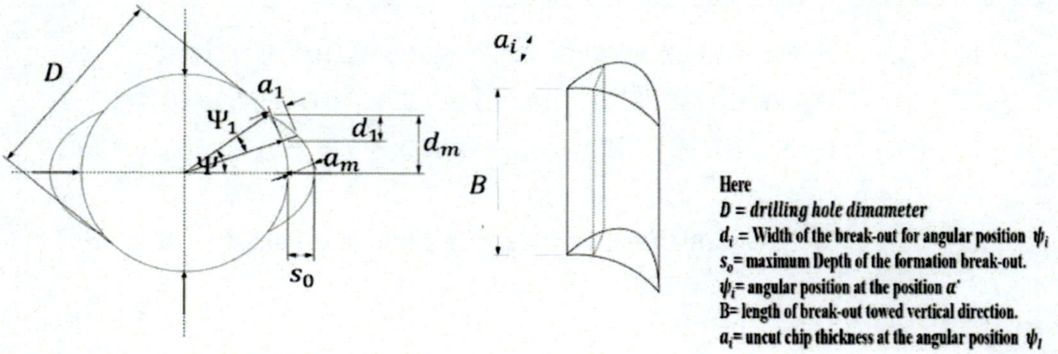

Fig. 1 Break-out as a chip cutting by milling machine

Compressive force acting perpendicular to the maximum principal stress breaking out of the formation is shown in Fig. 1. Now unbalanced force on the single side of the break-out is working as a shear cutting force by imaginary milling machine. Imaginary milling tool cut first-half of the break-out in a side and another tool cut second half of the break-out in opposite side. In this study, detailed analysis is done for the determination of unbalanced force and a new theory is proposed as a chip cutting by milling machine (Chattopadhyay, 2011).

2 Theory

Stress profile for strike-slip faulting regime is taken to understand a break-out theory. Stresses acting around the borehole as radial direction and tangential direction is given in Fig. 2a. In this figure, borehole circular periphery cut it out from the minimum horizontal stress point-A for better understanding of stress concentration.

Stresses acting on the point P on the periphery of the borehole is given in the Fig. 2b. Now unbalanced force acting on the point P due to the maximum horizontal (σ_H), minimum horizontal (σ_h) and radial stress (σ_r). Unbalanced force in the tangential direction (σ_T) is acting as a cutting force whenever unbalanced radial force (σ_R) is not responsible for the break-out but prevent the fall.

Now, value of σ_T and σ_R are given by $\sigma_T = 0$ at A, B, C and D and $\sigma_R = \sigma_r - \sigma_h$ at A and C and $\sigma_R = \sigma_r - \sigma_H$ at B and D point P in the first quadrant:

$$\sigma_T = \sigma_H \cos\alpha - \sigma_h \sin\alpha \tag{1}$$

$$\sigma_R = \sigma_H \sin\alpha + \sigma_h \cos\alpha - \sigma_r \tag{2}$$

For the second, third and fourth quadrant, equation of the tangential and radial force is changed with the changing of direction. Resultant force is given by $\sqrt{\sigma_T^2 + \sigma_R^2}$ and resultant angle with radial direction is $\tan^{-1}(\sigma_R/\sigma_T)$. Now when $\sigma_T = 0$, then point P given as P^* and the polar angle at that point is given by the half break-out angle (α^*) given as

$$\alpha^* = \tan^{-1}(\sigma_H/\sigma_h) \tag{3}$$

This σ_T is as force by the milling machine. Now dimension of cutting chip compared with the break-out given in the Fig. 1. As milling machine theory failure is

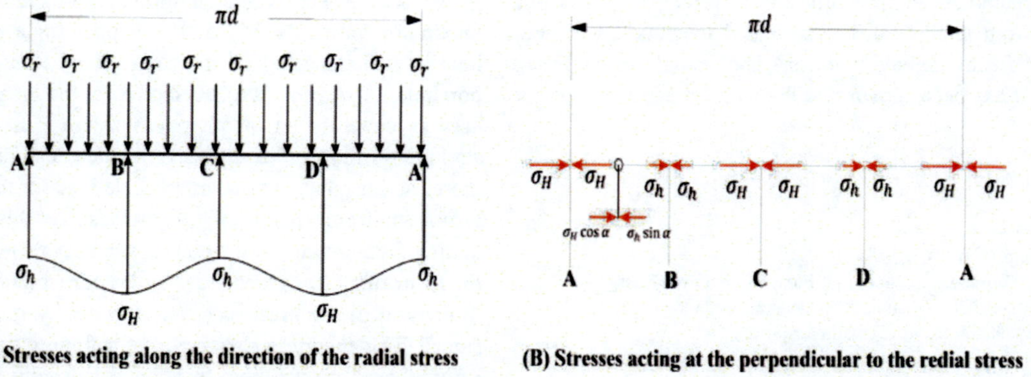

Fig. 2 Stress distribution at radial direction and tangential direction with polar angle

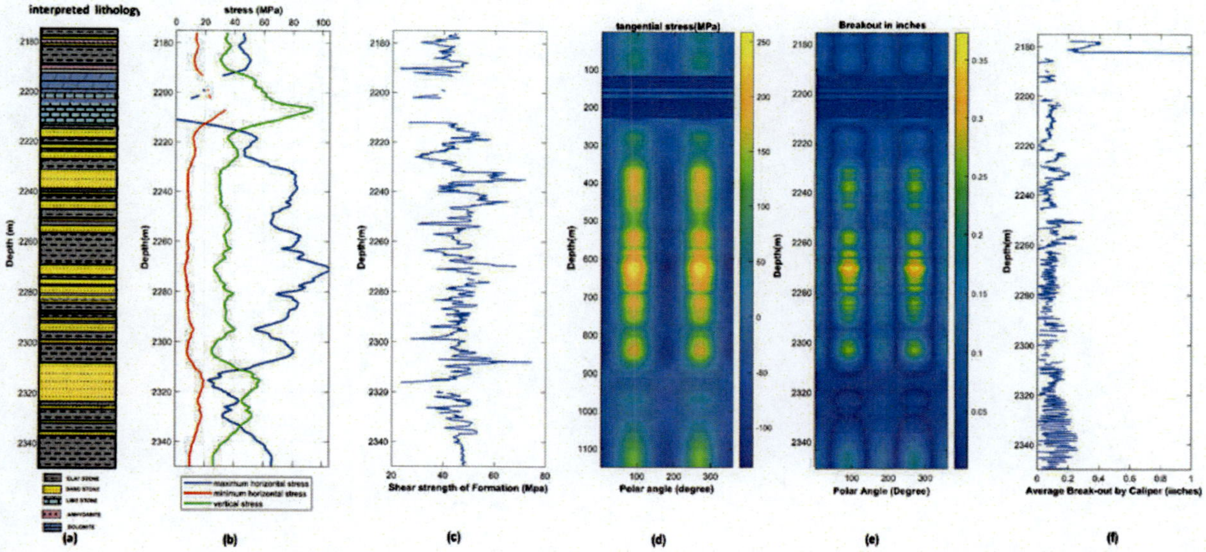

Fig. 3 Result on case study on Netherlands exploratory offshore gas well. **a** interpreted lithology, **b** stress profile, **c** shear strength of formation, **d** tangential stress, **e** predicted break-out, **f** actual break-out

$$P_{\text{ti}} = (A_i)P_s \quad (4)$$

P_s = specific strength of the formation = form factor (f)* shear strength (τ_s) and break-out thickness is

$$a_i = \frac{\sigma_T}{fB\tau_s} \quad (5)$$

From Mohr-Coulomb criterion (Labuz & Zang, 2012):

$$\tau_s = (\sigma_m - \tau_m \sin\phi)\tan\phi + c \quad (6)$$

Here: σ_m(mean effective stress) $= (\sigma_H + \sigma_h)/2$

$$\tau_m(\text{maximum shear stress}) = (\sigma_H + \sigma_h)/2$$

3 Results and Discussion

A case study of the exploratory gas offshore well at the Netherlands is taken with appropriate input as a gamma ray (gr), sonic log (*DTSM and DTCO*), porosity log (Φ), density log (ρ) and resistivity log (R_{deep}). The mechanical Earth model (MEM) is used to predict mechanical properties to find stress profiles (Mahetaji and Brahma, 2023), as shown in Fig. 3b. Lithology around the wellbore is given in track (a), shear strength is calculated with Mohr-Coulomb criteria track (c), track (d) is tangential stress and track (e) is predicted break-out as mud density is taken 1.5g/cc and calibrated and validated with track (f) of actual break-out driven from caliper log.

4 Conclusions

Break-out prediction as an imaginary cutting by milling machine is the new approach in geomechanics, as till date, no such literature is available for the prediction of break-out dimensions as a function of shear strength of the formation and stress profile. Further stress profile for the exploratory well is also estimated from the image log as break-out dimensions is also the relation of the unbalanced force acting on the borehole. This proposed mathematical model helps us to design a cost-effective and safe well in the geologically complex reservoir for exploitation and exploration of hydrocarbon. This theory is also applicable to maintain wellbore stability in real time.

References

Al-Ajmi, A. (2006). *Wellbore stability analysis based on a new true-triaxial failure criterion* (Doctoral dissertation, KTH).

Chattopadhyay, A. B. (2011). *Machining and machine tools (with CD)*. Wiley.

Labuz, J. F., & Zang, A. (2012). Mohr–Coulomb failure criterion. In *The ISRM suggested methods for rock characterization, testing and monitoring: 2007–2014* (pp. 227–231). Springer.

Mahetaji, M., & Brahma, J. (2023). Prediction of minimum mud weight for prevention of breakout using new 3D failure criterion to maintain wellbore stability. *Rock Mechanics and Rock Engineering*, 1–22. https://doi.org/10.1007/S00603-023-03679-4

Zhang, J. J. (2019). *Applied petroleum geomechanics*. Gulf Professional Publishing.

Real-Time Optimum Drilling Mud Weight Prediction by New 3D Criterion to Manage Wellbore Stability in Challenging Environments

Mohatsim Mahetaji, Jwngsar Brahma, and Rakesh Kumar Vij

Abstract

Drilling mud weight and wellbore orientation are convenient factors to manage wellbore stability. Improper selection of these convenient parameters causes the formation break-out or drilling-induced fracture. The main objective of this study is, to predict the optimum drilling mud weight with the concept of a new 3D failure criterion and drilling-exponent (D-exponent) factor for the proper maintenance of wellbore stability in challenging environments. The new failure criterion is derived as the function of principal stresses acting on the in situ rock. To prevent the formation break-out, optimum mud weight is calculated by the proposed criterion. Stresses around the wellbore acted in the cylindrical coordinate, so the kirsch equation converted all orthogonal stress into the cylindrical coordinate. A theoretical model calculates optimum mud weight by the proposed failure criterion and existing criteria like the Mohr-Coulomb criterion and the Mogi-Coulomb criterion. The D-exponent factor further corrects this suggested mud weight by real-time monitoring the drilling event. Mohr-Coulomb failure criterion overestimates the formation's strength as the criterion neglects the intermediated principal stress. Mogi-Coulomb criterion gave a dual result for optimum mud weight due to the quadratic equation of octahedral shear stress. Optimum mud weight prediction by the above methods is comparatively less accurate than our proposed model. The proposed criterion considers the effect of intermediate principal stress as a linear function of all stresses that make the mathematical model more effective and reliable for real problems. The predicted mud weight is updated in real-time monitoring by the D-exponent factor considering the drilling parameters like penetration rate (ROP) and tool cutting speed (RPM), weight on bit, and borehole diameter. This proposed model is also recommended for exploratory wells to predict and implement drilling parameters to maintain wellbore stability. The proposed new 3D failure criterion is an extended form of the Mohr-Coulomb criterion, where the roles of intermediate stress are also considered to predict optimum mud weight to keep the wellbore stable in a challenging environment. The proposed model is further helpful for maintaining wellbore stability in the exploratory well in high-pressure and high-temperature well, and environmentally challenged well by real-time monitoring.

Keywords

Wellbore stability • Geomechanics • 3D failure criterion • Mud weight prediction • D-exponent

1 Introduction

Drilling mud weight and wellbore orientation are convenient factors to manage wellbore stability. Wellbore instability in geomechanics is due to the break-out of the formation due to the compressive failure as lower weight selection or due to the drilling-induced fracture due to higher mud weight (Al-Ajmi & Zimmerman, 2005). Proper selection of these convenient parameters is important to maintain wellbore stability. Real time monitoring of the mud weight plays an important role for exploratory well where knowledge of the formation and uncertainty of reservoir. Drilling exponent is the useful tool to maintain wellbore stability as it continues prediction of the pore pressure (Das & Chatterjee, 2017; Eaton, 1975). Real-time pore pressure is implemented on the optimum mud weight selection using linear 3D failure criterion. The main objective of this study is, to predict the

M. Mahetaji (✉) · R. K. Vij
School of Petroleum Technology, Pandit Deendayal Energy University—PDEU, Gandhinagar, Gujarat, India
e-mail: mohatsimmehtaji@gmail.com

J. Brahma
School of Technology, Pandit Deendayal Energy University—PDEU, Gandhinagar, Gujarat, India

M. Bezzeghoud et al. (eds.), *Recent Research on Geotechnical Engineering, Remote Sensing, Geophysics and Earthquake Seismology*, Advances in Science, Technology & Innovation,
https://doi.org/10.1007/978-3-031-48715-6_54

optimum drilling mud weight with the concept of a new 3D extended Mohr-Coulomb failure criterion (Mahetaji et al., 2023) and D-exponent to maintenance of wellbore stability in challenging environments. A case study from the online dataset via the Dutch Oil and Gas Portal (NLOG) is taken to apply on actual data on this theory and then used to validate it (https://www.nlog.nl/datacenter/brh-overview; Maleki et al., 2014).

2 Theory

The new failure criterion is derived as the function of principal stresses acting on the in situ rock (Fig. 1). To prevent the formation break-out, optimum mud weight is calculated by the proposed criterion. Stresses around the wellbore acted in the cylindrical coordinate, so the kirsch equation converted all orthogonal stress into the cylindrical coordinate. A theoretical model calculates optimum mud weight by the proposed failure criterion and existing criteria like the Mohr-Coulomb criterion and the Mogi-Coulomb criterion (Ulusay, 2015; Van Adrichem Boogaert & Kouwe, 1993). The D-exponent factor further corrects this suggested mud weight by real-time monitoring the drilling event.

New 3D extended Mohr-Coulomb failure criterion with considering all three principal stresses acting around the bore hole is given as

$$\sigma_1 = \frac{2\gamma \sin c}{(2 - 2\sin\phi_m + \gamma)}\sigma_2 + \frac{2 + 2\sin\phi_m + \gamma}{(2 - 2\sin\phi_m + \gamma)}\sigma_3 + \frac{2(2+\gamma) * c * \cos\phi_m}{(2 - 2\sin\phi_m + \gamma)} \tag{1}$$

Here γ is weighting factor, c is cohesion, ϕ_m is the friction angle, and σ_1, σ_2 and σ_3 are principal stresses around the borehole.

This new failure criterion is applied in Mechanical Earth Model (MEM) for prediction of the optimum mud weight (Mahetaji & Brahma, 2023). Input required for this MEM is taken from the logging while drilling tool and wire line logging tool. Drilling exponent is the relation of the rate of penetration, speed of the tool (RPM), weight on bit (WOB), bit diameter (D), normal pressure (p_n), and mud weight (p_m) and is given as

$$D_C\text{exponant} = \frac{\log_{10}\left(\frac{\text{ROP}}{60\text{RPM}}\right)}{\log_{10}\left(\frac{12\text{WOB}}{D\times10^6}\right)}\frac{p_n}{p_m} \tag{2}$$

Now, as at the time of the drilling, based on the formation lithology and pressure profile drilling parameter changes and it impacts on the value of the D_C. Real-time value of the D_C predicts the stress profile around the borehole as new 3D failure criterion is used for the prediction of mud weight that has been corrected for maintaining wellbore stability.

3 Results and Discussion

For validation of this new theory, a case study of the exploratory gas offshore well at the Netherlands is taken with appropriate input as a real-time monitoring. Lithology around the wellbore is given mostly claystone, sandstone, limestone, anhydrite, and dolomite.

Wireline well log data (from 2175 to 2350 mTVD) is selected as an input to find out stress profile. MEM with appropriate empirical correlation with present lithology between input (Caliper, GR, newton-density and porosity and deep resistivity) and stress profile as a output (σ_h, σ_H, and σ_r). Optimum mud weight from above method on case study is given in Fig. 2. First track is giving the actual value D_C and normal value D_{Cn} at the time of the drilling. The value of D_{Cn} is given by the equation

$$D_{Cn} = 0.000175D_2 + 1.57 \tag{3}$$

Second track in Fig. 2 is the predicted value of the pore pressure and third track is showing the optimum mud weight based on the above method. Mud weight suggested based on this theory is around 2–4 g/cc that is higher than actual used

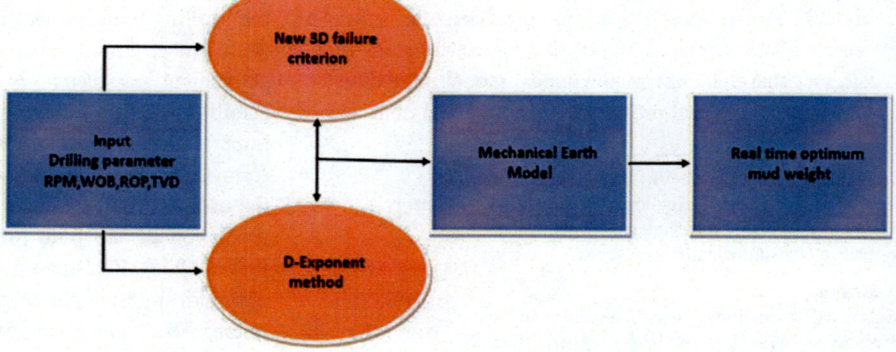

Fig. 1 Flow diagram of the methodology

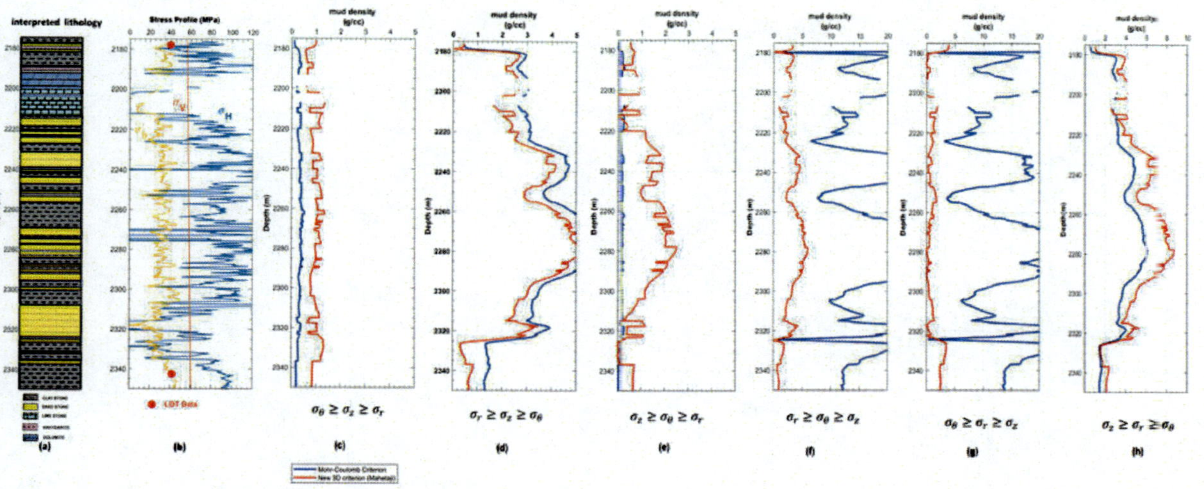

Fig. 2 Real-time minimum mud weight (g/cc) calculation with d-exponent by new 3D failure criterion on case study of Netherlands exploratory offshore gas well. **a** interpreted lithology, **b** stress profile, **c–h** minimum mud density for ($\sigma_\theta \geq \sigma_z \geq \sigma_r$); ($\sigma_r \geq \sigma_z \geq \sigma_\theta$); ($\sigma_z \geq \sigma_\theta \geq \sigma_r$); ($\sigma_r \geq \sigma_\theta \geq \sigma_z$); ($\sigma_\theta \geq \sigma_r \geq \sigma_z$); and ($\sigma_z \geq \sigma_r \geq \sigma_\theta$)

mud weight at the time of drilling that is 1.5 g/cc. Mud weight suggested is based on the drilling practice and lithology present at that point of drilling. Above method continuously improves drilling mud weight.

The formation's strength is overestimated by the Mohr-Coulomb failure criterion because it ignores the intermediated stress. Due to the quadratic equation of octahedral shear stress, the Mogi-Coulomb criterion produced two results for the ideal mud weight. The proposed criterion takes into account the impact of intermediate principal stress as a linear function of all stresses, improving the accuracy and dependability of the mathematical model for actual applications. The D-exponent factor updates the expected mud weight in real-time monitoring while taking drilling parameter like ROP, RPM, WOB, and borehole diameter into account. This proposed model is also recommended for exploratory wells to predict and implement drilling parameters to maintain wellbore stability.

4 Conclusions

This proposed novel 3D failure criterion is an extension of the Mohr-coulomb criterion, which develops considering the effects of intermediate stress on formation failure. Mud density calculation to prevent by new 3D linear failure criteria gave better results than the Mohr-Coulomb criterion as it neglects the effect of intermediated principal stress on the break-out of the formation. Wellbore stability should be maintained considering all six possible situations in high-pressure, high-temperature, and environmentally challenging wells. This hybrid approach improved minimum mud density by considering the drilling exponent for real-time monitoring.

References

Al-Ajmi, A. M., & Zimmerman, R. W. (2005). Relation between the Mogi and the Coulomb failure criteria. *International Journal of Rock Mechanics and Mining Sciences, 42*(3), 431–439.

Das, B., & Chatterjee, R. (2017). Wellbore stability analysis and prediction of minimum mud weight for few wells in Krishna-Godavari Basin, India. *International Journal of Rock Mechanics and Mining Sciences, 93*, 30–37.

Eaton, B. A. (1975). *The equation for geopressure prediction from well logs*. Fall Meeting of the Society of Petroleum Engineers of AIME.

https://www.nlog.nl/datacenter/brh-overview.

Mahetaji, M., & Brahma, J. (2023). Prediction of minimum mud weight for prevention of breakout using new 3D failure criterion to maintain wellbore stability. *Rock Mechanics and Rock Engineering*, 1–22.

Mahetaji, M., Brahma, J., & Vij, R. K. (2023). A new extended Mohr-Coulomb criterion in the space of three-dimensional stresses on the in-situ rock. *Geomechanics Engineering, 32*, 49–68.

Maleki, S., Gholami, R., Rasouli, V., Moradzadeh, A., Riabi, R. G., & Sadaghzadeh, F. (2014). Comparison of different failure criteria in prediction of safe mud weigh window in drilling practice. *Earth-Science Reviews, 136*, 36–58.

Ulusay, R. (Ed.). (2015). *The ISRM suggested methods for rock characterization, testing and monitoring: 2007–2014*. Springer.

Van Adrichem Boogaert, H. A., & Kouwe, W. P. F. (1993). Stratigraphic nomenclature of the Netherlands, revision and update by RGD and NOGEPA.

Earthquake Seismology

Joint Tomographic Inversion Using First Arrivals, Moho-Reflected Phases, Local and Teleseismic Events in the Region of Arraiolos (Portugal)

Ines Hamak, Piedade Wachilala, José Borges, Nuno Dias, Inês Rio, and Mourad Bezzeghoud

Abstract

For several years, geoscientists studied the seismicity of Arraiolos (central Portugal) which shows persistent but diffuse activity. Due to the slow Iberian intraplate deformation, it is difficult to observe a correlation between earthquakes and outcropping faults which induce a complexity in the interpretation of tectonic processes hidden behind the occurrence of the observed seismic activity. To overcome this issue, the study of the variation of velocity structures within depth in three-dimensions is obtained through seismic tomography. The crustal image will help to better constrain the seismicity. As it is generally unlikely to obtain a uniform distribution of rays sampling a medium, several approaches can be followed to enhance the tomographic imaging. In order to improve the ray coverage, we add reflected PmP and SmS phases. Following the Mw4.9 magnitude earthquake (January 15th, 2018), a temporary network of 34 seismic stations [21 short-period stations (Institute Dom Luiz, IDL) and 13 broadband stations (Institute of Earth Science, ICT)] were deployed recording a number of 437 earthquakes during a period of 6 months. The additional local and teleseismic events were detected on seismograms and relocated from a set of 12 broadband stations (DOCTAR network—Deep Ocean Test Array) operating between May 2011 and September 2012. In this study, a first inversion was performed from the Arraiolos aftershock sequence, generating crustal images that covered only a small part of the area and showing a poor ray density, mainly due to the narrow distribution of hypocenters. To significantly improve the resolution of the model, additional events and seismic phases are integrated in the process with the purpose of detecting the eventual presence of blind faults, correlating velocity layers with geological units and relocating the seismic events within the crustal model. This broader approach can be applied to any region of the world.

Keywords

Aftershock sequence • Seismic tomography • Later seismic phases • Arraiolos region

1 Introduction

Seismic tomography is a methodology used for decades to image deep layers of the inner Earth and figure out its geodynamics processes. The Mw4.9 magnitude earthquake occurred on the 15th of January 2018 in Arraiolos (Portugal) around 11.51 am at a depth of 11 km (located by the Instituto Português do Mar e da Atmosfera, IPMA) (Borges et al., 2018; Matias et al., 2019). This main shock has aroused the interest of geoscientists, who started to conduct seismological studies in order to understand the seismicity of the region (Matos et al., 2018). The outcome of each research has characterized the seismic activity of Arraiolos as diffuse but persistent. Additionally, it came out that none of the

I. Hamak (✉) · P. Wachilala · J. Borges · M. Bezzeghoud
Instituto de Ciência da Terra (ICT), University of Évora, Évora, Portugal
e-mail: hamak.ines@gmail.com

J. Borges
e-mail: jborges@uevora.pt

M. Bezzeghoud
e-mail: mourad@uevora.pt

P. Wachilala
Department of Teaching and Investigation of Exact and Natural Sciences, Instituto Superior de Ciências da Educação da Huíla (ISCED), Lubango, Angola

N. Dias
Department of Physics, Instituto Superior de Engenharia de Lisboa, Lisboa, Portugal
e-mail: nmdias@fc.ul.pt

N. Dias · I. Rio
Instituto Dom Luiz (IDL), University of Lisbon, Lisbon, Portugal
e-mail: irio@fc.ul.pt

M. Bezzeghoud et al. (eds.), *Recent Research on Geotechnical Engineering, Remote Sensing, Geophysics and Earthquake Seismology*, Advances in Science, Technology & Innovation,
https://doi.org/10.1007/978-3-031-48715-6_55

outcropping faults were active nor responsible for the seismicity of the area. Nevertheless, Araujo et al. (2020) have mapped two strike-slip faults Ciborro and São Gregorio faults that are intersecting in Aldeia da Serra, the highest point of the region, describing a reverse focal mechanism. He generated a hypothetical model that undoubtedly led to a better understanding of the tectonic mechanism hidden behind the observed seismic activity. However, additional investigations still need to be carried out (Fig. 1).

To clarify the unanswered questions regarding Arraiolos seismotectonic, a joint seismic tomography inversion was performed in this study, integrating global and local events as similarly as Moho-reflected phases (PmP and SmS) aiming the obtention of P and S waves distribution and the characterization of the crustal structure through all directions. By adding earthquakes and different ray trajectories to the study, a uniform distribution of hypocenters around the whole area will be obtained and consequently increase the ray density for a better resolution of the crustal model. Thus, the seismic activity will be well-constrained, and the tectonic processes clarified.

2 Methodology

A first inversion was performed using LOTOS code, created by Ivan Koulakov in 2009, based on the improvement of the Thurber approach (Koulakov, 2009). This seismic tomography has integrated the aftershock sequence recorded after the earthquake of Arraiolos occurred, comprising 437 events recorded by 34 seismological stations during a period of 6 months (Jan–June 2018) (Wachilala et al., 2019). This code is based on an iterative method that allows the simultaneous inversion of P and S waves and the source relocation within the 3D velocity model. The improvement of velocity distribution is observed iteration by iteration starting with the optimization of the hypocenter location within the 1D velocity model integrated at the beginning of the inversion process. As

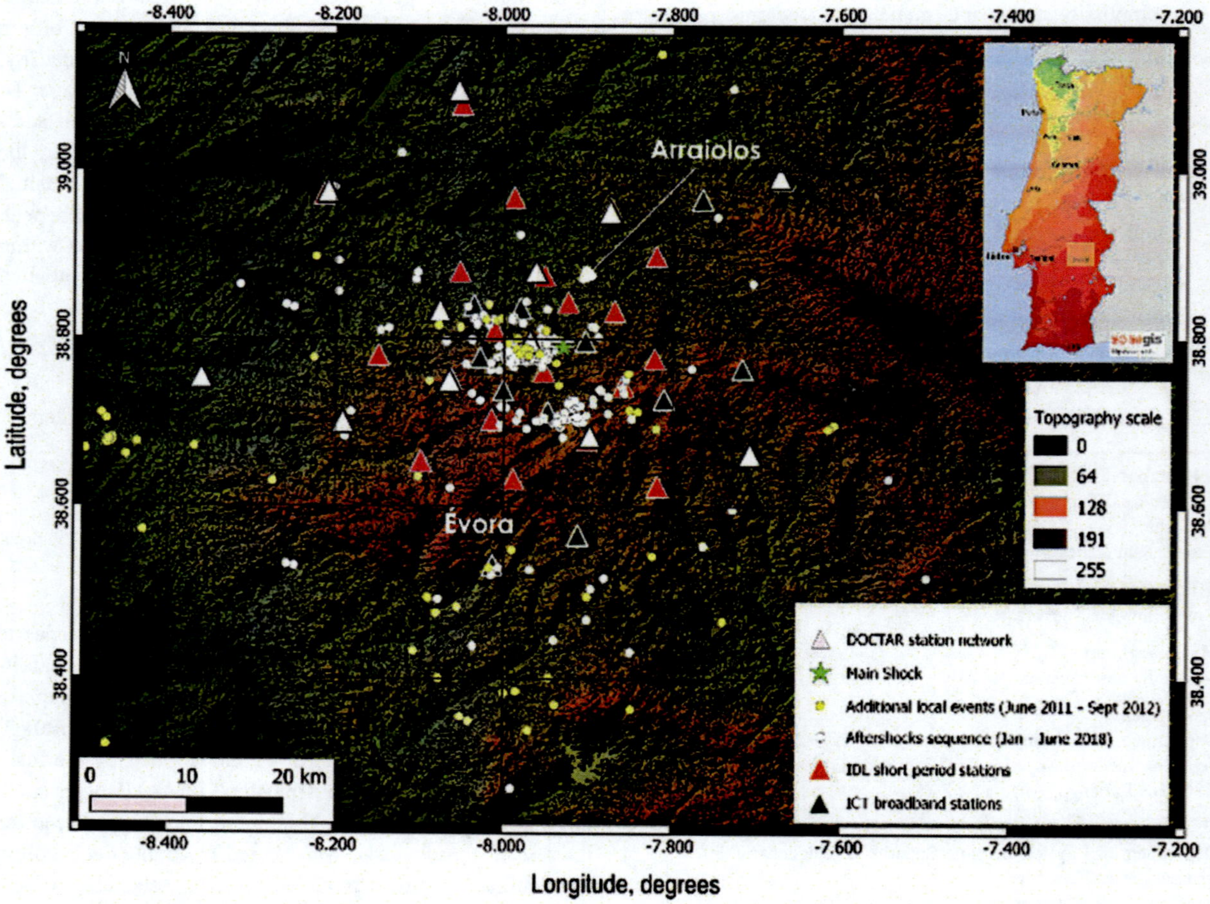

Fig. 1 Topographic map with the aftershock sequence that followed the main shock (2018) and the additional local events (May 2011–Sept 2012) used in this stage of study. The *triangles* represent the seismological stations operating for different periods of time belonging to three different local networks (DOCTAR, IDL and ICT)

the distribution of aftershocks used for the first inversion was narrowly concentrated in a local point within depth, it has generated a poor ray coverage enabling to image only a small part of the region (Fig. 2) (Hamak et al., 2020). To better constrain the tomographic images within a larger region, later phases (PmP and SmS), teleseismic events and additional local earthquakes are intended to be integrated to the inversion (Xia et al., 2011). Nevertheless, the approach proposed by LOTOS code was not suitable for this purpose and had to be abandoned for this stage of the work.

Picking PmP and SmS phases is one of the most challenging part of this study. Therefore, on a first step, the computation of PmP and SmS was conducted using predicted travel times, calculated from the 1D velocity model of the region on the purpose to obtain the eventual position of the arrival time of the phase. Then the visualization of amplitude changes on the rotated components (HHR, HHT) in the surrounding of the theoretical arrival time is carried out. Considering the approach of Nakajima et al. (Nakajima et al., 2002) the later phases were detected on the basis of the following characteristics: (1) the particle motions orientation being the same for P and PmP phases, as similarly as S and SmS, shows that they arrive at the station as compressional and shear waves, respectively (Fig. 4), (2) the large amplitude suggests that they are reflected in a sharp velocity discontinuity and (3) the difference between SmS and S arrival times is equal to 1.7 times the difference between PmP and P arrivals. However, a more precise methodology needs to be employed to predict the travel times of the reflected phases. This new approach is based on the use of the fast-marching code called FMTOMO (Rawlinson & Sambridge, 2005) that allows the integration of several types of phases (teleseismic and local events) to obtain the tomographic model.

3 Data and Results

3.1 Preliminary Seismic Tomography

The preliminary distribution of P and S waves velocities using LOTOS code are shown in Fig. 3. As the hypocenters were located in a narrow space, a uniform distribution of seismic rays was unlikely to be observed, which led to the generation of poorly resolved velocity images. The relocation of the aftershock sequence and the obtention of P and S velocity distribution were obtained with an RMS of 0.036 s for P waves and 0.051 s for S waves. The dataset used in this first stage of inversion was able to generate velocity contrasts for only a small region with a low resolution near the surface (Fig. 2).

3.2 Integration of Later Phases, Local and Teleseismic Events to the Seismic Tomography

In this study, a local station network named deep ocean test array (DOCTAR, University Potsdam, Portugal) was used, composed of 12 broadband stations operating from

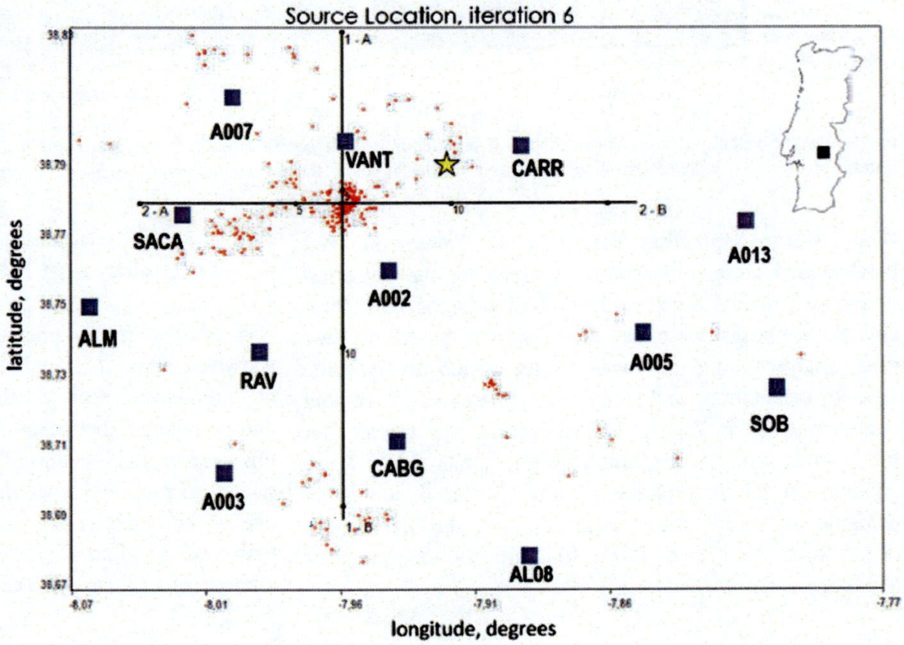

Fig. 2 Area of best ray coverage for Arraiolos region. *Red dots* represent the relocated epicenters after the obtention of the 3D velocity model and *blue squares* represent the station network. Two cross sections *1A–1B* and *2A–2B* were chosen to visualize the results. The *yellow star* represents the main shock

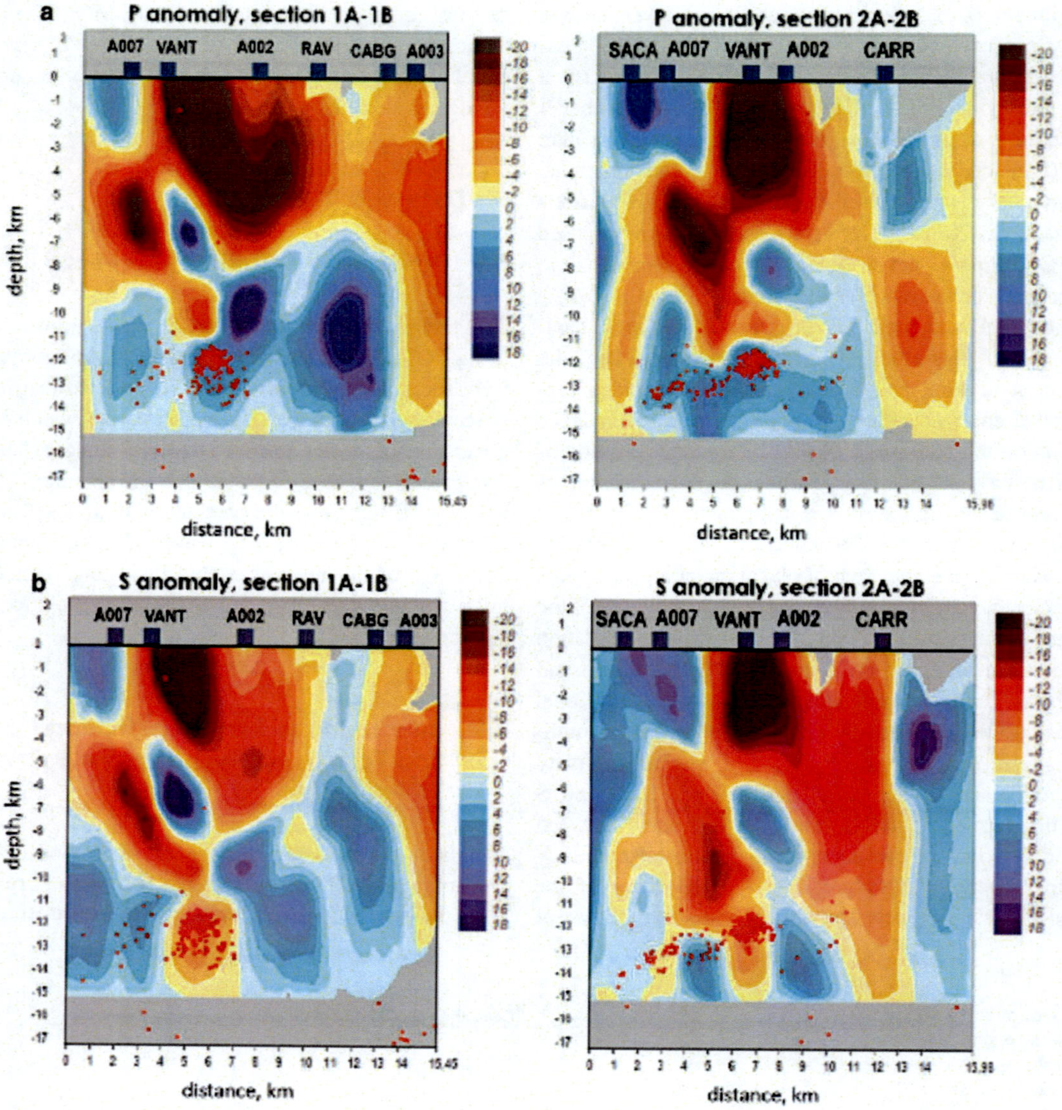

Fig. 3 P and S velocity anomalies distribution within depth beneath the two cross sections shown on Fig. 2. *Blue squares* represent seismological stations and *red dots* indicate aftershocks locations. On the right side the color variation represents the velocity anomalies values in percentage (%)

May 2011 to September 2012 (Fig. 1). Teleseismic and local events were relocated by performing the detection of P and S arrival times on DOCTAR seismograms. PmP and SmS phases were as well detected based on two main analyses, only on local events, which are: particle motion orientation and amplitude changes on rotated components as in Fig. 4. This detection was carried out for events with a magnitude higher than Mw1.7, to conduct an efficient phase picking. Figure 4 shows an example of PmP phase detection for the event that occurred on the 13th of July, 2012 with the magnitude of Mw2.5.

4 Discussion and Future Work

The velocity images obtained after the first inversion have shown a start of aftershock alignment that needs to be explored into more detail and on a wider scale. An anti-correlation between P and S waves anomalies is observed on the location of hypocenters, which is generally related to rock with specific properties or fluid content. As one of the objectives of this work is to detect the eventual existence of blind faults, it is important to increase the number of events, enhance the ray coverage and image a

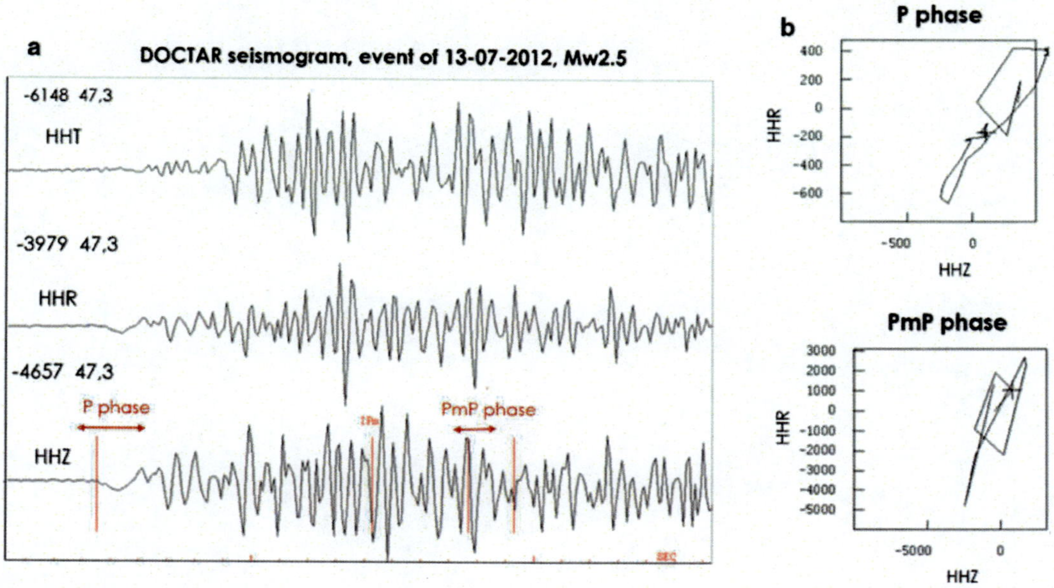

Fig. 4 **a** Seismogram of a local event recorded by DOCTAR station network (see in Fig. 1) on the 13th of July, 2012 having a magnitude of Mw2.5, visualized on rotated horizontal components (HHR, HHT). **b** Particle motions of P and PmP phases

larger area. By generating a velocity model of a wider region, the deformation mechanism hidden behind the observed seismic activity will be finally elucidated.

To reach this purpose, a deeper exploration of the FMTOMO code will be conducted during these months to hopefully generate first images of velocity contrast. In order to figure out the specific functioning of the software, a deep focus on each step of the inversion process needs to be carried out. Therefore, the new tomographic methodology used in this project aiming the additional integration of later phases and teleseismic events will be performed, providing reliable results. Finally, showing the efficiency of this new approach of seismic tomography will enable its use in any other regions of the world presenting the same problem.

Acknowledgements Thanks are due to FCT for the financial support to the ICT project (UID/GEO/04683/2013) with the reference POCI-01-0145-FEDER-007690 and to the IDL project (UIDB/50019/2020—IDL).

References

Araujo, A., Caldeira, B., Martins, A., Borges, J. F., Moreira, N., Fonseca Araújo, J., et al. (2020). Macrossismicidade associada ao sismo de Arraiolos do dia 15 de janeiro de 2018 com M = 4, 9 e eventuais implicações na geometria da rutura. *Comunicações Geológicas, 107*, 35–37.

Borges, J., Caldeira, B., Fontiela, J., Custódio, S., Dias, N., Waschilala, P., et al. (2018). The Arraiolos–Portugal–Moderate-Sized 2018 (M = 4.9) earthquake of January 15 and aftershocks: preliminary results. In *Proceedings of the 36th general assembly of the European seismological commission*

Hamak, I., Wachilala, P., Borges, J. F., Dias, N., Rio, I., Bezzeghoud, M. (2020). Three-dimensional crustal image of Arraiolos aftershock sequence, earthquake of M = 4.9, in Alentejo region, Portugal.

Koulakov, I. (2009). LOTOS code for local earthquake tomographic inversion: Benchmarks for testing tomographic algorithms. *Bulletin of the Seismological Society of America, 99*(1), 194–214.

Matias, L., Rio, I., Waschilala, P., Vales, D., Borges, J., Dias, N., et al. (2019). Investigação da sequência sísmica iniciada pelo sismo de Arraiolos (15/1/2018, ML= 4.9). Lisboa: IST, 14, 1–11

Matos, C., Custódio, S., Batló, J., Zahradník, J., Arroucau, P., Silveira, G., & Heimann, S. (2018). An active seismic zone in intraplate West Iberia inferred from high-resolution geophysical data. *Journal of Geophysical Research: Solid Earth, 123*(4), 2885–2907.

Nakajima, J., Matsuzawa, T., & Hasegawa, A. (2002). Moho depth variation in the central part of Northeastern Japan estimated from reflected and converted waves. *Physics of the Earth and Planetary Interiors, 130*(1–2), 31–47. https://doi.org/10.1016/S0031-9201(01)00307-7

Rawlinson, N., & Sambridge, M. (2005). The fast marching method: An effective tool for tomographic imaging and tracking multiple phases in complex layered media. *Exploration Geophysics, 36*(4), 341–350.

Wachilala, P., Borges, J. F., Matias, L., Rio, I., Fontiela, J., Oliveira, R., et al. (2019). Análise da distribuição espaço-temporal da sismicidade na zona de Arraiolos no período de janeiro-maio de 2018. *APMG 2019*

Xia, S., Qiu, X., Xu, H., Zhao, M., & Shi, X. (2011). Characteristics of PmP phases from earthquakes and their role in crustal tomography: An active volcanic area example, northeastern Japan. *Science China Earth Sciences, 54*(5), 640–646.

Evidence of Seismic Gap in Al Hoceima (Morocco): Implication for Seismic Hazard

Hamza Akka, Abdelilah Tahayt, Ismail Es-Sabbar, Hafid Ouammou, Nacer Jabour, and Elia d'Acremont

Abstract

This paper uses high-resolution seismic profiles of the complex active zone of Al Hoceima (Morocco) to illuminate the active fault system, acquired along a main, 10 km long transect cutting across the strands of an active fault system in Al Hoceima bay area. The investigation approached was to tackle and follow faults vertical offset. The final seismic sections led to identify faults on the profiles and to distinguish the likely-active ones, with visible wide deformation features, where normal strike-slip faults trend N–S and NW–SE, principally related to the major Al Idrissi strike-slip fault system. To aid interpretation of this complex setting, we complemented seismic images with seismicity catalogs and focal mechanism solutions to investigate the active tectonics. These faults are accommodating an important part of the complex seismotectonic movements. Seismicity catalog in the Al Hoceima area exhibits a moderate seismic release (Mw < 6), with shallow depths mostly $\leq$ 20 km, focal mechanisms suggest mainly a strike-slip regime with a subvertical $\sigma 3$ and subhorizontal $\sigma 1$ implying a transtensional regime in the area. The seismicity seems to be confined between two major faults: Trougout in the east and Rouadi in the west. However, this trend is no longer respected in the last 2 years as seismic activity is shifting toward the eastern side of the Trougout fault. Moreover, a seismic gap zone around this fault is evident considering the instrumental seismicity catalog of last 100 years. This fault system and related seismicity could represent a potential hazard for neighbor cities.

Keywords

Earthquake • Focal mechanism • Active tectonics • Al Hoceima bay • Rif • Morocco

1 Introduction

The Rif belt, located in northern Morocco, is widely considered as an active region. It is also recognized as a moderate seismic hazard zone (Cherif et al., 2017). Within this seismic context, Al Hoceima city is known as the most hazardous Moroccan city exposed to seismic risk, since it has experienced a series of major historical and instrumental earthquakes. The seismicity in this region results from the plate convergence of Africa toward Eurasia (Jolivet et al., 2009; López-Casado et al., 2014). The majority of this convergence is accommodated in the Rif–Betic–Alboran region (Dewey et al., 1998; Jolivet & Faccenna, 2000; Lis Mancilla & Diaz, 2015). Consequently, the Alboran region's seismicity particularly concenters along a NE–SW zone running through the Campo de Dalías region of Betics (Spain) at north, to the Al Hoceima region of the Rif (Morocco) at the south (Ammar et al., 2007). According to the seismic gap hypotheses, fault zones where no major earthquake has occurred recently are more probable than others to host the next major earthquake (Petrillo et al., 2022). The hypotheses validity assumes that it is possible to obtain indications about the time of the next major earthquake on the basis of the previous seismic history.

According to El Mrabet (El-Mrabet, 2005) the Al Hoceima region was subject to earthquake sequences reported in 1522, 1624, 1791 and 1800–1802. More recently, a series of seismic events have occurred with major earthquakes recorded on May 26, 1994 (Mw 6), February

H. Akka (✉) · A. Tahayt · I. Es-Sabbar · H. Ouammou
Geohazards and Georesources Group, FSTT,
Abdelmalek Essaadi University, Tétouan, Morocco
e-mail: hamza.akka@etu.uae.ac.ma

N. Jabour
National Institute of Geophysics, CNRST, Rabat, Morocco

E. d'Acremont
Sorbonne Universities, UPMC Univ Paris 06, CNRS,
Paris Institute of Earth Sciences (iSTeP), 4 Place Jussieu,
Paris, France

M. Bezzeghoud et al. (eds.), *Recent Research on Geotechnical Engineering, Remote Sensing, Geophysics and Earthquake Seismology*, Advances in Science, Technology & Innovation,
https://doi.org/10.1007/978-3-031-48715-6_56

24, 2004 (Mw 6.4) and January 25, 2016 (Mw 6.5) in Al Hoceima region and in the southern Alboran Sea area (Biggs et al., 2006; Jabour, et al., 2004; Kariche et al., 2018; Stich et al., 2020; Tahayt et al., 2009). These earthquake sequences took place at 10–20 km depth (Kariche et al., 2018). Moreover, the 1994 earthquake, (Mw = 6) occurred with a left-lateral strike-slip buried fault (Calvert et al., 1997), the 2004 (Mw = 6.4) earthquake event was believed to be a right-lateral strike-slip fault, striking NW–SE (Cakir et al., 2006). While, surface deformation inferred from InSAR (Tahayt et al., 2009) suggests 2004 event was induced by NNE–SSW and NW–SE-trending coupled strike-slip faults. Finally, the 2016 earthquake event occurred on NNE–SSW oriented rupture with a similar mechanism to the 1994 and 2004 events (Kariche et al., 2018). Thus, this paper shed light on the active faulting, seismicity and focal mechanisms, and aims to determine the recurring seismic faulting styles in Al Hoceima bay region, to provide a new picture of the present-day tectonics of the area, in order to get a better understanding of the kinematics, fault pattern and seismic hazard in the region.

In Sect. 2, a thorough examination of the data and the methods employed in its acquisition and analyzation will be presented. The findings from this analysis will be discussed in detail in Sect. 3. Finally, in Sect. 4, meaningful conclusions will be drawn from the results and a thought-provoking discussion regarding the implications of the findings will be conducted. This will bring closure to the article and provide insight into the significance of the research conducted.

2 Data and Methods

The seismic profiles used for this study were collected in May 2012, during the Marlboro-2 survey, which cover 190 km^2 from the eastern limits of Al Hoceima bay to 14 km westward of Ras El Abid Promontory continental shelf (d'Acremont et al., 2014; Lafosse, et al., 2018). Using a simple *P*-wave velocity model of 1550 m/s, the seismic profiles were acquired for the Quaternary deposits, with a constant sedimentation rate. (Lafosse et al., 2017; Martínez-García et al., 2013). The present work is focused on fault systems affecting the Quaternary of Al Hoceima bay. However, to characterize the activity of faults from the seismic profiles, we used OpendTect 6.6 attributes. Therefore, seismicity catalog of events $\geq$ 3 Mw during the period running from 1900 to 2020 provided by the Moroccan Institute of Geophysics ING CNRST along with the online database of the Spanish National Geographic Institute IGN were compiled. Finally, the focal mechanism solutions available for the most important events (4.1–6.4 Mw) in the area were extracted from the database of IRIS website (iris.edu) based on the GCMT (The Harvard Centroid Moment Tensor Catalog) additionally to the GFZ website catalog (gfz-potsdam.de). All the mentioned datasets and protocols were used to evaluate the seismotectonic characteristic in this region.

3 Results

All the three analyzed seismic sections display evidences of deformation mainly due to faulting and subordinately to small folds, affecting the bay Plio-Quaternary sediments successions, with a variable intensity, clarity and distribution. Within the Al Hoceima bay area, the seismic profile MAB 290 (Fig. 1a) reveals distinct characteristics of the tracked faults. Notably, these faults exhibit a prevailing N–S orientation and are marked by a discernible normal component. These faults dip, exceed 70 degrees, and are accompanied by evident vertical offsets, reaching up to 7 meters in certain faults. Additionally, some faults successively reach the seafloor ("outcropping faults"). Moreover, an important number of faults are buried under the sediment and do not reach the seafloor, mainly confined in the sediments under MRS3 (280 kyr), while above, between MRS2 (150 kyr) and MRS1 (25 kyr) faint deformation is observed. The seismic section MAB 301 was acquired almost back to the shore line (Fig. 1b). These lines, show related characteristics, in turn line MAB 290 (Fig. 1a) displays some particularities. The basement dips clearly from the westernmost part, faults are well-exposed and easy to track, the offsets on the seismic reflectors can reach 8 m in some cases.

Additionally, the vertical displacement of faults clustered under MRS3 (280 kyr) to MRS2 (150 kyr) displays that deformation is distributed throughout the Al Hoceima bay, since most of the faults have offset that may reach 12 m in some cases. Furthermore, most faults display normal component with possible strike-slip movement, few are offsetting the seafloor. Sediments are thinner toward the eastern side, indicating the asymmetric geometry of sediment fill. Moreover, an important number of faults are affecting the sediments section comprised between the MRS2 (150 kyr) and MRS1 (25 kyr) particularly in the central zone, while the region above MRS1 (25 kyr) is partially undeformed, where faults appear to disrupt the sediments above MRS1 (25 kyr). Finally, the section MAB 321 displays the basin geometry where the basement plunges to the NW toward the sea viewing from the SW to the NE. Offsets on the seismic reflectors illustrated the tectonic activity interpreted as piano-key faults affecting mainly the Quaternary sedimentary units under MRS3 (280 kyr) (Fig. 1c). On the other side, the seismicity map of Al Hoceima zone and its offshore areas for the last hundred years show two clusters of earthquakes notably observed, cluster 1 located in Alboran sea between 35.5°N; 36°N and − 3.5°W; − 4°W oriented NE–SW with,

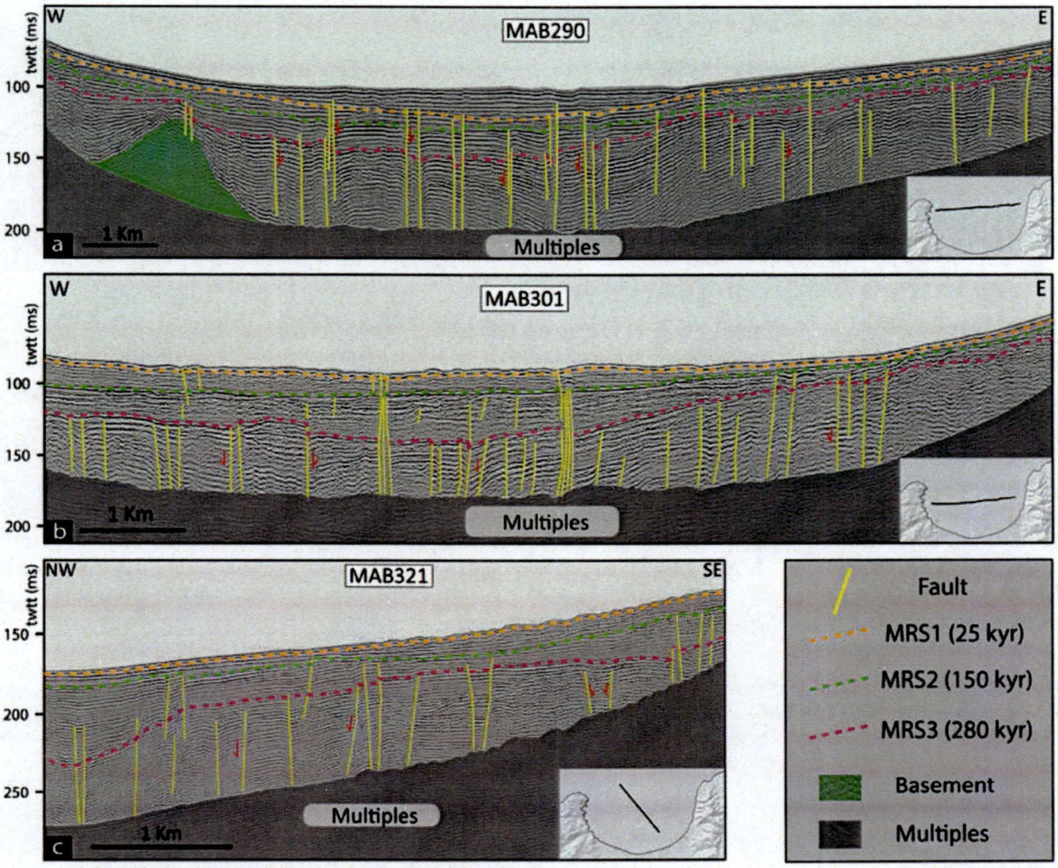

Fig. 1 Interpreted high-resolution seismic profiles that cross the Al Hoceima bay northern Morocco and southern Alboran basin. Yellow lines represent faults, dashed lines represent "marine regression surfaces"; pink (MRS3 280 kyr), green (MRS2 150 kyr), and orange (MRS1 25 kyr) represent seismic reflectors by ages. Red arrows show the most important vertical offsets

cluster 2 located et al. Hoceima onshore/offshore between 34.5°N; 35.3°N and − 3.5°W; − 4.3°W (Fig. 2). These clusters are separated by a seismic gap with ≈ 3 km wide, in the limit between the African and Eurasian plates.

4 Discussion and Conclusions

In the studied area of Al Hoceima bay Nekor basin, the acquired offshore high-resolution seismic data and seismicity catalog paired with focal mechanisms solutions and has provided new insights on the Quaternary tectonic evolution: (1) The Al Hoceima bay basin is tectonically active, mainly affected by piano-key faults trending NW–SE, and they are parallel to the Ajdir, Boujibar and Imzouren faults at onshore part, accommodating a part of the deformation within the bay. The focal mechanisms indicate a dominant strike-slip regime, with variation on the amount of normal or reverse mechanisms involvement. However, previous work has shown that the major faults bordering Al Hoceima bay display different types of movement, the left-lateral slip Nekor fault has a horizontal slip rate of 2 mm/year, while the N–S, onshore-offshore Trougout fault corresponds to a normal fault with a horizontal slip movement rate of 1.25–1.7 mm/year as reported by Poujol et al. (2014). These values are in concordance with general trend of the area. (2) Seismicity in the Al Hoceima area is moderate (Mw 6), with shallow hypocentral depths (mostly ≤ 20 km), focal mechanism for the area suggests mainly a strike-slip movement for the main earthquake events. The studied area is going under a subvertical $\sigma3$ and subhorizontal $\sigma1$ that imply a persistent transtensional regime. (3) The deformation and seismicity seem to be confined within the principal faults of the Rouadi and Trougout N–S trending faults. (4) The GPS block movement of Al Hoceima characterized by a particular direction, which can be rooted to the area tectonic features. Assuming the ages of movements and the induced recurrence times obtained by Poujol et al. (2014) for onshore faults [Trougout (N–S), Rouadi (N–S) and Boujibar (NW–SE)], we suggest that the NW–SE faults in Al

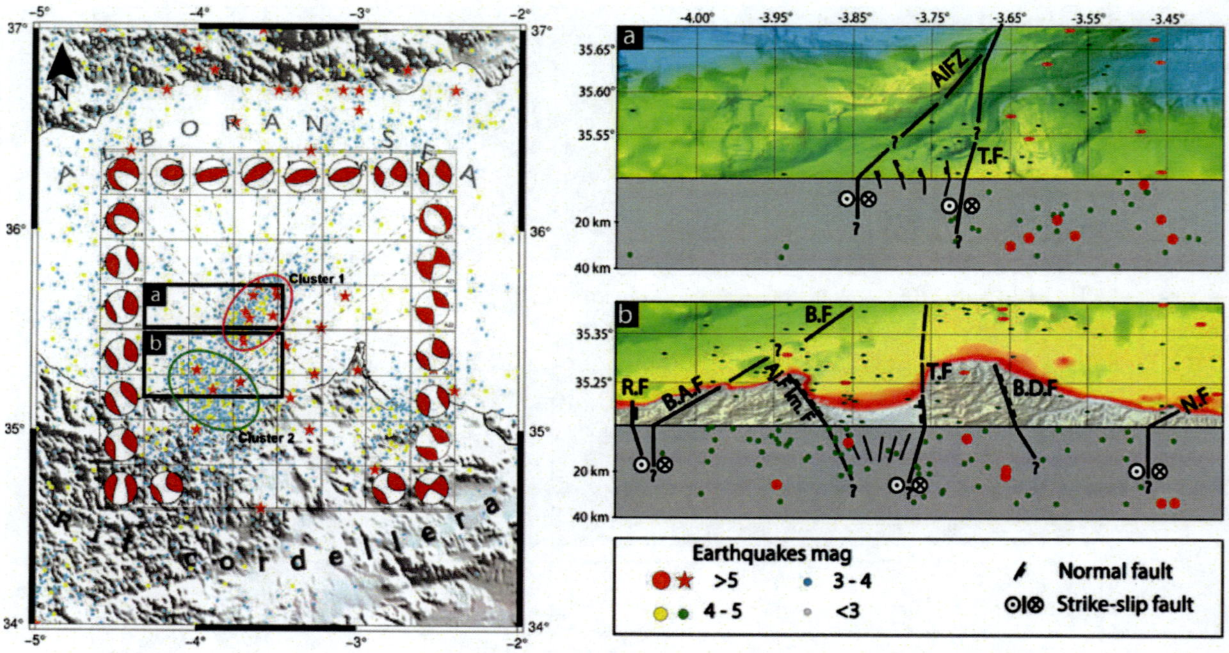

Fig. 2 Spatial distribution of earthquakes events with magnitudes ≥ 2 Mw (IGN Morocco catalogues 1900–2020) with two distinguished clusters **a** and **b** 3D blocks models of tectonic major faults in Al Hoceima region, accompanied by the earthquakes vertical and horizonal distribution. AIFZ—Al Idrissi fault zone, BF—Bokkoya fault, BAF—Bousekkour-Aghbal fault, AJF—Ajdir fault, RF—Rouadi fault, ImF—Imzouren fault, TF—Trougout fault, BDF—Boudinar fault, NKF—Nekor fault

Hoceima bay could play a role in the future rupture on Trougout fault. A possible scenario is to have a rupture on conjugate N–S and NW–SE faults which reminds the mechanisms of 1994, 2004 and 2016 earthquakes as revealed by the source time function analysis (Bezzeghoud & Buforn, 1999; Kariche et al., 2018; Stich et al., 2004). This led us to suggest that the seismicity is not linked to specific faults directions but to a various pattern playing contemporaneously.

Therefore, the earthquake catalogs show a clustering of seismic epicenters consistent with the main tectonic structures of the region, and confined mainly between the Rouadi fault to the west and the Nekor fault to the east. Moreover, when moving west to east of the study area (Fig. 2), it seems to be clustered on two main strips (Fig. 2), and a lack of significant earthquakes within the bay area. For this reason, we propose that Al Hoceima bay could be considered as a seismic gap. Therefore, we conducted a model based on the tectonic characteristics of the area as well as the seismicity catalog to explain the absence of significant earthquakes in the bay area (Fig. 2).

Acknowledgements The authors appreciate and acknowledge dGB Earth Sciences for proving academic license of OpendTect™ software to the Department of Earth Sciences, FST Tangier, Morocco. The MARLBORO-2 members are highly thanked for providing seismic datasets for pursuing this academic research. The authors would also like to express their great appreciation for reviewers.

References

Ammar, A., Mauffret, A., Gorini, C., & Jabour, H. (2007). The tectonic structure of the Alboran Margin of Morocco. *Revista De La Sociedad Geológica De España, 20*(3–4), 247–271.

Bezzeghoud, M., & Buforn, E. (1999). Source parameters of the 1992 Melilla (Spain, MW= 4.8), 1994 Alhoceima (Morocco, MW= 5.8), and 1994 Mascara (Algeria, MW= 5.7) earthquakes and seismotectonic implications. *Bulletin of the Seismological Society of America, 89*(2), 359–372.

Biggs, J., et al. (2006). Fault identification for buried strike-slip earthquakes using InSAR: The 1994 and 2004 Al Hoceima, Morocco earthquakes. *Geophysical Journal International, 166*(3), 1347–1362.

Cakir, Z., Meghraoui, M., Akoglu, A. M., Jabour, N., Belabbes, S., & Ait-Brahim, L. (2006). Surface deformation associated with the M w 6.4, 24 February 2004 Al Hoceima, Morocco, earthquake deduced from InSAR: Implications for the active tectonics along North Africa. *Bulletin of the Seismological Society of America, 96*(1), 59–68.

Calvert, A., et al. (1997). An integrated geophysical investigation of recent seismicity in the Al-Hoceima region of North Morocco. *Bulletin of the Seismological Society of America, 87*(3), 637–651.

Cherif, S., Chourak, M., Abed, M., & Pujades, L. (2017). Seismic risk in the city of Al Hoceima (north of Morocco) using the vulnerability index method, applied in Risk-UE project. *Natural Hazards, 85*(1), 329–347.

d'Acremont, E., et al. (2014). High-resolution imagery of active faulting offshore Al Hoceima, Northern Morocco. *Tectonophysics, 632*, 160–166.

de Lis Mancilla, F., & Diaz, J. (2015). High resolution Moho topography map beneath Iberia and Northern Morocco from receiver function analysis. *Tectonophysics, 663*, 203–211.

Dewey, J. F., Holdsworth, R. E., & Strachan, R. A. (1998). Transpression and transtension zones. *Geological Society, London, Special Publications, 135*(1), 1–14.

El-Mrabet, T. (2005). The great earthquakes in the Maghreb region and their consequences on man and environment. In *CNRS-LAG report, Rabat, Morocco*

Jabour, N., et al. (2004). The February 24th, 2004 Al Hoceima earthquake. In *Center of Sismology European-Mediterranean Seismological Centre* (p. 7). www.emsc-csem.org

Jolivet, L., & Faccenna, C. (2000). Mediterranean extension and the Africa-Eurasia collision. *Tectonics, 19*(6), 1095–1106.

Jolivet, L., Faccenna, C., & Piromallo, C. (2009). From mantle to crust: Stretching the Mediterranean. *Earth and Planetary Science Letters, 285*(1–2), 198–209.

Kariche, J., Meghraoui, M., Timoulali, Y., Cetin, E., & Toussaint, R. (2018). The Al Hoceima earthquake sequence of 1994, 2004 and 2016: Stress transfer and poroelasticity in the Rif and Alboran Sea region. *Geophysical Journal International, 212*(1), 42–53.

Lafosse, M., et al. (2017). Evidence of quaternary transtensional tectonics in the Nekor basin (NE Morocco). *Basin Research, 29*(4), 470–489.

Lafosse, M., et al. (2018). Late Pleistocene-Holocene history of a tectonically active segment of the continental margin (Nekor basin, Western Mediterranean, Morocco). *Marine and Petroleum Geology, 97*, 370–389.

López-Casado, C., Henares, J., Badal, J., & Peláez, J. A. (2014). Multifractal images of the seismicity in the Ibero-Maghrebian region (westernmost boundary between the Eurasian and African plates). *Tectonophysics, 627*, 82–97.

Martínez-García, P., Comas, M., Soto, J. I., Lonergan, L., & Watts, A. B. (2013). Strike-slip tectonics and basin inversion in the Western Mediterranean: The Post-Messinian evolution of the Alboran Sea. *Basin Research, 25*(4), 361–387.

Petrillo, G., Rosso, A., & Lippiello, E. (2022). Testing of the seismic gap hypothesis in a model with realistic earthquake statistics. *Journal of Geophysical Research: Solid Earth, 14*, e2021JB023542.

Poujol, A., et al. (2014). Active tectonics of the Northern Rif (Morocco) from geomorphic and geochronological data. *Journal of Geodynamics, 77*, 70–88.

Stich, D., Mancilla, F. L., Baumont, D., & Morales, J. (2005). Source analysis of the Mw 63 2004 Al Hoceima earthquake (Morocco) using regional apparent source time functions. *Journal of Geophysical Research: Solid Earth, 110*, 16.

Stich, D., Martín, R., Morales, J., López-Comino, J. Á., & Mancilla, F. L. (2020). Slip partitioning in the 2016 alboran sea earthquake sequence (western mediterranean). *Frontiers in Earth Science, 8*, 587356.

Tahayt, A., et al. (2009). (2009) The Al Hoceima (Morocco) earthquake of 24 February 2004, analysis and interpretation of data from ENVISAT ASAR and SPOT5 validated by ground-based observations. *Remote Sensors in Environmental, 113*(2), 306–316.

Source Parameters Spectral Modelling of the Aftershock Sequence of the 25th April 2012 Beni Haoua M_W 4.9 Earthquake and Its Relation with the El Asnam Fault, Northern Algeria

K. Abbes, M. S. Boughacha, M. Bezzeghoud, M. Y. Mehiaoui, H. Airouche, N. Benkaci, and Y. Bouhadad

Abstract

On 25th April 2012, a moderate M_w4.9 earthquake, followed by an aftershock sequence ($0.6 \leq ML \leq 4.3$), occurred in Beni Haoua, northern Algeria. To reveal the fault responsible for this earthquake and the likely associated subsurface deformation, a detailed study of the weak seismicity based on spectral modelling of source parameters has been performed. The earthquake affected the area located in the north-eastern extend of the El Asnam seismic zone and it appears likely related to the reactivation of a segment of the fault zone that caused the El Asnam M_w7.1 earthquake on 10th October 1980, one of the most destructive earthquakes recorded in northern Africa and western Mediterranean. The determination of the source parameters of these earthquakes is important because they are too small to be reported in a global catalogue. In this study, source parameters are determined for 35 associated aftershocks using spectral modelling of three components P and S waves, assuming the Madariaga 1976 model. This model is one of the most widely used for a singular crack, radially expanding at a constant rupture speed. In this study, we estimated the source parameters (seismic moment, size of the seismic source, source dislocation and stress drop) from data in the frequency domain and we show how the results depend on the model assumptions. The Q attenuation factors for P and S waves are estimated to be 97 (50–170) and 153 (73–242), respectively, with a ratio of $Qs/Qp = 1.62$. We analysed source spectra and stress drops of 35 micro-earthquakes and resolved significant variations in earthquake stress drop and apparent average source dimension R_p and R_s of about 17 and 75 m, respectively. From the local magnitude calculated for the studied aftershock sequence, we highlighted similar relationship between the moment magnitude and the local magnitude, $M_{wp} = 0.62\ M_L + 0.86$ and $M_{ws} = 0.63\ ML + 0.81$, for P and S waves, respectively. This type of study is very important since we have exploited low magnitude earthquakes to obtain information that can contribute to the seismotectonic analysis of active seismic zone.

Keywords

Beni Haoua • Source parameters • Spectral analysis • Quality factor • Algeria

1 Introduction

On 25th April 2012, the moderate M_w4.9 earthquake occurred in Beni Haoua, northern Algeria that was followed by a $0.6 \leq M_L \leq 4.3$ aftershock sequence (Abbes et al., 2015). The focal mechanisms within the damaged area and those of the El Asnam destructive M_w7.1 earthquake occurred on 10th October 1980, both seem associated to a same NE–SW oriented main fault zone (Fig. 1). In fact, the El Asnam seismic zone exhibits a significant seismic hazard and the 10th October 1980 event was one of the most destructive earthquakes recorded at northern Africa and western Mediterranean scale (Fig. 1).

The seismicity catalogues established for the Tellian zone (Ayadi & Bezzeghoud, 2015; Bezzeghoud & Ayadi, 2022;

K. Abbes · H. Airouche · N. Benkaci · Y. Bouhadad
National Centre of Applied Research in Earthquake Engineering (CGS), Rue Kaddour Rahim H. Dey Algiers, BP 252 Hussein Dey, Algeria
e-mail: kabbes@cgs-dz.org

M. S. Boughacha
Laboratoire de Géophysique, U.S.T.H.B, Bab Ezzouar, Algeria

M. Y. Mehiaoui
National School of Public Works, BP. 32, Rue Sidi Garidi, Kouba, Algeria

M. Bezzeghoud (✉)
Departamento de Física, Escola de Ciências E Tecnologia, ICT/IIFA, Universidade de Évora, Evora, Portugal
e-mail: mourad@uevora.pt

M. Bezzeghoud et al. (eds.), *Recent Research on Geotechnical Engineering, Remote Sensing, Geophysics and Earthquake Seismology*, Advances in Science, Technology & Innovation,
https://doi.org/10.1007/978-3-031-48715-6_57

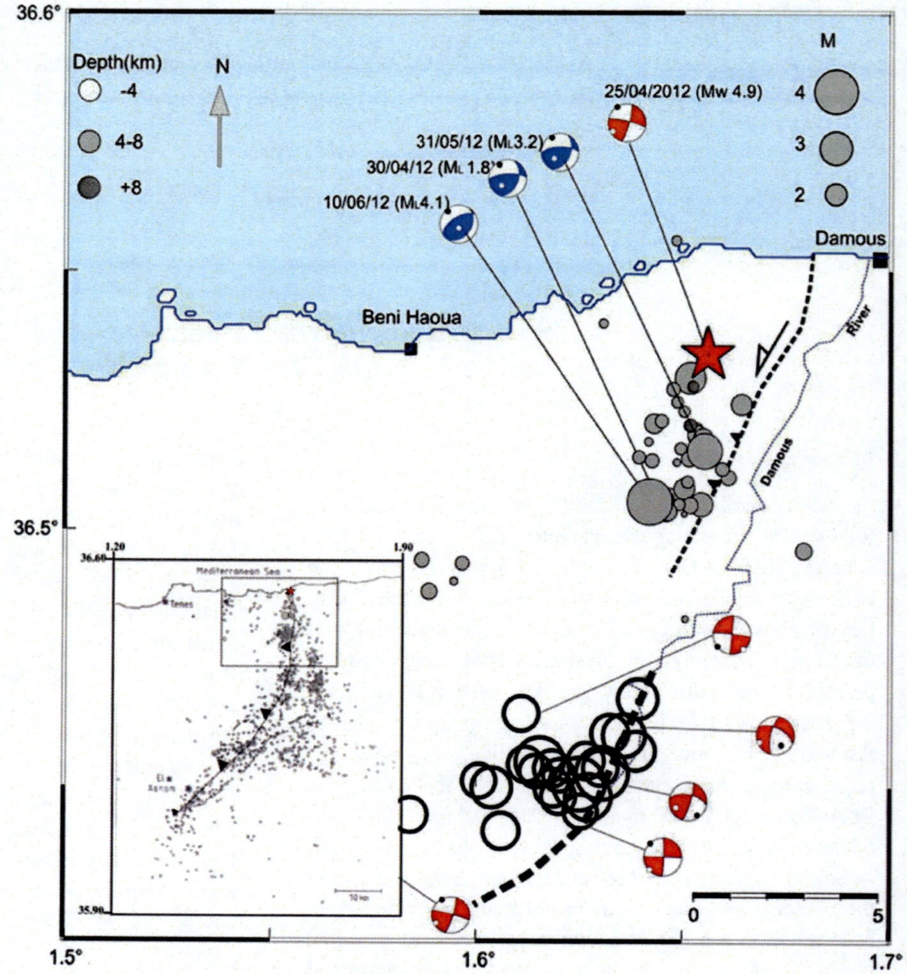

Fig. 1 Aftershock sequence in the Beni Haoua region which coincides with the northern end rupture of El Asnam earthquake (M_s7.3) in the Algerian Tellian Atlas shown in the inset. Filled circles show the aftershocks of the Beni Haoua main shock (red star). Open circles show the main aftershocks that followed the El Asnam 10th October 1980 earthquake in its northern part of the fault

Harbi et al., 2015; Mokrane et al., 1994) confirm the seismotectonic activity of the region with low to moderate seismic hazards especially in the vicinity of the Quaternary basins and along the coastal zone and active faults. In the study area, active faults are mostly sealed by Quaternary sediments (lack of surface ruptures) and structural analysis is not allowed to endeavour possible correlation with the seismotectonic activity and excludes the possibility of linking this activity to known faults with any certainty. Accordingly, the available wave packages of aftershock sequence of the 25th April 2012 earthquake event occurred in Beni Haoua region in northern Algeria have been alternatively processed and analysed in order to characterise i) the incipient seismic source parameters, ii) the tectonic effects of this earthquake in its local and regional context and iii) its relation with the El Asnam fault zone with emphasis of the structure of the crust through which the engendered seismic waves propagate during earthquake events.

In this present work, we study the Beni Haoua aftershock sequence (Abbes et al., 2015), using 321 P and S waveforms (35 × 2 × 3) CGS data, in terms of $M_{0p/s}$, $M_{wp/s}$, $R_{p/s}$, $\Delta\sigma_p/s$ and $Q_{p/s}$ using the Fourier amplitude spectrum analysis of P and S waves, which might provide information on the physical source process.

2 Methodology

The equation $M_w = 2/3\ (\log_{10}M_0 - 9.1)$ proposed by Hanks and Kanamori (1979) calculates the moment magnitude based on the physical model of the earthquake rupture. The latter makes possible to better characterise the magnitude of large earthquakes while maintaining consistency with other magnitude calculation methods for weak earthquakes. The scalar seismic moment M_0 [Nm] is related to the final static displacement after the occurrence of an earthquake and, hence, it is more related to the associated tectonic effects. Therefore, we have opted to calculate M_0 by using the displacement spectrum, Fourier transform of low-frequency (plateau) (Madariaga, 1976). Waveforms are easier to analyse in the frequency domain than in the time domain (Helffrich et al., 2013). The displacement spectrum geometry of the different emitted wave packages provides excellent information about the seismic source. The spectrum envelope is commonly presented in two main parts: a low-frequency *plateau* proportional to the seismic moment M_0, and a decreasing slope at high frequencies that corresponds to attenuation. The signal is firstly corrected for the instrument effect. Secondly, an average is deduced from the seismic trace. Finally, the resulted signal is integrated to obtain the Fourier displacement spectrum. The time window is chosen taking into account the first arrival of the P wave and ends before the arrival of the S wave on the three components Z, N and E as shown in Fig. 2.

In order to enhance seismic hazard assessment methods at regional scales, it is necessary to define realistic seismic motions appropriate to the context of each zone. So, we used the *ω-square model* (Kaneko & Shearer, 2014; Madariaga, 1976) to analyse the low-magnitude aftershock sequence that followed the Beni Haoua earthquake.

The *ω-square* model introduced first by Aki (1967) and then used by Brune (1970), is based on the observation of the spectral content of a large number of records. The seismic moment M_0 is defined by $M_0 = \mu \Delta U A = 2\mu E_r / \Delta\sigma$, in which μ is the medium shear modulus of the crust, ΔU is the coseismic displacement averaged over the rupture area of the fault $A = \pi R^2$, $\Delta\sigma$ is the static stress drop averaged over the rupture area, and E_r is the radiated seismic energy.

The source radius is given by $R = v_s K / 2\pi f_c$, in which v_s is the shear wave velocity at the source, K is a constant ($2.34/2\pi = 0.37$) and f_c is the corner frequency. The static stress drop, before and after the rupture, is given by $\Delta\sigma = 7M_0/16R^3$.

3 Data

For each event and for the same station we obtain average $Z_{p/s}$, $N_{p/s}$ and $E_{p/s}$ of the source parameters $M_{0p/s}$, $M_{wp/s}$ $R_{p/s}$, $\Delta\sigma_{p/s}$ and $Q_{p/s}$. The chosen Q, f_c and Ω_0 values were inferred by the best fitting Brune's theoretical ω^{-2} model (Brune, 1970; Hanks, 1972) with the observed displacement spectra. The seismic moment M_0 was estimated using the low-frequency displacement flat level Ω_0 of spectra defined by the following formula (Bezzeghoud et al., 1989):

$$M_0 = \frac{4\pi\rho v^3}{F\mathcal{R}} \Omega_0,$$

where ρ is the density, v is the velocity of the P or S wave, Ω_0 is the low-frequency spectral level, F *is* the free surface correction and $\mathcal{R}$ is the radiation pattern coefficient [$\mathcal{R}_P = 0.52$ and $\mathcal{R}_S = 0.63$] (Bezzeghoud et al., 1989; Boore & Boatwright, 1984). The free surface correction depends on the incident angle. However, due to the low-velocity layers near the surface, the incidence is not far from vertical, so $F = 2$ seems to be a good approximation (Bezzeghoud et al., 1989).

The displacement amplitude spectrum indicated by the low-frequency band plateau noted Ω_0 and the typical ω^{-2} spectral decay at high frequencies allows to estimate the corner frequency f_c. (Fig. 2b). In order to study the weak seismicity in detail, from the spectral analysis of P and S waves and based on the signal-to-noise ratio ($S/N \geq 3$), 35 relocated events (Abbes et al., 2015) occurred in the rupture zone with a magnitude that varies between 0.6 and 4.3 were selected (Fig. 1, Table 1). The $M_{0p/s}$, $M_{wp/s}$ and $\Delta\sigma_p/s$ were estimated using the frequency domain ground motion spectrum. We used the M_w scale (Hanks & Kanamori, 1979) to

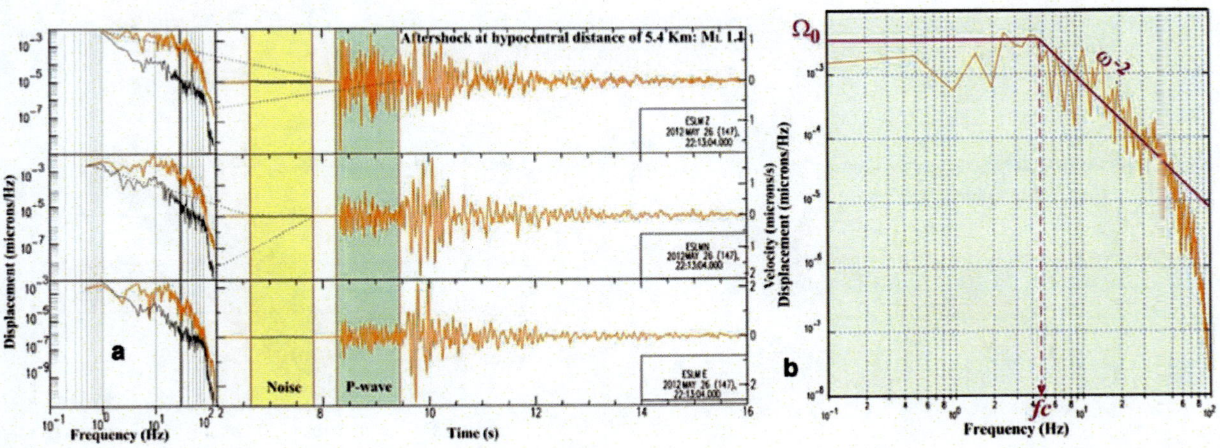

Fig. 2 **a** Three components P wave (orange) and noise (black) spectrum (left) and time history record of Beni Haoua aftershock. The time window depends on the magnitude of the earthquake. **b** Displacement amplitude spectrum indicated by the low-frequency band plateau noted Ω_0 and the typical ω^{-2} spectral decay at high frequencies allow to estimate the corner frequency f_c

Table 1 Number of waveforms per station, *sps* is the sample per second used in this study

Station ID	Amplifier gain	Number	sps
GOUR	10	18	250
DAMS	10	19	250
ESLM	10	24	250
MESL	1	9	250
TACH	1	20	250
BISA	1	13	250
ADEF	1	1	200
BCHG	10	13	250

process the three components of 35 events individually. Table 1 reports the recording stations parameters.

4 Results

The variations of the log of the seismic moment $M_0(p/s)$, measured in Nm, as a function of the local magnitude M_L (Abbes et al., 2015) were obtained from the regression of the data shown in Fig. 3a, c as $\log(M_{0p}) = 0.9M_L + 10.47$ and $\log(M_{0s}) = M_L + 10.24$, respectively, for P and S waves.

Figure 3b gives the variation of the seismic moment as a function of the source radius. Our results show that the variation of the source radius R_p and R_s ranges between 17 and 75 m with no clear correlation.

The moment magnitude M_w is calculated from the seismic moment M_0. The latter is estimated on the basis of the circular fault model (Hanks & Kanamori, 1979), using the following parameters: the velocity V_p is equal to 4.5 km/s, the density 2.4 g/cm^3 and the radiation coefficient R_p = 0.52.

The local magnitudes M_L of the aftershocks ($0.6 \leq M_L \leq 4.3$) estimated by Abbes et al. (2015) are compared with the calculated M_w, in order to define a scaling law according to the linear regression $M_{wp} = 0.62M_L + 0.86$, and the same goes for M_{ws} (Fig. 3c).

The stress drop is drawn as a linear variation with seismic moment. The inferred values of shear wave stress drop $\Delta\sigma_s$ and compressional wave stress drop $\Delta\sigma_p$ are between $0.02 < \Delta\sigma_s < 165$ MPa with an average value of 2.0 MPa, and between $0.02 < \Delta\sigma_p < 87$ MPa with an average value of 2.5 MPa, respectively (Fig. 3d). Using spectral analysis, (Allmann & Shearer, 2007; Bezzeghoud et al., 1989) analysed 2000 earthquakes worldwide with mb $\geq$ 5.5 and found that the estimated stress drops range from 0.3 to 50 MPa with a median value of 4 MPa. In this study, the estimated stress drops vary considerably for earthquakes of

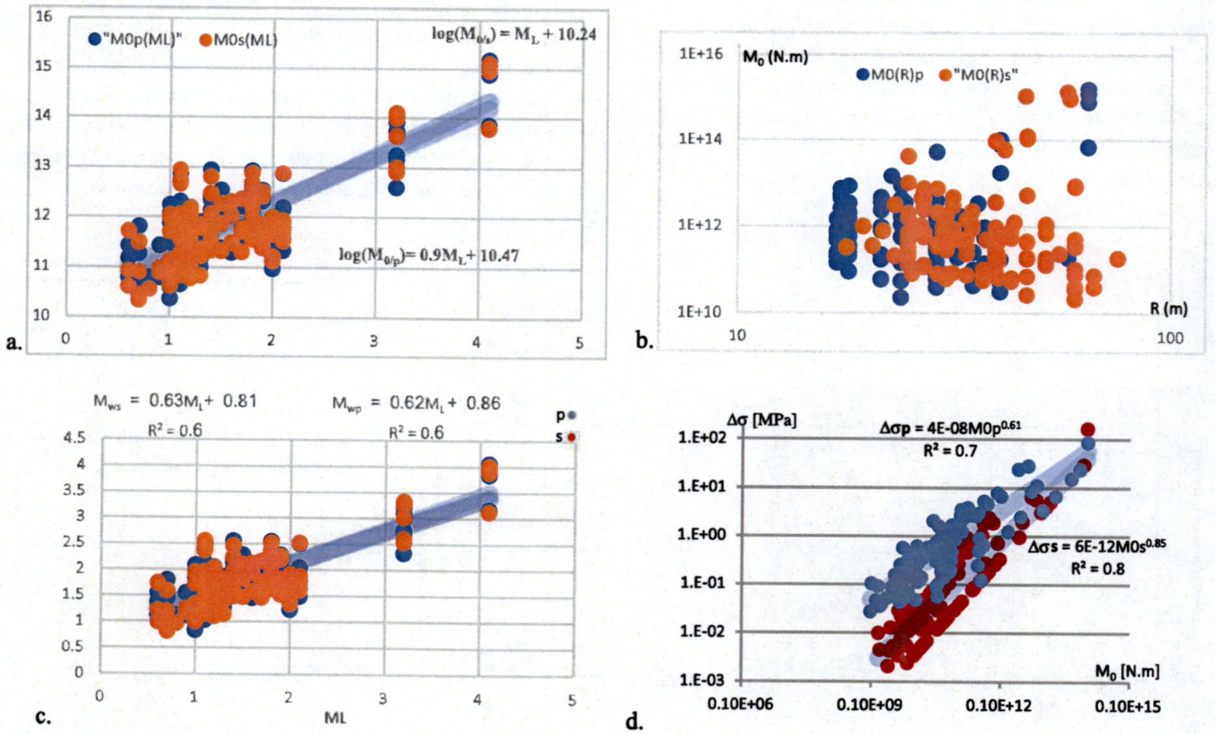

Fig. 3 Source parameters relationships determined in this study. **a** Seismic moment $M_0(p/s)$ *versus* local magnitude M_L; **b** $M_0(p/s)$ *versus* source radius *R*; **c** Moment magnitude *Mw(p/s)* versus local magnitude M_L; **d** *Stress drop versus seismic moment* M_0

the same magnitude. However, the median stress drops appear to be relatively constant with magnitude (Allmann & Shearer, 2007, 2009), which confirms our results.

Quality factor coefficients values for the studied aftershocks sequence of with hypocentral distances between 7 and 30 km show that Q_p and Q_s ranges, respectively, from 50 to 170 with an average of 97 and from 73 to 242 with an average of 153. In general, $Q_s/Q_p < 1.0$ corresponds to areas mostly outside the major sedimentary basins. However, in this case study $Q_s/Q_p > 1.0$ suggests partially fluid-saturated crust (Hauksson & Shearer, 2006; Klimentos, 1995).

5 Conclusion and Discussion

This work proposes a new contribution to incipient source parameters determination using Fourier spectrum analysis of weak earthquake related to an aftershock sequence occurred in the Beni Haoua region, along a narrow NE–SW left lateral strike-slip fault settled in the northern end rupture of El Asnam most destructive earthquake M 7.3 in the Algerian Tellian Atlas. The fact that the damaged area associated to the Beni Haoua earthquake lays towards the north-eastern extend of the El Asnam thrust fault, lets us to assume that both earthquakes are associated to a unique NE–SW transform zone that thrusted the northern extremity of the Plio-Quaternary Chellif basin.

The spectrum analysis of the aftershock sequence of Beni Haoua earthquake allowed to establish incipient seismic source parameters and to propose novel relationships among themselves. Firstly, it was easy to remark that Log $M_{0(p/s)}$ seismic moment [Nm] plotted against local magnitude; M_L, local magnitude versus moment magnitude and stress drop as a function of the seismic moment presents a functional dependence between these parameters as given by the equations shown in Fig. 3a, c and d. Secondly, considering the fact that the frequency content is different at different sites, as a consequence of the finiteness of the rupture velocity and the source dimensions, as shown in Fig. 3b the seismic moment parameter plotted against source radius does not reveal any clear dependence function; in particular, this could happen if the medium is not strongly attenuating and/or the epicentre of the event lies inside the area enclosed by the acquisition network, so that the attenuation term is almost the same at different stations (Modiano & Hatzfeld, 1982). The attenuation factor indicates that $Q_p < Q_s$ which means that the shear waves attenuate less than the compressional waves. The Q_s shows some dependence on the frequency f_c. This led to the following relationship: $Q_s = 49 f_{cs}^{0.36}$. On the other hand, Q_p varies between 50 and 170 with an average of 97 and does not allow to conclude on its f_{cp} corner frequency dependence. The average Qs/Qp ratio is 1.62, which is very close to the value determined by Lorenzo and Zollo (2003) estimated at 1.64. Additionally, the fact that tectonically active areas exhibit generally low Q values (< 200) (Benkaci et al., 2022) supports the hypothesis that this zone is a segment of the El Asnam thrusting fault.

Acknowledgements This work is a part of the daily research activities at the National Centre of Applied Research in Earthquakes Engineering (CGS, Algiers). The authors are thankful to the accelerograph network maintenance team of CGS. This work has been partially supported by the Portuguese Foundation for Science and Technology (FCT) through the projects UIDB/04683/2020-ICT (Institute of Earth Sciences) and SFRH/BSAB/143063/2018. We thank the editors and reviewers who shared their precious time to improve the manuscripts.

References

Abbes, K., Dorbath, L., Dorbath, C., Djeddi, M., Ousadou, F., Maouche, S., et al. (2015). The Beni Haoua, Algeria, M_w4.9 earthquake: source parameters, engineering, and seismotectonic implications. *Journal of Seismology, 20*(2), 655–667.

Aki, K. (1967). Scaling law of seismic spectrum. *Journal of Geophysical Research, 73*, 5359–5376.

Allmann, B. B., & Shearer, P. M. (2007). Spatial and temporal stress drop variations in small earthquakes near Parkfield, California. *Journal of Geophysical Research: Solid Earth, 112*(B4), 38–56.

Allmann, B. B., & Shearer, P. M. (2009). Global variations of stress drop for moderate to large earthquakes. *Journal of Geophysical Research: Solid Earth, 114*, B1.

Ayadi, A., & Bezzeghoud, M. (2015). Seismicity of Algeria from 1365 to 2013: Maximum observed intensity map (MOI 2014). *Seismological Research Letters, 86*(1), 236–244.

Benkaci, N., Airouche, A., Abbes, K., Mehiaoui, M. Y., & Bensalem, R. (2022). Attenuation of seismic coda-waves in Algeria: Algiers Vicinity and Mitidja Basin, 2022. *Pure and Applied Geophysics, 179*, 1011–1035.

Bezzeghoud, M., & Ayadi, A. (2022). The history of seismology and historical seismicity in Algeria: An overview. *Mediterranean Geoscience Reviews, 4*, 361–381.

Bezzeghoud, M., Deschamps, A., & Madariaga, R. (1989). Broad band P-wave signals and spectra on digitals stations. In R. Cassinis, G. Nolet, & G. F. Panza (Eds.), *Digital seismology and fine modeling of the lithosphere* (pp. 351–374). Plenum Publishing Corporation.

Boore, D. M., & Boatwright, J. (1984). Average body-wave radiation coefficients. *Bulletin of the Seismological Society of America, 74*, 1615–1621.

Brune, J. N. (1970). Tectonic stress and the spectra of shear waves from earthquakes. *Journal of Geophysical Research, 75*, 4997–5009.

De Lorenzo, S., & Zollo, A. (2003). Size and geometry of microearthquake seismic ruptures from P and S pulse width data. *Geophysical Journal International, 155*(2), 422–442.

Hanks, T. C. (1972). Wyss, M: The use of body-wave spectra in the determination of seismic-source parameters. *Bulletin of the Seismological Society of America, 62*, 561–589.

Hanks, T. C., & Kanamori, H. (1979). Moment magnitude scale. *Journal of Geophysical Research, 84*(B5), 2348–2350.

Harbi, A., Sebaï, A., Benmedjber, M., Ousadou, F., Rouchiche, Y., Grigahcene, A., Aïni, D., Bourouis, S., Maouche, S., & Ayadi, A. (2015). The Algerian homogenized macroseismic database (267–

1989): A deeper insight into the Algerian historical seismicity. *Seismological Research Letters, 86*(6), 1705–1716.

Hauksson, E., & Shearer, P. M. (2006). Attenuation models (Q_P and Q_S) in three dimensions of the southern California crust: Inferred fluid saturation at seismogenic depths. *Journal of Geophysical Research: Solid Earth, 111*, B05302.

Helffrich, G., Wookey, J., & Bastow, I. (2013). *The seismic analysis code: A primer and user's guide*. Cambridge University Press

Kaneko, Y., & Shearer, P. M. (2014). Seismic source spectra and estimated stress drop derived from cohesive-zone models of circular subshear rupture, 2014. *Geophysical Journal International, 197*, 1002–1015.

Klimentos, T. (1995). Attenuation of P- and S-waves as a method of distinguishing gas and condensate from oil and water. *Geophysics, 60*(2), 447–458.

Madariaga, R. (1976). Dynamics of an expanding circular fault. *Bulletin of the Seismological Society of America, 66*(3), 639–666.

Modiano, T., & Hatzfeld, D. (1982). Experimental study of the spectral content for shallow earthquakes. *Bulletin of the Seismological Society of America, 72*(5), 1739–1758.

Mokrane, A., Aït Messaoud, A., Sebaï, A., Menia, A., Ayadi, A., Bezzeghoud, M. (1994). Les séismes en Algérie de 1365 à 1992. In M. Bezzeghoud, & M. Benhallou (Eds.), *Centre de Recherche en Astronomie, Astrophysique et Géophysique, Alger-Bouzaréah*

Deterministic Tsunami Hazard Assessment for the Wadam As Sahil Coast, Sultanate of Oman

Issa El-Hussain, Zaid Al-Habsi, Rachid Omira, Ahmed Deif, Adel Mohamed, Maria Ana Baptista, and Yousuf Al-Shijbi

Abstract

The Wadam As Sahil port and facility (study area) is part of al Batinah region of northern Oman coast. This important coastal segment may be subjected to tsunami from two subduction zones in the Indian Ocean, namely Makran Subduction Zone (MSZ) and Sumatra Subduction Zone (SSZ) in the Indian Ocean. The MSZ is divided into the East Makran Subduction Zone (EMSZ) and the West Makran Subduction Zone (WMSZ) segments. The deterministic approach for tsunami hazard assessment is estimated using the scenarios of the historic events of 1945 (HMSZ), along with two scenarios featuring Mw 8.8 event in EMSZ and Mw 7.2 in WMSZ. Sumatra Subduction Zone (SSZ) poses a very low hazard to the coast of the study area due to shielding of the Indian subcontinent, therefore, it is not considered in this study. Tsunami numerical simulations are performed utilizing validated shallow water model over a high-resolution digital elevation model of the study area. A compilation of multisource elevation data from several sources was utilized to generate the digital elevation model (DEM) for the study area. The simulated maximum probable earthquake with 7.2 magnitude from the WMSZ resulted in tsunami hazard of more than 3 m wave height, about 2.8 m run up, and nearly 800 m inundation at the study area coast. The simulated maximum probable earthquake with 8.8 magnitude from the EMSZ resulted in tsunami hazard of less than 1.5 m wave height, about 2.0 m run up, and nearly 680 m inundation at the study area coast. The WMSZ Tsunamigenic source dominates the tsunami hazard due to its close proximity to the study area coast. The EMSZ poses less hazard to the study area due to orientation of the fault and it is distant from the study area coast. The last tsunami hazard is due to Makran 1945 earthquake due to its distance, strike orientation, and relative magnitude.

Keywords

Makran Subduction Zone • Tsunami hazard • Tsunamigenic source

1 Introduction

Tsunami is one of the most dangerous natural hazards affecting the Indian Ocean responsible for great loss of life and property throughout history. The impact of the 1945 tsunami was large in the Arabian Sea. It affected Pakistan, the western coast of India, Iran, and Oman, killing hundreds of people and causing great destruction. Reports of Tsunami waves arrived at some inhibited sites in northern Oman but no report from the Wudam As Sahil study area. Wudam As Sahil is among the many critical infrastructures that are now built for providing assistance for the rest of the country in case of calamity. The Deterministic Tsunami Hazard Assessment approach is utilized to assess the hazards for Wudam As Sahil coastal area where often used as a disaster relief and evacuation facility (Fig. 1).

I. El-Hussain (✉) · Z. Al-Habsi · A. Deif · A. Mohamed · Y. Al-Shijbi
Earthquake Monitoring Center, Sultan Qaboos University, Al-Khoudh, Muscat, Oman
e-mail: elhussain@squ.edu.om

R. Omira
Institute Dom Luiz, University of Lisbon, IDL-FCUL, Lisbon, Portugal

A. Mohamed
National Research Institute of Astronomy and Geophysics (NRIAG), Helwan, Cairo, Egypt

M. A. Baptista
Instituto Superior de Engenharia de Lisboa, Instituto Politécnico de Lisboa, R. Conselheiro Emidio Navarro, Lisbon, Portugal

M. Bezzeghoud et al. (eds.), *Recent Research on Geotechnical Engineering, Remote Sensing, Geophysics and Earthquake Seismology*, Advances in Science, Technology & Innovation,
https://doi.org/10.1007/978-3-031-48715-6_58

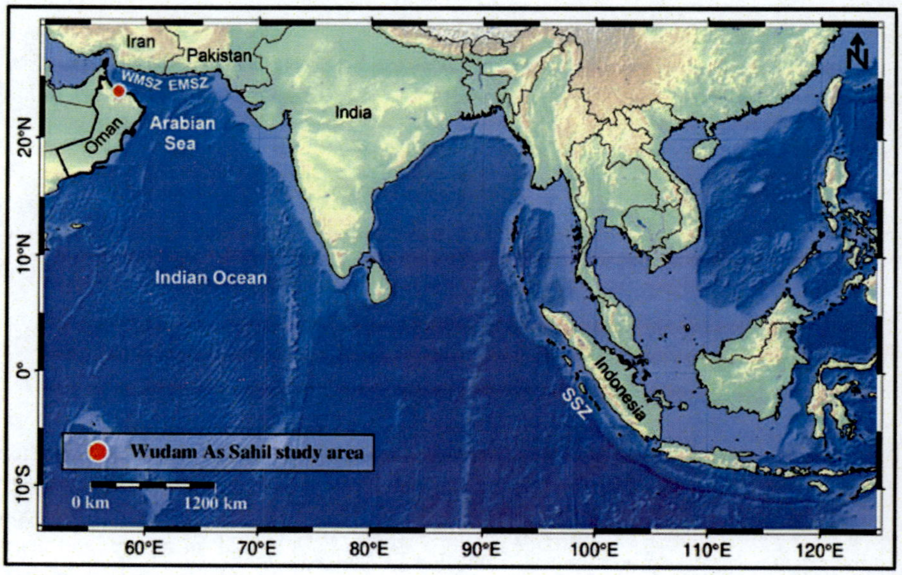

Fig. 1 Location of the study area in northern Oman for tsunami hazard assessment. MSZ = Makran Subduction Zone. SSZ = Sumatra Subduction Zone

2 Area Setting and Methods of Study

The Wudam As Sahil study area is located in the northern part of Al-Batinah coastal plain comprising piedmont and coastal zones dominated by late tertiary-quaternary alluvial deposits. The coastal zone is flat and comprises an extensive sand and gravel plain, housing several facilities and a port.

The deterministic tsunami hazard assessment (DTHA) approach uses a particular source scenario to calculate the tsunami impact at the coastal area of interest producing maps depicting the wave height, coastal inundation, and run-up heights. Two earthquakes were selected, one in each segment of the Makran Subduction Zone (EMSZ and WMSZ) to represent the worst-case tsunami scenarios (El-Hussain et al., 2021). Moreover, the historical Makran event (HMSZ) in 1945 was also considered (Table 1).

The accuracy of the tsunami simulation is critically dependent on the quality of data used in addition to the capability of the used numerical model. In this work, a DEM was generated with increasing resolution from the source area toward the target coast. A compilation of multisource height/depth data includes data from General Bathymetric Chart of the Oceans with 15 s arc resolution, the nautical maps, a local topographic elevation data with 5 m resolution obtained from the Omani National Survey Authority (NSA) were utilized to produce The DEM.

DTHA method has been applied by a number of scientific groups in various tsunami-prone coastal zones (Baptista et al., 2011; Omira et al., 2013; Tinti & Armigliato, 2003). Guidelines to perform this assessment have been published by Anwar et al. (2011) (El-Hussain et al., 2021). DTHA involves a number of steps outlined by Omira et al. (2013). In this work, these steps were adopted to produce the tsunami hazard maps, in terms of maximum wave height, run-up heights, and inundation depth, for the study area coast.

A non-linear shallow water model (NSWING-Nonlinear Shallow Water model with Nested Grids, Miranda et al., 2014) is used to simulate the tsunami propagation from the

Table 1 Fault parameters of the deterministic scenarios

Eastern Makran scenario						
L (km)	W (km)	Slip (m)	Dip (0)	Strike (0)	Rake (0)	Mw
461	110	11.1	7	263	90	8.8
Western Makran scenarios						
L (km)	W (km)	Slip (m)	Dip (0)	Strike (0)	Rake (0)	Mw
62	25	1.7	7	281	90	7.2
Makran historical event—November 27, 1945, earthquake						
L (km)	W (km)	Slip (m)	Dip (0)	Strike (0)	Rake (0)	Mw
150	70	6.6	7	246	90	8.1

source zone toward the coast. The code uses an explicit leapfrog finite difference scheme to solve the shallow water equations in spherical or Cartesian coordinates (Miranda et al., 2014).

3 Results

3.1 Results of the Western MSZ Mw 7.2 Scenario

Figure 2 shows maps of the near-shore mean water heights (MWHs), inundation flow depths, and maximum flow velocity at the coast of the study area with mean sea level (MSL) condition and mean high-high water (MHHW) tidal conditions, following the occurrence of the Mw 7.2 earthquake magnitude scenario in the western segment of the MSZ. Similar maps are produced for other eastern scenarios but are not presented here due to their lesser impact on the cost of interest, and their results are presented in the discussion section.

Results from modeling (Fig. 2) clearly demonstrate that the Mw 7.2 western scenario caused a high tsunami impact at Wudam As Sahil coast. The MWHs reach more than 3 m at some locations of the coast when combined with the MHHW tidal stage (Fig. 2c). Moreover, results show significant inundations of more than 800 m, along the Wadi channels close to the harbor of Wudam As Sahil. It is clear that the Mw 7.2 western scenario causes the worst tsunami impact among the selected scenarios for this particular coast, since this fault scenario lies very close and directly in front of the site.

4 Discussion

Figures 3 and 4 show the maximum run-up heights, flow depth, and maximum inundation lines for all deterministic tsunami scenarios considered in this study. Results from high-resolution modeling (Fig. 3) show that the Mw 7.2 western scenario caused a higher tsunami impact than the

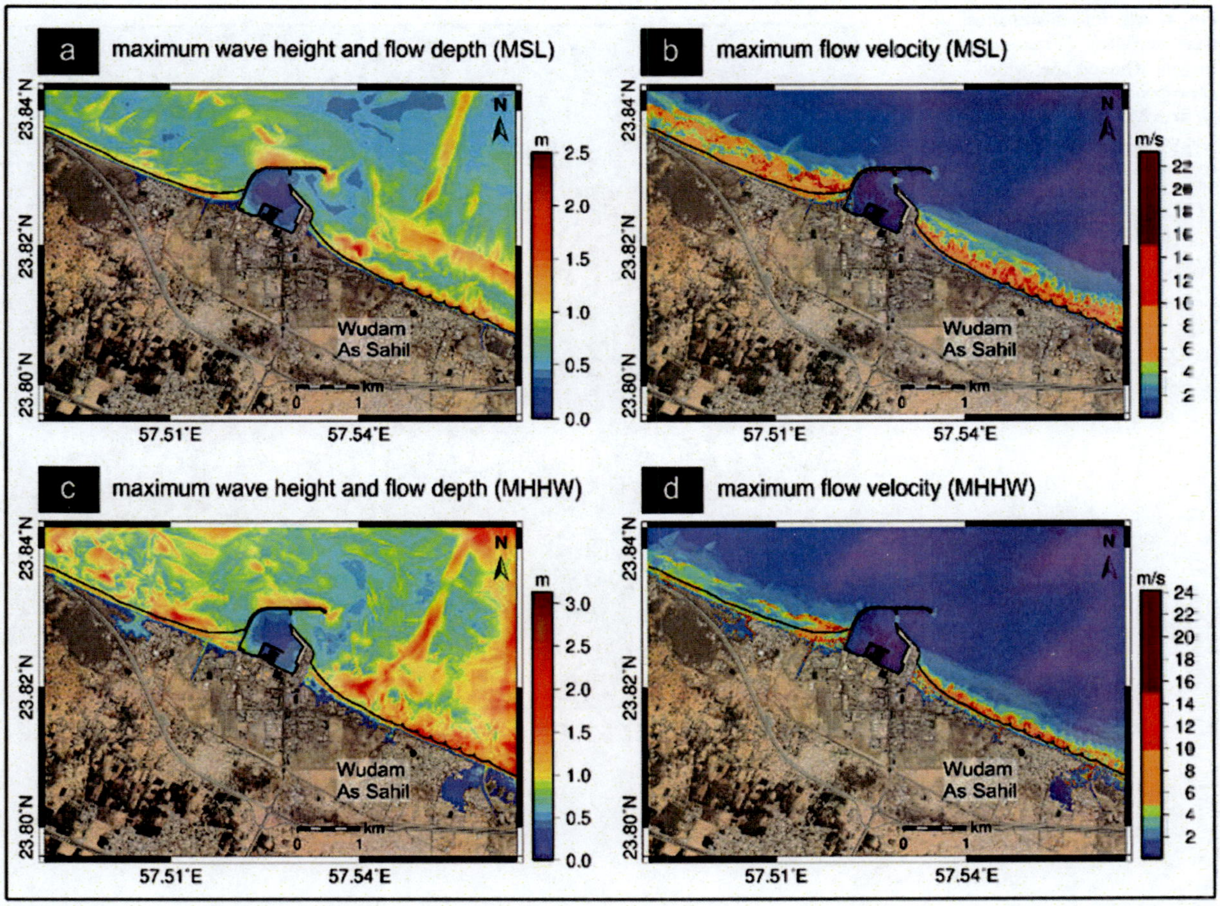

Fig. 2 Tsunami scenario at the coast for the Mw7.2 event in the western MSZ. **a** and **c** maps of maximum wave height and flow depth distribution and **b** and **d** maps of maximum flow velocity for different tidal conditions

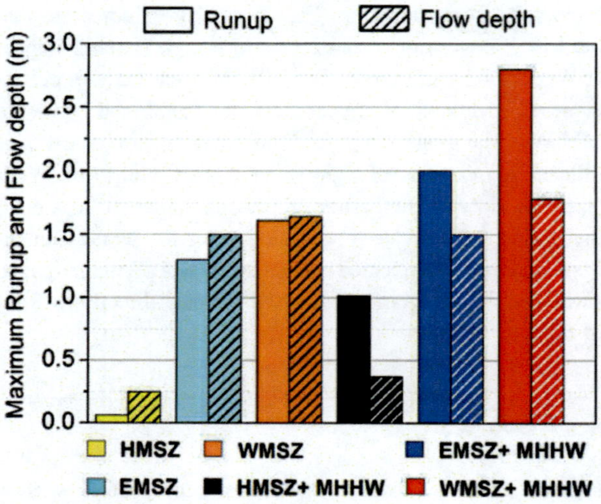

Fig. 3 All tsunami scenarios were considered in this study at the Wadam As Sahil area. The orange bars, which corresponds to the Mw 7.2 western MSZ scenario under high tide condition, represents the worst-case scenario considered at this particular site

Mw 8.8 eastern scenario at Wudam As Sahil coast. It is clear from the results that the Mw 7.2 western Makran scenario poses the maximum threat to this particular site as the maximum run-up height (about 2.7 m) and maximum flow depth (1.7 m) and the inundation limit map shows greater tsunami extent than from other scenarios (Fig. 4).

5 Conclusions

DTHA is conducted using credible earthquake scenarios and high-resolution numerical tsunami modeling. DTHA results showed that the site coastal area is under considerable threat of tsunami hazards. The maximum wave height near shore is about 3 m maximum flow depth is 1.7 m and maximum run-up of about 2.7 m at the coast. The maximum inland penetration in the low-lying areas is about 800 m in the study area. The hazard maps show higher hazard zones due to earthquake from WMSZ than that of EMSZ.

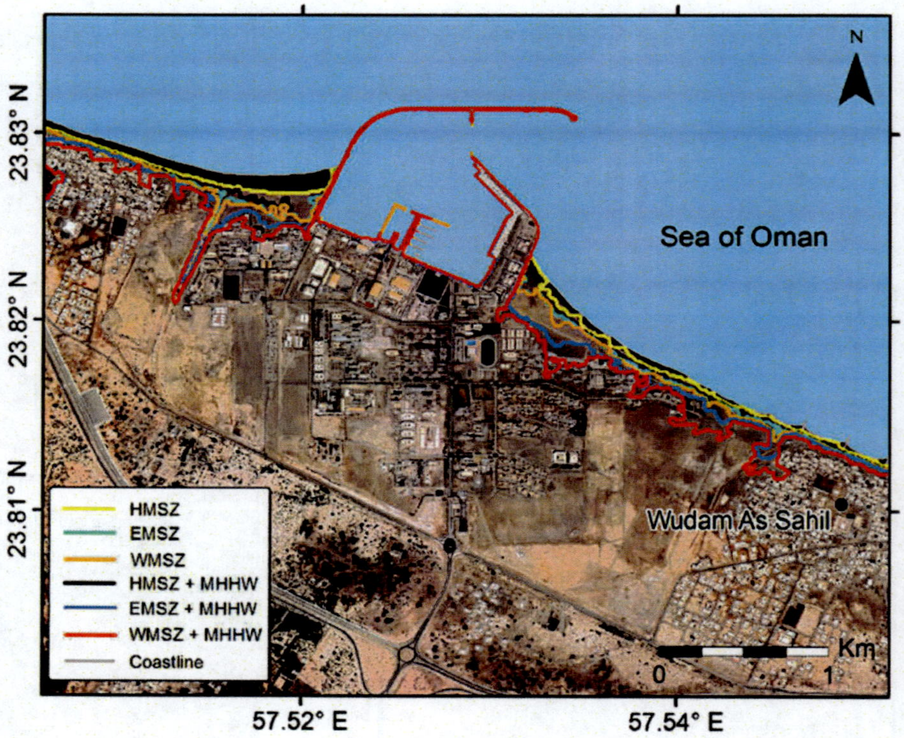

Fig. 4 All tsunami scenarios were considered in this study at the site. The red line, which corresponds to the Mw 7.2 western MSZ scenario under high tide conditions, represent the worst-case scenario considered at this site

References

Anwar H., Gebert N., Mueck., M., Muhari A., Post J., Stein E., Wegscheider S., Birkman J., Riedlinger T., & Strunz, G. (2011). *LIPI/DLR/UNU-EHS. Guidelines for tsunami risk assessment in Indonesia: scientific proposal for practioners and end-users* (132 p). Provided by Indonesian-German Working Group on Tsunami Risk Assessment.

Baptista, M. A., Miranda, J. M., Omira, R., & Antunes, C. (2011). Potential inundation of Lisbon downtown by a 1755-like tsunami. *Natural Hazards and Earth System Sciences, 11*, 3319–3326.

El-Hussain, I., Al-Habsi, Z., Al Bloushi, K., Omira, R., Deif, A., Baptista, M. A., & Mohamad, A. M. E. (2021). Site-specific deterministic and probabilistic tsunami hazard assessment for Diba-Oman and Diba-Al-Emirates. *Arabian Journal of Geosciences*, Special Issue, 1–14, https://doi.org/10.1007/s12517-021-07137-9.

Miranda, J. M., Luis, J., Reis, C., Omira, R., & Baptista, M. A. (2014). Validation of NSWING, a multi-core finite difference code for tsunami propagation and run-up. American Geophysical Union (AGU) Fall Meeting, San Francisco. Paper Number: S21A-4390. Session Number and Title: S21A, Natural Hazards.

Omira, R., Baptista, M. A., et al. (2013). Performance of coastal sea-defense infrastructure at El Jadida (Morocco) against tsunami threat—Lesson learned from the Japanese 11th March 2011, tsunami. *Natural Hazards and Earth System Sciences, 13*, 1–16.

Tinti, S., & Armigliato, A. (2003). The use of scenarios to evaluate the tsunami impact in southern Italy. *Marine Geology, 199*(3), 221–243.

Ionospheric Anomalies Associated with the Mw 6.0 and 6.4 South Iran Twin Earthquakes from GPS and Ionosonde Observations

John P. Pappachen, Hamdan Ali M. Hamdan, Rajesh Sathiyaseelan, Abdollah Masoud Darya, Abdallah Shanableh, and Moussa Leblouba

Abstract

Ionospheric anomalies associated with the Mw 6.0 and Mw 6.4 south Iran twin earthquakes were investigated through GPS and ground Ionosonde observations. The pre-seismic and the co-seismic ionospheric disturbances (CIDs) were observed in the GPS measured Total Electron Content (TEC) and the Ionosonde measured NmF2 and hmF2. The results show positive enhancement of TEC at near field stations ~2–6 days prior to the twin earthquakes. The estimated NmF2 and hmF2 values also show positive values in the pre-seismic phase compared to their mean values. These anomalies are attributed to seismic activity as the Planetary K index (Kp) and Disturbance Storm-time index (Dst) were quiescent during the entire observation period. Observation of sudden CIDs at near field stations immediately after the earthquakes explains the co-seismic energy propagation mechanism through the Lithosphere Atmosphere Ionosphere (LAI) coupling. Our observation on the differential NmF2 marks a dip in the peak electron density by -2.84 × 10^{11} e/m^3 during the time of earthquakes. The estimated TEC, NmF2, and hmF2 results show a consistent reduction in their values on the day of earthquakes. Moreover, their changes were similar in both pre- and co-seismic phases and hence they are qualified as potential parameters for earthquake precursors. Thus the pre- and co-seismic ionospheric anomalies detected in this study reveal the characteristic signatures of twin earthquake preparation processes and the co-seismic energy propagation.

Keywords

Ionospheric anomalies • Total Electron Content • GPS • Ionosonde • Earthquake precursors

1 Introduction

Earthquakes are one of the most destructive natural calamities caused by the sudden displacements of rock masses beneath the earth's surface. The process leads to sudden release of energy in the form of seismic waves which travel radially around the epicenter and cause a wide range of destructions from minor damages to loss of human lives. As we cannot prevent such natural events from happening, geoscientists have mainly been concentrating to understand the earthquake preparation processes with the goal to predict it and to significantly mitigate the level of hazard and its subsequent damage. Study of Ionospheric perturbations using GNSS observations is one of the recent advances in the earthquake prediction research. The changes in the Ionospheric Total Electron Content (TEC) prior to an earthquake is an important measurement which would provide a warning of the location, time, and magnitude of an expected strong event within a narrow period. It has been observed that earthquakes can create changes in the ionospheric Total Electron Content (TEC), peak electron density, and peak height before, during, and after the events (Gautam et al., 2018; Pulinets et al., 2003). In addition to this,

J. P. Pappachen (✉) · H. A. M. Hamdan · A. Shanableh · M. Leblouba
Research Institute of Sciences and Engineering (RISE), University of Sharjah, Sharjah, UAE
e-mail: jpappachen@sharjah.ac.ae

H. A. M. Hamdan
Department of Applied Physics and Astronomy, College of Sciences, University of Sharjah, Sharjah, UAE

R. Sathiyaseelan
Wadia Institute of Himalayan Geology, Dehradun, India

A. M. Darya
Sharjah Academy for Astronomy, Space Sciences and Technology (SAASST), Sharjah, UAE

A. Shanableh · M. Leblouba
Department of Civil and Environmental Engineering, College of Engineering, University of Sharjah, Sharjah, UAE

M. Bezzeghoud et al. (eds.), *Recent Research on Geotechnical Engineering, Remote Sensing, Geophysics and Earthquake Seismology*, Advances in Science, Technology & Innovation,
https://doi.org/10.1007/978-3-031-48715-6_59

activities related to volcanoes, nuclear explosions, and geomagnetic and solar storms can also cause atmospheric and ionospheric disturbances (Huang et al., 1985). There are numerous studies that reported the pre-seismic, co-seismic, and the post-seismic ionospheric anomalies associated with the earthquakes (Hammerstorm & Cornely, 2016; Heki, 2011; Pulinets et al., 2003). The variations in the ionospheric TEC prior to an earthquake are one of the important parameters which can give insight about the earthquake preparation processes and the energy transfer mechanism from the ground to the Upper atmosphere at an ionospheric height of ~300 to 350 km through a process called Lithosphere Atmosphere Ionosphere (LAI) coupling.

The TEC in the ionosphere can be measured at ground through the reception of GPS signals from the satellites to the ground GNSS antennas and it is expressed in TECU (1 TECU = 10^{16} electrons m^{-2}). The GPS transmits dual frequency signals of L1 (1.5 GHz) and L2 (1.2 GHz) which experience time delay and phase changes while passing through the ionospheric layers. This delay is a function of TEC along with its travel path. Earthquakes cause grating of rock masses along a fault plane and the polarization of mineral ions which generate electric fields. At ionospheric height, the TEC distribution undergoes perturbation and generates ionospheric disturbances due to the passage of ground surface Rayleigh waves that are generated by earthquakes. Both these processes lead to pre- and co-seismic ionospheric disturbances as the ground seismic energy gets transmitted through the immediate neutral atmosphere and then to the ionospheric layers.

The Mw 6.0 and 6.4 south Iran twin earthquakes are the recent significant Strong Magnitude earthquakes that occurred on November 14, 2021, about 65 km northwest of the port town of Bandar Abbas (USGS, 2022) (Fig. 1). The earthquakes occurred at the Zagros fold and thrust belt zone in a short interval of time and created widespread destruction like landslides in the region. These earthquakes can be referred as twin or doublet earthquakes by considering their close epicentral locations and short interval of their time of occurrence. Considering the importance of twin earthquakes, here we investigate the pre-seismic and co-seismic ionospheric TEC variations associated with these events to find out the possible pre-seismic signatures.

2 Data and Methodology

We have used GPS data from 6 continuously operating stations located around the earthquake epicenter for a period of 15 days (7 days before and after the earthquakes). GPS Rinex observation data along with the BRDC navigation file and the Differential Code Bias (DCB) data were used for the

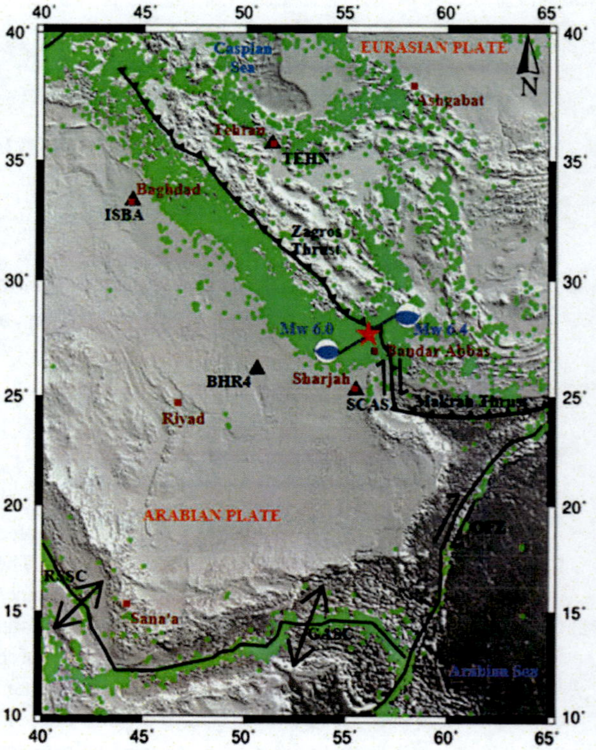

Fig. 1 Location map of the Mw 6.0 and 6.4 south Iran twin earthquakes. The red stars indicate the location of the twin earthquakes and the beachballs represent the focal mechanism of the earthquakes. The green dots indicate the seismicity of the region (USGS) and the black triangle shows the GPS station locations used for the study. Regional tectonic features namely, Zagros thrust, Makran Thrust, Owen Fracture Zone (OFZ), Red Sea Spreading Center (RSSC), and Gulf of Aden Spreading Center (GASC) are also marked

TEC calculation. We estimated the integrated TEC (ITEC) along the satellite receiver path from GPS measurements using the following formula:

$$\mathrm{ITEC} = \frac{1}{40.308}\frac{f1^2 f2^2}{f1^2 - f2^2}[L1\lambda 1 - L2\lambda 2] + \mathrm{Const} + \mathrm{nL}$$

where, '$L1$' and '$L2$' are the carrier phase measurements and '$f1$' and '$f2$' and '$\lambda 1$' and '$\lambda 2$' are their respective frequencies and wavelengths. '*Const*' is the unknown initial phase and the 'nL' represents the error in determining the phase path. In order to identify the noise in the TEC variation caused by solar and geomagnetic storms over the observation period, we analyzed the Planetary K index 'Kp' and the Disturbance Storm-Time 'Dst' indices. The diurnal effects in the estimated daily TEC of each station have been removed by subtracting the statistically estimated mean TEC over a period of 15 days. The resultant differential TEC or ΔTEC hence represents the noise-free TEC variation over the observation period.

To study the variation in ionospheric peak electron density (NmF2) and peak height (hmF2), we used data from the Canadian Advanced Digital Ionosonde (CADI) station installed in the University of Sharjah campus (25.285° N, 55.464° E). The Ionosonde operates at frequencies from 0.1 to 30 MHz. We have estimated the NmF2 and hmF2 values of the ionospheric F2 layer for the entire observation time and a manual scaling of the electron density profiles was performed using the polynomial analysis to ensure the accuracy.

3 Results and Discussions

In this study, we attempted to find out the anomalous changes in the Ionospheric electron content, Ionospheric peak density, and the peak height of electron density associated with the Mw 6.0 and 6.4 south Iran twin earthquakes as precursory signals.

3.1 Analysis of Total Electron Content (TEC) Anomalies

The ionospheric Total Electron Content (TEC) changes associated with the south Iran twin earthquakes are estimated from the continuous GPS data. The integrated TEC (ITEC) observations in the near field GPS stations show anomalous enhancement of ITEC prior to the twin earthquakes. Figure 2 shows the ITEC variations of the nearest GPS station SCAS situated within the earthquake preparation zone over the observation period. An enhancement of ~4 to 10 TECU is observed from its normal value (~15 TECU) around 2 to 6 days prior to the event. Similar enhancement in the ITEC is also observed near field stations namely, TEHN and ISBA. An anomalous increase in the TEC of ~4 to 8 TEC is observed at both TEHN and ISBA stations prior to the event from their normal value of ~11 TECU. This phenomenon was reported previously by (Zakharenkova et al., 2007) as a possible precursor of an impending earthquake. The ΔTEC over the observation period is calculated statistically by removing the average ITEC over the observation period from the daily ITEC. Hence the diurnal effects are removed, and it represents the noise-free TEC anomaly. The ΔTEC observation in the near field stations also supports the observation seen in the ITEC. The anomalous enhancement in the TEC is also well observed in the ΔTEC observations. A positive ΔTEC variation of ~2 to 10 TECU is observed at SCAS station, 3 to 7 days prior to the twin event. Similar positive anomalies of the range of ~2 to 4 TECU are observed in the TEHN and ISBA stations. Since the Kp and Dst indices do not show any prominent signatures of geomagnetic and solar storms over the observation period; hence, the enhancement in the local TEC can be a pre-seismically induced precursory effect of the twin earthquakes. From the observations, it is evident that maximum enhancement of TEC anomaly is at the closest station SCAS and then followed a reducing trend toward farther stations from the earthquake epicenter.

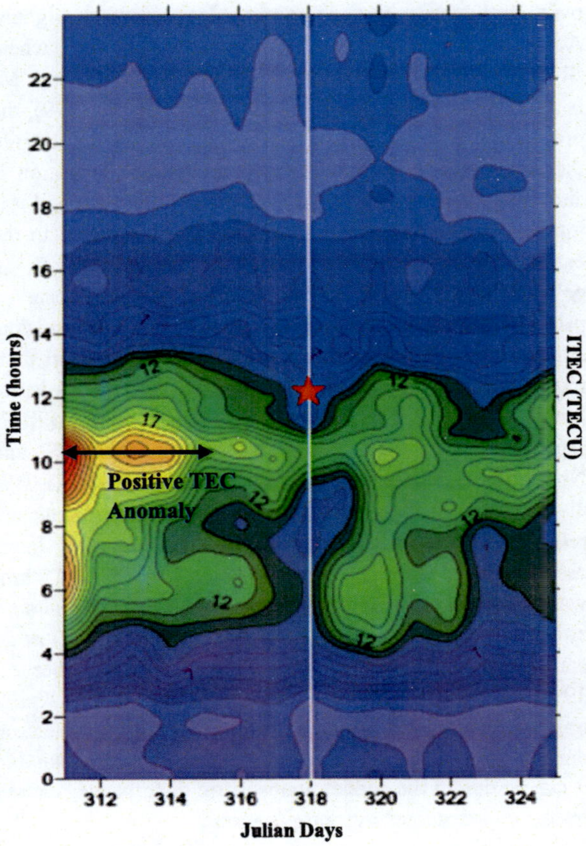

Fig. 2 Estimated ITEC anomaly values of the SCAS station over the observation period. The colored contour represents the integrated TEC values. The white bold line and the red star show the earthquake's day and time respectively

However, we observed a reduction in the daily TEC (~4 to 6 TECU) at near field stations on the day of twin events. But interestingly, a sudden co-seismic TEC enhancements of ~2 to 3 TECU are also observed at these near field stations immediately after the earthquakes due to the co-seismic ionospheric disturbances (CIDs) caused by the upward propagating atmospheric pressure waves generated during the earthquakes.

3.2 Ionospheric Peak Electron Density and Peak Height Observations

Our observations on ionospheric peak electron density (NmF2) and peak height (hmF2) changes associated with the twin earthquakes also show notable precursory signals. According to the Dobrovolsky formula, $R = 10^{0.43M}$, where 'R' is the radius of the earthquake preparation zone, and 'M' is the earthquake magnitude (Dobrovolsky et al., 1979), the preparation zone for an Mw 6 earthquake is 380 km around the earthquake epicenter. The ground Ionosonde station is situated well within the earthquake preparation zone and it is inferred that the observed anomalies will be related to the earthquake processes. As like in the estimated TEC, an enhancement in the NmF2 and the hmF2 was also observed prior to the twin earthquakes. The positive pre-seismic enhancements of both NmF2 and hmF2 values from their mean values are possibly attributed to the seismo-ionospheric coupling. Gupta and Upadhayaya (2017) report a similar kind of variation in ionospheric hmF2 and NmF2 values before the low-mid latitude earthquakes. It is also observed that pre-seismic ionospheric perturbations are relatively higher as compared to post-seismic period. Interestingly, a reduction in both NmF2 and hmF2 ($\sim$25 km) was observed on the event day as in the TEC. A similar effect in the NmF2 was reported by (Liu et al., 2004) as an earthquake ionosphere anomaly prior to the September 21, 1999, M 7.3 Chi-Chi earthquake. The differential NmF2 observations mark a prominent trough with a peak electron density of -2.84×10^{11} e/m^3 at the time of earthquake and it can be due to the sudden ionospheric disturbances caused by the co-seismic energy propagation.

4 Conclusions

We investigated the ionospheric anomalies associated with November 14, 2021, Mw 6.0 and Mw 6.4 south Iran twin earthquakes through GPS measured TEC and NmF2 and hmF2 observations from the ground based Ionosonde. Results from stations close to the epicenter show the development of a TEC-enhanced ionosphere at $\sim$2–6 days prior to the twin events. Since there are no prominent geomagnetic and solar activities during the observation period, the anomalies in the TEC are seismogenic in nature. A sudden increase in the ionospheric TEC is observed immediately after the twin earthquakes which indicates that the CIDs are created by the twin events through LAI coupling. Similarly, pre-seismic enhancement of NmF2 and hmF2 is also attributed to the preparation processes of the twin earthquakes. Interestingly, TEC, NmF2, and hmF2 regarded as potential precursors for earthquake show a reduction in their values on the day of twin earthquakes. Concurrent behaviors of these parameters in the pre- and co-seismic periods reveal their required scale variations to identify as potential earthquake precursors. Our study shows notable ionospheric anomalies in pre and co-seismic periods that signify the signatures of earthquake preparation and co-seismic energy propagation.

Acknowledgements We would like to thank the Director, the Research Institute of Sciences and Engineering (RISE), University of Sharjah, UAE, for providing the necessary facilities and encouragement to carry out this work. We acknowledge the Sharjah Academy for Astronomy, Space Sciences and Technology (SAASST), for the GPS and Ionosonde data. Author RS thanks the Director, Wadia Institute of Himalayan Geology, Dehradun, India, for providing the facilities to carry out this work. We also acknowledge the International GNSS service (IGS) (garner.ucsd.edu) for the GPS data and the United States Geological Survey [USGS] for providing the earthquake data.

References

Dobrovolsky, I. R., Zubkov, S. I., & Myachkin, V. I. (1979). Estimation of the size of earthquake preparation zones. *Pure and Applied Geophysics, 117*, 1025–1044.

Gautam, P. K., Chauhan, V., Sathyaseelan, R., Kumar, N., & Pappachen, J. P. (2018). Co-seismic ionospheric GPS-TEC disturbances from different source characteristic earthquakes in the Himalaya and the adjoining regions. *NRIAG Journal of Astronomy and Geophysics, 7*, 237–246.

Gupta, S., & Upadhayaya, A. K. (2017). Preearthquake anomalous ionospheric signatures observed at low-mid latitude Indian station, Delhi, during the year 2015 to early 2016: Preliminary results. *Journal of Geophysical Research: Space Physics, 122*, 8694–8719.

Hammerstorm, J. A., & Cornely, P. J. (2016). Total Electron Content (TEC) variations and correlation with seismic activity over Japan. *Journal of Young Investigators, 31*, 13–16.

Heki, K. (2011). Ionospheric electron enhancement preceding the 2011 Tohoku-Oki earthquake. *Geophysical Research Letters, 38*, L17312. https://doi.org/10.1029/2011GL047908

Huang, Y. N., Kang, C., & Chen, S. W. (1985). On the detection of acoustic gravity waves generated by typhoon by use of real time HF Doppler frequency shift sounding system. *Radio Science, 20*, 897–906.

Liu, J. Y., et al. (2004). Pre-earthquake ionospheric anomalies registered by continuous GPS TEC measurements. *Annales Geophysicae, 22*, 1585–1593.

Pulinets, S. A, Contreras, A. L, Bisiacchi, G, Ciraolo, L., & Singh, R. (2003). Ionospheric and thermal precursors of Colima earthquake of 22 January 2003. *Annual Meeting Mexican Geophysics, Union GEOS*, p. 170.

U.S. Geological Survey. (2022). *Earthquake lists, maps, and statistics*. https://www.usgs.gov/natural-hazards/earthquake-hazards/lists-maps-and-statistics. Accessed 15 March 2022.

Zakharenkova, I. E., Shagimuratov, I. I., Krankowski, A., & Lagovsky, A. F. (2007). Precursory phenomena observed in the total electron content measurements before great Hokkaido earthquake of September 25, 2003 (M = 8.3). *Studia Geophysica et Geodaetica, 51*, 267–278.

Comparison of Electromagnetic Signals Before an Earthquake Using the Radio Direction Finding Method. The Case of Po Plain Valley (Italy)

Valentino Straser, Daniele Cataldi, and Gabriele Cataldi

Abstract

In this study, we present the outcome of a trial of the Radio Direction Finding detection system, aimed at verifying the link between electromagnetic signals detected in Italian monitoring centers from those emitted in tectonically stressed areas. The area studied is in the Po Valley, in Northern Italy, in an area historically affected by seismicity with earthquakes rarely exceeding magnitude 6. The Radio Direction Finding system, designed to detect the direction of electromagnetic signals, confirmed the applicability of this method for areas subjected to crustal stresses that can evolve into seismic shocks. In the case of this experiment, the signals preceded the seismic events by about 24 h, and the intersection of the colorimetric lines, which appeared in the dynamic spectrograms, coincided with the future epicenter zone of the earthquakes. The seismic occurrence fell within a time window of three days, confirming a well-established trend, that overlapping with previously studied cases.

Keywords

RDF systems • Earthquake • Electromagnetic signals • Crustal diagnosis • Elf frequency

1 Introduction

After a pioneering phase and initial skepticism about the study of seismic precursors, expressed by the scientific community, today more and more research centers are taking up this area of seismology. Among the methods tested by various research teams on an international scale, this study presents the Radio Direction Finding (RDF) method, designed and tested in Italy since 2017 (Straser et al., 2017) by Radio Emissions Project (Rome, Italy), and discusses a successful application for two earthquakes, which occurred in Italy on February 9, 2022, with magnitudes between 4.0 and 4.3. Three RDF monitoring stations are set up in Italy: Lariano in Rome; Ripa-Fagnano in L'Aquila; and in Pontedera, near Pisa. The 24-h monitoring is based on the identification of different electromagnetic frequency bands, distinguishing natural-type signals that are emitted from the Earth's crust. The instrumentation is currently designed to identify the intensity, electromagnetic frequency, duration, and most importantly, the direction of arrival of the signals (azimuth) with respect to the geographical position of the station itself. The data, analyzed daily, have made it possible to ascertain that the electromagnetic signals come from areas under tectonic stress and historical seismicity (Cataldi et al., 2019; Straser et al., 2019). The archive of data extracted from the Italian RDF network shows common features regarding the occurrence of radio anomalies, associated with the azimuth of seismic epicenters. The detection of the anomalies occurs, instrumentally, 3 to 7 days before the seismic event, in which the signals decrease and then reappear a day before the earthquake. This repeated variation of discontinuous electromagnetic increments can be interpreted as increased tectonic stresses that usually occur in pre-earthquake phases.

2 Methods or Materials

The monitoring stations of the Italian RDF network are equipped with antennas connected to a radio receiver and interfaced with a computer, which is active 24 h a day. The device is equipped with two antennas aligned with each other orthogonally to the geographical poles. The variation of radio anomalies is detected in the VLF band in the range from 0.3-a to 96 kHz. The innovative aspect of the method is

V. Straser (✉)
UPKL, Rue de La Presse 4, Brussels, Belgium
e-mail: valentino.straser@gmail.com

D. Cataldi
Radio Emission Project, Lariano, Rome, Italy

G. Cataldi
Radio Emission Project, Albano Laziale, Rome, Italy

M. Bezzeghoud et al. (eds.), *Recent Research on Geotechnical Engineering, Remote Sensing, Geophysics and Earthquake Seismology*, Advances in Science, Technology & Innovation,
https://doi.org/10.1007/978-3-031-48715-6_60

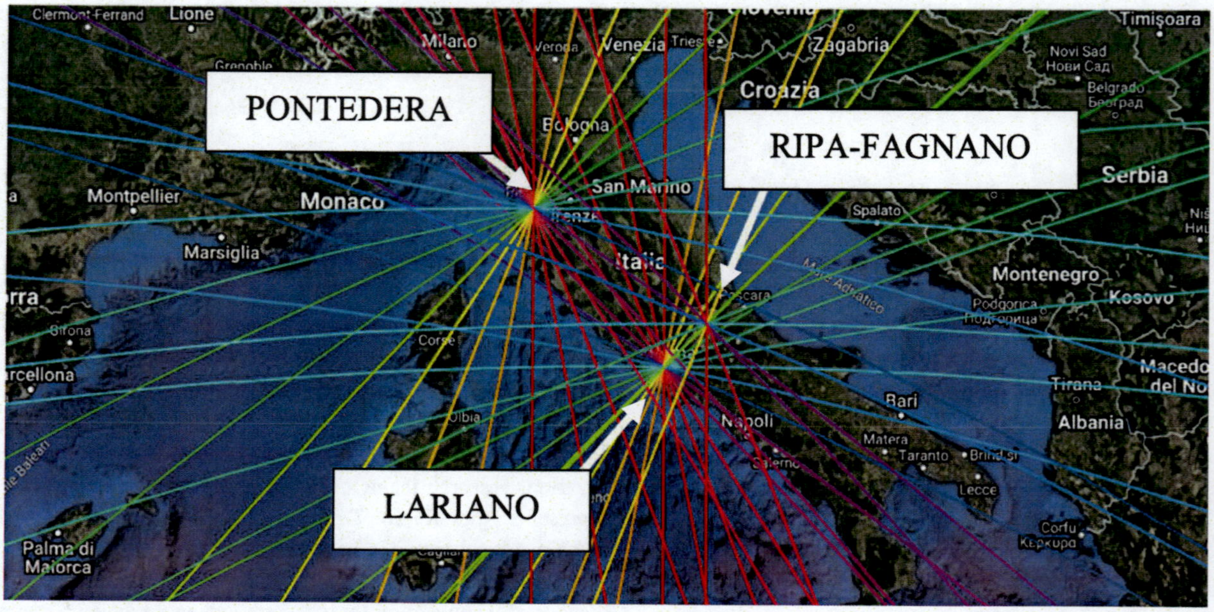

Fig. 1 RDF mapping of the three Italian stations. It shows the position of the RDF stations of Pontedera, Pisa, Italy; Lariano, Rome, Italy; and Ripa-Fagnano, L'Aquila, Italy. The colored lines identify the different azimuths of origin of the radio signals, recorded by each survey station. *Credits* Radio Emissions Project; Google My Maps

to associate the colors of the magnetic anomalies in the spectrogram (Fig. 1) with a map indicating the direction of the signals (azimuth).

The RDF stations are three and located at:

(1) Lariano, Rome, Italy (GPS: Lat. 41.728799 N, Long. 12.843205 E).
(2) Pontedera, Pisa, Italy (GPS: Lat. 43.672445 N, Long. 10.640100 E).
(3) Ripa-Fagnano, L'Aquila, Italy (GPS. Lat: 42.265709 N, Long. 13.583850 E).

RDF survey stations record electromagnetic monitoring, generating an archive of spectrograms that can be analyzed to understand the characteristics and evolution of natural-type radio signals. The data collected by the RDF monitoring stations are compared with earthquakes published on the website of the National Institute of Geophysics and Volcanology. The method used is the analysis, comparison, and interpretation of electromagnetic data compared with the azimuth of the signals and compared with known situations of previous earthquake events.

3 Results

The two earthquakes, which occurred on February 9, 2022, and were located at Bagnolo in Piano in the Po Valley in northern Italy (Figs. 2 and 3), were inferred from the website of the National Institute of Geophysics and Volcanology in Rome. The two seismic events had the same epicentral zone, spaced at the time level by a little more than an hour.

1. Ml 4.0 on 09-02-2022 at 18:55:12 (UTC), 44.7800, 10.7250, depth 6 km.
2. Ml 4.3 on 09-02-2022 at 20:00:57 (UTC) 44.7860, 10.7240, depth 7 km.

4 Discussion

The monitored area was chosen randomly and fell in the Po Valley in Northern Italy, identified based on historical and recent seismicity, to test the RDF method. The duration of monitoring was six months. The traces of radio anomaly signals to be considered in the experimentation are two: the color purple for the Lariano-Rome station and red for the Pontedera station. The frequency band considered for the study of these seismic precursor candidates is between 0.01 and 7 Hz. The colored lines appearing in the spectrogram represent the azimuth within which the natural electromagnetic signals are identified and detected by the RDF stations of Pontedera-Pisa and that of Lariano-Roma. On the day before the two seismic events, February 8, 2022, from 1:00 p.m. to 5:00 p.m., the Lariano-Rome station recorded violet signals from, therefore, the investigation area of the Po Valley (Fig. 4). Similarly, the RDF station in Pontedera, too,

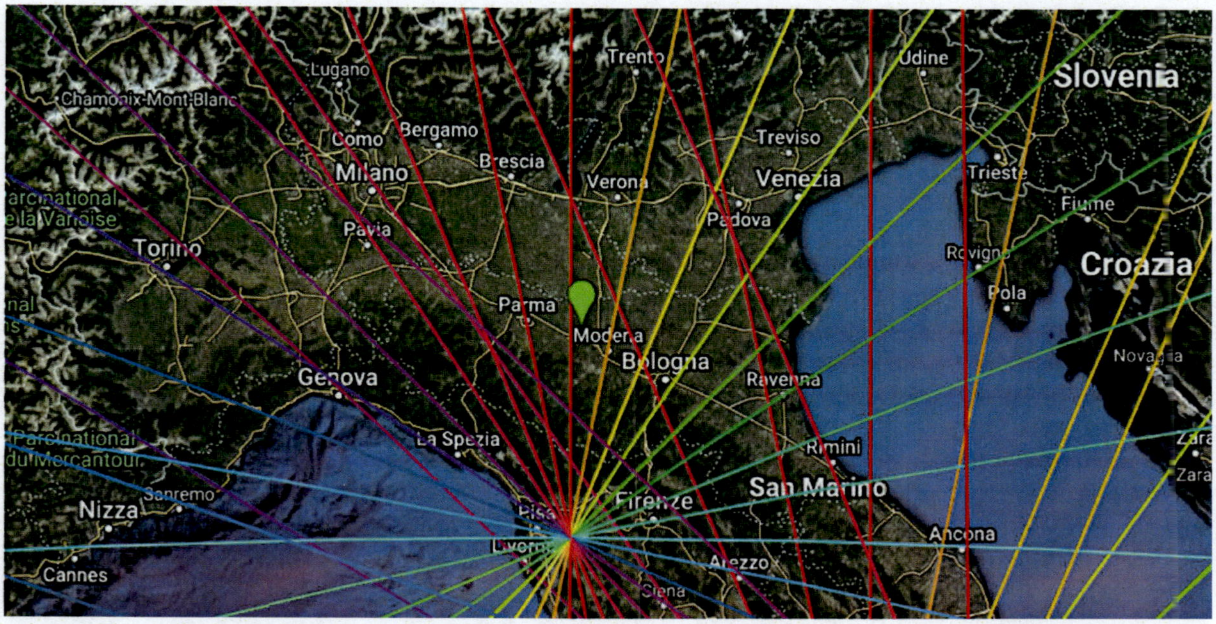

Fig. 2 Spatial position of the seismic epicenter with respect to the Italian territory which occurred in northern Italy on February 9, 2022, and of magnitude Ml 4.0. *Credits* Radio Emissions Project; Google My Maps

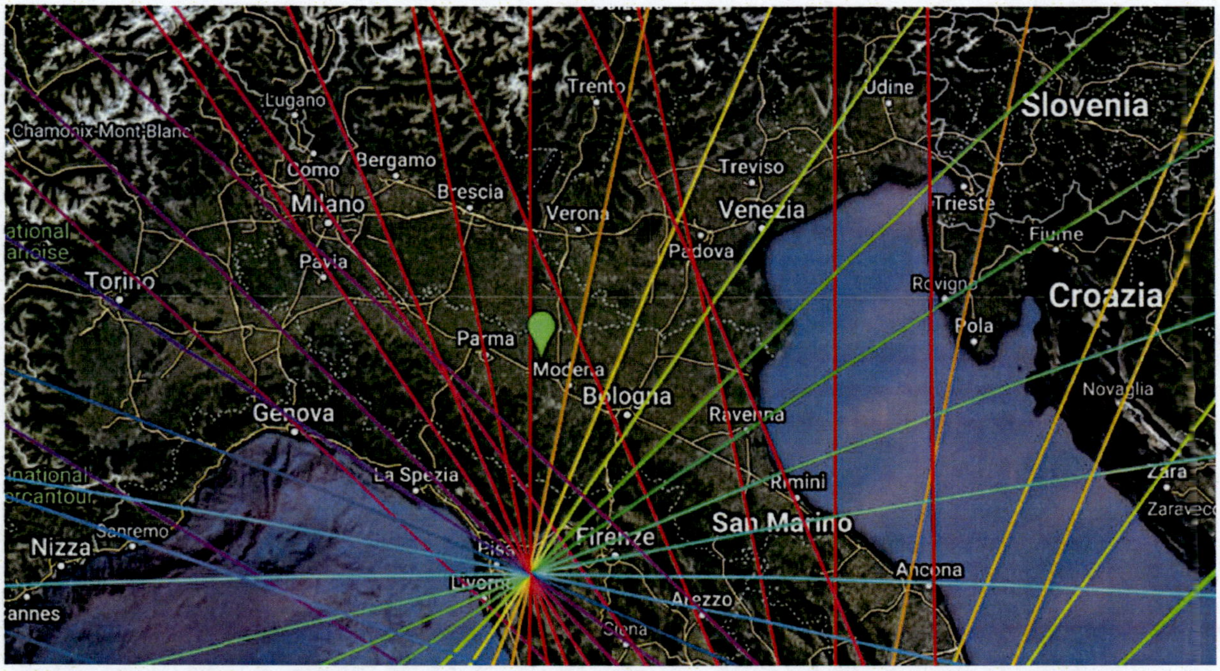

Fig. 3 Spatial position of the seismic epicenter with respect to the Italian territory which occurred in northern Italy on February 9, 2022, and of magnitude Ml 4.3. *Credits* Radio Emissions Project; Google My Maps

showed the appearance of a series of electromagnetic signals, which appeared suddenly and disappeared after a few hours. These signals presented a precise red-colored azimuth, indicating the geographical area of study. These emissions, clearly distinguishable from the natural geomagnetic background, identified a circumscribed geographical area, the location of future epicenters. By intersecting the directions of the two monitoring stations (purple and red),

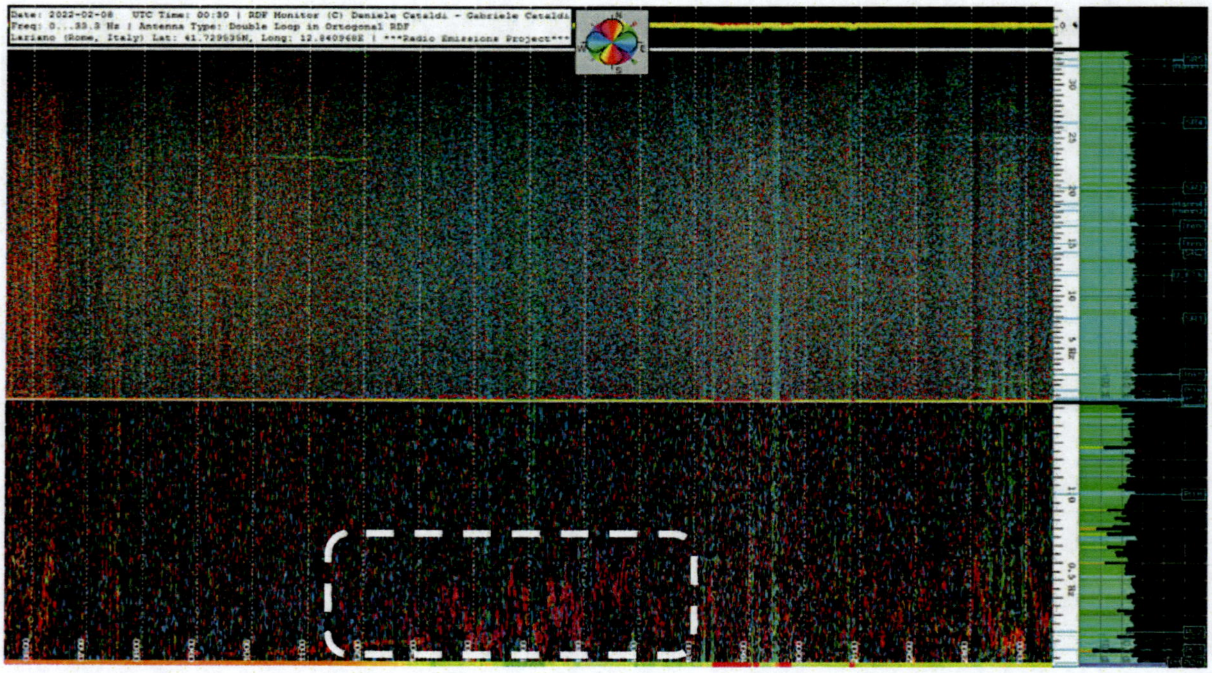

Fig. 4 Spectrogram produced by the RDF station in Lariano, Rome, Italy on February 8, 2022. The spectrogram shows the sudden appearance of a signal having as azimuth (purplish) the study area of the Po Valley. *Credits* Radio Emissions Project

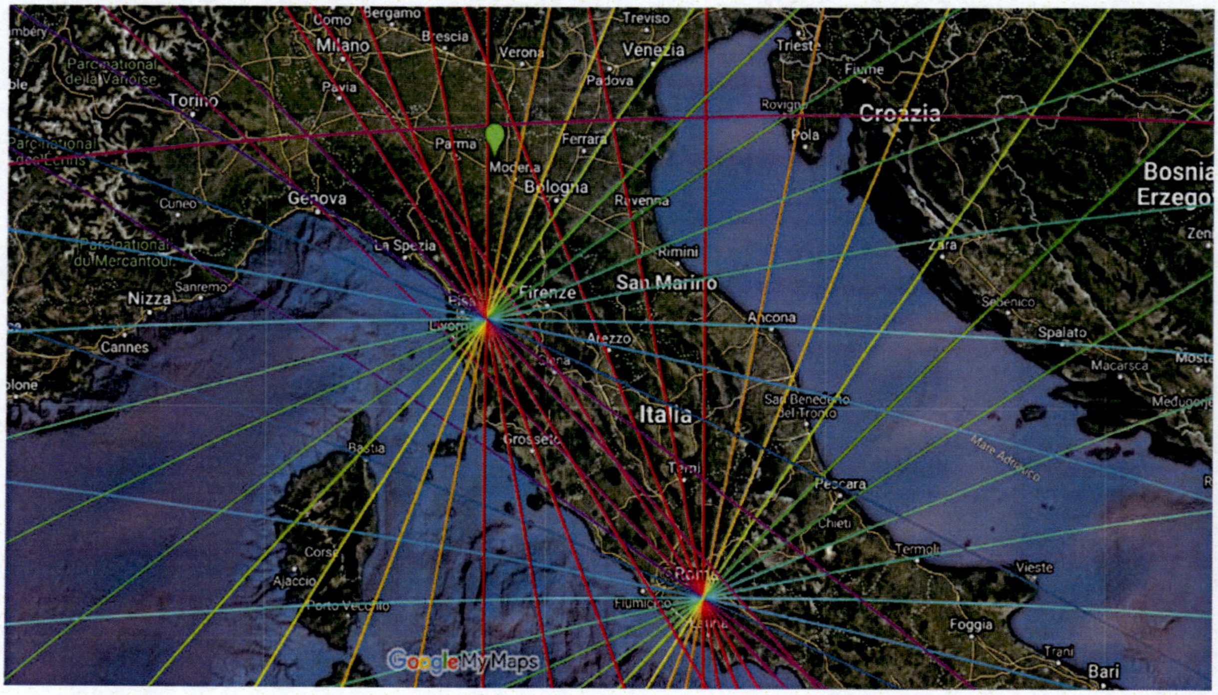

Fig. 5 Colorimetric map of the Italian RDF network showing the location of the epicenters of the earthquakes that occurred in northern Italy on February 9, 2022, of magnitude Ml 4.0 and Ml 4.3. The colored lines represent the azimuth within which natural electromagnetic signals are identified and depart from RDF stations. The purple line (RDF) at the top of the graph that crosses the word "Croazia" refers to the University of Malaya station, and does not refer to the Italian RDF stations, used in this study. *Credits* Radio Emissions Project; Google MY Maps

the area under crustal stress and the possibility of occurrence of a seismic event of magnitude 4 or greater was assumed. The colorimetric signals, associated with the map to identify the direction of the tectonic disturbance, usually appear within a three-day time window. In this case, however, the earthquakes occurred about 24 h after the appearance of the radio anomalies (Figs. 4 and 5).

5 Conclusions

We conclude that the occurrence of the radio anomalies, detected by the RDF Method, can be associated with instrumentally detectable tectonic stress conditions for a given seismic zone by comparison with the color scale and azimuth of the signals. The method of triangulation of electromagnetic signals carried out through the three Italian stations (Fig. 1), at low and very low frequencies, also made it possible to establish an element of coincidence between the intersection of the signals and the future epicentral zone, in the three-day time window. Such signals, for all cases studied in Italy, vary from three days to a few hours before the seismic event. Further steps of experimentation and an interdisciplinary comparison will confirm or not the reliability of the method.

References

Cataldi, D., Cataldi, G., & Straser, V. (2019). Radio Direction Finding (RDF)—Pre-seismic signals recorded before the earthquake in central Italy on 1/1/2019 west of Collelongo (AQ). *Geophysical Research Abstracts*, Vol. 21, EGU2019-3124, EGU General Assembly.

Straser, V., Cataldi, D., & Cataldi, G. (2017). Radio direction finding system, a new perspective for global crust diagnosis. *New Concepts in Global Tectonics Journal, 6*(2), 202–210.

Straser, V., Cataldi, D., & Cataldi, G. (2019). Radio Direction Finding (RDF)—Geomagnetic monitoring study of the Himalaya area in search of pre-seismic electromagnetic signals, *Asian Review of Environmental and Earth Sciences, 6*(1), 16–27.

Seismic Signal Discrimination Between Natural Earthquakes and Mining Explosions Using Moroccan Seismic Network

Ibrahim Ouchen, Oussama Arab, Younes El Fellah, and Mimoun Harnafi

Abstract

Some regions of Morocco are characterized by relatively moderate active seismicity and mining. The seismic activity in Morocco is related to the African and Eurasian plate's convergence. In this work, we examined several diagnostic techniques to identify and distinguish between earthquakes as natural events and explosions as artificial events in the northern part of Morocco. The methods used are based on seismic spectral amplitude ratio, moment magnitudes (M_w), the time duration of signals, and focal mechanisms. The discrimination has been checked regarding monitoring of local seismicity and chemical explosion using the data recorded from more than 80 seismic stations in the framework of several projects; PICASSO, Topo Iberia, Morocco-Array, SIMA, RIFSIS for the period from 2010 to 2012. It comes out that using the distribution of Amplitude/Frequency methods between 0.1 and 25 Hz frequency band, the separation appears clearly for the earthquakes in the range 1–3 Hz and explosions in the range 9–12 Hz. Therefore, the moment magnitude of natural earthquakes is greater than that of artificial explosions. The detectability of the local earthquake and chemical explosion is estimated by the near station to the mine. The magnitude of explosions and earthquakes is between 2 and 4. Our approach is to apply these methods to better separate the populations of earthquakes and explosions to help calibrate local discriminants, especially in the phosphate areas where explosions rich database is available. This may shed light on the seismicity pattern of the region.

Keywords

Discrimination • Earthquakes • Mining explosions • Method relative amplitude/frequency • Method M_W-M_L

1 Introduction

The discrimination between seismic events and explosions is one of the most studied sections in seismology. Moreover, with the Comprehensive Nuclear Test Ban Treaty Organization (CTBTO), the research is increasing in order to obtain efficient and powerful packages to characterize the sources of seismic signals captured by the seismometers.

In order to improve the performance of our seismic data recorded on the stations of our network, some discrimination methods have been applied on the seismic signals recorded on Moroccan network, with the aim of developing more relevant and adapted methods to our data in the future. The objectives are the distinguish between artificial events and natural events and to allow the correction of trajectory effects on seismic wave propagation in the different geological parts of Morocco (Michard et al., 2008).

2 Materials and Methods

The data selected for this work was collected from the Picasso, Topo Iberia, and Morocco-Array networks (Fig. 1). The stations used are broadband seismographs (BBS) type with three components, equipped with seismometers Nanometrics Trillium 120P/Streckeisen STS-2/ and Guralp. Eighteen small events were selected for this study in the magnitude range between 2.0 and 4.3, of which seven are small natural earthquakes, and eight are explosions induced by the RIFSIS (Gil, et al., 2014) and SIMA (Ayarza et al., 2014) projects. All events were recorded between 2010 and 2012, located between latitude 30–36 and longitude from − 3 to − 9.

I. Ouchen (✉) · O. Arab · M. Harnafi
Department of Geomorphology and Geomatics, Scientific Institute, Mohammed V University in Rabat, Rabat, Morocco
e-mail: ouchen.ibrahim@gmail.com

Y. El Fellah
Energy and Farm Machinery Department, Rural Engineering, Agronomic and Veterinary Institute Hassan II (IAV) Rabat, Rabat, Morocco

M. Bezzeghoud et al. (eds.), *Recent Research on Geotechnical Engineering, Remote Sensing, Geophysics and Earthquake Seismology*, Advances in Science, Technology & Innovation,
https://doi.org/10.1007/978-3-031-48715-6_61

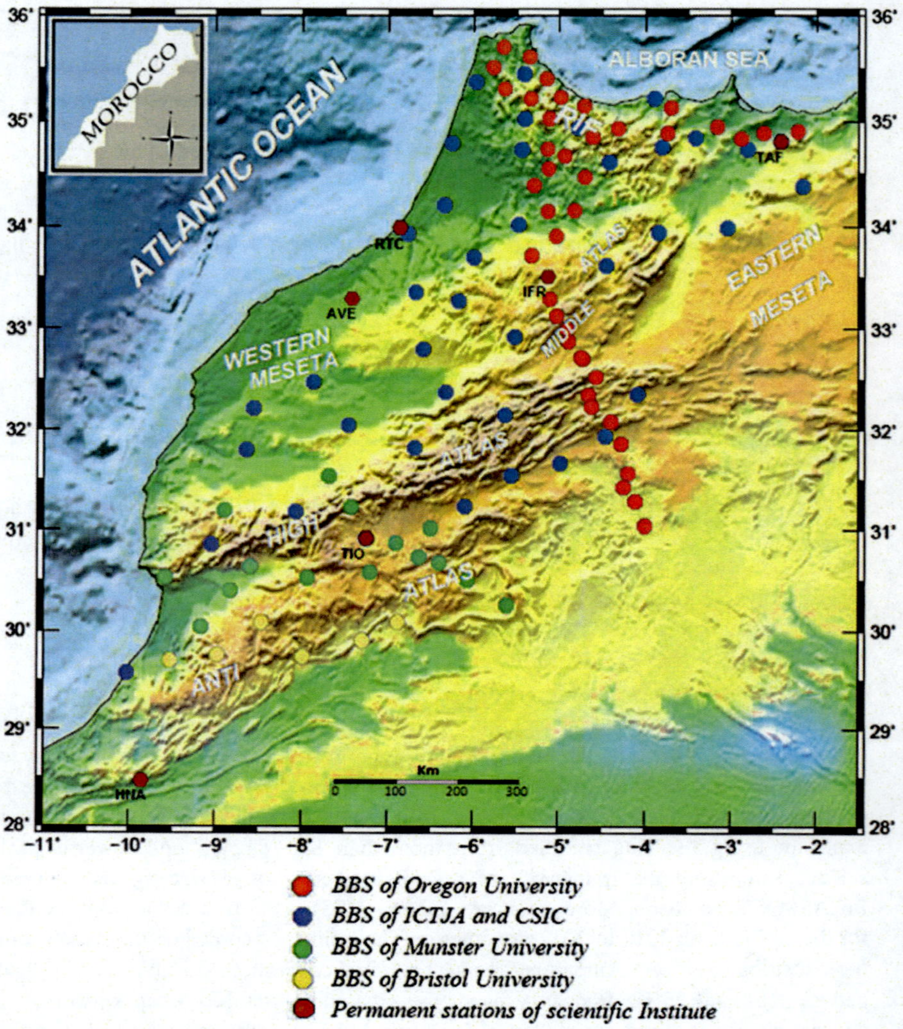

Fig. 1 Deployment of temporary and permanent seismic stations used in this study

A vast number of techniques have been used to discriminate between earthquakes and quarry explosions; in this work, we test some of these methods for the identification of chemical explosion signals from earthquakes. The main identification method previously used in discrimination is the duration of the signal, because for the majority of events, the duration is longer for earthquakes than for explosions (Li et al., 1995). The spectral analysis method is based on the distribution of the amplitude-frequency spectrogram in the time domain, and to study the spectral characteristics of signals and the frequency content of whole seismic waveforms (Plafcan et al., 1997). Finally, the moment magnitude M_W is the most accurate and modern quantification for measuring the size of an earthquake; however, magnitude scales have long been used for this purpose (Hanks & Kanamori, 1979). The following relation defines the moment magnitude M_W:

$$M_w = (\log_{10} M_0/1.5) - 6.07 \quad (1)$$

M_0 (N·m): is the seismic moment; it is a static parameter and is not related to the dynamic properties of the source.

In order to improve our study of discrimination, we calculate the ratio between the moment magnitude M_W and the local magnitude M_L using the maximum amplitudes on the Z channels (vertical) (Table 1). This method is based on a physical model of deformation of seismic events that allows the best calculation of the magnitude in both cases of large and small earthquakes.

3 Results and Discussion

Many papers have attempted to deal with the discrimination issue, including some similar to our study of small earthquakes and explosions (Allmann et al., 2008; Dahy &

Table 1 Moment magnitude and signal duration (SD) with corresponding M_L and Mc, for the studied earthquakes and explosions indicated by EV and SP/SR, respectively

Date	Events	H:M:S	Lat	Long	ELV (m)	Depth (m)	M_W	M_L	SD(s)	Mc
04/05/2010	SP1	16:00:03	33.67	− 5.20	381	60	2.2	1.6	23	0.6
06/05/2010	SP3	16:59:59	32.86	− 4.97	1525	30	2.0	1.3	34	1.0
07/05/2010	SP4	12:01:00	32.30	− 4.49	1380	30	2.3	1.6	35	1.0
07/05/2010	SP5	17:00:00	31.79	− 4.22	1030	30	2.3	1.6	24	0.6
08/05/2010	SP6	12:00:00	30.99	− 3.96	714	30	3.1	1.8	31	0.9
11/10/2011	SR2	17:01:00	35.10	− 5.30	496	30	2.6	1.7	40.5	1.2
12/10/2011	SR3	16:00:00	34.49	− 5.19	276	30	2.6	1.9	31.4	0.9
13/10/2011	SR4	14:00:00	34.74	− 3.87	1368	30	2.4	1.8	30.5	0.9
17/02/2012	SPY1	14:09:28	32.22	− 8.39	390	< 100	2.4	0.9	34	1.0
28/02/2012	SPY2	15:13:00	32.22	− 8.39	390	< 100	2.2	1.2	34	1.0
28/02/2012	SPB	15:04:00	32.32	− 7.86	560	< 100	2.4	1.0	34	1.0
21/01/2010	EV1	16:57:07	34.78	− 5.73	317	18,000	4.3	3.4	166	2.9
19/02/2010	EV2	04:13:09	33.43	− 4.14	2232	0	2.9	2.5	101	2.3
05/08/2010	EV3	18:54:14	32.17	− 5.84	2383	7000	4.3	4.2	212	3.3
01/03/2011	EV4	10:26:26	35.13	− 4.69	123	76,000	3.2	2.5	92.7	2.3
28/02/2012	EV5	06:26:30	32.32	− 6.96	360	400	2.8	2.5	112	2.4
02/06/2010	EV6	16:35:42	34.71	− 3.89	1157	3000	3.4	3.0	111	2.2
29/02/2012	EV7	04:11:52	32.19	− 4.84	1631	0	2.9	2.7	115	2.5

Hassib, 2009). This work indicates that one of the most effective methods for distinguishing artificial events from natural events is the use of relative amplitude and frequency.

Applying this ratio between the moment magnitude M_w and the local magnitude M_L discussed in Sect. 2, we can see that the distinction between them is very clear (Fig. 2a); the discrimination is easier when the moment magnitude M_w is > 2.8. However, the local magnitude of the explosions is smaller than that of the earthquakes.

We deduce from Fig. 2b that the correlation between the duration and coda magnitude (Mc) works as an important factor in discriminating between earthquakes and explosions, the two populations appear separated and there is no confusion between them. The results of the spectral distribution in the different stations (Table 1), represented in Fig. 2c show the general view of the average of the spectrograms of mining explosions and earthquakes in station. This demonstrates the relationship between the relative amplitude and frequency detected in the Moroccan seismic network. As the graph reveals, discrimination in the frequency domain is observed between 9 and 12 Hz for earthquakes and between 1 and 3 Hz for explosions.

4 Conclusions

In this paper, we have applied some methods for discriminating between earthquakes and explosions in Morocco. We work then to quantify the performance of these methods as sufficient discriminants in the case of small earthquake magnitudes and artificial events. The dataset used in this study was collected from 2010 to 2012 with more than 80 seismic stations distributed over the Moroccan territory.

The relative amplitude-frequency ratio used for the totality of the signals recorded at the Moroccan network shows that explosions have a relatively high amplitude-frequency ratio compared to earthquakes. We conclude that the best frequency band filter to apply for earthquakes is 1–3 and 9–12 Hz for explosions. It should also be noted that to complete our study of discrimination, we need to explore the effect of other methods. It is clear from the results obtained in this work that an additional research is needed for regional explosions recorded via teleseism in our seismic network to provide valuable information regarding discrimination for future studies.

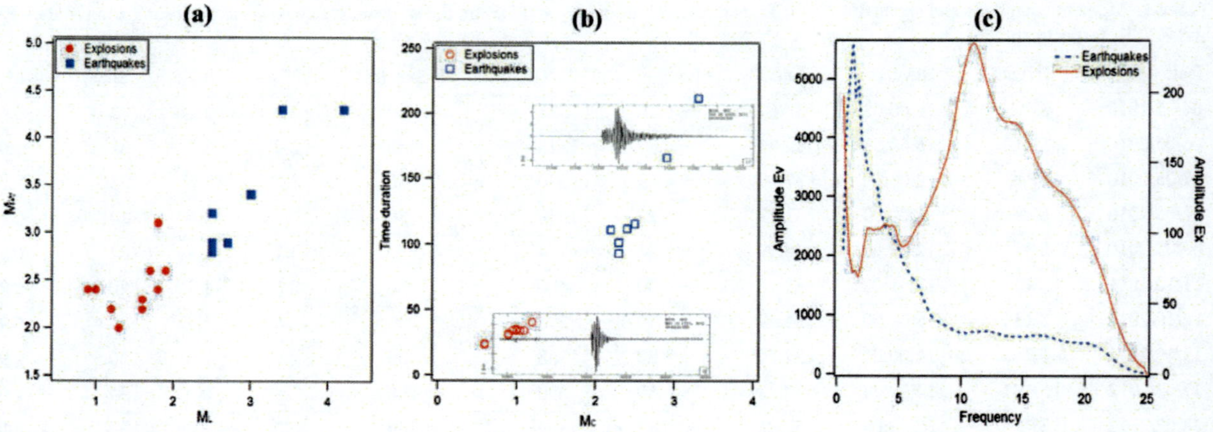

Fig. 2 **a** The M_W/M_L ratios for the studied events **b** Relationship between time duration (second) and coda magnitude **c** Amplitude-Frequency average spectra from all events studied in this work. Explosions are in Red and earthquakes are in Blue colour

References

Allmann, B. P., Shearer, P. M., & Hauksson, E. (2008). Spectral discrimination between quarry blasts and earthquakes in southern California. *Bulletin of the Seismological Society of America, 98*(4), 2073–2079.

Ayarza, P., et al. (2014). Crustal thickness and velocity structure across the Moroccan Atlas from long offset wide-angle reflection seismic data: The SIMA experiment. *Geochemistry, Geophysics, Geosystems, 15*(5), 1698–1717.

Dahy, S. A., & Hassib, H. G. (2009). Discriminating nuclear explosions from earthquakes at teleseismic distances. *European Journal of Applied Sciences, 1*(4), 47–52.

Gil, A., et al. (2014). Crustal structure beneath the Rif Cordillera, North Morocco, from the RIFSIS wide-angle reflection seismic experiment. *Geochemistry, Geophysics and Geosystems, 15*(12), 4712–4733.

Hanks, T. C., & Kanamori, H. (1979). A moment magnitude scale. *Journal of Geophysical Research: Solid Earth, 84*(B5), 2348–2350.

Li, Y., Rodi, W., Toksöz, M. N. (1995). Discrimination of earthquakes, explosions, and mining tremors using the empirical Green's function method. In *Massachusetts Inst of Tech Cambridge Dept of Earth Atmospheric and Planetary*

Michard, A., Saddiqi, O., Chalouan, A., & de Lamotte, D. F. (2008). Continental evolution: The geology of Morocco. In *Structure, stratigraphy, and tectonics of the Africa-Atlantic-Mediterranean triple junction*. Springer

Plafcan, D., Sandvol, E., Seber, D., Barazangi, M., Ibenbrahim, A., & Cherkaoui, T.-E. (1997). Regional discrimination of chemical explosions and earthquakes: A case study in Morocco. *Bulletin of the Seismological Society of America, 87*(5), 1126–1139.

Quick Determination of Bolide Explosion Locations Using Seismic and Optic Data

Jordi Díaz, Josep M. Trigo-Rodríguez, Mar Tapia, and Mario Ruiz

Abstract

Bright bolides entering the atmosphere announce the entrance of meter-sized meteoroids often surviving as meteorites reaching the ground. To be able to find them, it is necessary to locate quickly the position of the final explosion. This is usually done using optical sensors, but the bolides produce shock waves that can be detected by infrasound stations and, after coupling to the ground, also by seismic stations. The increasing availability of temporal seismic station deployments and the gathering of the records on permanent seismic networks can become a powerful tool to complement optical data, in particular for diurnal events or explosions located far from optical stations. We present here two examples of bolides exploding over NE Iberia that were reported in the Spanish Fireball and Meteorite Network (SPMN) catalog with magnitudes − 12 and − 17 and recorded by temporary seismic arrays deployed near their entry trajectory. The location of the final explosions has been derived from the seismic data using a simplistic hypothesis and the codes used systematically for earthquake location. Although locating accurately the origin point of sonic waves is a complex task, our approach allows getting a quick location of the explosion location, with errors estimated in few kilometers that appears as a useful tool to better define the search zone for eventual meteorites.

Keywords

Bolide • Meteoroid • Seismic data • Optical data • Quick location

1 Introduction

In order to find possible meteorites resulting from bolide fragmentation, it becomes necessary to locate quickly the position of the explosion. This is usually done using optical sensors, which allow to identify the trajectory and final explosion of the bolide. However, as the number of such stations is limited, the distance between the explosion and the nearest optical stations might be large and the determination of the meteoroid disruption location can sometimes be not enough accurate. In addition, optical instruments cannot be used in the case of daytime meteoroids.

With the densification of the permanent seismic networks and the increasing availability of temporal seismic station deployments, seismic data can become a powerful tool to complement optical data, as the sensors detect clearly the ground coupling of acoustic waves generated by the explosion. We present here two examples of bolides exploding over NE Iberia, for which the available seismic data provides useful information.

2 Data and Methods

Obtaining a precise location of the fireball explosion using seismic waves is a complex task, as the propagation of sonic waves through the atmosphere is affected by multiple aspects, including the vertical temperature profile, the direction of the prevailing winds at each height level, humidity, and ground topography (Arrowsmith et al., 2010). We propose in this contribution the use of an oversimplified location strategy, based on the methods used routinely for

J. Díaz (✉) · M. Ruiz
Geosciences Barcelona (CSIC), E08023 Barcelona, Spain
e-mail: jdiaz@geo3bcn.csic.es

J. M. Trigo-Rodríguez
Institute of Space Sciences (CSIC), Cerdanyola del Vallés, E08193 Barcelona, Catalonia, Spain

J. M. Trigo-Rodríguez
Institut d'Estudis Espacials de Catalunya (IEEC), 08034 Barcelona, Catalonia, Spain

M. Tapia
Laboratori d'Estudis Geofísics Eduard Fontserè, Institut d'Estudis Catalans, Barcelona, Spain

M. Bezzeghoud et al. (eds.), *Recent Research on Geotechnical Engineering, Remote Sensing, Geophysics and Earthquake Seismology*, Advances in Science, Technology & Innovation,
https://doi.org/10.1007/978-3-031-48715-6_62

earthquake location, to get a quick approximative location of the bolide explosion point. We use the classical Hypo71 (Lee & Lahr, 1975) algorithm, included as a plug-in in the SeisComP data processing package (Helmholtz-Centre Potsdam, 2008), we assume that the signal is generated during the final bolide explosion (source point) and we define an homogeneous model of the atmosphere, with a speed set at 340 m/s. It is important to note that the accuracy of this approach, as always in seismic location problems, is strongly dependent on the number and geographical distribution of the available seismic stations.

3 Results

On September 30, 2019, a fireball entered the atmosphere above NE Iberia at 08:50 UTC. Despite occurring during daylight hours, the entry of the fireball was witnessed by a significant number of observers and its fragmentation generated a burst widely noted by the population. The Spanish Fireball and Meteorite Network identified this fireball with the code SPMN300919, assigning it an absolute magnitude of − 12 ± 2. This magnitude places it among the ten fireballs of greater magnitude detected by the network during the year 2019, being the only daytime event cataloged during that year. The visual observations available indicate that the fragmentation of the bolide occurred along an east–west trajectory, over the area located between the towns of Solsona and Coll de Nargó/Oliana (NE Iberia) (yellow line in Fig. 1a).

This event was recorded by the permanent seismic stations located in NE Iberia (Institut Cartogràfic i Geològic de Catalunya 2000) and some stations located in the northern side of the Pyrenees. However, the most relevant contribution arises from the data acquired by the SANIMS temporary network (Diaz et al., 2023), including 25 broad-band instruments deployed over the NE termination of the Pyrenees.

The location procedure results in a source point located to the SW of the zone covered by the dense seismic network. The height of this source is calculated to be 57 km, but it is poorly constrained. As it is estimated that most fireball fragmentation do occur between 20 and 45 km (Moreno-Ibáñez et al., 2015), we consider that our data does not allow us to constraint properly the explosion height and we decide to fix this value to 35 km to obtain a more accurate location. The final results indicate that the explosion origin time was 08:48:25 and it was located at 41.988°N and 1.511°E, very close to the city of Solsona and approximately 10 km southwards from the trajectory estimated from visual witnesses, with formal errors in latitude and longitude of ± 3 km. It is worthy to note that seismic signals were recorded to distances over 80 km to the north and the east of the origin zone, but were not recorded to the south and west, probably due to the effect of the dominating winds.

On May 23, 2022, a very brightly fireball was detected over NE Iberia and reported by a large number of witnesses, in despite of occurring at 02:45 official time. The Spanish Fireball and Meteorite Network identified this fireball with the code SPMN230522, assigning it an absolute magnitude of − 17 ± 2. This is the brightest object detected by the network since January 2021 and has been observed all around central and southern Iberia. Several of the optical observatories have recorded the final explosion of the bolide, and shown the later trajectory of several fragments. From the optical measurements, it has been stated that the bolide started to bright above the coastline 15 km SW of Barcelona and followed a SSW-NNE trajectory, with the fragmentation point located 75 km further north, near the city of Vic (yellow line in Fig. 2). This event has been recorded by up to 32 seismic stations distributed over NE Iberia, including 10 stations deployed in high schools of the Barcelona metropolitan area in the framework of the SANIMS project. The classical location procedure described above has provided an origin zone located at 41.822°N and 2.911°E, with

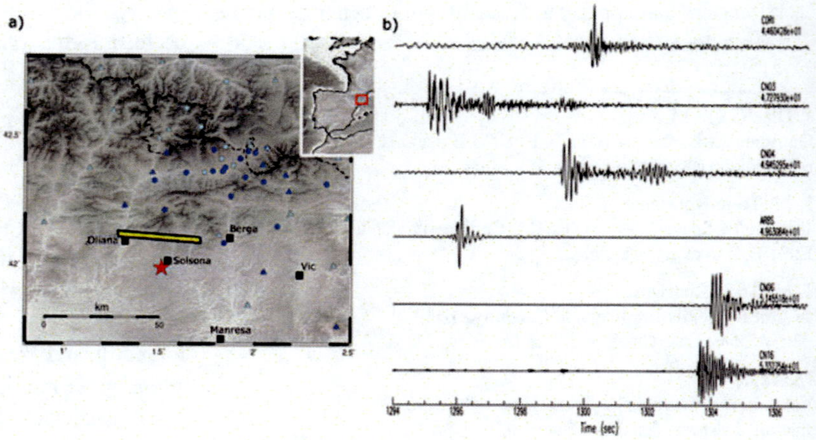

Fig. 1 **a** Map showing the temporary (dots) and permanent (triangles) seismic stations around the 30/9/2019 bolide explosion (red star). Dark blue colors show the stations with clear arrivals. Yellow line shows the trajectory inferred from witnesses. Black squares show the location of nearby towns. **b** Seismic traces at representative sites

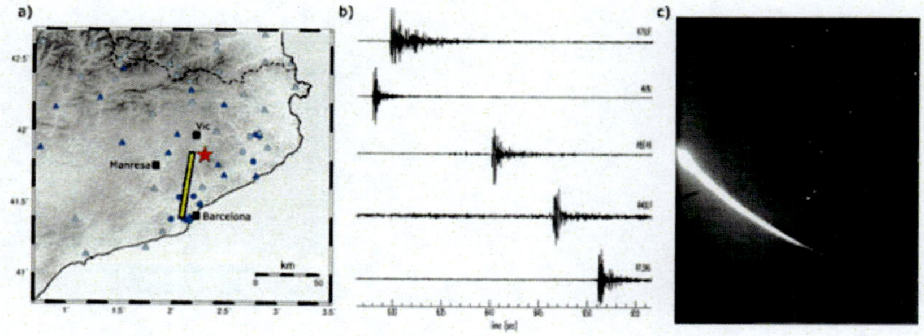

Fig. 2 **a** Map showing the temporary (dots) and permanent (triangles) seismic stations around the 23/5/2022 bolide explosion (red star). Dark blue colors show the stations with clear arrivals. Yellow line shows the trajectory inferred from optical observations. Black squares show the location of the nearby towns. **b** Seismic traces at representative sites. **c** Image from the bolide taken by the all-sky camera from the Observatori Astronòmic del Montsec

a height of 35 km. The formal errors in latitude and longitude are of ± 0.85 km, while for the height the error is estimated in ± 3.2 km. Comparing the results from the optical and seismic locations, their latitude is very similar, while there is a difference of about 15 km in longitude. In this case, the seismic signals are recorded at distances of 130 km by stations distributed all around the explosion, hence resulting in lower error bars. Station located to the SW does not record seismic arrivals associated to the explosion.

4 Conclusions

Our data prove that even if using simplistic hypothesis, seismic data can be a useful tool to quickly locate the explosion point of bolides entering the atmosphere, hence allowing researchers to constrain the zone where eventual meteorites could be found. As we propose the use of a location method implemented in the data managing software packages routinely used by the seismic community, the localization of bolide explosions can be implemented as a standard procedure, as done for example with quarry blast. However, this approach relies on the availability of a large number of seismic instruments around the entry area.

Regarding the two events inspected here, for the September 30, 2019, the seismic data provided key information, as the event occurred in broad daylight without optical recordings. On contrary, the May 23, 2022, bolide was recorded by 21 optical sensors allowing to reconstruct the bolide trajectory and explosion zone, which is consistent with the seismic location, hence confirming the validity of our approach.

Acknowledgements The authors want to acknowledge the Geo3Bcn-CSIC LabSis Laboratory to provide data from temporary deployments. We are grateful to all the meteor observers for sharing their observations through the citizen science project Red de Investigación sobre Bólidos y Meteoritos (SPMN-CSIC): https://www.ice.csic.es/news/citizen-science/2-uncategorised/319-spmn-network-meteorite. We thank Observatori del Montsec IEEC personnel for allsky camera maintenance. This research has been supported by the Spanish Ministerio de Ciencia, Innovación y Universidades grants numbers RTI2018-095594-B-I00 and PGC2018-097374-B-I0.

References

Arrowsmith, S. J., Johnson, J. B., Drob, D. P., & Hedlin, M. A. H. (2010). The seismoacoustic wavefield: A new paradigm in studying geophysical phenomena. *Reviews of Geophysics, 48*, 4.

Diaz, J., Ventosa, S., Schimmel, M., Ruiz, M., Macau, A., Gabàs, A., Akin, O. & Vergés, J. (2023). Mapping the basement of the Cerdanya Basin (Eastern Pyrenees) using seismic ambient noise. *Solid Earth, 14*, 499–514. https://doi.org/10.5194/se-14-499-2023

Helmholtz-Centre Potsdam. (2008). GFZ German Research Centre for Geosciences and gempa GmbH. The SeisComP seismological software package. GFZ Data Services

Institut Cartogràfic i Geològic de Catalunya. (2000). Catalan seismic network. International Federation of Digital Seismograph Networks. Other/Seismic Network. https://doi.org/10.7914/SN/CA

Lee, W. H. K. & Lahr, J. C. (1975). HYP071 (revised); A computer program for determining hypocenter, magnitude, and first motion pattern of local earthquakes. https://doi.org/10.3133/ofr75311

Moreno-Ibáñez, M., Gritsevich, M., & Trigo-Rodríguez, J. M. (2015). New methodology to determine the terminal height of a fireball. *Icarus, 250*, 544–552.

On the Effect of COVID-19 Lockdown on Seismic Detection Capability

Sergio Arévalo, Mario Ruiz, and Jordi Díaz

Abstract

During 2020, people around the world lived a lockdown of different stages consequence of COVID-19 pandemic, switching almost every daily life activity to a remote version. As evidenced by previous studies, the human activity reduction resulted in lower levels of background seismic vibrations mostly for frequencies ranging between 2 and 20 Hz. As this is also the characteristic frequency band for the body-wave arrivals of local events, we could expect an increase on the signal-to-noise ratio during lockdown, resulting in a better detection of such phases, for seismic stations located in urban environments. The deployment in the city of Barcelona of a large seismic network during the lockdown period provides a unique opportunity to check this hypothesis. A manual picking of the local earthquake arrivals reported in the Institut Cartogràfic i Geològic de Catalunya ICGC catalog has been carried out on the waveforms recorded by the permanent and temporary seismic stations located within the city. We have reviewed the seismic events occurring between December 2019 and September 2020, to compare the detection level in the pre-lockdown, lockdown, and post-lockdown time periods. By quantifying the number of picked P-waves on each station and classifying this information according to the hour of day, it has been possible to analyze the correlation between the human common activity and the capacity of seismic stations to record seismic events. First results indicate that the reduction of seismic noise during lockdown resulted in a higher number of pickings at labor hours. However, the limited number of local events and the occurrence of a small seismicity cluster close to Barcelona during the lockdown, makes it difficult to be conclusive on the influence of the seismic noise reduction on the detection capability.

Keywords

Seismic detection capacity • COVID-19 lockdown • Seismic noise

1 Introduction

During the lockdown period due to the COVID-19 pandemic, a temporary seismic array including 14 short-period stations was deployed within the city of Barcelona (Spain), in the framework of the SANIMS project (RTI2018-095594-B-I00). This temporary network added to the 3 accelerometers and the 2 broad-band stations from the ICGC network (Institut Cartogràfic i Geològic de Catalunya, 2000) located in or close to the city center (Fig. 1a). Therefore, up to 19 stations were operating in the city before, during, and immediately after the lockdown period, hence providing a unique opportunity to check the influence of the low level of human activity in the identification of seismic arrivals.

Lockdown measures in Barcelona started on March 15, 2020, and were reinforced between the 20 March and the 12 April. On May 4, the plan to ease the lockdown began, and commercial and industrial activities were gradually resumed. Finally, the State of Alarm expired at midnight on Sunday, 21 June. Figure 1b shows the power amplitude in the 2.0–20.0 Hz range at the ICJA broad-band station, evidencing an abrupt lowering following the lockdown measures. Similar results have been reported for locations around the world (Lecocq et al., 2020; Pérez-Campos et al., 2021). An increase in the perception of seismic events by the population has been reported in areas as Central America (Arroyo-Solórzano et al., 2021).

S. Arévalo
Univ. Barcelona, E08023 Barcelona, Spain

M. Ruiz · J. Díaz (✉)
Geosciences Barcelona (CSIC), E08023 Barcelona, Spain
e-mail: jdiaz@geo3bcn.csic.es

M. Bezzeghoud et al. (eds.), *Recent Research on Geotechnical Engineering, Remote Sensing, Geophysics and Earthquake Seismology*, Advances in Science, Technology & Innovation,
https://doi.org/10.1007/978-3-031-48715-6_63

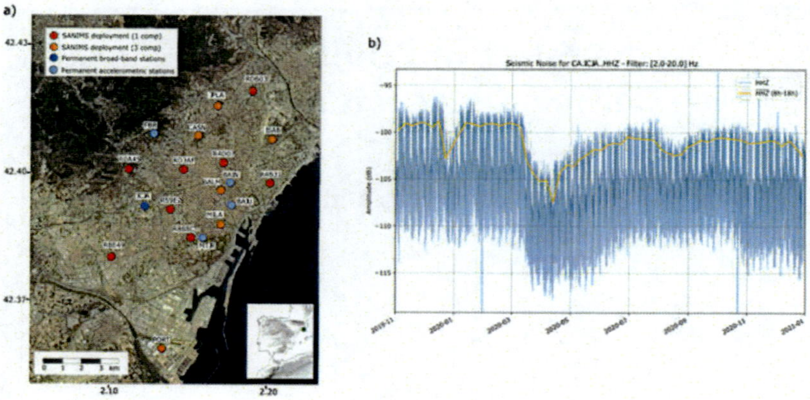

Fig. 1 **a** Map of the stations deployed within Barcelona. **b** Seismic power amplitude in the 2–20 Hz band between 11/2019 and 01/2021 measured every 30 min at ICJA station. Blue line shows the values measured each 30 min, while orange line shows the weekly mean values recorded between 8 and 18:00 UTC (adapted from Diaz et al. 2021)

2 Data and Results

2.1 Seismicity Patterns

In a first step, we reviewed the local seismic activity reported in the ICGC catalog to compare the detection level during three 100-day periods representing the pre-lockdown (December 2019–March 2020), lockdown (March–June 2020), and post-lockdown (June–September) time periods. Figure 2, shows the seismicity maps and the number of events per day during these 100 days periods, evidencing that there is not an overall change in seismicity between them. The local seismicity level in NE Iberia is low, and it tends to concentrate in clusters of irregular occurrence, hence affecting strongly statistical studies on seismic occurrence. As an example, during the lockdown period, a small cluster in front of the coast of Tarragona (1.4°E) did occur, while another cluster, located NE of Barcelona has been active during the whole period, with larger events during the lockdown and post-lockdown periods.

We have also verified if the magnitude threshold for the ICGC was lowered during the lockdown period, taking as a reference a large time period starting in 2010, without observing any effect. Therefore, seismic catalogs in this region are not a good option to infer the potential impact of the human activity in seismic detection. This is not surprising, as permanent seismic stations tend to be installed as far as possible of populated area, and the seismic noise of anthropogenic origin, characterized by their high-frequency content, attenuates at short distances, in the orders of 100 m.

2.2 Seismic Arrivals Observations in Barcelona Stations

The next, and more relevant, step has been to focus in the stations located within the city. In the first stage, we have focused on the events reported in the ICGC catalog with magnitude > 1.0. For those events, the waveforms of the 19 Barcelona stations have been extracted and inspected manually to pick P-wave arrivals. In a second step, we have also inspected the event with magnitude $0.0 < m < 1.0$ and epicenters close to Barcelona to check if additional arrivals could be identified.

We have reviewed 262 events, but only for 71 of them we have been able to identify clear P-wave arrivals. Regarding individual traces, more than 4500 have been inspected, but only 385 have been retained. Figure 3 shows the number of valid readings identified each day in the Barcelona stations during the pre-, lockdown, and post-lockdown 100 days long periods. As observed, this number is highly variable, as depends directly of the number, magnitude, and epicentral distance of the earthquakes. The lockdown period (gray shading) does not seem to have a pattern different from the rest of the investigated periods (Fig. 3a). For days with detections, the number of readings is usually ranging between 12 and 25, proving that relative large earthquakes are detected in the network independently of the background noise level. However, during the lockdown period up to 10 days show reading values around 5, indicating that only a few of the 19 stations provided clear lectures. We interpret this as denoting that the background noise reduction during lockdown made it possible to detect, on a limited number of the Barcelona stations, week arrivals that would not be

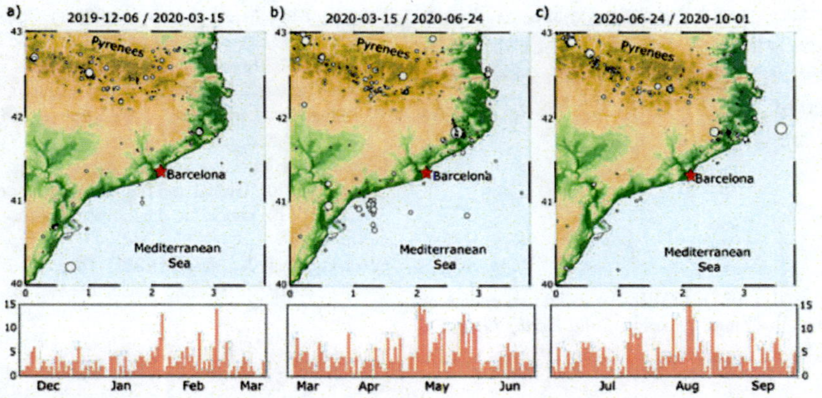

Fig. 2 Seismic activity in NE Iberia as reported by the ICGC catalog. Upper panels show the location of the event occurring before (**a**), during (**b**), and after (**c**) the lockdown period. The size of the dots is proportional to the assigned magnitude. Lower panels show the number of per-day events

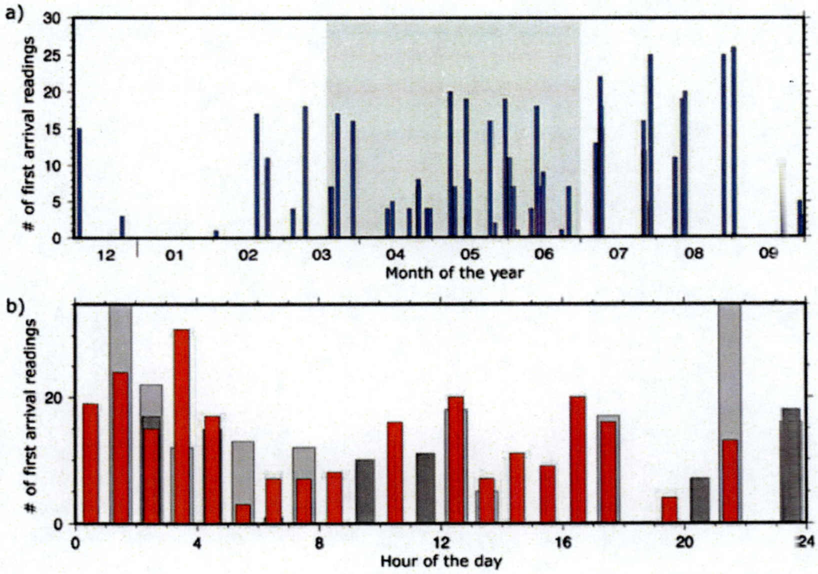

Fig. 3 **a** Daily number of seismic arrival lectures in the stations located in Barcelona between November 2019 and December 2020. The lockdown period is marked by a gray background. **b** Bars show the number of phase arrival readings per hour in the seismic stations located in Barcelona, during three 100-day long intervals before (light gray), during (red), and after (dark gray) the lockdown period

recorded otherwise. This is therefore our first evidence of an enhancement in the detection capability of seismic stations during the COVID-19 lockdown.

Next, we have quantified the number of picked P-waves on each station and classifyied this information according to the hour of day (Fig. 3b). We consider here as labor hours the period between 8:00–20:00 official time. During the pre- and post-pandemic 100-day long intervals, a total of 126 valid readings have been done during labor hours, in front of a total of 385 readings (32%). During the lockdown period, 121 readings were identified during labor hours, out of a total 195 (62%). Therefore, there is a clear increase in the ratio of phase arrivals identified in Barcelona during the working hours of the days with lockdown measures, that we interpret as resulting from the lower level of seismic background noise due to the decrease in human activity.

3 Conclusions

From the inspection of the ICGC catalog in the NE Iberia region, there are no indications of an enhancement of the seismic detection capability related to the background noise lowering following the COVID-19 lockdown measures. However, by quantifying the number of picked P-waves on the large number of seismic stations deployed within the city of Barcelona it has been possible to evidence a significant increase in the number of phase arrivals detected during labor hours. Additionally, more events providing a scarce number of lectures were detected during lockdown times. This means that their weak signals were only detectable in the best Barcelona stations due to the general background noise reduction. However, the effect of lockdown must be considered as moderate, as we have not been able to identify previously undetected seismicity. Similar studies should be carried out on other regions to check if our conclusions regarding the effect of COVID-19 lockdown on seismic detection capability are of general application.

References

Arroyo-Solórzano, M., et al. (2021). COVID-19 lockdown effects on the seismic recordings in Central America. *Solid Earth, 12*, 2127–2144.

Diaz, J., Ruiz, M., & Jara, J. A. (2021). Seismic monitoring of urban activity in Barcelona during the COVID-19 lockdown. *Solid Earth, 12*, 725–739.

Institut Cartogràfic i Geològic de Catalunya (2000) *Catalan Seismic Network. International Federation of Digital Seismograph Networks*. Other/Seismic Network

Lecocq, T., Hicks, S. P., Noten, K., van Wijk, K., van Koelemeijer, P., et al. (2020). Global quieting of high-frequency seismic noise due to COVID-19 pandemic lockdown measures. *Science, 369*, 1338–1343.

Pérez-Campos, X., et al. (2021). The effect of confinement due to COVID-19 on seismic noise in Mexico. *Solid Earth, 12*, 1411–1419.

Hydro-Seismicity Triggered by Heavy Rainfall

Behrooz Bazargan Sabet, André Burnol, and Antoine Armandine Les Landes

Abstract

Seismic hazard assessment in stable continental regions ("intraplate domain") involves specific difficulties, due to the lower recurrence of events compared to active zones, and to the fact that the triggering processes are particularly complex and still poorly understood. Several recent studies emphasize the role of local processes such as erosion, deglaciation, fluid circulation, etc. The concept of "hydro-seismicity" (Costain, Geol Soc Lond Spec Publ 432:91, 2017) explains a possible mechanism, the triggering of intraplate seismicity by the pressure increase in the interstitial fluid, decreasing the effective stress within the fault plane. Although climate change may not be perceptible at the level of the annual volumes of precipitation, the alternation of drought periods with intense rainy episodes may constitute a significant transient water supply, reaching shallow faults by infiltration into the subsoil and inducing the phenomenon of hydro-seismicity. In this context, we have shown that the local magnitude 5.4 earthquake of November 11, 2019, in Le Teil (France) could have been triggered by the transient increase in hydraulic pressure following heavy rainfall in the days preceding the event, while the trigger of the event was initially attributed to the activities of a quarry located near the earthquake zone. Using soil moisture data acquired by the SMOS satellite and modeling the height variations of the unsaturated zone by the ComPass code developed by the Bureau de Recherches Géologiques et Minières (BRGM), we show that water infiltration has induced hydraulic overpressures. The maximum value is located at the intersection of a system of faults at 1200 m depth; the hypocenter of the earthquake. The impact of this overpressure has been estimated to be several times greater than the cumulative effect of mechanical stress release due to the exploitation of the quarry over the two past centuries.

Keywords

Intraplate earthquake • Hydro-seismicity • Triggered seismicity • Hydraulic modeling • Mechanical modeling

B. B. Sabet (✉) · A. Burnol · A. A. L. Landes
Bureau de Recherches Géologiques Et Minières (BRGM), Orléans, France
e-mail: b.bazargan-sabet@brgm.fr

1 Introduction

On November 11, 2019, at 11:52 a.m., the Le Teil village in the Rhône Valley was violently shaken for 5–6 s by an earthquake of local magnitude 5.4 (Mw 4.9). The first investigations have shown that the earthquake is due to the reactivation of the La Rouvière fault, a shallow fault that originated from the Oligocene period, which was not considered as active so far. This reverse fault caused an average ground displacement of 10 cm. The earthquake hypocenter was estimated at around 1 km depth.

While the presence of the LafargeHolcim quarry in the epicentral zone, exploited since 1833, raised the question of possible anthropogenic origin of this earthquake (Larroque et al., 2020), we put forth another potential triggering source that is the hydraulic effect related to heavy rainfall after drought period, on a network of faults under long-term tectonic stress loading.

2 Methods

We rely on a local updated 3D geological model of the area built by compiling existing cartographic data, deep drilling data, 240 km of seismic line including 8 km of new acquisition and surface traces of local faults network studied through the differential SAR interferometry (DInSAR) analysis. The seismic interpretation and the construction of the 3D geological model reveal the presence of a positive

M. Bezzeghoud et al. (eds.), *Recent Research on Geotechnical Engineering, Remote Sensing, Geophysics and Earthquake Seismology*, Advances in Science, Technology & Innovation,
https://doi.org/10.1007/978-3-031-48715-6_64

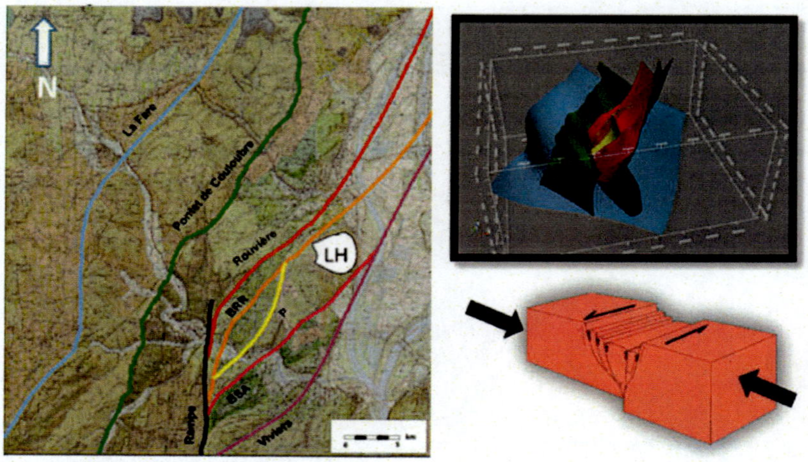

Fig. 1 Geological 3D model faults position

flower structure, limited to the north by the Rouvière fault, dipping south-east, and to the south by the Viviers fault (not detected before) dipping steeply to the northwest. Inside this blooming structure, there are steeply dipping secondary faults, from the NW to the SE: Bayne Rocherenard (BRR), Paurière (P), and Bayne Saint-Alban (BSA), delimiting different blocks (Fig. 1).

Based on the 3D geological model and using the BRGM ComPass code, we build a 3D hydrogeological model. The hydrostatic equilibrium with the Rhône River, located in the eastern part of the Le Teil, allows setting the boundary condition. The initial calculation defines the unsaturated zone of the model. The result of the equilibrium is then used to simulate, over the period 2010–2019, the dynamic recharge linked to the re-saturation of the vadose zone according to the variations in surface soil humidity acquired by the SMOS satellite. The main input data are shown in Fig. 2. The results of these simulations allow evaluating, for different geological configurations and boundary conditions, the order of magnitude of the pressure variations along the Rouvière fault, which is identified as the origin of the earthquake.

The results of our hydraulic model are compared with the results obtained from 3D mechanical model built on the same geometry, to assess the effect of mass removal during the quarry exploitation, by evaluating the Coulomb stress change. The calculations are performed using Itasca 3DEC code.

3 Results

3.1 Results of Hydraulic Model

According to the "hydro-seismicity" concept, the intersections between two or multiple faults are the most probable location zones for the hypocenter of an earthquake triggered by hydraulic recharge (Costain, 2017). The results of the 3D hydraulic simulations show that the variation in hydraulic pressure in the fault system between 2018 and 2019 reached a minimum value on September 24, 2019, (on the eve of intense rainy episode) after a drought period of 6 months; the longest period since 2015. After the recharge period, the differential pressure (ΔP) reaches the

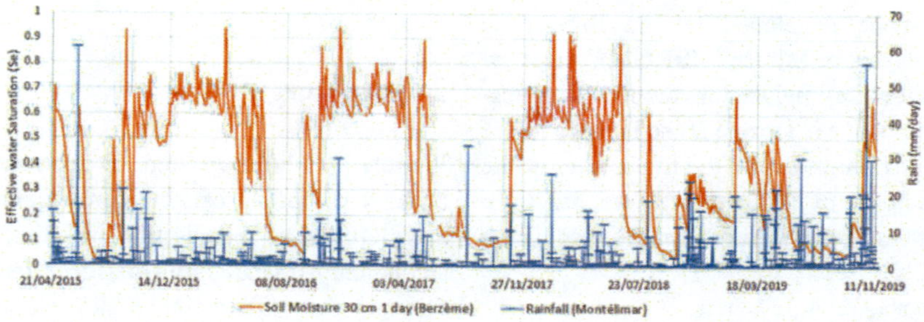

Fig. 2 Input data: Precipitation (rainfall) in blue line. In red line, effective saturation (Se) calculated from *in situ* soil moisture at 30 cm depth acquired every 3 days

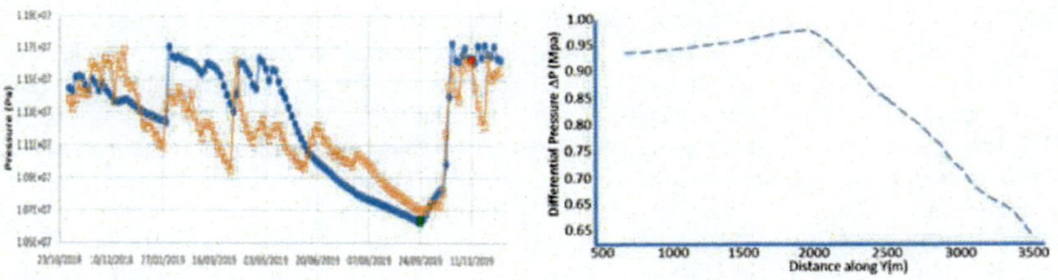

Fig. 3 Hydraulic pressure on the faults system (left). Maximum ΔP along Y axis of the hydraulic 3D model (right). Maximum value at faults intersection

maximum value of 0.98 MPa, near the junction of Rouvière, Bayne Rocherenard et Paurière faults system, at 1200 m depth (Fig. 3).

3.2 Results of Mechanical Model

We use the distinct element method to assess the impact of mass removal. Our results show the maximum Coulomb stress change of 0.25 MPa, a value of the same order as what is obtained with Boussinesq's analytical solution (Ampuero, et al., 2019). A secondary maximum value is located near the intersection of the Rouvière fault with Bayne Rocherenard fault at 1.3 km depth (Fig. 4). To deal with the uncertainties, we perform a parametric study taking plausible values for each parameter. We show that the uncertainty associated with our calculations affects the results within a range of < 10%.

4 Discussion

The origin of the earthquake of November 2019 is not well known and in particular, tectonic origin is not excluded. However, if the earthquake is "triggered" by an external event, it is more likely that the trigger mechanism be linked to the hydraulic loading due to an intense rainfall event than to the mechanical effect due to the mass removal during the quarry exploitation. In fact, the hydraulic overpressure linked to the rainfall is several times greater than the mechanical stress change due to the quarry excavation. Furthermore, the hydraulic overpressure reaches its maximum just before the earthquake following a dynamic pattern while the mass extraction is a quasi-static process acting over two centuries.

Another clue is the position of the earthquake hypocenter, which has been relocated using multiple approaches (Delouis et al., 2021). This position is seemingly very close to the location of the maximum ΔP obtained from our hydraulic simulations. The accordance between the seismological analyses and our modeling results (Burnol, et al., 2022) suggests that the intersections between secondary faults might have played an important role for the nucleation process of the Le Teil earthquake.

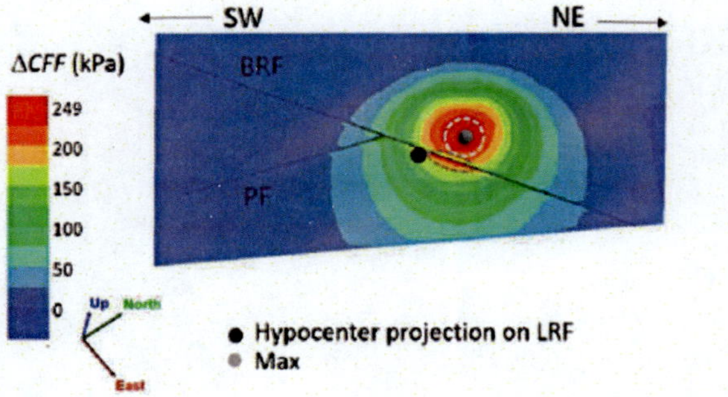

Fig. 4 Mechanical model and the maximum value of Coulomb stress. Black point represents the hypocenter position, estimated by Ritz et al. (Ritz et al., 2020)

5 Conclusions

Two numerical models have been developed to simulate the potential hydraulic vs. mechanical triggering factor of the Le Teil earthquake. Our results suggest large increase in fluid pressure occurring during a shift from long drought period to heavy rainfall, as the triggering origin of the event. With climate change, it is now clear that such an abrupt transition scenario will occur more and more frequently in the future. It is therefore necessary to re-evaluate the intraplate seismic risk by taking into account the possible reactivation of shallow faults, which until now have been classified as slightly or no active.

References

Ampuero, J. P., et al. (2019). *Rapport d'évaluation du groupe de travail (GT) CNRS-INSU sur le séisme du Teil du 11 novembre 2019 et ses causes possibles*

Burnol, A., et al. (2022). Double surface rupture and hydraulic recharge of a three-fault system during the Mw 4.9 earthquake of 11 November 2019 at Le Teil (France)

Costain, J. K. (2017). Groundwater recharge as the trigger of naturally occurring intraplate earthquakes. *Geological Society, London, Special Publications, 432*, 91.

Delouis, B., et al. (2021). Constraining the point source parameters of the 11 November 2019 Mw 4.9 Le Teil earthquake using multiple relocation approaches, first motion and full waveform inversions. *Comptes Rendus Géoscience, 353*, 1–24.

Larroque, C., Ampuero, J.-P., Delouis, B., & Cornou, C. (2020). Aux origines du séisme du Teil. *La Recherche, 561*, 94–97.

Ritz, J. F., Baize, S., Ferry, M., et al. (2020). Surface rupture and shallow fault reactivation during the 2019 Mw 4.9 Le Teil earthquake. *France. Communications Earth and Environment, 709*(1), 10.

Effect of Earthquake-Induced Pounding on Adjacent Building Structures with Unequal Number of Stories

Pedro Folhento, Rui Carneiro de Barros, and Manuel Braz-César

Abstract

Structural pounding among insufficiently separated buildings constitutes one of the main reasons for damage and the possible collapse of structures during severe earthquakes. Building pounding is still not properly understood, as it involves several complex phenomena, leading to unclear patterns for the dynamic behaviour of the colliding structures. Hence, this study considers floor-to-floor collisions between fixed-base reinforced concrete buildings under earthquake excitations. The different number of stories and separation distances are considered to analyze the results in terms of story shear forces and interstory drifts. Finite element models are constructed, considering a fibre-based approach to model the inelasticity at the ends of structural members. Structural pounding was simulated through the use of a viscoelastic impact model, the Kelvin-Voight model. Results showed that the stories above the points of impact experienced 50% to almost 100% increases in interstory drifts and shear force demands, which may lead to brittle failures of the columns. The number and magnitude of pounding forces were higher for upper stories. The case with no gap size did not always lead to the worst scenario, and a separation distance close to the no-pounding case still presented pounding forces of significant magnitude, emphasizing the importance of proper separation of building structures. In conclusion, the buildings in the scenarios with an unequal number of stories were more vulnerable to pounding forces, experiencing higher increases in story shear and interstory drift.

Keywords

Seismic analysis • Building pounding • Impact models • Non-linear inelastic behaviour • Parametric investigation

1 Introduction

The influence of earthquake-induced structural pounding among buildings is a paramount subject in the seismic analysis and design of structures. This phenomenon leads to unclear patterns or trends of the colliding structure's dynamic responses which explain contradictions in research results (Mahmoud et al., 2019). The recognition of such phenomenon has been growing in the last decades, contributing to a better understanding of building structural pounding and mitigation of its negative consequences. Constituting the main goals in the study of such occurrences in seismic events. Several mitigation measures and techniques have been proposed over the years in the literature to attenuate or eliminate the negative consequences of such phenomenon (Folhento et al., 2020).

This study comprises an investigation of the effect of earthquake-induced structural pounding on the response of adjacent reinforced concrete (RC) structures with variable gap distance and number of stories. Pounding is considered to happen among floors, and five scenarios for the RC buildings in terms of the number of stories will be assessed. This will allow the understanding of how different number of stories and the separation of structures may influence the structures' behaviour, important for the safety of human lives and services during a seismic event in the occurrence of pounding.

P. Folhento · R. C. de Barros (✉)
CONSTRUCT, Faculdade de Engenharia, Universidade do Porto, Porto, Portugal
e-mail: rcb@fe.up.pt

P. Folhento
e-mail: up201811645@edu.fe.up.pt

M. Braz-César
CONSTRUCT, Instituto Politécnico de Bragança—ESTiG, Bragança, Portugal
e-mail: brazcesar@ipb.pt

M. Bezzeghoud et al. (eds.), *Recent Research on Geotechnical Engineering, Remote Sensing, Geophysics and Earthquake Seismology*, Advances in Science, Technology & Innovation,
https://doi.org/10.1007/978-3-031-48715-6_65

2 Problem Formulation

This parametric investigation is based on five scenarios of adjacent RC structures with variable separation distance: L3R3, L3R4, L3R5, L4R3, and L5R3, in which “L” stands for building 1, “R” for building 2, and the number indicates the number of stories of the buildings. Building 1 presents a more flexible layout compared with building 2.

The six different fixed-base RC structures were designed according to the Eurocodes, using class C25/30 concrete and steel rebars of S500. The seismic design was performed following capacity design considerations, assuming Portimão, Portugal’s seismic region.

The finite element models are built in OpenSees (McKenna, 2011) using the fibre model with finite-length lumped plasticity at the critical sections of plastic hinge formation.

The pounding forces generated from the collisions between the adjacent structures are calculated using the Kelvin-Voight impact model (Anagnostopoulos, 2004) as shown in Fig. 1.

3 Results

Non-linear time-history analyses were carried out, considering three seismic actions adjusted to the seismic region of Portimão, Portugal: El Centro (RSN6), Loma Prieta (RSN811), and Northridge (RSN949) (PEER, 2020). The results are presented in Figs. 2 and 3 in terms of the number and magnitude of collisions, interstory displacements and story shear forces, for the five pounding scenarios and across all the values of the gap size.

4 Discussion

Observing Fig. 2, it can be verified that a scenario that presents a greater difference in the number of stories will be more susceptible to more collisions and of a larger magnitude than a scenario with an equal number of stories. The number of collisions and their magnitude naturally depend on the gap size. Nevertheless, a zero-gap size does not always present the highest magnitude of pounding forces,

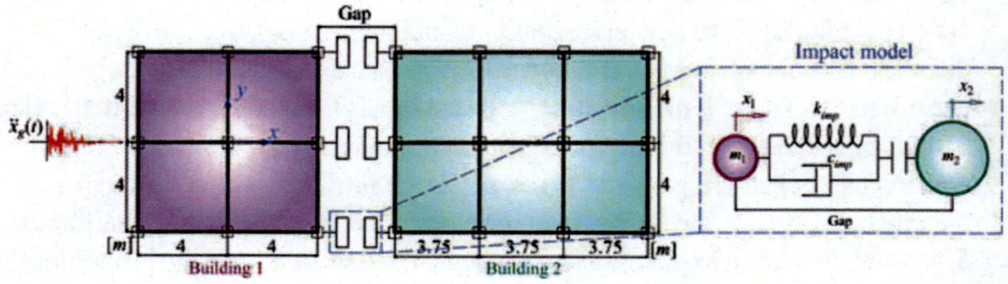

Fig. 1 Plan view of the adjacent RC building structures and the impact model

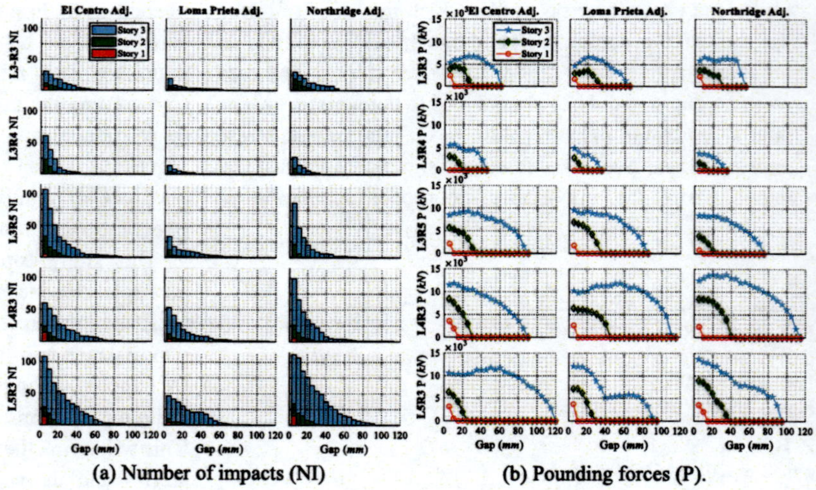

Fig. 2 Results regarding the number and magnitude of pounding forces

Fig. 3 Normalized (N) results of building 1 concerning the no-pounding case

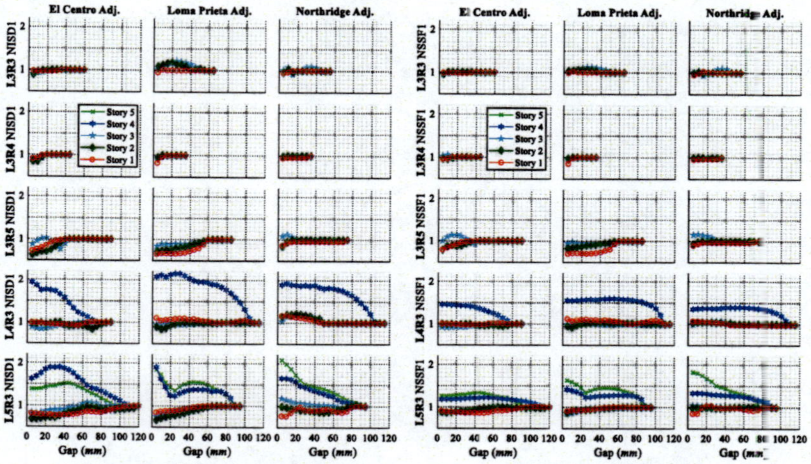

(a) Interstory displacements (ISD) (b) Story shear forces (SSF).

though it is always the case with a greater number of impacts. Figure 3 corroborates with these results, showing that interstory displacements and story shear forces can be amplified, particularly, in scenario L5R3, in which the taller building was the most flexible in the original configuration (L3R3). When building 2 is the taller structure, interstory displacements and story shear forces of building 1 are reduced for some gap sizes. Building 2 presented smaller variations in interstory displacements and story shear forces (not shown due to limited space) and fewer amplifications due to pounding forces.

5 Conclusions

In conclusion, it was verified that scenarios with a different number of stories and more differences in dynamic behaviour are more susceptible to pounding effects (number and magnitude of collisions). Top stories above the point of impact suffered bigger amplifications of the responses studied, which reached twice the case of no-pounding. The range of the gap size in which pounding occurs was naturally wider for the cases of scenarios with an unequal number of stories, and gap sizes closer to the no-pounding case still presented pounding forces of significant magnitude that could influence the structures' behaviour, emphasizing the importance of proper separation of buildings.

Acknowledgements This paper is within the scope of the first author's Ph.D. degree in progress, financially supported by the Portuguese Foundation for Science and Technology (FCT) through the PhD grant reference SFRH/BD/139570/2018 under the programme POCH (N2020–P2020) and subsidized by the European Social Fund (FSE) and national funds from MCTES. This work was financially supported by: Base Funding—UIDB/04708/2020 of the CONSTRUCT—Instituto de I&D em Estruturas e Construções—funded by national funds through the FCT/MCTES (PIDDAC).

References

Anagnostopoulos, S. (2004). Equivalent viscous damping for modelling inelastic impacts in earthquake pounding problems. *Earthquake Engineering and Structural Dynamics, 33*(8), 897–902.

Folhento, P., Barros, R., & Braz-César, M. (2020) Mitigation of earthquake-induced structural pounding between adjoining buildings—State-of-the-Art. In J. Gonçalves, M. Braz-César, & J. Coelho (Eds.), *CONTROLO 2020. Lecture Notes in Electrical Engineering*, Vol. 695. Springer

Mahmoud, M., Choong, K., & Jankowski, R. (2019). Seismic pounding between adjacent buildings: Identification of parameters, soil interaction issues and mitigation measures. *Soil Dynamics and Earthquake Engineering, 121*, 135–150.

McKenna, F. (2011). OpenSees: A framework for earthquake engineering simulation. *Computing in Science and Engineering, 13*(4), 58–66.

PEER. (2020). Pacific Earthquake Engineering Research Center (PEER) strong ground motion data base. https://peer.berkeley.edu/peer-strong-ground-motion-databases

Check for updates

Evaluation of Liquefaction Potential Through Geophysical Tests Case: Earthquake in Lagunas, Peru 2019

Jorge Alva, Carmen Ortiz, Joao Rengifo, and Grover Riveros

Abstract

On May 26, 2019, at 02:41 h (Peruvian time), the Mw 8.0 Sismo Lagunas 2019 earthquake occurred. It had its epicenter at the coordinates: − 05.74°, − 75.55°, 60 km south of the town of Lagunas (Loreto), at a depth of 135 km with Mw 8.0. It caused a lot of damage, including the liquefaction phenomenon reported in the woody village of Tamarate. The maximum horizontal acceleration recorded was at the UNTRM Station (0.098 g) in Chachapoyas with an epicentral distance of 292.21 km. In December 2019, geophysical tests were conducted in the woody village center of Tamarate, the area where the soil liquefaction event was evidenced. The reasons for choosing to use multi-channel surface wave and ambient vibration tests are based on (i) the fact that the area is only accessible by river. And (ii) the used geophysical equipments are practical and easy to handle and transport compared to the usual equipment such as the Standard Penetration Test and Cone Penetration Test for liquefaction potential evaluation. Multi-channel surface wave and ambient vibration analysis tests were performed to determine the shear wave velocity profile and the fundamental period of the ground. Seismic profiles and average wave velocities were Vs30 (163–168 m/s). The predominant period values obtained in the study area were 0.5 and 0.7 s. The occurrence of soil liquefaction (FSL) was evaluated using the deterministic method as the ratio of the minimum strength normalized cyclic shear stress producing liquefaction or "Cyclic Resistance Ratio" (CRR) divided by the normalized cyclic shear stress induced by the earthquake or "Cyclic Stress Ratio" (CSR). The CRR parameter was determined at each depth up to 30 m. With this evaluation, the actual minimum acceleration value that occurred in Tamarate was found between 0.1 and 0.12 g. A soil liquefaction was generated up to a depth of 13.2 m.

Keywords

Liquefaction • Acceleration • Geophysical test • Microtremor

1 Introduction

A seismic event occurred on May 26, 2019, in Alto Amazonas (Loreto-Peru with coordinates − 05.74°, − 75.55°), named "Lagunas Earthquake." The epicenter was 60 km south east of Lagunas with a depth of 135 km and a magnitude of 8 Mw. The maximum horizontal acceleration recorded at UNTRM Station (0.098 g) from the city of Chachapoyas, which is located at an epicentral distance of 292.21 km.

The Sismo Lagunas 2019 earthquake caused the soil liquefaction in the woody village of Tamarate, located 112 km northwest of the epicenter. In fact, liquefaction is a secondary seismic hazard that can be triggered in water saturated sandy soils and results in several damage of anthropogenic and natural habitats. Hence a detailed assessment of the soil liquefaction in Tamarate is needed for prevention and mitigation purposes.

According to De la Cruz Bustamante et al. (1999), Tamarate is 100 m above sea level on the banks of Huallaga River and has a geology belonging to the Cenozoic era, Quaternary system, Holocene series with alluvial deposits defined as seasonal floodplain, these deposits are masses of sandstones and gravels in a sandy matrix.

Accessibility to Tamarate and Lagunas is by river from the city of Yurimaguas in ∼ 7 h by speedboat. This was one

J. Alva · C. Ortiz (✉) · J. Rengifo · G. Riveros
CITDI of Universidad Nacional de Ingeniería, Lima, Peru
e-mail: cortiz@uni.edu.pe

M. Bezzeghoud et al. (eds.), *Recent Research on Geotechnical Engineering, Remote Sensing, Geophysics and Earthquake Seismology*, Advances in Science, Technology & Innovation,
https://doi.org/10.1007/978-3-031-48715-6_66

of the main reasons why it was decided to evaluate the liquefaction potential by means of geophysical tests using heavy equipment.

2 Methods

The deterministic method described in Seed and Idriss (1971) and (Andrus & Stokoe, 2000) and Multi-channel Analysis of Surface Waves (MASW) tests were performed. Four microtremor tests allowed obtaining the profiles of shear wave velocities (Vs) and the predominant periods of the soil (T) to determine the dynamic parameters of the study area (Fig. 1a).

The H/V Spectral Ratios obtained from the measurement of the microtremor tests were compared with the H/V of the Rayleigh waves for the fundamental mode of the profiles obtained from the MASW tests. Some test pits were excavated in the studied site to identify the depth of the water table, which was found to be 2.3 m and to sample soil in order to determine the percentage of fines content and classify the samples using SUCS. To determine the liquefaction potential, we relied on the relationship between the Cyclic Resistance Ratio CRR and the Cyclic Stress Ratio CSR presented in Eq. (1):

$$FS_L = \frac{CRR_M}{CSR} \tag{1}$$

To determine CRR_M, Eq. (2) presented by Andrus and Stokoe (2000) was applied.

$$CRR_M = \left(a\left(\frac{Vs_1}{100}\right)^2 + b\left(\frac{1}{Vs_1^* - Vs_1} - \frac{1}{Vs_1^*}\right)\right) MSF \tag{2}$$

To determine the CSR value, the Eq. (3) formulated by Seed and Idriss (1971) was applied.

$$CSR = \frac{\tau_{av}}{\sigma'_{v0}} = 0.65 * \left(\frac{a_{\max}}{g}\right) * \left(\frac{\sigma_{v0}}{\sigma'_{v0}}\right) * r_d \tag{3}$$

3 Results

The two MASW tests showed that the shear wave velocities Vs in the first 30 m are on average 163 and 168 m/s. Figure 1b shows the profile obtained from the conjoint inversion using both the dispersion curve and the H/V spectrum microtremors. The microtremor's H/V spectral ratio performed in the studied site estimates the spectral ratio performed in the studied site estimate the spectral ratios of 0.5–0.7 s.

Likewise, it is observed that the predominant period in the area where the liquefaction phenomenon occurred in Tamarate is 0.63 s (Fig. 1c). With the Vs results obtained initially, Eqs. (2), (3), and (1) were applied to obtain the values of CRR, CSR, and FSL at each depth where there is a record.

However, it was necessary to previously determine the densities of the strata for the effective stress values, for which the correlations presented by Anbazhagan et al. (2016) were applied. Finally, for the seismic values requested to determine CSR, data from the Mw 8 Lagunas 2019 earthquake presented by Tavera (2019) and acceleration data recorded at stations of the UNI Accelerographic Network (UNI, 2016) were used. The results are shown in Table 1.

4 Discussion

The soil Liquefaction is considered if the FSL value is < 1. In the studied site, FSL values are < 1 under a depth of 13.2 m, which coincides with the inspection carried out after the Lagunas earthquake.

On the other hand, considering that in the study area, there was evidence that the liquefaction occurred, the acceleration values used for this investigation were analyzed. It was found that the minimum value of acceleration in Tamarate, that can trigger the soil liquefaction was

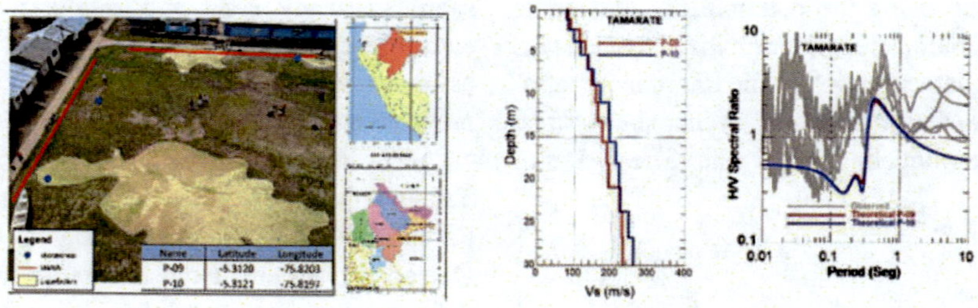

(a) (b) (c)

Fig. 1 Location of tests performed and areas where liquefaction occurred (a). Shear wave velocity profile (b) and observed and theoretical H/V spectral ratio (c)

Table 1 Assessment of liquefaction potential with acceleration $a = 0.12$ g at P09

Depth (m)	Vs (m/s)	CRR	CSR	FSL
1.1	82.07	0.0406	0.063	0.643
2.3	89.11	0.0411	0.063	0.656
3.7	113.65	0.0635	0.081	0.783
5.3	128.51	0.0768	0.094	0.819
8.9	145.30	0.0831	0.107	0.775
11.0	158.25	0.0948	0.106	0.897
13.2	163.85	0.0929	0.102	0.913
15.6	175.74	0.1040	0.096	1.085

0.12 g. It is observed that the acceleration value directly affects the CSR value and the FSL. It is worth mentioning that data was recorded up to 30 m, but results are presented up to the first 15.6 m because under this depth the FSL result increases.

Likewise, when comparing the H/V Spectral Ratios obtained from microtremors with the H/V of the Rayleigh waves of the inverted profiles of the MASW tests (Fig. 1c), a similar fundamental period of the soil was observed. It also allowed to validate the information used from the soil profiles. Also, the obtained predominant period of 0.63 s confirms that the study area is associated with flexible soil deposits.

5 Conclusions

The evaluation of soil liquefaction using shear wave velocity measurements presents reliable results and it is very useful in areas of difficult accessibility, allowing for the evaluation of a greater area and greater depth.

The FSL values obtained are < 1 under a depth of 13.2 m, which coincides with the inspection carried out in Tamare. These results allowed the generation of a set of knowledge and experiences on the dynamic properties of the soil to the scientific community worldwide.

The minimum acceleration value in Tamarate, which caused the liquefaction phenomenon was 0.12 g. The predominant period obtained of 0.63 s confirms that the study area is filled with flexible soil deposits.

References

Anbazhagan, P., Uday, A., Moustafa, S. S., & Al-Arifi, N. S. (2016). Correlation of densities with shear wave velocities and SPT N values. *Journal of Geophysics and Engineering, 13*(3), 320–341.

Andrus, R. D., & Stokoe, K. H. (2000). Liquefaction resistance of soils from shear-wave velocity. *Journal of Geotechnical and Geoenvironmental Engineering ASCE, 126*(11), 1015–1026.

De la Cruz Bustamante, N. S., Fídel Smoll, L., Lipa Salas, V. R., Zuloaga Gastiaburu, A. D., Cavero Loayza, R., Ticona Turpo, P., et al. (1999). Geología de los cuadrángulos de Lagunas

Seed, H. B., & Idriss, I. M. (1971). Simplified procedure for evaluating soil liquefaction potential. *Journal of the Soil Mechanics and Foundations Division ASCE, 97*(9), 1249–1273.

Tavera, H. (2019). *Sismo de Lagunas del 26 de mayo del 2019 (M8.0): Aspectos sismológicos*. Instituto Geofísico del Perú

UNI. (2016). *Red Acelerográfica UNI Homepage*. http://www.red-acelerografica-peru.uni.edu.pe/es/page/report/371. Accessed 16 Aug 2022

Residential Buildings Made with Local Materials and Their Classification on the Basis of a Field Experiment

Zhanybek Mamatov, Sagynbek Orunbaev, Yzat Sydykov, and Nurlan Shamshiev

Abstract

The high seismicity of the territory imposes special requirements on the construction industry of the Kyrgyz Republic; it is necessary to develop and implement a long-term and targeted state policy in order to increase the level of seismic resistance of existing and newly constructed buildings and structures. This research presents the classification of low-rise buildings from local materials and the analysis of a series of experiments on models of low-rise buildings from local materials. The local material is clay. As a local type of clay used earthenware clay. Owing to their excellent insulation properties and their heat storage capacity, clay building materials help to reduce the energy consumption of buildings. A series of experiments were carried out to analyze seismic resistance and to increase insulation properties. Their intensity can reach 8–9 balls by the MSK64 scale or even more. The Medvedev–Sponheuer–Karnik (MSK) scale of seismic intensity was widely used in former Soviet Union countries starting in 1964 and it is based on typical masonry structures. The last series of experiments showed that the model of a house built from local materials and reinforced with a wooden frame of the "Synch" type has the following properties: does not collapse at different weights of coatings, even with intensity 9 by MSK64 scale (in terms of acceleration) earthquakes; of all the options, the most acceptable in terms of efficiency and seismic resistance were houses with a wooden frame "Synch" filled with clay materials. Wooden-framed houses filled with clay materials are called «synch» houses. The Synch is a timber frame structure filled up with earthenware clay.

Keywords

Classification • Simulation • Experiment • Local materials • Model • Wooden • Ferroconcrete framework

1 Introduction

The main part of the territory of the Kyrgyz Republic is in the seismically active zone. Such location creates certain risks when buildings and structures may be affected by seismic impacts. Earthquakes occur very frequently in the Kyrgyz Republic. Their intensity can reach 8–9 balls by MSK64 scale or even more. The Medvedev–Sponheuer–Karnik (MSK) scale of seismic intensity was widely used in former Soviet Union countries starting in 1964 and it is based on typical masonry structures. The high seismicity of the territory imposes special requirements on the construction industry of the republic; the development and implementation of long-term and targeted State policy is necessary to increase the seismic resistance of existing and newly constructed buildings and structures (Orunbaev et al., 2019; State Programme, 2011).

More than 65% of the population of the Kyrgyz Republic live in rural areas, where 90% of the residential houses were built and are still being built with local materials. It should be mentioned that construction of individual residential houses is traditionally made of clay materials not only in rural areas but also in the Bishkek and Osh cities.

The local materials are clay. As a local type of clay used earthenware clay. Owing to their excellent insulation properties and their heat storage capacity clay building materials help to reduce the energy consumption of buildings. A series of experiments was carried out to analyze seismic resistance and to increase insulation properties.

Z. Mamatov · Y. Sydykov · N. Shamshiev
Kyrgyz Civil Engineering Institute named after. N. Isanova, Kyrgyz State Technical University named after I. Razakova, Bishkek, Kyrgyz Republic

S. Orunbaev (✉)
American University of Central Asia, Bishkek, Kyrgyz Republic
e-mail: orunbaev_s@auca.kg

M. Bezzeghoud et al. (eds.), *Recent Research on Geotechnical Engineering, Remote Sensing, Geophysics and Earthquake Seismology*, Advances in Science, Technology & Innovation,
https://doi.org/10.1007/978-3-031-48715-6_67

2 Methods and Materials

Depending on the region, certain construction methods have traditionally been established, based on the type of house construction and on the technology of the production of materials. Therefore, there is a need to develop seismic-resistant structural and planning solutions for residential buildings using local construction materials.

When technical task was developed, it was decided to use simulation as the experimental-theoretical investigation method of the object. It consists of the following operations: construction of model, investigation of model parameters under the given conditions or effects, and transfer of the obtained data to simulated object. For this purpose, the centuries of experience in construction of houses from low-strength materials inherited from our ancestors were studied including earthquake-proof measures as well as reviving the methods and technologies of their construction (Mamatov & Chymyrov, 2010; Shapanov et al., 2008). After studying and synthesizing the existing practice in the Central Asian region (Mamatov, 2017; Rashidov & Mamatov, 2019), houses made from local materials, their structural concepts, and basic structure were classified into 4 (four) types for Kyrgyz Republic (Mamatov, 2015):

1. Houses with walls built of raw brick or blocks of correct shape;
2. Wooden-framed houses filled with clay materials called «synch» houses. The Synch is a timber frame structure filled up with earthenware clay;
3. Houses with walls built of clay material or from various clay pieces of irregular shape («Sockmo»);
4. Houses made of reinforced concrete frame filled clay material.

According to the results of the classification, the direction of mechanical modeling was chosen—research on models process with the purpose of transferring the research results to the full-scale structure. In this case, the scale of the model, its elements, and the mechanical characteristics of the materials are selected according to certain laws of similarity (Mamatov, 2019).

3 Results

A series of experiments were conducted on the seismic platform in the laboratory «Seismic Construction». This laboratory tested 4 types of houses and made a comparison between them for seismic resistance (Mamatov, 2019).

3.1 Sochkmo and Pakhsa Construction

The construction of the walls of "Sockmo" or "Pakhsa" is the most common method of wall construction from materials on the territory of the Kyrgyz Republic—clay technology. When erecting a clay wall, specially made shapes are used, or scaffolding with a shield height of 50–60 cm, is used. The laying is like that of raw brick, but the joints are not strong enough. Clay was used in the construction of this type, although in practice the soil extracted from the site or surroundings of the house is most commonly used.

Pakhsa is a part of a clay wall 50–60 cm high, erected layer by layer with special formwork. To make it, the clay mass is mixed with the straw which is pre-coated with cloth. Next, the mass is wrapped in matter, rolled, and placed in a cast formwork («gualyac») along the entire perimeter of the building in rows. In other words, each layer should be laid only after the previous layer of the wall is able to bear the weight of the next round. The results of the simulation have shown that houses constructed by the «Sockmo» type completely break down already at intensity 5–6 MSK64 scale (according to the acceleration index—188 cm/c2) earthquake (Fig. 1).

3.2 Raw Bricks and Blocks Construction

House structures with walls built of raw brick and blocks of correct shape are very often found in the countryside and new buildings in the suburbs of the city of Bishkek. The peculiarity of this type of wall is that due to the fit of bricks and blocks in the corner joints, the walls do not settle.

Masonry of this type is almost the same as brickwork. The thickness of the walls of raw brick can be taken in 1; 1.5 or 2 bricks, and from blocks of correct shape—1 or 1.5 blocks. This type of wall without adequate reinforcement, as

Fig. 1 General view of the model of the house «sockmo» after the experiment

Fig. 2 General view of the model of the house, made of brick raw, after the experiment

well as the previous one, is not earthquake resistant and needs to be reinforced. The main feature of strengthening this model is to strengthen the inside and outside of the wall with nets of cords with surface gunning sand-cement mix. According to the results of the experiment, it was discovered that the model of a house built of raw brick with simple reinforcement has a much higher resistance to a house of simple «sockmo» (Fig. 2).

In these figures, you can see a lot of useful sides of the reinforced model, such as hauling a net of cords and splashing sand-cement mixture under high pressure, with the result that the plaster clings to the wall. With an intensity of 6–7 by MSK64 scale (in terms of acceleration) earthquake, the plaster remained stuck to the wall and did not fall off, which proves its relative resistance to earthquake.

3.3 Wooden-Framed Houses Construction

Wooden-framed houses filled with clay materials «synch» houses are very often used in the most populated areas of Central Asia, for example in the Batken region of the Kyrgyz Republic (Mamatov, 2017; Mamatov & Chymyrov, 2010; Rashidov & Mamatov, 2019). The wooden frame is built to carry the load from the roof, wood ceiling, and insulating materials. «Gualyak» is mainly used to fill the frame. Like «synch» homes can be recommended as a simple and reliable construction with local materials. This type of house can be used as a model in seismic zones, as well as housing for victims of natural disasters. The cross sections of the supporting pillars of the lumber should be dimensions—100 × 50 mm, 50 × 50 mm, struts—50 × 30 mm (Fig. 3), and the sill—150 × 150 mm; 150 × 100 mm. Another option, the supporting pillars of the logs must be of circular cross-section diameter of 60–80 mm, planed on all sides, to square, to make it even, so that they were attached to the base and mounted on top of the other logs.

Fig. 3 The process of construction of a model house of wooden frame "synch"

When filling the frame, it is necessary to ferment (real) clay to prepare shuttering the right size, add clay, and various additives, such as straw, pour the dough around the perimeter, tightly tamping. When filling the wall, wait, let the clay dry and solidify the solution to a certain extent. Houses of similar design are considered very resistant buildings for earthquake compared to other buildings, and it is recommended to build such houses in seismically hazardous areas. Results of the experiments have shown that the intensity 7–8 by MSK64 scale (in terms of acceleration) of an earthquake, filling collapsed, but the wooden frame was durable and resistant (Fig. 4). It should be noted that in the event of severe earthquakes, these houses do not have the possibility of tragic consequences, and the collapsed filling is not difficult to recover in the future.

Fig. 4 General view of the model house of «synch» after the experiment

3.4 Reinforced Concrete Frame Houses Construction

The house of «reinforced concrete frame» with the filling of the raw bricks mainly are used in new built quarters around the city of Bishkek and in other regions. In the construction of houses of such construction, first, when the foundation is filled with concrete, it is necessary to lay reinforcement upright columns at the corners by four rods in each corner, in the openings of doors and windows (two bars). Secondly, at the intersections of the walls where there is a switch-reinforced, it should continue to the level beams. However, every 60 cm in height should go at least two rods at the joint between the wall supports. Thirdly, in the soaked clay, straw should be added, and a scaffold should be installed, which makes one row of walls 60 cm high. In the same way, the following rows are arranged up to the level of the trimmer beam.

Next, the two rebar rods that protrude from the foundation along the edges of the doors and windows should be rolled into the band belt, and the concrete should be covered (Fig. 5). This element is called a monolithic or reinforced concrete core.

From the results of the experiments, we can see that on intensity 7–8 by MSK64 scale (in terms of acceleration) earthquake, in a house built with respect to all construction technologies and methods, there are only cracks in the plastering of the house and in the joints of the clay walls with concrete elements at the level of the windowsills and lintels have appeared (Fig. 6).

Obviously, such a house construction is resistant to earthquakes, compared to the other listed three types of

Fig. 5 The process of construction of model homes from the «reinforced concrete frame»

Fig. 6 A general view of the "reinforced concrete frame" model of the house, after the experiment

structures above. Houses with reinforced concrete supports are much more durable than houses made of clay wall blocks. This fact must always be considered.

The main components of seismic protection are the three main parts of the building (foundations, walls, and roof), which must be bonded in order to function as a unit during an earthquake. Poorly built seismic belt may even exacerbate the devastating effect of the earthquake. There are also buildings with columns, cores, and seismic belts, with low concrete quality and too thin rebar, which cannot ensure the building's earthquake resistance. Therefore, developers should apply to architectural authorities and relevant design organizations for a detailed project and strictly comply with design requirements in construction.

4 Discussion

Based on the results of experimental studies of the buildings presented, it is possible to make a conditional pedestal of seismic stability, considering the economic component, construction technology, and earthquake resistance:

- in the first place—houses with "reinforced concrete frame"—the frame of houses of such structure will assume all the burden if the recommended building technologies (techniques) are met.
- in the second place—houses of the type «synch»—each element of such houses is bound with other elements and works with them in unity. This type of house is built easily and quickly in comparison with others and does not require excessive transport costs; therefore, they can be recommended for erection in earthquake risk zones.
- in the third place—houses with walls constructed of raw brick or blocks of correct shape, as well as pasha houses—a feature of the walls of this type of house is that the

corners of the brick or brick blocks are adjusted to each other crosswise, thus preventing the walls from splitting apart and the stability of the brick walls is much greater than that of other materials, made of clay. Pakhsa houses retain relative stability due to the fact that the whole perimeter is laid with various binding additives, which creates a single monolithic structure.

- in the fourth place—houses with walls erected from mud-brick material or different ground (clay) pieces of irregular shape.

5 Conclusions

The reasons for the frequent destruction of such houses are the unbound structure, the absence of anti-seismic measures and, of course, the improper exploitation of the houses.

Given that the Kyrgyz Republic is in a seismically active zone and more than 93% of its area is mountainous, the delivery of materials is a major obstacle. Therefore, we recommend the use of local materials, namely a wooden frame with clay filling or "Synch" constructions, as they are more earthquake resistant. Also, the construction of the houses must be carried out strictly in accordance with the project executed by the authorized project agency. The specific conditions of the construction should be considered, and acceptance carried out in accordance with the requirements of the construction (Imanbekov et al., 2002; SNIP KR 20-02, 2018).

References

Imanbekov, S. T., Kosivzov, G. V., Troshkin, V. N., Gron, A. G., & Begaliev, U. T. (2002). Construction of individual dwellings with local materials in the Kyrgyz Republic, Bishkek. *KNIPIS, 64*, 40.

Mamatov, Z. Y., & Chymyrov, A. U. (2010). The construction of earthquake-resistant houses with local materials in Kyrgyzstan. In *Proceedings of the international forum "Natural disasters and the security of the construction of buildings and structures"*. UNECF Committee on Housing and Land Management

Mamatov, Z. Y. (2015). Modelling and experimental analysis of residential buildings from local materials/Science, technology and life-2015. In *Proceedings of materials the international scientific conference* (pp. 131–143)

Mamatov, Z. Y. (2019). Modelling damage and destruction of low-rise buildings» annual report SRW.

Mamatov, Z. Y. (2017). Ways to build safe homes and reinforce homes" textbook: B. *Polygraph Album Resource, 14*, 164.

Orunbaev, S. Z., Mendekeyev, R. A., Moldobekov, B. D., & Rodkin, M. V. (2019). Microseismic and vibroseismic testing of a house: Comparative study of results with the example of a typical privately owned residential house in Bishkek, Kyrgyzstan. *Seismic Instruments, 55*, 92–100.

Rashidov, T. R., & Mamatov, Z. Y. (2019). Results of the assessment of the degree of seismic damage to private residential developments. *Eurasian Union of Scientists, 6*(63), 33–37.

SNIP KR 20-02. (2018). *Earthquake engineering. Design standards*. State Agency for Architecture and Construction of the Kyrgyz Republic

State Programme. (2011). Seismic security in the Kyrgyz Republic for 2012–2019

Shapanov, A. T., Tolegenov, M. N., & Mamatov, S. (2008). Design and construction of buildings made of clay in seismic regions. *CGAP Newsletter, 3*(21), 23–28.

Wavelet Analysis of Seawater Seismic Reflection Image

Jun Chen, Xinyan Wang, and Haibin Song

Abstract

Using seawater seismic reflection images has certain advantages for studying small and mesoscale physical ocean features. In addition to various physical phenomena such as internal waves and vortices, there may also be various interference factors in seawater seismic reflection images, which have different characteristics in seismic reflection images. Analyzing and extracting these different features is of great significance for image-denoising interpretation, analysis, and recognition. The wavelet analysis method is widely used in image processing, which can effectively separate different frequency/wavenumber components in the image to achieve image-denoising and specific frequency/wavenumber feature extraction. The effect of wavelet processing is greatly affected by the wavelet base function. Only selecting the appropriate wavelet base can have ideal results for different types of images. Here, three bases, Haar, db4, and bior4.4 are selected to process the seawater seismic wave images. Comparing the decomposition results of three different wavelet bases, the decomposition results of the Haar wavelet can clearly show the horizontal characteristics of the water layer in the image. In contrast, the decomposition results of DB4 and bior4.4 are relatively weak. The specific final decomposition order should be determined according to the purpose and demand of actual image interpretation.

Keywords

Seawater seismic reflection image • Wavelet analysis • Wavelet base • Image-denoising • Feature extraction

J. Chen (✉) · X. Wang · H. Song
School of Ocean and Earth Science, Tongji University, Shanghai, 200092, China
e-mail: junxianchen@tongji.edu.cn

1 Introduction

The seawater layers have different acoustic velocity and density due to different temperature, salinity, and pressure properties, and then forms some reflection events on the seismic reflection section. These reflections consist of layered, elliptical, wavy, and high-angle reflectors. These waveforms can be used to explain traditional physical oceanographic phenomena, such as the thermohaline fine structure of seawater, internal wave, vortex/eddy, front (Holbrook et al., 2003; Song et al., 2009), and nepheloid layer (Vsemirnova et al., 2012). Compared with traditional physical oceanography methods, seismic oceanography has a high lateral resolution, which can image the seawater column in full depth and four dimensions in a short time, and monitor the temporal and spatial changes of seawater (Song et al., 2021). The use of seawater seismic reflection images has certain advantages for studying small and mesoscale physical oceanographic characteristics.

2 Wavelet Analysis Method

At present, the wavelet analysis method is widely used in image processing, which can effectively separate different frequency components in the image to achieve image-denoising and specific frequency feature extraction. There are a lot of achievements of wavelet analysis in stratigraphic seismic image processing, but there are not many applications for seawater seismic image processing. Li et al. (2019) used wavelet transform to analyze the scale characteristics of the plume model migration section. Wavelet analysis decomposes the signal into two parts: detail and approximation. Detail represents the local high-frequency characteristics of the signal, and approximation represents the low-frequency background information of the signal. After N times of decomposition, the signal S can be expressed as:

M. Bezzeghoud et al. (eds.), *Recent Research on Geotechnical Engineering, Remote Sensing, Geophysics and Earthquake Seismology*, Advances in Science, Technology & Innovation,
https://doi.org/10.1007/978-3-031-48715-6_68

$$S = A_N + \sum_{i=1}^{N} D_i \quad (1)$$

A_N is approximation after N times of decomposition. D_i is the details of the ith time decomposition. The approximation can be further decomposed to get new large-scale details and approximation.

$$A_j = A_N + \sum_{i=j+1}^{N} D_i \quad (2)$$

The effect of wavelet processing is greatly affected by the wavelet basis function. In other words, wavelet analysis can perform well in image-denoising and information extraction, but the analysis results of different wavelet bases are quite different.

Only selecting the appropriate wavelet base can have ideal results for different types of images. There are some differences in the transform characteristics of different wavelet functions. Therefore, before using a wavelet for image processing, selecting the appropriate wavelet function is necessary.

3 Results

Figure 1a is a seawater seismic reflection image from the northern South China Sea. The water layer presents chaotic reflection at a water depth of 500–600 m and above the seafloor of CMP6400-8800 (CMP = Common Middle Points). Above the seafloor of CMP9600-11600, the water layer presents a blank reflection. An internal solitary wave is developed in the water layer on the right/NE side of the image, and a vortex/eddy is developed at the position of CMP8800-11000. The complex ocean dynamic action forms the chaotic and blank reflection structure. In addition, at the water depth of 225–375 m, CMP6800-7800, a triangular wedge-shaped reflection structure is observed. The upper layer of the reflection event is inclined to the sea surface, and the lower layer is inclined to the seabed.

To extract the features of the image, wavelet analysis is applied. Since the selected wavelet bases are likely to affect the processing results, we processed it (Fig. 1a) with three different wavelet bases. Figures 1b–d show the horizontal details of the first-order wavelet after wavelet calculation. Figure 1b is the result of the Haar wavelet base. The horizontal structure of the water layer can be clearly seen in the high-frequency horizontal details, the boundary of seawater structure can be clearly delineated, and the structures such as vortex/eddy and internal solitary wave can be distinguished. Figure 1c is the result of the db4 wavelet base. It is found that compared with the result of the Haar wavelet, the result of the db4 wavelet has more high-frequency detail noise, and cannot see the reflection structures such as the water layer and the seafloor. Figure 1d is the result of bior4.4 wavelet base. There are no reflection structures such as the water layer and the bottom shape of the seafloor. Compared with the result of Haar wavelet base, there is more noise.

Figure 2 shows the first-order wavelet decomposition results with the Haar wavelet base. Figure 2a is the horizontal wavelet details, Fig. 2b is the vertical wavelet details, Fig. 2c is the diagonal wavelet details, and Fig. 2d is the low-frequency approximation. On the high-frequency horizontal details, the horizontal structure of the water layer can be clearly seen, such as the boundary of the seawater structure, vortex/eddy, and internal solitary wave. High-frequency vertical details highlight the vertical structure of internal solitary waves. The features of high-frequency diagonal

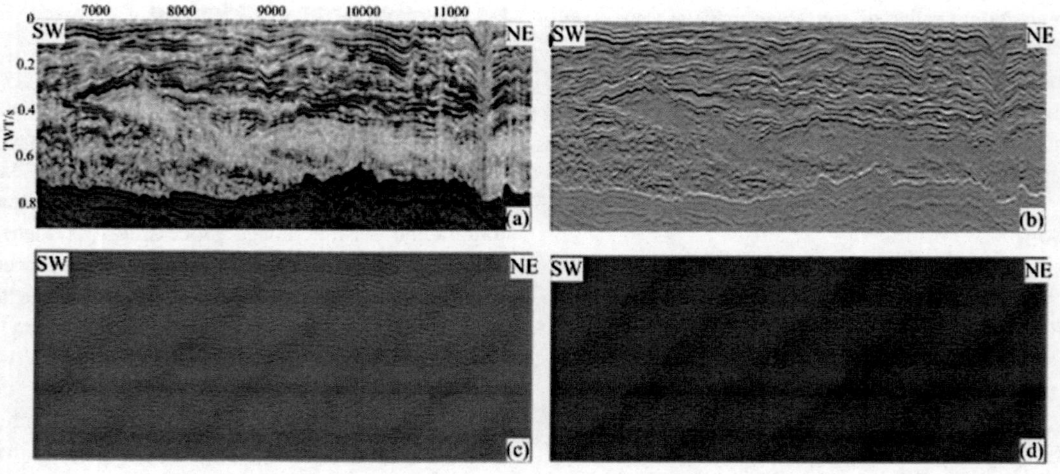

Fig. 1 Sketch map of horizontal detail of first-order wavelet after decomposition with different wavelet bases. **a** original image, **b** result of Haar base, **c** result of db4, **d** result of bior4.4

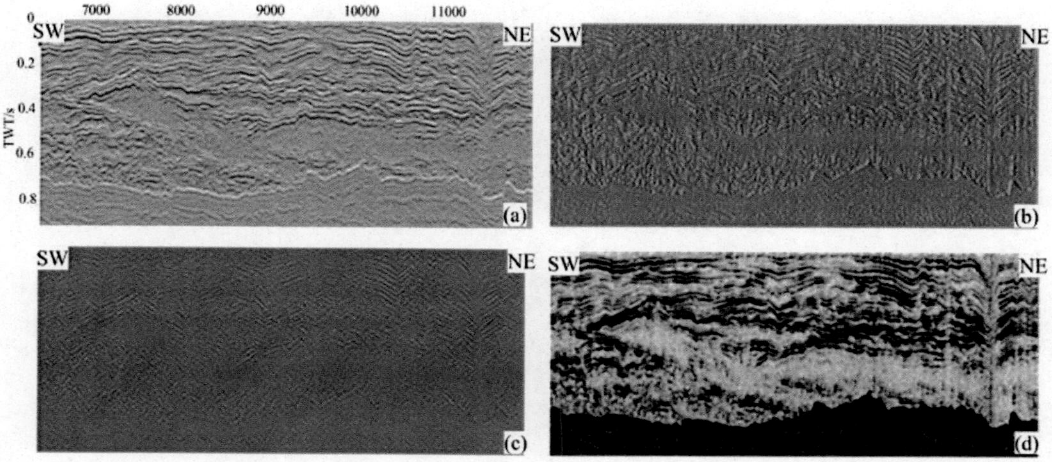

Fig. 2 Results of first-order wavelet decomposition with Haar wavelet base. **a** Horizontal wavelet details, **b** vertical wavelet details, **c** diagonal wavelet details, and **d** approximation

details are not obvious. The low-frequency approximation reflects the sketch/outline of the image, and the image contrast is higher than that of the original image (Fig. 1a). The background of the image can be observed clearly.

In addition to wavelet details and approximation, wavelet modulus maps with different scales are also obtained by wavelet transform. After wavelet decomposition, the components on different scales are processed by superposition, threshold processing, enhancement attenuation and reconstruction, to achieve the purposes of sea water seismic image-denoising, feature extraction, and image enhancement. For space reasons, it is not shown here.

4 Conclusions

We choose Haar, db4, and bior4.4 wavelet bases to process the seawater seismic wave images. Comparing the decomposition results of three different wavelet bases, the decomposition results of the Haar wavelet can clearly show the horizontal characteristics of the water layer in the image. In contrast, the decomposition results of db4 and bior4.4 are relatively weak. Therefore, it is considered that for the wavelet processing of seawater seismic wave reflection image, the effect of using Haar wavelet base is better. At the same time, the essence of multi-level wavelet decomposition is to continuously decompose the low-frequency approximation. The specific final decomposition order should be determined according to the purpose and demand of actual image interpretation.

Acknowledgements The authors thank the reviewers for their feedback on the article.

References

Aslanov, L. F. (2016). Reflected wafes from bored or CFA piles of large section in the offshore soils. *Neft. khozyaystvo, 7*, 112–116.

Holbrook, W. S., Páramo, P., Pearse, S., et al. (2003). Thermohaline fine structure in an oceanographic front from seismic reflection profiling. *Science, 301*(5634), 821–824.

Li, C. P., You, J. C., Gou, L. M., et al. (2019). Analyzing scale characteristics of migration sections of plume model by wavelet transform. *Progress in Geophysics, 34*(3), 1192–1198.

Song, H. B., Chen, J., Pinheiro, L. M., et al. (2021). Progress and prospects of seismic oceanography. *Deep-Sea Research Part I Oceanographic Research Papers, 177*, 103631.

Song, H. B., Pinheiro, L., Wang, D. X., et al. (2009). Seismic images of ocean meso-scale eddies and internal waves. *Chinese Journal of Geophysics, 52*(11), 2775–2780.

Vsemirnova, E. A., Hobbs, R. W., & Hosegood, P. (2012). Mapping turbidity layers using seismic oceanography methods. *Ocean Science, 8*(36), 11–18.